T0258413

Emulating Natural Forest Landscape Disturbances

Emulating Natural Forest Landscape Disturbances

Concepts and Applications

Edited by

AJITH H. PERERA,

LISA J. BUSE, AND

MICHAEL G. WEBER

COLUMBIA UNIVERSITY PRESS

NEW YORK

Columbia University Press
Publishers since 1893
New York   Chichester, West Sussex
© 2004 Columbia University Press
All rights reserved.

LIBRARY OF CONGRESS CATALOGING-IN-PUBLICATION DATA

Emulating natural forest landscape disturbances : concepts
    and applications / edited by Ajith H. Perera, Lisa J. Buse,
    and Michael G. Weber.
        p.   cm.
    Includes bibliographical references (p.).
    ISBN 0-231-12916-5 (cloth : alk. paper)
    1. Forest management—United States.   2. Forest
ecology—United States.   3. Ecological disturbances—
United States.   4. Forest management—Canada.   5. Forest
ecology—Canada.   6. Ecological disturbances—Canada.
I. Perera, Ajith H.   II. Buse, Lisa J.   III. Weber, Michael G.

SD143.E5 2004
634.9′2—dc22   2003068813

Columbia University Press books are printed on durable and
acid-free paper.

Printed in the United States of America

10 9 8 7 6 5 4 3 2 1

CONTENTS

CONTRIBUTORS

JAMES A. BAKER
Ontario Forest Research Institute
Ontario Ministry of Natural Resources
1235 Queen Street E
Sault Ste. Marie, ON P6A 2E5

DAVID J. B. BALDWIN
Ontario Forest Research Institute
Ontario Ministry of Natural Resources
1235 Queen Street E
Sault Ste. Marie, ON P6A 2E5

YVES BERGERON
Industrial Chair in Sustainable Forest
    Management
Universite du Quebec en Abitibi-
    Temiscamingue
C.P. 8888, Succ Centre-ville
Montreal, QC H3C 3P8

DEN BOYCHUK
Aviation and Fire Management Branch
Ontario Ministry of Natural Resources
70 Foster Drive, Suite 400
Sault Ste. Marie, ON P6A 6V5

LISA J. BUSE
Ontario Forest Research Institute
Ontario Ministry of Natural Resources
1235 Queen Street E
Sault Ste. Marie, ON P6A 2E5

TOM CLARK
CMC Ecological Consulting
R.R. 1, 1150 Golden Beach Road
Bracebridge, ON P1L 1W8

THOMAS R. CROW
North Central Research Station
U.S. Department of Agriculture Forest Service
1831 Highway 169 E
Grand Rapids, MN 55744

WILLIAM DIJAK
North Central Research Station
U.S. Department of Agriculture Forest Service
1831 Highway 169 E
Grand Rapids, MN 55744

PIERRE DRAPEAU
Department of Biological Sciences
Université du Québec à Montréal
C.P. 8888, Succ Centre-ville
Montreal, QC H3C 3P8

DAVID L. EULER
Birchpoint Enterprises
Box 6, Site 4, R.R. 4
Echo Bay, ON P0S 1C0

GLENN FOX
Department of Agricultural Economics and
    Business
University of Guelph
Guelph, ON N1G 2W1

SYLVIE GAUTHIER
Laurentian Forestry Centre
Canadian Forest Service
1055 du P.E.P.S., Box 3800
Sainte-Foy, QC G1V 4C7

DANIEL W. GILMORE
Department of Forest Resources
University of Minnesota
115 Green Hall, 1530 North Cleveland Avenue
St. Paul, MN 55108

PIERRE GRONDIN
Forest Research Branch
Quebec Ministry of Natural Resources
2700 rue Einstein
Sainte-Foy, QC G1P 3W8

ERIC J. GUSTAFSON
North Central Research Station
U.S. Department of Agriculture Forest Service
5985 Highway K
Rhinelander, WI 54501

ALTON S. HARESTAD
Department of Biological Sciences
Simon Fraser University
8888 University Drive
Burnaby, BC V5A 1S6

HONG S. HE
School of Natural Resources
University of Missouri–Columbia
203 ABNR Building
Columbia, MO 65211

DARYLL HEBERT
Encompass Strategic Resources
R. R. 2, Highway 21 S
Creston, BC V0B 1G2

BRENDAN HEMENS
BioForest Technologies
527 Beaverbrook Court, Suite 416
Fredericton, NB E3B 1X6

CHRIS HENSCHEL
Forest Programs
Canadian Parks and Wilderness
    Society–Wildlands League
401 Richmond Street W, Suite 380
Toronto, ON M5V 3A8

PAUL F. HESSBURG
Pacific Northwest Research Station
U.S. Department of Agriculture Forest Service
1133 N Western Avenue
Wenatchee, WA 98801

AMY E. HESSL
College of Forest Resources
University of Washington
Box 352100
Seattle, WA 98195-2100

ROBERT E. KEANE
Rocky Mountain Research Station
U.S. Department of Agriculture Forest Service
Box 8089
Missoula, MT 59807

J. P. KIMMINS
Department of Forest Sciences
University of British Columbia
2424 Main Mall, #3041
Vancouver, BC V6T 1Z4

DENNIS H. KNIGHT
Botany Department
University of Wyoming
Laramie, WY 82071

ALAIN LEDUC
Université du Québec à Montréal
C.P. 8888, Succ Centre-ville
Montreal, QC H3C 3P8

CHAO LI
Northern Forestry Centre
Canadian Forest Service
5320 122 Street, Room M076
Edmonton, AB T6G 2E9

DAVID A. MACLEAN
Faculty of Forestry and Environmental
    Management
University of New Brunswick
Box 44555
Fredericton, NB E3B 6C2

DANIEL MCKENNEY
Great Lakes Forestry Centre
Canadian Forest Service
1219 Queen Street E
Sault Ste. Marie, ON P6A 2E5

DONALD MCKENZIE
College of Forest Resources
University of Washington
3200 SW Jefferson Way
Corvallis, OR 97331

JOHN G. MCNICOL
JGM Forestry Consulting
1470 Highway 61
Thunder Bay, ON P7J 1B7

AL MUSSELL
The George Morris Centre
102–150 Research Lane
Guelph, ON N1G 4T2

THUY NGUYEN
Sustainable Forest Management
Universite du Quebec en Abitibi-
    Temiscamingue
C.P. 8888, Succ Centre-ville
Montreal, QC H3C 3P8

ETSUKO NONAKA
Department of Forest Science
Oregon State University
321 Richardson Hall
Corvallis, OR 97330

BRIAN J. PALIK
North Central Research Station
U.S. Department of Agriculture Forest Service
1831 Highway 169 E
Grand Rapids, MN 55744

RUSSELL E. PARSONS
Rocky Mountain Research Station
U.S. Department of Agriculture Forest Service
Box 8089
Missoula, MT 59807

AJITH H. PERERA
Ontario Forest Research Institute
Ontario Ministry of Natural Resources
1235 Queen Street E
Sault Ste. Marie, ON P6A 2E5

DAVID L. PETERSON
FRESC Cascadia Field Station
University of Washington
Box 352100
Seattle, WA 98195-2100

FRED PINTO
Forest Research and Development Section
Ontario Ministry of Natural Resources
3301 Trout Lake Road
North Bay, ON P1A 4L7

KEVIN B. PORTER
Atlantic Forestry Centre
Canadian Forest Service
Box 4000, Regent Street
Fredericton, NB E3B 5P7

SUSAN PRICHARD
College of Forest Resources
University of Washington
Box 352100 Cascadia Field Station
Seattle, WA 98195-2100

KEITH M. REYNOLDS
Pacific Northwest Research Station
U.S. Department of Agriculture Forest Service
3200 SW Jefferson Way
Corvallis, OR 97331

MERRICK B. RICHMOND
Pacific Northwest Research Station
U.S. Department of Agriculture Forest Service
1133 N Western Avenue
Wenatchee, WA 98801

MATHEW G. ROLLINS
Rocky Mountain Research Station
U.S. Department of Agriculture Forest Service
Box 8089
Missoula, MT 59807

STEPHEN ROMANIUK
Southcentral Science and Information Section
Ontario Ministry of Natural Resources
3301 Trout Lake Road
North Bay, ON P1A 4L7

WILLIAM H. ROMME
Department of Forest Sciences
Colorado State University
104 Forestry Building
Fort Collins, CO 80523-1470

R. BRION SALTER
Pacific Northwest Research Station
U.S. Department of Agriculture Forest Service
1133 N Western Avenue
Wenatchee, WA 98801

FRANK SCHNEKENBURGER
Ontario Forest Research Institute
Ontario Ministry of Natural Resources
1235 Queen Street E
Sault Ste. Marie, ON P6A 2E5

STEPHEN R. SHIFLEY
North Central Research Station
U.S. Department of Agriculture Forest Service
202 Anheuser-Busch Natural Resources
    Building
University of Missouri
Columbia, MO 65211-7260

THOMAS A. SPIES
Pacific Northwest Research Station
U.S. Department of Agriculture Forest Service
3200 SW Jefferson Way
Corvallis, OR 97331

ROGER SUFFLING
School of Planning
University of Waterloo
200 University Avenue W
Waterloo, ON N2L 3G1

IAN D. THOMPSON
Great Lakes Forestry Centre
Canadian Forest Service
1219 Queen Street E
Sault Ste. Marie, ON P6A 2E5

DANIEL B. TINKER
Department of Environmental Resource
    Management
Western Carolina University
College of Arts and Science
Cullowhee, NC 28723

MONICA G. TURNER
Department of Zoology
University of Wisconsin–Madison
430 Lincoln Drive
Madison, WI 53706

KEVIN WEAVER
Geographic Information Systems
Sault College of Applied Arts and Technology
443 Northern Avenue
Sault Ste. Marie, ON P6B 4J3

MICHAEL G. WEBER
Great Lakes Forestry Centre
Canadian Forest Service
1219 Queen Street E
Sault Ste. Marie, ON P6A 2E5

MICHAEL C. WIMBERLY
Warnell School of Forest Resources,
    Room 4-228
University of Georgia
Athens, GA 30602-2152

DENNIS YEMSHANOV
Great Lakes Forestry Centre
Canadian Forest Service
1219 Queen Street E
Sault Ste. Marie, ON P6A 2E5

JOHN C. ZASADA
North Central Research Station
U.S. Department of Agriculture Forest Service
1831 Highway 169 E
Grand Rapids, MN 55744

FIGURES AND TABLES

Figures

Tables

PREFACE

Our motivation for assembling this book came from the escalating debate in both scientific forums and the public media about the need to emulate natural disturbance in forest management. The participants in this debate are diverse, ranging from forest managers to policymakers, legislators, nongovernment organizations, researchers, students, and forestry stakeholders. Our goals are to draw the attention of those involved in the debate to an array of basic questions on this topic and summarize the state of our knowledge from a North American perspective. The basic questions we raise are many. They include:

- What is a *natural* disturbance?
- What is meant by *emulating* natural disturbance?
- Why would one *want* to emulate natural forest disturbance?
- How well do we *understand* natural disturbance?
- *How* might one emulate natural disturbance?

The views and information presented here represent a broad range of geographic perspectives from across Canada and the United States, ranging from the Pacific Northwest through the Rocky Mountains, Midwest, Northeast, and Maritimes. Although natural disturbances and their emulation encompass a range of spatial and temporal scales, our focus in this book is largely upon the broader scale of the forest landscape.

The contents of this book are organized into three sections. The chapters in the first section examine the concepts behind emulating natural forest disturbances, addressing the questions of what emulating natural disturbance involves and what ecological reasoning substantiates it. These include a broad overview (chapter 1), a detailed review of emerging forest management paradigms and their global context (chapter 2), and an examination of the ecological premise for emulating natural disturbance (chapter 3). In addition, the current understanding of natural disturbance regimes is explored, including a general overview (chapter 4) and detailed reviews that focus on two of the most prevalent disturbance regimes in North America: fire (chapter 5) and insects (chapter 6).

The second section contains case studies from a wide geographical range that address characterization of natural disturbances and development of applied templates for their emulation through forest management. The preponderance of studies that focus on fire regimes (chapters 7–10) in this section is not intentional; rather, it represents the greater focus that has traditionally been placed on understanding and managing fire compared with other forms of disturbance (chapter 11). The lessons learned from historical disturbance patterns are addressed from several viewpoints (chapters 12–14).

The final group of chapters addresses another aspect of applications: expectations for and feasibility of emulating natural disturbance through forest management. The expectations of forest stakeholders are expressed from two divergent perspectives: those of conservation (chapter 15) and of utilization (chapter 16). The feasibility of embracing this management paradigm is also explored from economic (chapter 17), silvicultural (chapters 18, 19), regional-scale planning (chapter 20), and policy development (chapter 21) points of view. Finally, we reflect upon the current state of knowledge of this topic as well as on the immediate challenges and future directions (chapter 22).

The intended readership of this book is broad. Although forestry professionals (practitioners, policymakers, and researchers) are the primary audience, anyone with a general forestry background will find it useful reading. We did not presume that this book would provide an authoritative statement on every aspect of the complex and evolving topic of emulating natural forest disturbance. However, we do expect that it will provide the reader with a good foundation for clarifying the ambiguities inherent in the concept, a broad perspective from which to grasp its intricacies, an appreciation of the divergent expectations that practitioners face, and a balanced view of the promises and challenges associated with the application of this emerging forest management paradigm.

Ajith H. Perera, Lisa J. Buse,
and Michael G. Weber
March 2003

ACKNOWLEDGMENTS

We are grateful for the funding and institutional assistance to produce this book provided by the Ontario Ministry of Natural Resources and the Canadian Forest Service. Connie Bouffard, Fraser Dunn, and Geoff Munro deserve special thanks for their sustained support.

All chapters were extensively peer reviewed, thanks to the generous efforts of many colleagues. On behalf of all authors, we thank reviewers James Agee, Dave Archibald, Jim Baker, William Baker, David Baldwin, Ken Baldwin, Wayne Bell, Dennis Boychuk, Terry Carleton, Joe Churcher, Dave Cleland, Bill Crins, Tom Crow, Dan Dey, Peter Duinker, Glen Dunsworth, Dave Euler, Richard Fleming, Jerry Franklin, Eric Gustafson, Bob Keane, Maureen Kershaw, Will Kershaw, Dan Kneeshaw, Alain Leduc, Chao Li, Marty Luckert, Dan McKenney, Don McKenzie, Tom Nudds, David Peterson, Bruce Pond, Reino Pulkki, Tim Schowalter, Susan Smith, Brian Sturtevant, Roger Suffling, Kandyd Szuba, Michael Ter-Mikaelian, Carolyn Whittaker, Jeremy Williams, Michael Wimberly, and Dennis Yemshanov. For sharing their views on emulating natural forest disturbances, thereby enriching the concluding chapter, we thank Joshua Breau, Joe Churcher, Barry Davidson, Dave Deugo, Rod Gemmel, George Graham, Rick Groves, Chris Henschel, Brian Hillier, Monte Hummel, Faye Johnson, Janet Lane, Rob MacLeod, Glen Niznowski, Lauren Quist, Peggy Smith, George Stanclik, Kandyd Szuba, Eric Thompson, and Kent Virgo.

Finally, we gratefully acknowledge those who assisted us in publication of the book: Terri Weaver for administrative support; Megan Holder and Agnes Ouellette for formatting the text and illustrations; Trudy Vaittinen for developing the cover design; and Robin Smith and Khristine Queja, our liaisons at Columbia University Press. Our sincere thanks to Geoff Hart, for improving the readability and clarity of the chapter manuscripts.

PART I

Concepts

CHAPTER 1

▪▪ Emulating Natural Disturbance in Forest Management
An Overview

AJITH H. PERERA and LISA J. BUSE

Sustainability has been a goal in forest management for several decades. However, as the concept of forest sustainability expanded beyond its original focus (that of sustained timber yield) to include other values such as biodiversity, various paradigms have evolved as possible management approaches (Brooks and Grant 1992; Hunter 1993). Rather than contrasting these approaches, we have chosen in this chapter to focus on one: the use of natural disturbance as a guide for forest management, a paradigm that has grown in popularity over the past decade. Given the elusive nature of the goal of forest sustainability, the emulation of disturbance has emerged as a surrogate goal (e.g., Hunter 1990; Attiwill 1994). This emergence coincided with the increased recognition of periodic disturbance as an integral process in forest landscapes (e.g., Turner 1987). In addition, public awareness of the existence and potential severity of natural forest disturbance and the forest's ability to recover from dramatic natural disturbances began increasing after the 1988 Yellowstone fires and the eruption of Mount St. Helens in 1980.

Hunter (1990) has also advocated using large natural disturbances as benchmarks in forest management and as a practical means of applying the *coarse filter* and *fine filter* approaches to conserving biodiversity (Noss 1987). These approaches consider landscape- and regional-scale factors first, followed by stand- and site-scale factors. Consequently, land managers in many regions of North America began to examine how the emulation of natural disturbance could guide their approach to forest management. This concept garnered special attention in regions where large forest disturbances are frequent, such as the northwestern (e.g., Cissel et al. 1994; DeLong and Tanner 1996) and northeastern (e.g., Hunter 1993; Bergeron and Harvey 1997) regions of Canada and the United States. For example, in Ontario (Canada), the approach of emulating natural forest disturbance gained enough public acceptance to receive legislated status (Statutes of Ontario 1995) and was subsequently translated into policies and prescriptive guidelines for forest management planning (Ontario Ministry of Natural Resources 1996a, 2002a). Other provincial governments, the national network of model forests, and the forest industry in Canada have included the goal of emulating natural disturbance in their strategies and guidelines and are testing aspects of the concept operationally at various scales (e.g., British Columbia Ministry of Forests 1995a; Government of Alberta 1997; Bonar 2001). The United States has yet to legislate the use of this approach. In recent years, the increased scientific interest in emulating natural forest disturbance has been reflected in the added attention devoted to this topic in the literature; for example, special issues on the subject have appeared in *Ecological Applications* (volume 9, 1999), *Forest Ecology and Management* (volume 155, 2002), and *Silva Fennica* (volume 36, 2002). In this chapter, we provide a conceptual overview of the paradigm of emulating natural forest disturbance.

The Concept of Emulating Natural Forest Disturbance

What Is the Meaning of *Emulating Natural Forest Disturbance?*

The broad concept of emulating natural forest disturbance in management has been referred to using many terms. Some examples include *ecological forestry* (Seymour and Hunter 1999), *emulation silviculture* (McRae et al. 2001), *natural disturbance models* (Armstrong 1999), *historical range of variability* (Morgan et al. 1994), and *natural variability* (Landres et al. 1999). All these terms imply that natural forest disturbance and its variability may be used as a general guide in developing forest management strategies. More specifically, natural disturbance can be used as a *null model* (a baseline against which to compare and evaluate alternative strategies) that can serve as a guideline for designing appropriately modified management practices such as harvesting, fire suppression, and prescribed burning.

The various terms that have been used for this and related concepts do, however, differ in specific intent. For example, ecological forestry and natural disturbance models refer to the use of natural disturbance as a general null model in the development of forestry practices. Emulation silviculture implies a more specific instance, in which the natural structures and processes are managed to emulate disturbance at the stand scale. Managing within the bounds of variation implies even more specific target setting, in which compositional and structural targets for the forest or landscape that fall within a predetermined range are set based on understanding natural variability or the historical range of variability of natural disturbance. The complete abandonment of forest manipulation in favor of wilderness development (in places where the actual natural disturbances themselves are essential, and cannot be emulated by forestry practices) can also be considered a form of emulating natural disturbance, albeit an extreme one.

To clarify the terminology, we propose the following broad and inclusive working definition of emulating natural disturbance:

> Emulating natural disturbance is an approach in which forest managers develop and apply specific management strategies and practices, at appropriate spatial and temporal scales, with the goal of producing forest ecosystems as structurally and functionally similar as possible to the ecosystems that would result from natural disturbances, and that incorporate the spatial, temporal, and random variability intrinsic to natural systems.

The Ecological Premise behind the Paradigm

What makes emulating natural disturbance so appealing to scientists, forest managers, and the public? The fundamental ecological premise of this paradigm appears to have two parts. First, it has been long known that periodic disturbances are inherent to forest ecosystems at all spatial and temporal scales (e.g., Pickett and White 1985; Turner 1987; Turner et al. 1997a). All forest ecosystems undergo periodic disturbances and recover from these disturbances because of the ecosystems' inherent resilience (Holling 1981). Second, it has been argued that forest ecosystems and their species assemblages have evolved and adapted in such a way that they can persist amidst these disturbances (e.g., Mutch 1970; Bunnell 1995). In this context, the goal is to apply a form of forest management that matches as closely as possible the natural disturbances that gave rise to the current forest ecosystems and will thus ensure the sustenance of these ecosystems.

Although many definitions have been proposed for *ecosystem resilience,* we have chosen the definition of Holling (1973) as the basis for this discussion: resilience implies the "persistence of relationships within a system and is a measure of the ability of these systems to absorb changes . . . and still persist." Figure 1.1 presents a simplified general view of this concept. An ecosystem at stage *a* is at a dynamic equilibrium (not necessarily a steady or stable state) when a disturbance occurs. The ecosystem's structure (e.g., species richness, biomass) and function (e.g., productivity, nutrient cycling) are affected by the disturbance, and respond. As long as the effect of the disturbance is less than the resilience of the ecosystem, the ecosystem's state (expressed in terms of its structural, functional, and other attributes) changes to stage *b* and then recovers to stage *c*, which approximates the predisturbance equilibrium. At this point, the original level of resilience has been restored and the ecosystem is once again able to withstand the recurrence of similar disturbances and thereby sustain itself.

Although this is the default (null) expectation, other scenarios are also possible. For example, if the effect of the disturbance exceeds the ecosystem's resilience, the response may be for the ecosystem to change to level *d*. From this point, the ecosystem may either recover slowly but fully to *f* or may only recover to *e*, which represents a dynamic equilibrium different from the original conditions. Alternatively, as Paine et al. (1998) illustrated, if a disturbance recurs during the recovery period (e.g., from b to c), or if a series of compounded disturbances occurs coincidentally, the ecosystem may be altered so that it can no longer return to its predisturbance state.

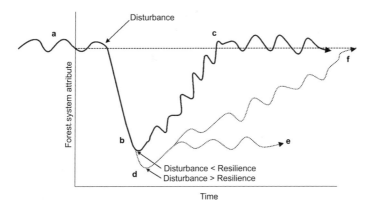

FIGURE 1.1. A conceptual illustration of how a forest ecosystem might respond to disturbance. Ecosystem states represented in the diagram: a, predisturbance state; b, state immediately after the disturbance (when the disturbance does not exceed the ecosystem's resilience); c, state after recovery from the disturbance; d, state immediately after the disturbance (when the disturbance exceeds the ecosystem's resilience); e and f, alternate states of the ecosystem after recovery (when the disturbance exceeds the ecosystem's resilience).

Although many exceptions and qualifications apply to the premise of resilience, it nonetheless provides a conservative and simplified tenet to inform forest management: interventions must be maintained within the limits imposed by the ecosystem's resilience, and this can be accomplished by avoiding disturbances that go beyond the resilience inherent in an ecosystem. Uncertainty and imperfection in our knowledge of ecosystems and their manipulations cause some to advocate emulating disturbance as a precautionary principle in the interest of conserving biodiversity. Logically, the emulation of natural disturbance would ensure sustainability, but many questions about the practical applicability of the approach remain unanswered.

Moving from Concept to Practice

The most important prerequisite for developing a framework for real-world applications that emulate natural disturbance is an adequate understanding of forest disturbance: what it is, the relative importance of different disturbance types, their characteristics, and the effects or outcomes of the disturbance.

What Is a Natural Disturbance?

Even though defining *natural* seems simple in principle, it is much more complex in practice. In the context of emulating disturbance, *natural* can be interpreted to mean either *inherent* (an integral process) or *normal* (a process observed commonly). Even at a given spatial and temporal scale, forest ecosystems inherently experience disturbances of widely different magnitudes, frequencies, intensities, and causal factors. Although some disturbances are *autogenic* (arising from within), others are *allogenic* (arising from outside the system). Some disturbances are minor and others, even though inherent, are so significant that they have been called *disasters or catastrophes* (Harper 1977) or (less subjectively) *large and infrequent disturbances* (Turner and Dale 1998). Common disturbance regimes occasionally have *extreme events* (Alvarado et al. 1998). Although such disturbances are no less inherent than less drastic disturbances, which of them can be considered sufficiently "normal" that they should be considered in emulation strategies?

Those grappling with the concept of emulating natural disturbance (e.g., Parsons et al. 1999) have focused mainly on historical disturbances, which are assumed to shed light on what is natural (e.g., Landres et al. 1999). The most commonly used historical benchmark in North America is the pre-European settlement condition (e.g., Morgan et al. 1994). This approach implies the existence of natural conditions before settlers arrived and assumes that any effects caused by aboriginal peoples are a part of nature.

In this context, naturalness may not be an absolute term; instead, it may represent a position along the continuum between absolute exclusion of human-caused disturbance and increasingly dominant anthropogenic effects. As Dale et al. (1998) illustrated, the same disturbance may occur both naturally and through human intervention, and distinguishing between the two based solely on their effects may be difficult at times. However, exclusive reliance on historical disturbance, whether natural or not, may not be an adequate guide for management practices because forest ecosystems are always in a state of flux and are affected by many factors that change (some randomly) with time. The inevitability of increased human population and increased impacts, combined with the perceived likelihood of global climate change, will further modify the context for forest ecosystems and their inherent disturbance regimes, whether or not these broadscale changes are themselves natural. To emulate natural disturbance in forest management, some agreement about what is natural in relation to forest disturbance is necessary.

What Can Be Emulated?

Studies that attempt to understand forest disturbance regimes at substand, stand, landscape, regional, and continental scales have proliferated in the past decade. These studies have addressed disturbances caused by tree falls, fire, insects, pathogens, wind, floods, hurricanes, and even volcanoes. Assisted by this information, practitioners must determine which disturbance regimes are important in their management area, based on both past occurrences and predictions for the future. In addition, the spatiotemporal scales of the disturbance regimes must be well understood and be matched to the corresponding scale of forest management. Once the suite of relevant disturbance regimes in an area has been selected, a framework must be developed for each of the disturbance regimes to guide the emulation of natural disturbance in management strategies. This framework should include explicit *emulation criteria,* which are those characteristics of disturbance events that quantitatively describe a disturbance regime and could thus be used as an explicit guide in developing management strategies and practices. Some emulation criteria that could be used for broad-scale applications of emulated natural disturbance include the structural attributes of disturbances, such as the rate of occurrence, the spatial and temporal probabilities of occurrence, the intensity of disturbance, the spatial patterns of disturbance, and the variability in each of these factors (table 1.1). Also included in the criteria are measurable consequences of disturbance: compositional, structural, and process descriptors for the post-disturbance forest landscape. This suite of emulation criteria is also useful in setting minimum standards for developing methods and tools that describe disturbance regimes and thereby help emulate natural disturbance.

Another aspect that needs clarification to develop strategies for emulating natural disturbance in forest management is the degree to which disturbances are emulated. The *degree of emulation* is the extent to which the emulation criteria are used in modifying forest management practices. The degree of emulation indicates how closely the actual values of the emulation criteria are mimicked by the management

TABLE 1.1. Examples of Disturbance Attributes and Emulation Criteria for Developing Strategies that Emulate Natural Disturbance

Disturbance Attributes	Example Emulation Criteria
Nature of disturbance	
Overall rate for a large region and its variation	Fire-return interval, annual rate of defoliation
Spatial tendency and its variation	Spatial probabilities of wind damage, burns, floods
Temporal pattern and its variation	Intervals between fires, insect epidemics, floods
Intensity and its variation	Severity of fire or defoliation and their spatial and temporal patterns
Geometry	Sizes, shapes, and adjacency of patches of disturbance
Consequences of disturbance	
Spatial and temporal patterns in landscape composition	Patterns in residual vegetation, postdisturbance succession
Spatial and temporal patterns in landscape age structure	Patterns in age-class distribution, species–age patterns
Ecological processes	Nutrient, carbon, and hydrological dynamics

practices, and thus represents the depth and intensity of the emulation strategies. The degree of emulation can range from 100%, with all emulation criteria mimicked absolutely, to nearly zero, where selected emulation criteria are used only as a very general guide to develop management strategies.

Expectations from Emulating Natural Disturbance

In reality, practical goals for emulating natural disturbance lie between these two extremes. In addition to the suite of ecological emulation criteria, implementation tactics are also tempered by economic and social criteria. For this reason, emulating natural disturbance is not a single strategy or tactic; instead, it acts as an umbrella under which any number of practices could be used to meet the chosen criteria or attain the specified objectives. Our discussion of what are essentially semantic considerations is intended to clarify expectations for the results of emulating natural disturbance. This is one of the rare forest management paradigms that appears to be equally appealing to scientists, forest resource managers, policymakers, special interest groups, and the public at large. This diverse group of stakeholders may have different implicit expectations that make emulating natural disturbance intuitively logical as a concept but may lead to disagreement at the strategic and tactical levels. Explicit agreement on the criteria and degrees of emulation of natural disturbance in management strategies a priori may help reconcile disparate expectations.

The general expectation of emulating natural disturbance is sustainability of the forest, but as mentioned by Hunter (1993), expectations related to sustainability can also differ. Citing Callicott's review (1990), Hunter points out that forest sustainability can be addressed from three different perspectives: Gifford Pinchot's resource-based perspective, which argues for sustaining the supply of forest resources; Aldo Leopold's evolutionary perspective, which argues for sustaining ecosystems; and John Muir's preservationist perspective, which argues for sustaining wilderness. These views parallel the three types of expectations for the results of emulating natural disturbance: economic, ecological, and social. Economic expectations from emulating natural disturbance may include a continued resource supply, whether the resource is timber, tourism, or wildlife for hunting. Ecological expectations may include the conservation of biodiversity and minimizing the environmental effects of forest management. Social expectations may include preservation of a suite of sociocultural values such as wilderness, forest esthetics, and iconic representations of cultural heritage. Although all these expectations are realistic, each must be explicitly addressed in terms of their congruence with specific emulation goals when developing forest management strategies. We see this as one of the most significant challenges in the practice of emulating natural disturbance, because it requires that some agreement be reached about priorities: it is unlikely that all expectations can be met simultaneously.

The Essence of Emulating Natural Disturbance

Emulating natural disturbance appears to be an intuitive and logical approach to forest management, in which the intrinsic disturbance processes and their effects on forest ecosystems are used to guide management strategies. Many forest management practices could conceivably be included under this umbrella. At this early stage of its evolution, emulating natural disturbance enjoys a rare universal popularity among forest resource managers, policymakers, special interest groups, scientists, and the public. However, it should not be seen as a panacea that will inevitably meet the multitude of expectations placed on forest management by all stakeholders. As with any management paradigm, this approach must first prove its practicality and effectiveness. At the same time, the body of scientific knowledge in support of emulating natural disturbance must be expanded, particularly in the areas of understanding disturbance regimes and testing the effectiveness of strategies for emulating natural disturbance in comparison with alternative management strategies. In short, emulating natural disturbance both as a concept and a practical approach must still mature. Nonetheless, in the absence of perfect knowledge of all forest ecosystem processes and their long-term responses to human intervention, emulating natural disturbance provides a strong conservative basis on which to develop management strategies for ensuring forest sustainability and a logical null model against which to measure the results of our efforts.

CHAPTER 2

:: Emulating Natural Forest Disturbance
What Does This Mean?

J. P. (HAMISH) KIMMINS

The Need for a New Forestry Paradigm

The world's population is predicted to increase by 50% (three billion people) in the time required for trees to grow to economic maturity over most of Canada (International Institute for Applied Systems Analysis 2001; Lutz et al. 2001). As a result, concerns over the environment are not likely to diminish soon. Many of the global environmental issues of the past four decades, during which the human population doubled from three to six billion, can ultimately be traced to population growth unaccompanied by appropriate changes in per capita resource use, pollution, and the rate and extent of ecosystem alterations. And relief may not necessarily come with a cessation in population growth. The demographic and economic transitions leading toward population stabilization or decline are linked to increased personal wealth and a correspondingly increased per capita "ecological footprint" (Wackernagel and Rees 1996).

A new relationship must be established between humans and the world's forests. Forestry has evolved at various times and in various places (Winters 1974; Baker et al. 1999) in response to the negative consequences of unregulated exploitation. Early cultures frequently practiced sustainable exploitation, using a resource without deliberate plans to ensure its renewal, but at a lower rate than the natural renewal rate for that resource (Kimmins 2002). Exploitation by small human populations with little technology is generally sustainable (Salim and Ullsten 1999), but as human numbers rise and the power of their technology increases, resource removal and ecosystem alterations exceed nature's capacity for unassisted recovery. The resultant resource depletion, or perceived threat thereof, has always been the progenitor of forestry.

Early forestry was usually political, legalistic, and bureaucratic in nature (i.e., *administrative forestry*), in which generalized regulations and inflexible practices did not respect ecological and biological diversity. As a result, this approach generally failed to achieve sustainability. Forestry then developed an ecologically based stage, but historically, this change has focused on timber production and has not necessarily sustained a variety of nontimber values. This stage in turn led to *social forestry,* in which ecologically based practices and policies are employed to sustain a wide variety of values. This overall sequence can be identified, in part or whole, in the history of forestry of many countries, although wars, social unrest, natural disasters, and inappropriate beliefs about forests have frequently interrupted and sometimes reversed the sequence.

Progress toward the social stage requires a forestry based on respect for nature, as well as on economic and noneconomic social values and emotional considerations (Kimmins 1993, 1999, 2000). This respect, which should be the foundation for a new forestry paradigm, must include an understanding of the ecological role of disturbance of the forest ecosystem in sustaining natural and desired ecosystem conditions (Attiwill 1994; Kimmins 1996). In this chapter, I explore the emulation of natural forest disturbance as such a paradigm, or its inclusion as a component of other possible paradigms.

Paradigms for a New Approach to Forestry

In designing ecologically based or social forestry, several different approaches or paradigms could be employed (Kimmins 2002), as discussed in the following sections.

Ecosystem Management and Ecosystem-Based Management

Ecosystem management is perhaps the most important paradigm discussed here. Although there are many interpretations and definitions for ecosystem management (Boyce and Haney 1997), fundamentally, it means managing ecosystems as integrated systems. This differs from the current practice of using fragmented and frequently uncoordinated management policies for trees (or even individual tree species), wildlife, water, esthetics, and other elements in the same ecosystem. Ecosystem management requires a single, coordinated plan that includes all desired values and is implemented by a single agency. This plan must be area based (must have a defined forest area) and must cover ecologically appropriate spatial and temporal scales. This approach demands a dramatic overhaul of present tenure systems in many parts of the world.

Ecosystem management requires a clear definition of the desired forest at various times in the future and must respect all forms of diversity (including temporal diversity) and the ecology of sustainability and natural disturbance. The approach thus requires a focus on selecting the trees and other components to leave undisturbed, selecting the processes to be sustained, and defining the sequence of conditions to create after harvesting, rather than focusing on what trees to take and regulating what not to do. It is results-based rather than regulations-based. Ecosystem management differs from ecosystem-based management in managing an entire forest as a system under a single plan executed by a single agency. In contrast, ecosystem-based management respects the ecology of individual components or values of the ecosystem, but generally manages each resource separately, with other resources constraining the objectives established for the focal resource. Ecosystem-based management generally involves separate plans and agencies in the management of the different resources.

Tenure systems for public forests generally license foresters to manage only the timber values for economic return, which impedes implementation of ecosystem management: the institutional arrangements do not encourage management of the forest over a defined area as an ecosystem. Even in private forests, many ecosystem values—water, air, migratory fish, and terrestrial animals—are considered common property and are not considered marketable by the forester or agency involved in forest management. This may prevent these nontimber values from being adequately represented in an ecosystem management plan and may result in a suboptimal overall balance of values. In consequence, ecosystem-based management is much more common than ecosystem management.

Ecosystem management and ecosystem-based management can both be applied as stand-level or landscape-level paradigms because the ecosystem concept does not inherently define spatial boundaries.

Adaptive Management

Adaptive management, which applies to many different management paradigms, accepts the uncertainty arising from the incomplete state of our knowledge and from risk. Adaptive management considers all forest management to be essentially an experiment, because most management practices have not been applied long enough or widely enough for us to be confident about their outcomes (Walters 1986). Each management intervention is thus an experiment from which we should learn by monitoring the outcomes. This monitoring requires a statement of the anticipated (hypothesized) outcomes against which the actual outcomes can be evaluated. Monitoring the environmental and social consequences of management in the absence of predicted outcomes provides little or no basis on which to judge the success of management; it may constitute little more than ecological and social "stamp collecting."

Forecasting outcomes generally requires the use of process-based models (Korzukhin et al. 1996), because we generally lack sufficient experience with new and untested management practices and policies. The failure of short- and medium-term experience to accurately forecast longer-term outcomes emphasizes the need for such models (e.g., Lundmark 1977, 1986). Adaptive management thus involves synthesizing existing knowledge and experience into predictive systems, trying different management strategies in different places and at different times, monitoring the results, comparing them with anticipated outcomes, and improving practices or forecasting systems when the outcomes are not as predicted. The forecasting systems also play a role in selecting the policies and practices to be tested, and in exploring alternative possible and desired future forests that may be unimaginable without such tools.

If adaptive management is accepted as the forestry paradigm, one cannot have rigid, regulation-based forestry, because such an administrative approach implicitly assumes we know everything we need to know and can always accurately predict the outcomes of our actions. Neither is true.

Adaptive management is particularly suitable for (and was developed for application in) resource management systems in which the results of changes in policy and practice can be measured over relatively short times, as is the case for agriculture or fisheries. In forestry, we frequently cannot measure the full impacts of changes in management for many decades (e.g., Lundmark 1977, 1986) or even for several tree crop rotations—perhaps centuries. As a consequence, adaptive management without adequate forecasting systems may not solve critical problems in forest management over time scales of decades. Forecasting for adaptive management in fisheries or agriculture has involved relatively simple models, and although more complex models are undoubtedly needed, some simple models have nonetheless proven useful. In contrast, adaptive management in forestry requires the use of relatively complex, multiple-value, ecosystem-management models spanning spatial scales from tens to hundreds of thousands of hectares, and from a few years to decades or centuries; simple models are not adequate (Seely et al. 1999). Appropriately complex models have not yet replaced simpler traditional growth and yield models (although the replacement has begun), and this will significantly constrain the application of adaptive management until forecasting becomes more advanced. Adaptive management has equal relevance at the stand and landscape scales.

Zonation

Doing things differently in different places (i.e., matching policy and practice to geographical differences in ecosystem characteristics, as well as to local cultural and social circumstances) is a cornerstone of the zonation paradigm, as well as of adaptive management. Examples of multiple-use and integrated resource management approaches from the past provide ample evidence of the need for spatial or temporal separation of certain values—it is simply not possible to provide all values at all times from a given stand-level ecosystem. Zonation provides spatial separation of different types and intensities of timber management, as well as of different timber and nontimber values across the forest landscape.

Theoretical considerations suggest that the world's demand for wood could be satisfied by intensively managing 20% or less of the world's forested area (Binkley 1997; Sedjo and Botkin 1997; Sedjo 2001). There are many reasons why a "20–80" solution, in which 20% of forests are managed intensively for timber and 80% are managed for a balance of values (including nontimber values and preservation) may fail to satisfy everyone. These reasons range from social and cultural considerations to political and sovereignty issues. However, the concept clearly suggests that there is insufficient global demand to justify intensive timber management on all or even most of the world's forested land. Zonation of the "working forest" into areas with more- and less-intensive timber management makes good economic sense and can also be good for nontimber values. Thus, some level of zonation also makes good environmental management sense.

Zonation is principally a landscape-level paradigm, but the concept of managing different types of forest ecosystems differently and for different values applies equally to stand-level variations in ecosystem function, structure, and composition along local gradients of soil moisture and fertility. This variation is captured by various ecological site classifications.

Variable Retention

The different values in a forest have different "ecological rotations" (Kimmins 1974; see below for a full definition and discussion of the term)—that is, they require different periods for their renewal. For example, trees suitable for wildlife (e.g., trees large enough to support eagle nests or provide winter dens for bears) and large decomposing logs require longer for their renewal than do trees suitable for commercial wood production. These different values can be sustained in different areas by means of zonation, but there is merit in sustaining values with different renewal periods in the same area by retaining elements of older forest within a matrix of younger forest zoned for timber production. This is achieved by retaining individual "wildlife trees," patches of wildlife trees, and snags within more intensively managed timber stands. Where there is a spatially variable retention strategy that accounts for different values, site types, management economics, worker safety, and other values, this paradigm is referred to as *variable-retention (VR) forestry* (Franklin et al. 1997). Variable retention is a subset of *structure-management silviculture* based on conceptual models of natural disturbance and stand development (Franklin et al. 2002). Although the approach was developed with wildlife and biodiversity issues in mind, it is applied as much for reasons of esthetics and public acceptance as for conservation purposes. As with the other paradigms, it can be used in combination with ecosystem management, adaptive management, and zonation.

Despite the many merits of variable-retention forestry systems, there can be significant insect, disease, windfirmness, biodiversity, and economic constraints on their design, and these may limit VR application in some forest ecosystems. Variable retention should be adapted for specific sites and situations rather than becoming an inflexible "one size fits all" rule. Variable retention and structure-management silviculture are both stand-level paradigms.

Emulation of Natural Disturbance and the Natural Range of Variation

A popular paradigm with some environmental groups and many natural disturbance ecologists is the emulation of natural disturbance (Attiwill 1994; Lieffers et al. 1996; Angelståm 1998; Y. Bergeron et al. 1999; Spence et al. 1999; McRae et al. 2001). This paradigm may also be discussed in terms of the concept of natural range of variation, which is the outcome of natural disturbance. The paradigm focuses on the landscape- and stand-level spatial patterns of ecosystem conditions that have resulted from past natural disturbance. The natural range of variation accounts for the historical, present, and possible future range of variation.

The natural range of variation can provide a very useful template for the management of such forests as boreal forests, which have experienced relatively frequent stand-replacing natural disturbances. It describes both the historical and the present range of disturbance patch sizes, the landscape-level pattern of disturbance patches and events, and the range in disturbance severity, and thus describes the variability in plant community composition and structure. Social, cultural, and economic considerations are then used to select the acceptable portion of this range. A comparison of the natural range of variation with the socially selected range of desired conditions alerts managers to the neglected portion of the natural range of variation (and its associated values).

The natural range of variation is less useful in ecosystems that have experienced infrequent, very large-scale, or very severe disturbances. Society is unlikely to accept the emulation of such disturbance regimes, and this constraint severely limits the utility of management designs based on the natural range of variation in such forests. Similarly, if the natural range of variation is applied to forests in which natural disturbance occurs at small spatial scales with infrequent stand-replacing disturbances, the natural range of variation may imply a level and frequency of intervention that is inconsistent with controlling tree disease, soil compaction, road density, forest composition, and other management considerations, and inconsistent with desired levels of wildlife and other measures of biodiversity.

Another problem with the emulation of natural disturbance and the natural range of variation is that long-term climatic variations, the invasion of nonnative species, air pollution, and the threat of rapid, human-induced climate change render some past disturbance regimes and the resultant natural range of variation inapplicable in the future. It also poses the question of which historical period should provide the baseline for the natural range of variation and whether human impacts on forests should be excluded from the definition of *natural*. These issues have led many experts to reject the paradigm of the natural range of variation based on the historical and present data in favor of managing for a desired future forest condition and pattern (i.e., the possible future range of variation), explored through process-based or hybrid ecosystem-simulation models.

The natural range of variation nonetheless remains a useful concept that is applicable in some areas, although it is probably not universally applicable. The natural range of variation has been developed largely as a paradigm for landscape-level patterns, but incorporates the stand-level patterns dealt with by the variable-retention paradigm.

As noted above, one way in which forests could be maintained in conditions close to those that have prevailed historically would be to design management-related disturbances or to combine "natural" and management disturbances so that their effects on the forest ecosystem's structures and processes are similar to those of disturbance regimes during the period selected as the baseline for comparison (e.g., Bengston et al. 2000). In British Columbia, interest in emulating natural disturbance has been expressed in the form of a natural disturbance type classification in the Biodiversity Guidelines of the British Columbia Forest Practices Code (British Columbia Ministry of Forests 1995a). However, this useful first step has not yet been developed in sufficient detail to constitute a basis for emulating natural disturbance. Much more detail is required to clarify the definition of disturbance, the natural range of variation in disturbance and resultant ecosystem conditions in particular forest types, what portion of this natural range is socially acceptable, and what approach to forest management would result in a socially acceptable and nondeclining pattern of change in forests.

Definition of Disturbance and Emulation of Natural Disturbance

Before we can discuss the emulation of natural disturbance, we must agree on what this phrase means. There has been considerable discussion of what *disturbance* means, and it is clear that the definition depends on the level of organization (suborganism, organism, population, community, or ecosystem) being considered, the spatial scale involved, and the time over which the definition is to be applied (White and Pickett 1985). White and Pickett defined disturbance as "any relatively discrete event in time that disrupts ecosystem, community, or population structure and changes resources, substrate availability, or the physical environment" (p. 7). They differentiate *perturbation, disaster,* and *catastrophe* as rare, occasional, or human-induced events that affect ecosystems. However, their discussion makes it clear that semantic problems persist in the debate over the meaning of these terms and that it remains important to define how the terminology is being used. Rogers (1996) noted that most ecologists focus on the disruption of the functions of ecosystems when they think about disturbance. His review provides a useful entry into the diversity of thought about ecological disturbance.

Arguably, one might abandon any attempt to define *disturbance.* Just as terms such as *clearcutting* and *old growth* pose formidable difficulties in their definition, and a single definition is rarely satisfactory in all circumstances, so the definition of disturbance will always depend on the continuum of temporal, spatial, and severity scales being considered and on the object, value, condition, or process being disturbed. Nevertheless, some definition is necessary before we can discuss, plan, and apply forestry practices that emulate natural disturbance and reproduce either a natural or historical range of ecosystem variation or a socially acceptable subset thereof.

Before presenting the working definitions of *natural disturbance* and its emulation that I use in this chapter, it is worthwhile examining the meaning of individual words.

Disturbance

There are many different interpretations of *disturbance. Webster's New Twentieth Century Dictionary,* second edition (McKechnie 1978), defines disturbance as "interruption of a settled state of things." But Edwards et al. (2000) noted that most forest ecosystems have no settled order, and that White and Pickett's (1985) definition may be more appropriate. However, it is unclear what "disrupts" ("to break or burst apart," in Webster's definition) in the White and Pickett definition (op. cit.) means in the context of an ecosystem that is constantly being changed by nondisruptive processes.

Paehlke (1995) provided no entry under *disturbance*—somewhat surprising for an encyclopedia of conservation and environmentalism. Similarly, a recent book on ecosystem management (Boyce and Haney 1997) gave no definition of disturbance, and presented no detailed treatment of the topic. There was also a notable absence of a discussion of disturbance in Costanza et al. (1992), despite the obvious importance of disturbance in defining ecosystem health.

Natural

Webster's dictionary defines *natural* as "forming part of, or arising from, nature; what is found or expected in nature," and "in a state provided by nature, without man-made changes; wild, uncultivated." But these definitions are confounded by Webster's definitions of *nature:* "the sum total of all things in time and space," "the primitive state of man," or "not cultivated or tamed; wild; savage," which either do not deny that humans are part of nature or which explicitly include humans. Evenden, quoted in Paehlke (1995), notes that *nature* is so ambiguous as to make it one of the most obscure terms in the English language. He points out that in the modern environmental movement, *nature* has been synonymous with *wilderness,* an etymological shift that would substantially change the word's traditional meaning.

One of the best treatments of *natural* is that of Peterken (1996), who noted that Henry David Thoreau, that "apostle of the wild," believed that mankind and nature form an ecological unity." Peterken believes that the definition of *natural* depends on personal philosophies and belief systems, but that strict adherence to definitions that exclude humans would exclude most terrestrial ecosystems from membership in the class *natural,* a view supported by Botkin (1995).

Emulation

Webster's defines *emulation* as "to try to equal or exceed," whereas *mimic* is "to copy closely, to imitate accurately, to simulate," and *copy* is "to duplicate, to fully reproduce, to imitate." There has been considerable discussion over whether forest management mimics or copies natural disturbance (e.g., McRae et al. 2001), and the word *emulation* was introduced as a more neutral term that better recognizes the inevitable differences between management-induced and nonanthropogenic effects. The essence of emulation, then, is to try to equal the overall effects of natural disturbance through well-designed management disturbance (Führer 2000).

A Working Definition

From this discussion, it seems that although we have a good grasp of the meaning of *emulation,* it is far from clear what *natural* means, and there is an urgent need to understand what *disturbance* means. As a working definition for this chapter, I define the emulation of natural disturbance as:

> Management over ecologically significant temporal and spatial scales that attempts to emulate the ecosystem effects of physical (allogenic) or biotic (biogenic) disturbance events, the frequency and/or severity of which have been changed by human action but which have historically determined the potential pathways, patterns, and rates of autogenic successional development in the ecosystem in question. Such emulation aims to maintain the historical range of variation, or a socially acceptable subset thereof, in desired ecosystem conditions and functions over defined spatial and temporal scales.

This is similar to the definition proposed by Perera and Buse (chapter 1, this volume). The rest of the present chapter explores the implications of my working definition, which interprets *disturbance* as *allogenic or biogenic events that change the prevailing rate, direction, and pathway of autogenic succession.* Disturbance is an event that either returns the ecosystem to an earlier seral stage, reinitiates the present seral stage (thereby temporarily arresting autogenic succession),

accelerates the transition to a subsequent seral stage (compared with the rate expected from autogenic succession), or alters the successional pathway (the sequence of seral stages) that the autogenic succession had been following. The normal autogenic processes of invasion, colonization, competition-related mortality, environmental alteration, and displacement of existing species, as in the stand-development phases of Oliver and Larson (1990) or Franklin et al. (2002), and their associated processes, are not considered ecosystem disturbances. I propose this definition in the full knowledge that disturbance is a continuum along axes of scale, severity, and frequency, and that, as a consequence, no single definition will satisfactorily serve all discussions of the management of disturbance.

Succession, Stand Dynamics, and the Ecological Role of Ecosystem Disturbance

The propensity to change over time is one of the key attributes of a forest ecosystem (Kimmins 1997a). This characteristic results from the ubiquity of allogenic and biogenic disturbances in the world's forests, and from the inevitability of the processes of stand dynamics and autogenic succession following disturbance. Allogenic and biogenic disturbances alter the overstory or understory species composition or the canopy cover; moreover, they cause changes in soil surface conditions that are sufficiently greater in their rates, spatial scale, and severity than would have been produced by autogenic processes that alter the direction, rate, or pathway of autogenic succession. The key to defining and understanding ecosystem disturbance, and thus to designing management interventions that can emulate it, is thus an understanding of autogenic succession and its relationship to allogenic and biogenic events.

According to the monoclimax theory of Clements (1916),[1] the absence of disturbance lets autogenic succession take a forest ecosystem through a series of seral stages to reach a self-perpetuating soil condition and biotic community dominated by the most shade-tolerant tree species in the region. Seedlings of the shade-tolerant canopy trees are able to become established and grow in the canopy gaps created by

---

[1]Many ecologists reject the simple Clements model of succession and recommend that the concepts of climax, successional convergence, and predictable seral sequences be abandoned, but I disagree. For many forest ecosystems in which the combined frequency and severity of retro-

gressive disturbance exceeds the rate of autogenic succession, the Clements model seems inadequate. However, where disturbance frequency and severity are low and autogenic succession is relatively rapid, the Clements model has some value.

the death of overstory dominants. Continuous occupation of the site by the same species may allow sufficient buildup of pathogens or parasites of these trees to promote gap formation through the death of individual trees or small groups of trees by largely autogenic processes. These biological agents may lower the likelihood of stand-replacing allogenic disturbance, replacing this process with autogenic events that create canopy gaps or small patches. The monoclimax theory suggests that the successional convergence of soil conditions that results from the accumulation, in the long-term absence of disturbance, of thick forest floors with increasing quantities of decayed wood results in climatic control of the self-perpetuating climax stand. Rich sites become nutritionally poorer, poor sites become richer, dry sites become more moist, and wet sites drier. The climax forest, reproducing itself by means of gap processes, is seen as the apex of ecosystem development in the absence of disturbance.

This tree-centric view of autogenic succession leading inevitably to a self-perpetuating, closed, or relatively closed forest stand in the long-term absence of disturbance is probably wrong for many of the world's forests, especially cold forests dominated by winter snow packs and very humid forests, such as many coastal, northern, and high-elevation forests in Canada. In such forests, the ubiquity of closed forest is the result of periodic disturbances at a frequency, scale, and severity that gives trees a competitive advantage over other life forms. Deviations (either too much or too little disturbance, or the wrong disturbance regime) from the critical range of disturbance regimes that favor closed-canopy forest communities will promote open forest, woodland, or grass, forb, shrub, or bryophyte communities that will become the climax community. Although such nonforest communities have traditionally been considered disturbance-induced seral stages, they may also be the ultimate climatic climax in the very-long-term absence of disturbance in the types of forest ecosystem noted above. This successional interpretation suggests that the disturbance-related polyclimax model of succession (Gleason 1927, 1939; Phillips 1935; Tansley 1935) provides a better explanation for the extent of closed forest than the monoclimax model.

One corollary of this hypothesis that this discussion is based on is that many forests, if not most, require a regime of periodic disturbances at a particular frequency, severity, and scale if they are to persist as tree-dominated communities within their historical range of variation. Another corollary is that as humans alter the type, severity, scale, and frequency of allogenic and biogenic disturbances, emulating the effects of the missing disturbances becomes not only acceptable but also necessary if we wish to sustain forests within their historical range of variation. Where human activity has increased the combined impact of the severity, scale, and frequency of disturbances and led to undesirable changes in forest condition, forest management should be designed to return disturbance to within its natural range of variation. Conversely, where we have reduced the overall impacts of ecologically appropriate, nonanthropogenic disturbance, management should replace the missing effects. Herein lies the rationale, and a template, for the emulation of natural disturbance.

Many forestry practices, such as planting in disturbed areas and stand density control (thinning), can emulate the population-level processes of stand dynamics: stand initiation, stem exclusion, and understory reinitiation, sensu Oliver and Larson (1990). Uneven-aged or variable-retention management can create forest structures similar in the living tree component to the "old growth" phase of a particular seral stage and with some elements of the dead tree component. There are notable differences, of course. Stand management, whether based on clearcutting, partial harvesting, or "continuous forest cover" and on even-aged or uneven-aged forestry, generally takes much or most of the stemwood away, whereas natural disturbances generally leave most of it in the forest. Planting may introduce nonnative tree species, or reestablish native species in an area where there is no natural seed source because past disturbances or autogenic succession have excluded the species from the area. Early thinning (precommercial thinning or spacing) can alter the tree species composition of the stand, which is sometimes the treatment's purpose. In fact, spacing operations in British Columbia generally have had a far more profound effect on tree species composition in the subsequent stand than do planting or harvesting methods. This is because early postharvest natural regeneration enriches the tree species composition of planted sites in naturally mixed-species forests as long as seed sources are available (i.e., as long as harvest openings are not larger than seed-dispersal distances). In spite of such differences, appropriately designed management can nevertheless emulate much of the natural stand dynamics.

Both forestry practices and natural disturbance affect the physical and chemical environment and biological processes: the microclimate and the soil are both altered. Practices such as clearcutting often emulate the microclimatic effects (e.g., light, humidity, wind) of severe, large-scale natural disturbances more closely than does partial harvesting, although the standing dead stems that remain after wildfire sometimes modify the microclimate significantly during the few years critical for tree seedling establishment after the disturbance. The microclimates resulting from less severe natural disturbances are better emulated by partial harvesting systems. Depending on the type of harvesting equipment used and the season of harvesting, timber removal may create more or less soil disturbance but more soil compaction than does natural disturbance, although the comparison again depends on the type, severity, and scale of the natural disturbance.

Public response to the visual impact of forest management has caused a trend toward requiring less soil and visual disturbance. As a consequence, forest harvesting is becoming a less severe disturbance than some natural events, and the resulting problem of desynchronization of soil and microclimatic conditions is likely to become more common in the future. Emulating natural disturbance sometimes requires more soil disturbance than is currently publicly acceptable.

What management has frequently failed to do is emulate the sequence of postdisturbance seral stages produced by autogenic succession. Early seral stages of herbs, shrubs, or unmarketable hardwoods have often been truncated by site preparation, "weed" control, and planting. Harvest-related disturbance and stand management have tended to perpetuate a particular seral stage, or even a subset of the stand development phases of that stage. At the stand level, this aspect of the emulation of natural disturbance requires attention. At the landscape scale, much remains to be done to alter spatial harvesting patterns so that they emulate the character and variability of the natural landscape more closely, if this is desired. In some cases, this will require the aggregation of disturbances in patterns that are the opposite of those suggested by esthetic and some other values.

The debate over the comparability of natural and harvest-related disturbance has focused on stand structure and composition, or on relatively temporary soil and microclimatic alterations. However, wildlife ecologists, hydrologists, stream ecologists, and fish biologists generally agree that the greatest ecological impact of timber harvesting is often caused by roads, and roads are probably the most "unnatural" aspect of harvesting disturbance. Forest management systems with frequent entries, partial harvesting, and low levels of disturbance generally require higher road densities, more frequent road use, a more permanent road system, and more frequent harvesting activity than low-frequency, more-complete harvesting systems such as clearcutting or shelterwood cuts. Illegal hunting, wildlife harassment, habitat fragmentation, visual impact, effects on disease organisms, the hydrological and fish habitat aspects of forest roads, and the risks of human-caused forest fires must all be considered in comparing natural and management-induced disturbance. If a system for the emulation of natural disturbance based on the natural range of variation in landscape patterns and stand characteristics requires more roads and more frequent activity, the total environmental impact of management may be greater than that of natural disturbance, and greater than in a system of timber management and harvesting that requires fewer roads and management activities but that is otherwise less similar to natural disturbance. The following discussion of successional management does not include these important considerations.

Figure 2.1 presents a model of natural disturbance of moderate to high severity interacting with autogenic succession in a hypothetical mesosere with the following communities: a deciduous broadleaved pioneer forest community; early, mid-, and late seral conifer communities; and a shrub plus herb woodland climax. In such a mesosere, where Egler's initial floristic composition pathway (Egler 1954) or the tolerance pathway of Noble and Slatyer (1977) is the expected pattern of ecosystem development, autogenic succession can undergo any one of a variety of postdisturbance sequences. The wide variety of possible successional consequences of allogenic and biogenic disturbance shown in this figure reflects differences in the type and severity of disturbance and the different seral stages in which the disturbance can occur.

Figure 2.1 also shows the possible successional consequences of allogenic and biogenic disturbances of low to moderate severity. In contrast to

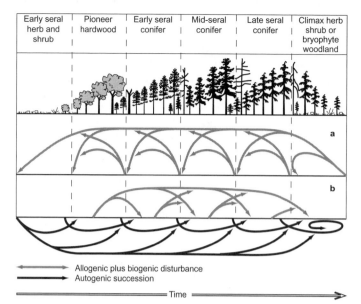

| Early seral herb and shrub | Pioneer hardwood | Early seral conifer | Mid-seral conifer | Late seral conifer | Climax herb shrub or bryophyte woodland |

Allogenic plus biogenic disturbance
Autogenic succession

Time

FIGURE 2.1. Possible pathways of successional retrogression resulting from disturbance of various types and spatial scales with (a) moderate to high severity or (b) low to moderate severity, occurring in various seral stages of a hypothetical mesosere.

the regressive effects of moderate to severe disturbances, low to moderate severity disturbances can accelerate autogenic succession. The wide diversity of possible regressive and progressive effects of disturbance on ecosystem conditions and dynamics lies at the root of the difficulty in making useful general statements about the effects of disturbance in forest ecosystems.

Figure 2.2 diagrams management-induced (biogenic) disturbance designed to emulate allogenic (nonmanagement) or biogenic disturbance to achieve a variety of seral objectives. Figure 2.2a presents successional diagrams for six different management-disturbance objectives: conversion of economically nonproductive early seral stages (e.g., by chemical, mechanical, fire, biotic, or manual vegetation management techniques) or late seral stages (by fire or mechanical disturbances) to an economically productive forest community, and disturbance-recovery sequences that maintain four different seral conditions. Figure 2.2b depicts the use of silvicultural strategies that employ disturbance to convert (accelerate) various seral stages into late, closed-forest seral conditions and then maintain these conditions. Figure 2.2c shows a variety of disturbances of late seral forest to initiate new seral sequences of various lengths, followed by management-induced low-severity disturbances to accelerate stand and seral development. Figure 2.2d contrasts silvicultural disturbance regimes that sustain (1) early seral or (2) mid- to late seral sequences. Achievement of these different successional pathways and outcomes requires the use of mixtures of several traditional silvicultural systems, modified in various ways (e.g., variable retention; Franklin et al. 1997) so as to emulate natural disturbance and recreate the desired portion of the historical range of variation in forest ecosystem conditions.

The Ecological Rotation Concept as a Framework for the Emulation of Natural Disturbance

An *ecological rotation* (Kimmins 1974) is the time required for an ecosystem or a particular ecosystem attribute (e.g., composition, structure, function, complexity, diversity) to return to its predisturbance condition or to reach some desired new condition following disturbance. The rotation is a function of the degree of disturbance-induced change interacting with the rate of autogenic recovery from that change (a measure of ecosystem resilience; figure 2.3).

The concept of an ecological rotation can provide a conceptual framework for designing regimes for the emulation of natural disturbance. It provides a way of taking the successional models in figures 2.1 and 2.2 and converting them into a management scenario. Figure 2.4 presents a hypothetical example that contrasts several alternatives:

- A conventional approach of repeated clearcutting and planting with an intermediate, 60-yr rotation (system 1 in the figure) or a short, 30-yr rotation (system 2);

FIGURE 2.2. Diagrammatic summary of various strategies for managing ecosystem disturbance to achieve a wide variety of successional objectives. (a) Ecosystem disturbance to convert either early seral herb-shrub communities or late seral herb-shrub woodland to productive closed forest, and disturbance and recovery sequences to maintain particular desired seral stages. (b) Acceleration of autogenic succession by selective harvesting combined with natural regeneration or underplanting to create late seral stand compositions and structures. (c) Management disturbance of late seral conditions to create different seral stages, and management to produce seral-stage sequences on the time scale of management rather than of succession. (d) Management-induced disturbance and accelerated recovery sequences to sustain either (1) early or (2) mid- to late seral sequences.

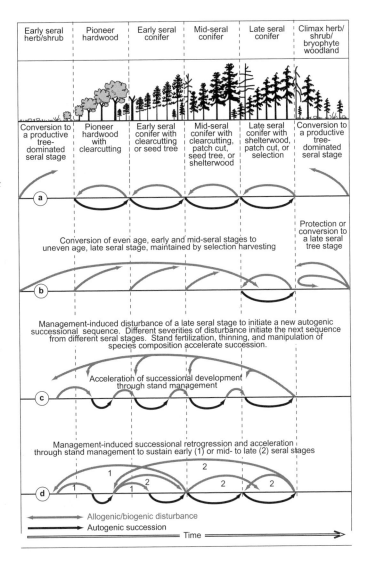

- A "soft touch," low severity, frequent entry disturbance regime (system 3); and

- One possible scenario for emulating natural disturbance (system 4).

In system 4, an infrequent high-severity, stand-replacing disturbance is followed by a period of stand development and succession accelerated by periodic lower-severity disturbance (stand density reduction by means of commercial thinning), followed by several partial harvests and underplanting. This system would cycle the ecosystem through a range of seral stages over time. The clearcutting systems (systems 1 and 2), with no thinning, sustain early to mid-seral conditions with a fairly small range of seral variation; the size of this range and the seral stage that

is maintained depend on the rotation length, which also determines how many phases of stand development (Oliver and Larson 1990) will occur before clearcutting. Without thinning, this system would sustain mid- and early seral species, respectively, in the stand-initiation and stem-exclusion phases of stand development. The understory reinitiation phase would only occur if thinning was undertaken to accelerate the stand's structural development.

The "soft touch," partial-harvest system depicted by system 3 (a harvest every 10 yr, or six entries in the same time as one rotation of the conventional clearcutting system) accelerates autogenic succession toward late successional conditions. The stand would exhibit many aspects of those conditions by the end of the first

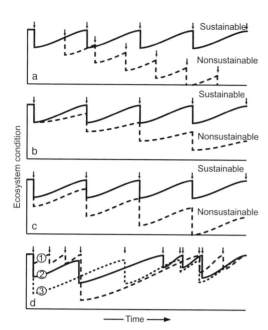

FIGURE 2.3. Relationship between ecosystem disturbance events (indicated by arrows), the frequency of disturbance, and the rate of ecosystem recovery (the slope of the recovery line). Sustainable and nonsustainable trends as a result of variation in (a) frequency, (b) rate of recovery, and (c) severity. (d) Variations over time in both severity and frequency of disturbance for three different disturbance regimes that might represent emulation of natural disturbance. Although the three scenarios in (d) differ greatly, they all return to the same starting condition. Panels a–c are based on Kimmins (1974).

60-yr period if silviculture manages the stand structure, and would sustain a late seral stage thereafter, assuming that a seed source of late seral species exists or that underplanting of shade-tolerant, late seral tree species occurs. This system would support only mid- to late seral species within a narrow range of seral conditions after the first 60–80 yr. Most early seral and many mid-seral species would be excluded.

The scenario for emulating natural disturbance (system 4) produces the widest range of successional conditions over time and supports the widest range of species, including early, mid-, and late seral species. It would also support the greatest range of forest products to allow for uncertain future market demands, would be the most adaptable in the face of accelerated global climate change, and would likely have the fewest problems with diseases, insects, and soil nutrition.

Figure 2.4 is only a hypothetical example. For most northern forests, the time scale in this fig-

ure is unreasonably short; the development of such ecosystems through the seral sequence requires longer periods even with thinning and underplanting. However, the ranges of seral conditions supported by the different systems appear to be valid, and the seral development under silvicultural management would certainly be much faster than in unmanaged stands. Figure 2.4 also implies an inevitable, Clementsian sequence of seral stages. In reality, a much greater variety of pathways would occur, as shown in figure 2.1.

As noted above, the successional management diagrams presented in this chapter have focused on stand structure, species composition, and seral stage. They have not addressed such important issues as landscape patterns of seral stages, the character of the edges of disturbance patches, and the many impacts of the road systems that are an inevitable consequence of forest management. Helicopter logging and long-line cable yarding can reduce road densities, and new techniques have rendered these technologies capable of some types of partial harvesting. Although they pose their own set of environmental impacts (e.g., high use of fossil fuels by helicopters), technological developments in timber harvesting are making a wider variety of the successional management options just examined much more feasible than in the past, and are thus increasing our ability to emulate natural disturbance.

Comparison of Natural and Harvest-Induced Disturbances in Western Canada and Alaska

A key question in the emulation of natural disturbance is how closely management disturbance emulates natural disturbance (cf. Keenan and Kimmins 1993; Kimmins 1997; McRae et al. 2001). This question must be evaluated at both the stand and landscape scales. At the stand scale, there are questions of the degree (severity) of ecosystem alteration and the frequency of disturbance. At the landscape scale, there are questions of pattern and of the ratios of different stand ages and seral stages that reflect the frequency and scale of stand-level disturbance. Frelich (2002) provides a useful review of forest disturbance in temperate forests.

Stand Level

Disturbance in humid and cool or cold northern forests

The major disturbance that maintains ecosystem productivity in interior Alaska is fire, whereas

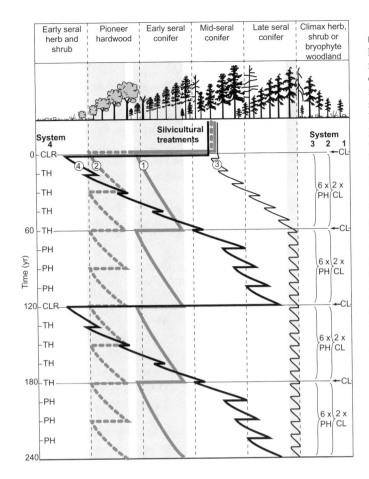

| Early seral herb and shrub | Pioneer hardwood | Early seral conifer | Mid-seral conifer | Late seral conifer | Climax herb, shrub or bryophyte woodland |

FIGURE 2.4. Seral sequences in a hypothetical mesosere that might result from conventional (system 1, corresponding to line 1) and short-rotation (system 2, line 2) even-aged silviculture with repeated rotations based on clearcutting followed by planting; a low-disturbance, frequent-entry, partial-harvest system with natural regeneration or under-planting of shade-tolerant species (system 3, line 3); and a system for the emulation of natural distur-bance that alternates moderate to severe harvest disturbance every 120 yr with periods of partial har-vesting and incorporates variable retention and silviculture for man-aging stand structure (system 4, line 4). The ranges of seral condi-tions sustained are far greater for the scenario with emulation of nat-ural disturbance than for the other systems. The time scales shown in this hypothetical example are prob-ably unreasonably short for many northern forests and shorter than it would take an unmanaged eco-system to develop through such a sequence of seral stages. The pat-terns are thought to be valid, however.

in coastal Alaska and northern coastal British Columbia it is windthrow and landslides.

The progressive decline in soil temperatures caused by the development of a thick forest floor and dense coniferous canopy in the cold boreal forest of the Alaskan interior causes nutritional stress that leads to the breakup of the spruce for-est and the development of a *Sphagnum* climax community (Crocker and Major 1955; Heilman 1966, 1968; Ugolini and Mann 1979; Banner et al. 1983). Fire or landslides remove the insulating layers, warm the soil or underlying mineral ma-terial, and initiate a new round of succession. Clearcutting without disturbing the mineral soil and minor vegetation (e.g., winter logging) may warm the upper soil horizons somewhat, but unless the operations break through the moss and organic mat, harvesting can accelerate the transition to late seral soil conditions that even-tually lead to a *Sphagnum* bog. Partial harvesting with no disturbance of soils and relatively minor disturbance of the vegetation would have a sim-ilar effect. Clearcutting and burning with the

retention of patches of live trees, snags, and scat-tered wildlife trees would probably emulate the key features of natural disturbance by fire most closely, but there is a strong public antipathy toward these practices, and the declining use of prescribed fire is a major challenge to the emula-tion of natural disturbance in these forests.

In cool and humid northern coastal forests, soil temperature is less important than the de-composition of the heavy aboveground litterfall and root litterfall, which results in the produc-tion of abundant humic and fulvic acids. Under the prevailing acidic forest-floor conditions, or-ganic colloids leach into the mineral soil, where they form organic coatings on the soil particles in the B-horizon. As these deposits accumulate, they reduce pore space, hydraulic conductivity, and gas exchange. The reduced aeration and drainage eventually restrict rooting and promote boggy conditions, which in turn impede nutri-tion and lead to the development of open canopies and *Sphagnum* communities. However, in many areas, windthrow and landslides prevent this sequence by stirring, loosening, and aerating the soil; exposing the organic-rich B-horizon to oxidation and decomposition; and maintaining soil aeration and drainage. Long before this pedogenic influence on the plant community develops, invasion by ericaceous shrubs, such as salal (*Gaultheria shallon* Pursh) and *Vaccinium* species, increases light and nutrient competition for tree seedlings and lowers nitrogen availability, contributing to the reduction in canopy cover and promoting the trend toward open, low-productivity forest. Clearcutting with minimal disturbance of soils and minor vegetation can promote the development of ericaceous shrubs and other minor vegetation and the regeneration of western redcedar (*Thuja plicata* Donn), and does little to promote the regeneration and growth of western hemlock (*Tsuga heterophylla* [Raf.] Sarg.). This "soft touch" clearcutting does little to emulate the dominant natural disturbances that occasionally disrupt the surface mat and mix and aerate the upper mineral soil.

The maintenance of closed forest with high levels of net primary productivity and tree growth in these ecosystems appears to require periodic disturbance by fire, stand-replacing windthrow (e.g., Prescott and Weetman 1994; Bormann et al. 1995), or landslides to reinitiate autogenic succession and prevent the systems from developing toward treeless climax communities of *Sphagnum* or ericaceous shrubs. Low intensity tree harvesting that fails to disturb the soil and minor vegetation can accelerate succession toward the treeless or open woodland condition, as will low severity allogenic or natural biogenic disturbance (see figure 2.1).

Clearcutting in which some proportion of the trees are pushed over would be the closest emulation of natural disturbance by wind in these coastal forests, but is technically infeasible, generally uneconomical, and visually unacceptable, and would pose strong risks of erosion, slope instability, and sedimentation of streams if conducted on slopes. Research on northern Vancouver Island suggests that surface scarification, at least of spots for planting or natural regeneration, would improve rooting, drainage, and soil aeration (Prescott and Weetman 1994). If combined (where needed) with fertilization to speed closure of the tree canopy and shading out of competing minor vegetation, this approach may offer the best system of forest renewal and the closest emulation of windthrow effects that is socially acceptable. The past practices of clearcutting and burning slash reduced but did not eliminate early problems with ericaceous shrubs. They did not duplicate the soil stirring caused by windthrow, and they reduced available site nitrogen sufficiently (by burning off the relatively nitrogen-rich surface of the forest floor and the fine fuels in the slash) to pose a threat to the longer-term growth and competitiveness of the trees if these practices were repeated over short rotations without nitrogen fertilization. The removal of up to 1600 $m^3 \cdot ha^{-1}$ of stemwood from clearcut productive stands in harvest areas constitutes a substantial difference between harvesting and windthrow. The long-term nutritional consequences of this removal are not yet clear, but are being explored using the FORECAST model (Kimmins et al. 1999a; Brunner and Kimmins 2003).

Disturbance in high-elevation forests in interior British Columbia and Alberta: The Engelmann spruce-subalpine fir biogeoclimatic zone

Although relatively undisturbed mixed-species stands can be found (Antos and Parish 2002), and small-scale disturbance and gap dynamics do play a role in boreal, northern, and subalpine forests (McCarthy 2001), most closed forest stands in this ecological zone are the result of infrequent severe fires (Veblen et al. 1994). Bark beetles (*Dendroctonus rufipennis* Kirby) have also periodically played an important role (Veblen et al. 1991; Parish et al. 1999). Postfire areas within the Engelmann spruce-subalpine fir biogeoclimatic zone often regenerate slowly—up to 100 yr is required on some sites in some subzones to reach commercial stocking levels (Jull 1990) and produce a forest of lodgepole pine (*Pinus contorta* Dougl. var. *latifolia* Engelm.) or Engelmann spruce (*Picea engelmanni* Parry), depending on the size and severity of the fire, seed sources, slope and aspect, and the climatic subzone. The more shade-tolerant subalpine fir (*Abies lasiocarpa* [Hook.] Nutt.) may colonize sites at the same time or after the establishment of pine and spruce, but its relatively slow early growth generally puts it in a subordinate crown position. The initial cohort of pine disappears from these stands within 250 yr and does not regenerate without stand-replacing disturbances, leaving a dominant spruce overstory, an occasional large subalpine fir, and a subordinate canopy layer consisting mainly of subalpine fir with a few spruce. Susceptibility to bark beetles and root rots prevents most of the fir from reaching great ages and large sizes, giving this species a reverse-J size distribution in comparison to the relatively flat age distribution for the spruce in stands older than 300 yr (see Clark 1994). In stands that escape stand-replacing disturbances for very long periods, subalpine fir may entirely replace spruce, but small-scale disturbance or regeneration of spruce on decaying logs generally sustains a spruce component in these stands (Antos and Parish 2002).

Simply clearcutting and planting late seral "gappy" forests in this zone, with minimal disturbance of surface soils and of the herb and shrub layer, has frequently resulted in unsatisfactory regeneration and failed to closely emulate the effects of wildfire or beetle epidemics. The best seedling growth often occurs on disturbed soil at the edge of skid trails, where the soil has not been severely compacted. Exposed mineral soil mixed with forest-floor material and freed from herb and shrub competition produces superior growth compared with undisturbed areas. If a closed tree canopy with a minimal shrub and herb understory had existed prior to logging, regeneration can be successful (if slow) in the absence of soil disturbance. However, where a significant component of herb- or shrub-filled gaps existed prior to logging and there is little disturbance to this vegetation, regeneration may be largely restricted to road edges and areas previously occupied by closed forest or clumps of trees. The undisturbed patches of herbs and shrubs appear highly resistant to tree invasion and may have persisted through several cycles of low to medium severity disturbance (Caza 1991). Fungal pathogens of seeds in the forest floor also greatly limit the recruitment of tree seedlings (Zhong and van der Kamp 1999).

A hot slash burn or spot treatments using herbicides or scarification can reduce these problems and permit the establishment (albeit slow) of new trees in such clearcuts, thereby more closely emulating infrequent stand-replacing natural disturbance. Low-disturbance partial harvesting and reliance on natural regeneration can potentially accelerate the loss of forest cover and the expansion of herb and shrub communities (see figure 2.1). In contrast, partial or small-patch harvesting with disturbance of soils and minor vegetation, combined with planting and natural regeneration, can be a successful regeneration strategy. Moreover, frost and problems from the pressure of deep snow can occur in clearcuts. Frost may be somewhat ameliorated in small

patch cuts and is greatly reduced in partial harvesting systems, but it may still be necessary to control minor vegetation that was abundant in the preharvest stand.

Disturbance on northern Vancouver Island

As was the case for northern coastal forests, clearcutting does not closely emulate the effects of stand-replacing windthrow in natural "second growth" western hemlock-amabilis fir (*Abies amabilis* [Dougl.] Forbes) stands on northern Vancouver Island (Prescott and Weetman 1994). In these stands, windthrow can periodically deposit up to 1600 m^3·ha^{-1} of stemwood on the ground; in contrast, harvesting removes much of this wood from the site. The past practice of burning slash in clearcuts—in this area, where natural fire is infrequent—also failed to closely emulate natural disturbance, but the major difference between disturbances by wind and harvesting on northern Vancouver Island is often the degree of physical soil disturbance. Clearcutting and burning in old growth western redcedar-western hemlock stands in this area has no natural disturbance analog, because these stands have a low probability of stand-level windthrow. However, clearcutting with mechanical disturbance or burning and planting of redcedar can result in redcedar-salal (a dominant ericaceous shrub) stands that become increasingly similar to natural redcedar-hemlock stands over time, except for the lower size and age of the redcedar and the reduced frequency of large decaying redcedar logs.

Although clearcutting does not emulate several of the effects of wind disturbance, it does mimic stand replacement in response to windstorms more closely than does any other silvicultural system. Not all windstorms fell all the trees in the area affected by this type of disturbance, so a mixture of clearcutting, clearcutting with reserves, variable retention, irregular shelterwood and group-selection cuts would probably emulate landscape-level effects more closely than ubiquitous clearcutting. However, the latter three systems would more closely emulate the processes that lead to dwarf mistletoe (*Arceuthobium campylopodium* f. sp. *tsugensis* Rosend. [Gill]) infestations in developing stands, and this result would probably not be socially acceptable. Also, recent experience with variable retention in these western forests suggests a low survival probability for isolated small groups or individual reserved trees in wind-exposed locations. Many become snags or coarse woody debris as a result of wind damage on such sites.

Based on the results of the SCHIRP research project (Prescott and Weetman 1994), it appears that replacement of western hemlock-amabilis fir stands in the absence of windstorms is best accomplished by clearcutting followed by planting and the application of nitrogen fertilizers, which results in stands very similar to those that develop after natural wind disturbance, except for the lower quantities of decaying logs. As noted above, there is no natural disturbance analog for harvesting very old western redcedar-western hemlock-salal stands. The current practice of clearcutting followed by spot scarification, planting, and fertilization appears to be an effective way to establish productive stands on these sites, but this approach does not emulate natural disturbance.

*Forests of the dry interior and
dry coastal Douglas-fir zones*

Millennia of relatively frequent stand-maintaining underburning and occasional stand-replacing fire, epidemics of bark beetles and defoliating insects, and scattered outbreaks of root decay fungi have created a diversity of stand structures in the forests of the interior Douglas-fir (*Pseudotsuga menziesii* [Mirb.] Franco) zone of British Columbia, ranging from dense, even-aged stands to open, savanna-like stands and multi-age stands. Emulating this natural range of structural and compositional variation and the range of disturbance that produced it would require a diverse mixture of silvicultural systems, including clearcutting, clearcutting with retention, VR systems, patch cutting, shelterwood cuts, and group selection. Anywhere there is a wide range of types, severities, and scales of disturbance, stand-level emulation of natural disturbance requires the use of a variety of silvicultural systems to reflect the variety of stand characteristics produced by natural disturbance.

In the summer-dry forests of the coastal Douglas-fir zone of British Columbia, underburning by indigenous peoples appears to have been a major source of disturbance that maintained overstory dominance by relatively fire-resistant Douglas-fir (except on moist, fertile sites). It prevented the development of mixed stands of Douglas-fir, western redcedar, western hemlock, and grand fir (*Abies grandis* [Dougl.] Lindl.) on medium (zonal) and drier sites. Underburning periodically opened up the Douglas-fir overstory canopy to varying degrees and eliminated the understory and regeneration, creating suitable light and seedbed conditions on these sites for regeneration of the relatively shade intolerant Douglas-fir. Regeneration of redcedar, western hemlock and grand fir also occurred if there were seed sources, but regeneration of these fire-sensitive species was periodically removed by subsequent fires, and they failed to dominate these stands. However, if these sites are partially harvested without underburning, the shade tolerant species persist. In the understory, they are protected from the summer heat and the moisture stress that accompanies periodic summer droughts in this zone. But if the Douglas-fir overstory is removed in a subsequent partial harvest, they are exposed to drought and heat stress to which they are not adapted and damage or death is common. There is considerable local climatic variation within this area due to variations in rain-shadow effects, drainage of cold air from the nearby mountains, and slope and aspect; as a result, there is considerable local variation in natural disturbance, the natural and historical range of variation, and stand types.

Clearcutting without burning at locations in this zone with cool, moist characteristics (including aspect) does not accurately emulate historical disturbance patterns either, as it releases the understory of shade-tolerant species. Some of these will be killed by the periodic summer droughts, but many will survive in poor physiological condition, subject to attack by insects and diseases. However, the Douglas-fir will eventually dominate these sites, as this species is the best adapted to the climate and tolerates the drought stress. Shelterwood cuts with the retention of some overstory Douglas-fir will emulate the structural aspects of historical underburning over the short term better than clearcutting, but will reduce Douglas-fir populations in the long run, because this approach favors the shade-tolerant species unless the understory's species composition is controlled manually. Low-impact, selective removal of canopy Douglas-fir ("high-grading") will accelerate the loss of Douglas-fir, the growth of shade-tolerant species in the understory, and the development of a physiologically maladapted forest. Group selection and narrow strip cuts have the same effect. Thus, clearcutting with retention or shelterwood cuts, both with underburning, would most closely emulate natural disturbance. Because the burning of slash or understory vegetation is not socially acceptable in this area, clearcutting with retention (essentially a seed-tree system) and precommercial spacing to regulate the understory composition could be substituted.

At hotter, drier locations in the coastal Douglas-fir zone, regeneration of shade-tolerant species is much reduced. The lower leaf area of the canopy on such sites casts less shade, so Douglas-fir regeneration is more successful in partial harvesting systems than it is in cooler, wetter areas in this zone. Combinations of shelterwood cuts, small patch cuts, and group-selection systems, all with permanent retention of some large old trees, would emulate the range of variation in stand structure and composition that has resulted from historical disturbance patterns on these hotter and drier sites.

Landscape Level

Different disturbance types in different forest types produce a wide variety of landscape patterns in terms of forest age, species composition, and stand structure. Different disturbance frequencies (return intervals) and severities result in diverse age-class distributions and variations in space and time in the relative areas of different forest conditions. Emulation of these aspects of disturbance diversity requires differences in the rates, types and spatial distributions of timber harvesting. The network of roads in the landscape represents another major, often neglected distinction between natural and management-related disturbances. Roads affect wildlife movement and natural predation, can cause some types of fragmentation, generally increase illegal hunting, and can affect a landscape's hydrology, streams, and fish populations.

Natural disturbance produces a distribution of disturbance sizes that usually includes a large number of small patches, fewer medium-sized patches, and a small number of large patches. In terms of area, the opposite trend occurs, with most of the disturbed area accounted for by large patches and the many small disturbances accounting for a small proportion of the total disturbed area (figure 2.5a). Each forest type has a unique relationship between the frequency of disturbance sizes and areas, and emulating natural disturbance requires that the socially acceptable portion of these frequency distributions becomes the basis for the design of cut blocks.

Conventional, administrative forestry has generally resulted in most cut blocks being of a similar size—close to the maximum size permitted, so as to reduce harvesting and road construction costs (figure 2.5b). With the advent of helicopter logging, it has become possible to use greater

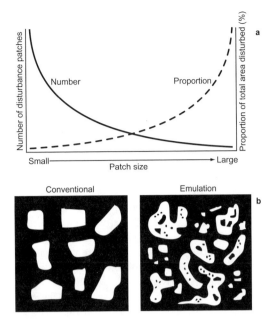

FIGURE 2.5. (a) A generalized frequency distribution for the sizes of disturbance patches and the proportion of the total disturbed area in different patch sizes. (b) Conventional clearcutting (at left) produces a regular pattern of relatively rectangular openings (disturbance patches) across the landscape, whereas emulating natural disturbance (at right) would produce a variety of patch sizes and shapes, with green retention patches or individual residual trees within larger patches.

numbers of small blocks in mountainous areas without the negative consequences of increased road construction. However, social pressures generally limit the upper end of the frequency distribution for the size of harvesting areas, leading to a strange situation in which wildlife managers and ecologists are often arguing to permit a number of significantly larger cut blocks, while some conservation groups are trying to prevent this in the name of "biodiversity" or "stewardship"—the "beauty and the beast" or "small is always good" syndrome (Kimmins 1999). Emulation of natural disturbance requires a much larger range in cut block size than current public opinion or (until recently) the British Columbia Forest Practices Code have allowed. Having all the harvested area in the form of small patches can fragment wildlife habitat, reduces the amount of forest interior, increases total road length, and leads to increased harassment of sensitive wildlife species. However, aggregating all harvesting into a few large patches excludes the habitat features associated with small openings and reduces the ecotone (edge) habitat that some species require.

In addition to frequency distributions of patch sizes, emulation of natural disturbance must also consider the spatial distribution of the patches. Natural disturbance caused by diseases generally produces scattered small patches, as do landslides. However, windstorms, insect epidemics, and fires generally produce a range of disturbance patches in the overall area of disturbance. In any given region, there is a frequency distribution of disturbance event sizes, as well as a frequency distribution of patch sizes within events. Landscape-level emulation of natural disturbance would use harvesting to produce a range of disturbance sizes and patch sizes within each disturbance similar to those observed in nature (figure 2.6). Many natural disturbances occur over days or weeks, whereas harvesting the same area might take months or years. Nevertheless, the resulting landscape patterns would be similar. Roads in these areas can be deactivated after harvesting is complete, leaving the area as young forest-pocket wilderness until the next harvest.

A final aspect of emulating natural disturbance is the nature of edges and the connectivity of forest stands. Some natural and most management-induced disturbances produce "hard" edges in which the transition from maximum to minimum disturbance occurs over a short distance (i.e., less than a tree's height). These hard edges produce numerous edge effects with positive or negative impacts on various abiotic and biotic characteristics of the landscape (Chen et al. 1992; Murcia 1995; Harper and Macdonald 2001). For example, a review of these edge effects found evidence for increased nest predation, greater herbivory, higher species richness, and accelerated plant mortality at edges compared with intact forest (Murcia 1995). Hard edges have also been shown to channel animal movements along rather than through the edges, thereby changing the flow of animals within the landscape and creating predator corridors that provide easy access to prey (Forman 1995).

In contrast, many natural disturbances create "soft" or diffuse edges, with the transition covering considerable distances (i.e., several to many tree heights). The communities that develop in and around these soft edges may be qualitatively different from those that develop in the forest's interior, within openings, or along hard edges due to the increased horizontal and vertical complexity of the transition (Fleming and Giuliano 2001; McIntire 2003). Early evidence suggests that some of the negative impacts associated with hard edges are less common than with more diffuse, naturally created edges (Suarez et al. 1997; Fleming and Giuliano 1998, 2001; Saracco and Collazo 1999). The complex spatial pattern of natural regeneration from intact mature forest into naturally disturbed stands also increases ecosystem complexity compared with structurally simpler hard edges. If emulating natural disturbance is the management goal, the character and variability of patch edges must be defined, understood, and emulated.

Emulation of Natural Disturbance as an Ecosystem-Level Strategy

Much of the debate over the paradigm of emulating natural disturbance has been based on population- or community-level attributes (e.g., stand structure) or simply on visual appearances. Many people object to clearcutting because it does not appear to resemble natural disturbance; they want things to "look natural" and believe that "soft touch," low-impact partial harvesting is a "kinder, gentler, more natural forestry." In contrast, others have claimed that clearcutting emulates natural disturbance because it creates broad-scale stand replacement, as do severe fire and wind disturbance. The truth often lies somewhere in between (Keenan and Kimmins 1993; Kimmins 1997; McRae et al. 2001).

Clearcutting often fails to emulate many aspects of natural disturbance. Failure to retain standing dead trees and scattered large, live individuals or patches of trees generally differentiates past clearcutting from natural disturbance,

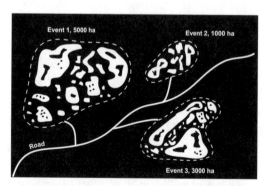

FIGURE 2.6. Emulating landscape-scale patterns of natural disturbance requires the concentration of harvest patches into events, each of which has its own patch-size distribution. Based on D. W. Andison (pers. comm.) and Saskatchewan Environment and Resources Management (2002).

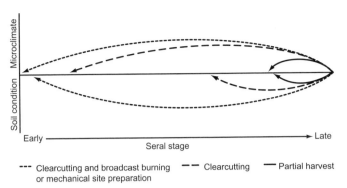

FIGURE 2.7. Desynchronization of the microclimatic and soil components of a seral stage by low-impact clearcutting, and resynchronization by soil-disturbance treatments, compared with the impacts of partial harvesting.

as does the removal of most of the stem volume. Failure to kill or suppress herbs and shrubs, remove moss or lichen mats, and prepare suitable seedbeds are some of the other ways in which some clearcutting fails to emulate some natural disturbances. In many instances, clearcutting is a less severe disturbance than fire, windstorms, or landslides. Although it generally produces an early seral microclimate, clearcutting often produces much less regression to earlier seral soil, seedling bank, and minor vegetation conditions than does stand-replacing natural disturbance. This is why foresters have frequently combined clearcutting with burning of slash or mechanical soil disturbance to resynchronize the microclimatic and soil components of the seral stage produced by harvesting (figure 2.7).

Disturbance that returns stand microclimate to that of an earlier seral stage without a similar effect on soil conditions can desynchronize the soil and atmospheric components of a seral stage, making it difficult for some plants to become reestablished. Species adapted to early seral microclimates are usually not well adapted to establishment on a late seral forest floor, especially one exposed to an early seral microclimate. This is one reason why burning slash and scarification can be an ecologically appropriate treatment for a clearcut that has a late seral forest floor.

If the ecological or other effects of soil disturbance are unacceptable and desynchronization impedes prompt ecosystem recovery, a partial harvesting system may be needed rather than clearcutting. However, compared with more severe disturbances, some types of partial harvesting may exclude early seral species or reduce their abundance. This would affect, for example, flowering plants important for insects, berry-producing shrubs important for wildlife, and shade-intolerant deciduous hardwoods important for maintaining and improving soil fertility, long-term site productivity, and many measures of biodiversity.

Clearly, the design of management-induced disturbance must take a multiscale, multivalue ecosystem approach and must include an assessment of the long-term consequences of different disturbance strategies. "Looking nice" today is simply not a sufficient basis for such designs, even though visual forest values are important (Kimmins 2001).

In the absence of experience over multiple rotations, long-term evaluations require the use of stand-level models, such as those of Kimmins et al. (1999a,b) and Seely et al. (1999) or landscape-level models (e.g., Mladenoff and Baker 1999; Gustafson et al. 2000; Pennanen and Kuuluvainen 2002; the models described in other chapters of this volume). For a recent review of modeling tools in forestry, see Messier et al. (2003).

The Role of Modeling in Sustainable Forest Management

Experience has been the traditional basis for making predictions in forestry. The complexity of forest ecosystems and their responses to disturbance has exceeded either our understanding of ecosystem processes or our ability to synthesize this understanding into accurate and reliable predictors. However, the rapidly increasing power of computers and improved techniques for synthesizing our expanding knowledge of ecosystems, as well as escalating public demands for a different kind of forestry, have rendered purely experience-based forecasting unsuitable for many of the challenges we face today (Kimmins 1990; Korzukhin et al. 1996; Kimmins et al. 1999b).

Modeling forest ecosystems has traditionally focused on either the stand or the landscape level, but rarely both. Stand-level growth and yield models dealing with harvestable volumes were the earliest type of forestry model (Assmann

1970) and are still the dominant type. Landscape-level models were initially nonspatial aggregates of stand-level growth and yield models used to calculate allowable harvesting rates and identify age-class gaps in the "normal" forest. More recently, landscape-level timber supply models have become spatial and have revealed the short-comings of nonspatial timber supply models with respect to sustainable multivalue forest management.

Stand-level ecosystem management models are relatively recent; their main progenitor is the JABOWA gap model developed in the late 1960s (Botkin et al. 1972). The International Biological Program of the 1970s and 1980s stimulated the growth of process modeling, which produced a wide variety of detailed physiologically based stand-level models of populations, communities, or (less commonly) ecosystems. Although these models have made great contributions to re-search and education, they have proven difficult to use in stand or landscape management, al-though some have been applied in regional and global environmental assessments (e.g., Running and Gower 1991; Running and Hunt 1993). How-ever, there is a growing acceptance of the need to incorporate ecophysiological and ecological processes into management models (Korzukhin et al. 1996; Morris et al. 1997; Proe et al. 1997; Seely et al. 1999), and the trend toward certifica-tion of sustainable forestry requires increasing use of hybrid (experience plus knowledge-based) ecosystem management models (e.g., FORECAST; Kimmins et al. 1999a).

The emulation of natural disturbance has also tended to focus on either stand-level succes-sional processes and results or on landscape-level disturbance, the resulting patterns, and the relationship of these patterns to landscape-scale processes. There is a new trend developing to link these two scales in modeling. By combining stand-level, hybrid models (e.g., FORECAST) with landscape-level forest models such as HORIZON (Kimmins et al. 1999b) or ATLAS (Nelson 2000, 2001; Messier et al. 2003), one can scale up the ecophysiological and stand-level detail to the landscape level, thereby incorporating this local ecosystem information into the representation of landscape-level processes. Simulation of the emulation of natural disturbance requires this cross-scale linkage. Purely stand-level perspectives fail to place the emulation of natural disturbance into the landscape context, whereas purely land-scape-level perspectives fail to adequately repre-sent the stand-level processes that drive much of the landscape-level processes.

Such modeling will probably become a stan-dard component of sustainable forest manage-ment based on emulating natural disturbance, on another approach, or (more likely) on a combi-nation of approaches. For the predictions of such models to be meaningful to professional for-esters and the public, they will increasingly have to be linked with visualization systems (Shep-pard and Harshaw 2001). This visualization will be needed for spatially explicit, individual-tree models of complex stands (e.g., FORCEE) and spatially explicit models of complex cut blocks (e.g., LLEMS), as well as for more conventional stand- and landscape-level models (Kimmins 2001).

Conclusions

Forestry is first and foremost about people—not ecology, biodiversity, or other aspects of the natural sciences. Forestry is the art, skill, prac-tice, science, and business of managing forested landscapes to sustain the balance of values, con-ditions, and environmental services that are eco-logically possible and desired by humans. Bio-diversity is desired in developed countries because its environmental and social values are now rec-ognized. Biodiversity conservation is *not* an objective of forest management when people do not understand its importance, or when other necessities of life and survival are more urgent.

Because forestry is people-centric, the choice of one or more paradigms for forestry is always strongly influenced by such social considera-tions as employment, worker safety, recreation, aesthetics, spiritual values, resource supply, eco-nomics, and the creation of wealth. These con-siderations are always balanced against the pre-dominantly biophysical considerations of most biophysical scientists and environmentalists. As choices are made between the applicability of different paradigms, including emulating natu-ral disturbance, the full range of social, man-agerial, and economic consequences must be valued against the full range of ecological, envi-ronmental, and biodiversity considerations. The outcome will undoubtedly be that emulating natural disturbance becomes the primary tem-plate for the design of sustainable forest manage-ment in some situations, but becomes relatively unimportant or only one of several approaches in others. Emulating natural disturbance is one tool in the toolbox of sustainable forest management; it must not be permitted to become a religion (see Kimmins 1993). The complexity of this approach requires the use of multiscale ecosystem-management models that address a wide variety of environmental and social values, and not only ecological values and conditions.

In his review of disturbance ecology and forest management, Rogers (1996) noted the paradigm shift that has taken foresters from thinking of disturbance as a negative event to recognizing that natural disturbance regimes or emulation of their ecological consequences is essential for the maintenance of long-term ecosystem function within the historical range of variation—which is interpreted to mean a "healthy" ecosystem. He also noted the convergence between ecology and the social sciences. Unfettered emulation of large-scale and severe natural disturbance is not socially, politically, and economically acceptable in a world whose human population has doubled over the past 40 yr and in which an anticipated three to four billion more people will be added to the population within the period covered by a commercial rotation for northern coniferous trees. Nonetheless, acceptance of the role of disturbance suggests that we should not control all fires, insect outbreaks, or diseases; nor should we try to salvage all trees threatened by these disturbances. Where we exerted such control in the past, we should now attempt to emulate natural disturbance by replacing some of the missing ecosystem components and conditions that resulted from disturbance.

Managers will have to establish a balance between emulating nature slavishly (where we know what "nature" is or has been) and what is politically and socially acceptable. Although much remains to be done to describe and understand past natural disturbance, at least as much work remains to be done to inform the public about the ecological role of and necessity for appropriate disturbance regimes, so that foresters can obtain a social license to practice some level of emulation of natural disturbance. The past and current campaigns of various environmental groups against such disturbances as clearcutting, and the insistence by these and other groups that forestry should promote late seral conditions by implementing low-disturbance management everywhere (see, e.g., figure 2.4, system 3) pose as great a threat to the values we wish to sustain in many of our forests as the nearly ubiquitous application of clearcutting in a way that failed to emulate the variability of natural disturbance.

In defining systems for the emulation of natural disturbance, foresters, forest ecologists, and other scientists must form a partnership with social scientists and public interest groups to promote an understanding of the ecological role of disturbance and develop a social license to practice emulation of natural disturbance in an economically and environmentally sustainable manner. This activity should be guided by the thoughts of Aldo Leopold (1953; cited in Leopold 1966) in his essay *The Land Ethic:*

> The evolution of a land ethic is an intellectual as well as emotional process. Conservation is paved with good intentions that prove to be futile, or even dangerous, because they are devoid of critical understanding either of the land, or of economic land use.

Occam's razor, annunciated by William of Occam (1285–1349), suggests that the simplest explanation for a phenomenon should be preferred. This philosophy has become a basic tenet of science, and the classical scientific method of hypothesis and deduction has encouraged reductionism and simplification as the basis for understanding. However, a literal translation of Occam's razor is that "plurality should not be posited without necessity." In other words, the razor has two edges: *as simple as possible but as complex as necessary.* One of the basic problems in debating stewardship and other environmental issues in forestry arises from oversimplification and overgeneralization. Unless we recognize and understand nature's spatial and temporal diversity, we will be unable to respect nature in our forest management. Successful emulation of natural disturbance requires "a critical understanding of the land" (i.e., ecosystems) and of "economic land use" (i.e., the relationship between humans and forests).

There is little doubt that forestry is, and should be, undergoing a paradigm shift. We have a variety of new paradigms to choose from, many of which are not mutually exclusive and will probably be used in combination. One of these is the emulation of natural disturbance. In the absence of adequate experience and knowledge, the coarse filter approach to forest stewardship and conservation has many merits. This approach is based on the notion that "nature knows best"—that historical ecosystem disturbance regimes have resulted in nature as we know it today and continue to produce the many values we want from forests. According to this concept, sustaining these values requires that the ecosystem effects of these historical disturbance regimes be sustained by permitting natural disturbance to occur, or, where this is socially unacceptable, by emulating this disturbance through management.

Acknowledgments

I thank Robin Duchesneau and Eliot McIntire for their contributions to this chapter, Maxine Horner for typing repeated drafts, and Christine Chourmouzis for the artwork.

CHAPTER 3

The Ecological and Genetic Basis for Emulating Natural Disturbance in Forest Management
Theory Guiding Practice

IAN D. THOMPSON and ALTON S. HARESTAD

Concerns about biodiversity loss, old forests, and ecological processes have given rise to the development of new approaches to forest management. Forests support the majority of terrestrial biodiversity, and wherever timber harvesting is conducted, questions arise about whether these forests can recover the biological diversity that existed prior to logging. There is global concern over the scale at which forestry is practiced and the long-term effects on ecosystems and biodiversity (e.g., Sala et al. 2000). As part of a broad-based response to these concerns, forest management has recently begun a paradigm shift away from sustained yield toward silviculture and planning that is "close to nature," ecologically based and sustainable, or that emulates natural disturbances at scales from forest stands to large landscapes (e.g., Harris 1984; Hunter 1990; Attiwill 1994).

The concept of emulating natural disturbance resulted from concern over the sustainability of all forest resources, including biodiversity; from new concepts of complexity, disturbance regimes, thresholds, and scale; and from the nascent sciences of landscape ecology and conservation biology. Managers accept that forests should be used sustainably, that ecosystems are complex and have multiple states, and that management should avoid going beyond the ecosystem's resilience (e.g., Perry 1994). Natural systems are dynamic, but can be so substantially altered by management that unintended consequences can result. This change in thinking about forest management followed a flood of papers in the late 1970s that underscored the importance of disturbance in forest ecosystems from the tropics (Whitmore 1975) to the boreal regions (Heinsel-man 1973; Van Wagner 1978). These studies indicated that temporal variability was the norm in all forests, and were summarized in the important book by Pickett and White (1985).

Fire is a natural disturbance in forests across much of North America and helps determine the characteristics of forests from fine to broad spatial and temporal scales. However, human-caused fires and fire suppression may have changed the magnitude and size distribution of the area burned in a given year compared with natural disturbance regimes (Johnson 1992; Li 2000c; Granström 2001). Furthermore, forest management may be altering the processes required to maintain forest function and species at many scales. The effects in all forest types and under all management regimes may vary for most species, but current knowledge suggests that some ways of managing forests are better than others (Haila et al. 1994; Franklin et al. 1997; Niemela 1999). Emulating natural disturbance is a management model that attempts to protect biodiversity in the long term, at multiple spatial scales, by providing stands and landscapes with structural conditions to which species have adapted (e.g., Hunter 1990). Forest management has changed landscape patterns and forest age classes compared with what would be present (and had been present prior to commercial forestry) under a natural disturbance regime (e.g., Harris 1984; Mladenoff et al. 1993; Perera and Baldwin 2000).

Forestry practices can alter habitats at spatial and temporal scales important to the survival of populations of plants and animals (Bunnell and Huggard 1999). Long-term declines in some species may have resulted from the loss of mature and old forests, reductions in the contiguous area of forest, and reduced stand complexity (Hansen et al. 1991; Virkkala et al. 1994). In Finland, for example, 43% of the endangered species have been affected by forest management, and most are associated with older forests of various types (Rassi et al. 1991, cited in Niemela 1999). Concerns over such changes relate to ecological genetics, dispersal, the effects of biodiversity on forest productivity, the continued capacity for stable communities to develop, and habitat loss and change. Forest managers must learn whether the forest communities that result from harvesting will resemble those that would result from natural disturbance, thereby maintaining the associated biodiversity and providing the same goods and services. In other words, is it possible to use forests while maintaining their ecological integrity and ensuring their stability (e.g., in terms of age structure, species composition, physical structures, and processes) within the known bounds of variation? What harvesting characteristics (e.g., size, rate, legacies, dispersion) can best reconcile the preharvest and eventual postharvest stands?

These questions require a detailed ecological understanding of the basis for forest management at multiple scales, and much of the basic science is not yet known; considerable research, some of it long term, remains to be done. Important basic principles of population genetics, population ecology, conservation biology, and theoretical ecology lie at the heart of the issue, and must be understood before managers can plan to emulate natural disturbance. Managers often inadvertently constrain their options by attempting to reduce the risk of detrimental effects on the future forest. Unfortunately, these constraints can affect factors beyond landscape structures and exacerbate some problems; for example, for the same total area harvested, small cut blocks produce more roads and forest edge than do large cut blocks.

This chapter expands on the ecological basis for emulating natural disturbance presented by Attiwill (1994). It focuses primarily on the effects of forestry on wildlife and trees and their habitats, and draws on mostly North American examples. To describe the ecological basis, we have reviewed the theoretical and empirical information pertaining to species within ecosystems, and the reasons why, beyond ethical considerations, it is important to maintain species. Most arguments for emulating natural disturbance ultimately depend on theories of population genetics and systems ecology and on the need to manage forests to maintain species and their genetic diversity in the face of environmental variability and uncertainty.

Miller (1999) noted that attention to genetic diversity during management planning would improve our ability to achieve long-term conservation objectives. These arguments, based on genetics theory, "scale up" to considerations of how to manage patches across a landscape in time and space so as to maximize the probability of maintaining genetic diversity and species survival—that is, the probability of conducting sustainable forest management.

Disturbance theory is an underlying theme throughout this chapter, but is not explicitly discussed. Numerous discussions of disturbance ecology and theory are available, including Pickett and White (1985), Hunter (1990), Frelich (2002), and Kimmins (chapter 2, this volume). Our current knowledge of disturbance confirms that variability is a normal feature of forest ecosystems and that these ecosystems accommodate this variation; this in turn provides guidance to managers on whether emulating natural disturbance is a good model or perhaps is even necessary. Arguments that support the emulation of natural disturbance fall into three types: arguments based on species adaptation, ecosystem processes, and conservation biology. We discuss the latter two together because they are closely related and stem from similar fundamental principles. When integrated, the three arguments form a compelling theoretical basis for emulating natural disturbance. However, direct empirical evidence is scant, and managers should design and implement this new paradigm conservatively.

Disturbance Emulation and Species Adaptation

Concerns about Forestry Based on Ecological and Population Genetics

Loss of genetic diversity within species reduces the capacity of individuals to respond to environmental variation

Genetic diversity is the basis for evolution and for the adaptive response of individuals to environmental change and other selection pressures. Genetic diversity is evident at various scales, including within and between populations. Factors that reduce genetic variation are thought to reduce the capacity of a species to adapt to changes and can lead to localized extinction. However, these effects may not always be expressed. Populations challenged by different environmental conditions undergo local selection. Isolation of populations can lead to divergence, and when the populations reconnect, the frequency of alleles may differ in the joined populations. The existing genetic diversity is the product of challenges, selection, isolation, and the joining of populations at fine to broad spatial and temporal scales. Forest management activities may simulate some or all of these factors (Ledig 1992; Miller 1999; Mosseler et al. 2000), but there has not yet been enough research to confirm the nature or magnitude of their effects (Hedrick and Miller 1992).

An important and well-documented concern is that inbred individuals are less productive and adaptable to change, and possibly more prone to extinction than individuals with higher levels of heterozygosity (Van Delden 1994; Friedman 1997; Mosseler et al. 2000, 2003; Brook et al. 2002). The Florida panther (*Puma concolor coryi*) is one vertebrate that has suffered this inbreeding depression or fixation (i.e., homozygosity in one or more alleles) through genetic drift related to habitat loss (Hedrick 1995; Maehr et al. 2002; references in W. E. Johnson et al. 2001). This species now experiences low survival of its young (W. E. Johnson et al. 2001). Such genetic issues are also a concern because future environments are unlikely to be similar to current conditions. Loss of genetic diversity could reduce the capacity of a species to respond to environmental variation, and species with low genetic diversity may become extinct.

Forest management may reduce genetic diversity within species

Ledig (1992), Van Delden (1994), Friedman (1997), and Miller (1999) reviewed possible forestry and other effects on genetic diversity and raised a number of concerns with respect to human impacts on evolutionary processes. First, high-grading, deforestation, forest fragmentation, loss of old-growth forest, and plantation forestry may all reduce local and regional within-species genetic diversity. The mechanisms include direct loss of genes through excessive removal of a species locally and regionally (e.g., mahogany species throughout Central and South America, especially in the Caribbean), inbreeding among remnant individuals, selfing in trees, and enhanced genetic drift (random changes in gene frequencies within populations of a species). The evidence that selective removal of trees directly affects the genetic diversity of commercial tree species is equivocal; adverse effects have been reported in some studies (e.g., Bergmann et al. 1990; Buchert et al. 1997; Lee et al. 2002; Rajora et al. 2002), but no effects have been reported in others (e.g., Rajora et al. 1998; Glaubitz et al. 2000). Some geneticists question the validity of some of the former studies, which do not necessarily measure genes relevant to adaptive characteristics and provide only a small window on total genetic variation (Hedrick and Miller 1992; A. Mosseler, pers. comm.). Outplanting studies under controlled environments would be required to demonstrate physiological responses to these changes. Moreover, such studies may fail to detect rare alleles that may confer adaptability under certain environmental conditions.

Environmental alterations (e.g., landslides, changing water tables), changes in the forest's age structure over large areas, and accidental or intentional movement of organisms by humans may cause species invasions, disease or pest problems, hybridization, and altered selection pressures, all of which can reduce genetic variation. Ledig (1992) and Mosseler et al. (2003) stated that extensive hybridization between black spruce (*Picea mariana* [Mill.] B.S.P.) and red spruce (*P. rubens* Sarg.) after large-scale logging in New Brunswick and Nova Scotia was reducing the viability of populations of the latter species.

Landscape configurations produced by forest management may create barriers to dispersal for some species, while facilitating movements of others. In the former case, this can isolate populations; in the latter, it may lead to hybridization or altered selection pressures from invasive competitors. The widespread facilitation of dispersal by roads or other linear corridors may benefit mobile generalist species such as coyotes (*Canis latrans*), but may only rarely benefit habitat specialists (Sutherland et al. 2000), although the concomitant effects of habitat loss may complicate interpretation of the results of studies done on the specialists.

Rare alleles are of particular concern, because enough individuals must possess the trait to ensure that it remains in the population. Yanchuk (2001) estimated that conserving an allele with a frequency of 0.01 (i.e., that occurred in only 1% of the population) would require a total population of 280,000 individuals, presumably under relatively constant environmental conditions. He further suggested that five hundred individuals would be sufficient to maintain an allele with a frequency of 0.05. To maintain an adaptive balance between genetic drift and mutations, populations should be at least five hundred to one thousand individuals (Franklin 1980; Lynch 1996).

Alteration of gene frequencies or the biological processes that contribute to these frequencies could potentially diminish a genotype's ability to persist. Such theoretical estimates depend on their underlying assumptions, but nonetheless help assess the difficulties of retaining rare traits in a population. Clearly, doing so is not a simple matter of saving isolated habitats, and the theory helps set realistic expectations for management initiatives. The Kermode bear (*Ursus americanus kermodei*), which is the white phase of the more familiar black bear (*U. americanus*), is found on islands off the northern coast of British Columbia, Canada. This white phase accounts for 10–20% of island populations and remains at these levels primarily due to isolation from the mainland; however, its persistence could be at risk if forestry practices increased the rate of immigration of black bears from the mainland (Marshall and Ritland 2002). Rare genes may confer no advantage to a species under current environmental conditions, but may become important when conditions change. Managing forests to maintain rare genes (and genetic diversity as a whole) may thus help a species adapt to future environments.

The aspects of ecological genetics discussed in this section provide a means of assessing management practices by identifying potential problems and the actual problems that resulted from past forestry practices. Species have adapted to the habitats, climates, and disturbances within their local and regional environments; thus, management should maintain the genetic diversity within and among populations that resulted from this adaptation. Current research and management initiatives are primarily directed at vertebrates, and scant attention has been given to maintaining viable populations of most other species, especially groundcover vegetation and species with known roles, such as decomposers. Yet these species contribute greatly to the ecological processes that support forests and the wildlife they contain. Genetic theory suggests that outcrossing must be encouraged by preventing the isolation of populations, which could deplete local populations to the point that negative consequences arise. Rajora et al. (2002) provide an example based on inbreeding in isolated populations of white pine (*Pinus strobus* L.).

Genetic theory also suggests that managers should prefer local stock for reforestation and should adhere to zoning models that maintain local genomes, but should also include some non-local trees to enhance genetic diversity. Moreover, they should make special provisions for rare species, endangered species, and rare habitats. Of course, forests must be managed for more than just genetic conservation, and at each higher scale of organization of the forest, managers must consider the system's emergent properties (Allen and Hoekstra 1992). For example, rare habitats express a spatially rare suite of environmental conditions within a larger area defined by different environmental conditions, regardless of scale. Such properties at the site and stand levels contribute to the variation inherent in ecosystems at the landscape level. Forest management must thus maintain or create the full range of ecological conditions in managed forests—including rare habitats—to prevent the loss of biodiversity and the ability to support this diversity.

Dispersal is the mechanism that enables outcrossing

Population genetics assumes that dispersal is the key mechanism for gene flow (Emlen 1972) and defines the ability of a species to maintain its genetic variability over time and across space through outcrossing. Animals and plants disperse as part of a process to improve their individual fitnesses by locating habitats that are optimal for breeding and that minimize competition for resources. Although evidence on the effects of forest management on species dispersal is limited, species that inhabit a particular region have generally adapted to the landscape pattern that developed in response to the region's typical natural disturbances (Hunter 1990; Niemela 1999). Some forestry practices may hinder dispersal for some species but eliminate barriers to dispersal for others. For example, white pine blister rust (*Cronartium ribicola*) and chestnut blight (*Cryphonectria parasitica*) dispersed rapidly in anthropogenically fragmented landscapes, but were less capable of dispersing in the continuous forests that existed at the time of European settlement of North America (Perry and Amaranthus 1997). However, dispersal of Florida panthers is limited by the lack of available habitat (Maehr et al. 2002).

Connectivity of habitat is important to dispersing individuals

Organisms differ in their dispersal distances based on their size, life history, and capacity to move (vagility). Small soil Collembola cannot, on their own, disperse across long distances, and the dispersal of propagules of sessile organisms is obviously affected differently by forest management than that of vagile organisms. However, the lack of empirical information on dispersal distances and mechanisms is a major limitation to understanding whether forest management disrupts movements of animals and plants (Greene et al. 1999; Sutherland et al. 2000) and to the development of population models. Metapopulation models (i.e., models of groups of distinct populations) can only be quantified by knowing the numbers of individuals involved, their dispersal rates, and the range of distances crossed: managers need this information to understand the effects of altered landscapes on species survival (Beissinger and Westphal 1998). An important question with respect to the dispersal of juvenile vertebrates in forests is the relative hostility of each habitat type (i.e., the risk of an individual's death) for each species (Harestad and Sutherland 2001). For example, annual trapping and predation rates for furbearers such as marten (*Martes americana*) were significantly greater in second-growth forests at least 40 yr old than in expanses of old natural forest during the marten's fall-winter juvenile dispersal (Thompson and Colgan 1987). However, the duration of this hostile effect in a given forest type as it ages is not known; it may depend on recruitment of new forest structures, for example.

Sutherland et al. (2000) modeled the dispersal distances of mammals and birds. They attributed the observed variation to taxon, body size, and diet type, with carnivores dispersing two orders of magnitude farther than herbivores or omnivores of the same body size. In contrast, many smaller or sessile forest organisms have a limited capacity to disperse by their own movements. Such species either rely on other organisms or wind to carry their propagules over long distances, or disperse their propagules only over short distances. Some of these species survive as "legacies" (Hansen et al. 1991) after disturbance and repopulate an area only gradually as the forest regrows. Wilkinson (1997) noted that for many tree species, even wind dispersal is confined to distances of only a few meters. Large, open cut blocks created by logging may eliminate species with low dispersal capability from an area, possibly removing them from the pool of local species and decreasing genetic diversity within the species. In tropical and neotropical forests, large-scale forest clearing has reduced populations of some birds and mammals and hence diminished their role as seed-dispersal agents, preventing some plant species from re-

colonizing open areas (e.g., Guariguata and Pinard 1998). Forest management that attempts to emulate natural disturbance patterns, such as creating small openings in tropical forests rather than expansive clearcuts, would be expected to reduce such effects.

Forest Landscape Patterns Can Affect the Species Assemblage

Multiple factors determine the number and abundance of species in a forest

In individual stands, the communities derive from a species pool that occurs over a broader spatial scale (e.g., landscape, region). At this scale, species do not occur uniformly; hence, even among similar forest patches, the particular species occupying each patch may differ. Habitats are nonuniform as well, so communities also vary at landscape scales. Local species richness may be controlled ultimately by top-down, extrinsic, broad-scale factors such as climate (Currie 1991; Thompson 2000). These factors can lead to appreciable differences in species richness across a forest landscape. Furthermore, species assembly depends to a large degree on history, stochastic factors, availability of suitable habitats, environmental gradients, and the recruitment of species through immigration. Forest management may disrupt the processes that historically led to a given species assemblage by affecting the capacity of the regional species pool to occupy habitats and disperse.

Numerous processes may affect the ability of an individual species to occupy a forest stand, including competition for resources and demographic factors. At a local scale, variation in species diversity is affected by numerous proximal biotic and abiotic variables, as well as by annual variation in the species pool (e.g., Haila et al. 1994). When studying the effects of change in forests, the alternative states in similar forest types must be considered. Furthermore, a patch of habitat may or may not be occupied by a particular species at various times (Pulliam and Danielson 1991). For example, the absence of a given species (perhaps an indicator species) may suggest problems or changes in the ecosystem, but that absence can only be conclusively demonstrated by means of a monitoring program that recognizes spatial and temporal variability in abundance. Although a given community's structure will vary over both short and long periods, the most successful competitors will almost always be present in the absence of massive disturbance. However, some seral species can be so common in early stages that other species that will eventually dominate the old forest may be difficult to detect. Although the presence of these dominant species may indicate challenges to that ecosystem's function, their absence suggests a loss. Similarly, the absence of uncommon species may reveal changes in ecosystem function or simply demonstrate the system's inherent variability. Neither presence nor absence necessarily reveals how close a system may be to a threshold in species richness.

Species composition can be altered by harvesting

The alteration, simplification, and homogenization of forest characteristics from stand to landscape scales reduces the variety of resources and leads to an influx of generalist species that can dominate communities in the succeeding forests or produce atypical predation (e.g., Crooks and Soulé 1999). For example, the spread of coyotes from the western-central United States across North America, including into the boreal, Acadian, and Great Lakes–St. Lawrence forests, has resulted largely from the capacity of the species to occupy open and altered habitats after logging and land-clearing, with associated road construction and a consequent decline in numbers of wolves (*Canis lupus*) (Harrison and Chapin 1998; Crooks and Soulé 1999). Similarly, Amur tigers (*Panthera tigris altaica*) have declined in Russia as a result of increased access to their habitats combined with habitat fragmentation and loss (Kerley et al. 2002).

Managed forest ecosystems are not necessarily converging on their prelogging spatial patterns (Mladenoff et al. 1993; Gluck and Rempel 1996; Elkie and Rempel 2000; Perera and Baldwin 2000), plant communities (Johnston and Elliott 1996; Carleton 2000; Lindenmayer et al. 2002; Rees and Juday 2002), or associated animal communities (Schieck et al. 1995; Hobson and Schieck 1999; Imbeau et al. 1999; Niemela 1999; Thompson et al. 1999; Voigt et al. 2000). Much of this evidence comes from the northern boreal forests. At least among animal communities, differences seem related to habitat alteration following logging, but no effort has been made to determine how facilitation, competition, or predation among species could affect the community structures of second-growth forests. However, in the southern boreal forest, convergence of the tree and groundcover communities between logged and burned stands has been re-

ported (Reich et al. 2001) in jack pine (*Pinus banksiana* Lamb.), black spruce, and trembling aspen (*Populus tremuloides* Michx.) forests in Minnesota. Reich et al. cited the "patchy" nature of logged and burned areas that apparently was sufficient to enable local communities to reassemble after disturbances and eventually approach their prelogging state.

Often, history and serendipity are responsible for the reassembly of communities, and ecological indeterminism (i.e., variability in the communities that appear on a given site type) may be common, because nature is heterogeneous in time and space. For example, McCune and Allen (1985) found that only 10% of the variance among tree communities in nearby canyons was explained by environmental variables, and the rest was attributable to history and stochastic factors. Frelich (2002) notes that multiple disturbances within a short period create a different forest state than that which develops after a single disturbance. Although variation among nearby communities is not surprising, many studies indicate sufficient similarity among the vertebrate communities in boreal forests to suggest that these communities do not arise solely by chance (e.g., Voigt et al. 2000). Perhaps forest managers can only attempt to ensure that all species have the opportunity to become part of the postharvest communities at some stage. This does not mean that managers should accept defeat, but rather that we must recognize our inability to accurately predict the future at all scales or to assert full control over the ecosystem, and must compensate for these inabilities by using a range of management regimes.

Species may be adapted to disturbance regimes

The diversity of wildlife species is positively associated with the diversity of forest habitats (Boecklen 1986; Ruggerio et al. 1991; Bunnell 1995; Welsh and Lougheed 1996; McKinnon 1998). One premise behind managing to emulate natural disturbance is that many species are thought to be regionally adapted to the stand and landscape patterns and structures created by natural disturbance (Mutch 1970; Noble and Slatyer 1980; Hunter 1990; Bunnell 1995). Although there is some evidence to the contrary (Titterington et al. 1979; Coates 2002), most research suggests that certain species found in naturally disturbed forest stands are absent or significantly less abundant in previously logged stands of similar types and ages (Hutto 1995;

Drobyshev 1999; Hobson and Schieck 1999; Simila et al. 2002; Simon et al. 2002), even 40–50 yr after logging.

Bunnell (1995) reports correlations between fire regimes and the proportions of species that, for example, breed in early seral stages or nest in cavities. These relations were interpreted as evidence that the birds and mammals have adapted to particular fire regimes, but the gross (mean) characteristics of disturbance may not be the features that actually determine the habitat opportunities for these species. Such relations may depend more on extremes and the specific disturbance features that create habitat. The character of the landscape is determined more by a few large disturbances than by many small disturbances (Johnson 1992; Perera and Baldwin 2000). Furthermore, Bunnell's (1995) relations are difficult to interpret, because proportions of the total number of species were used as correlates, and hence although proportions may increase, the total number of species could decrease.

Rather than assuming that species have adapted to disturbance, it is simpler to assume that species have different requirements that are not equally available in the various seral stages and forest types. The habitat requirements for such species as deer mice (*Peromyscus maniculatus*) are broadly available across seral stages. However, the requirements for other species are confined to specific seral stages; for example, pileated woodpeckers (*Dryocopus pileatus*) excavate their nests in large diameter, decaying trees and require large old trees for foraging (Harestad and Keisker 1989; Flemming et al. 1999; Bonar 2000), and marbled murrelets (*Brachyramphus marmoratus*) nest in old trees with large diameter branches that form suitable nesting platforms (Manley et al. 1999; Meyer and Miller 2002). The correlations reported by Bunnell (1995) may simply arise from the availability of structures that themselves depend on forest age and thus correlate indirectly with a given disturbance regime.

Habitat Loss and Fragmentation in Continuous Forests

Habitat loss and fragmentation reduce species richness

Habitat loss and fragmentation are key mechanisms behind population declines and extinction, especially for habitat specialists, when habitat patches become small and isolated (e.g., Wilcox and Murphy 1985; Lovejoy et al. 1986; Bierre-gaard et al. 1992; Tilman et al. 1994). In these situations, reducing forest habitat to about 30% or less of the landscape leads to disproportionately large declines in species populations and richness (Andren 1994). Similar declines have also been observed for biomass and tree species richness in fragmented tropical forests (Laurance et al. 1997). There is experimental evidence of the negative effects of fragmentation and the positive effects of connectivity between habitats for invertebrates (e.g., Gilbert et al. 1998; Partridge et al. 1998), but as yet, little experimental evidence exists for vertebrates. The processes leading to extinction in fragmented landscapes may vary, but include inbreeding, increased predation rates or an altered suite of predators, competition from invading species, changes in microclimate, loss of food plants, and increased levels of parasitism (Andren 1994). In highly fragmented landscapes where forests are now interspersed with agriculture, the land between forest patches will remain unforested. However, fragmentation in continuous forests is usually ephemeral and may be of concern for only a short period and in a relatively small portion of the landscape (Thompson and Angelstâm 1999; Boutin and Hebert 2002; Schmiegelow and Monkkonen 2002). After a short period, the managed area regenerates sufficiently to again allow dispersal of most organisms, although exceptions may occur, especially for species with limited capacity to disperse and those susceptible to high levels of predation (see the section of this chapter entitled *Connectivity of habitat is important to dispersing individuals*).

Boreal and temperate forests are naturally heterogeneous across a landscape, and a certain level of isolation always exists among patches of identical or similar ecosystems, in part imposed by natural disturbance. However, patches of specific forest types or habitats, such as old-growth forest or a rare microhabitat, may become much more fragmented under intensive forest management, where rotation age is short and harvest rates high compared with the natural rate of stand replacement. Situations could develop in which patches of the old forest habitat required by various species become rare across a landscape. The result would combine the impacts of habitat loss (e.g., logging of an old forest type faster than its rate of creation) and fragmentation (increased distance among patches and small patch size), causing an Allee effect (in which fitness decreases disproportionately for small populations). This effect has been well documented for larger animals (Schaller 1976; Hedrick 1995; Beier 1996; Robertson and van Shaik 2001) and some species of forest birds in Scandinavia (Virkkala et al. 1994), particularly as a result of the loss of old forest habitats. However, these studies reported the combined effects of the habitat loss and its spatial configuration (expressed as fragmentation) rather than isolating the effects of the two factors.

Habitat availability is the most important factor affecting populations

Most studies of fragmentation have focused on the characteristics of individual fragments, but forestry problems and their solutions must be assessed at broader scales. Ecological processes also occur at broad scales, and effects observed at fine scales are not necessarily expressed at broader ones. Hence, focusing on the larger landscape is necessary. In particular, forest management often causes habitat loss through the conversion of mature forests to young stands or new forest types, the elimination of old forests, the replacement of natural forest by plantations, and the loss of stand or landscape structures. These changes result in multiple secondary effects, such as invasions by exotics, conversion of endemic diseases to epidemics, and alteration of herbivore communities, as well as increasing the effects of access by recreationists, trappers, and hunters.

Based on both empirical and theoretical considerations, Thompson (1987), Harrison and Bruna (1999), and Boutin and Hebert (2002) argued that habitat availability, not patch size or fragmentation, ought to be the primary management objective for maintaining biodiversity in boreal forests. In particular, Andren (1994) and Boutin and Hebert (2002) suggested that as long as habitat availability is above the critical threshold for a species (typically about 30%), then the landscape configuration in boreal forests is of little consequence. This coarse filter approach, in which general requirements of most species are met, neglects the effects on species for which less than 30% of the landscape normally provides habitat, referred to as "micro-ecosystems" by Haila (1994). Such rare ecosystems include isolated habitats with rare plants, or the habitat of certain amphibians in an old forest's riparian areas. In tropical forests, local species richness is high and individual abundance is usually low, so Boutin and Hebert's (2002) 30% rule would be unsuccessful as a conservation tool, because individuals of a species may be far apart and thus, effectively isolated. Nevertheless, the concept of managing to emulate natural disturbance in tropical forests could still apply if the approach involved small patch cuts and selection harvests to maintain habitat (i.e., reducing habitat loss).

It is important to maintain high quality habitats for species

Animals and plants can tolerate a range of habitat conditions. Some conditions increase individual fitness, whereas others decrease it. Metapopulation theory recognizes that source habitats (habitats where reproduction exceeds mortality) are better than sink habitats (habitats where mortality exceeds reproduction) in one or more aspects of habitat quality, and that the key feature may be something as simple as patch size. However, in terms of survival and reproduction, habitats that supply sufficient food, breeding sites, shelter, and cover from predators are preferred to those that provide insufficient amounts of any of these resources. A major legacy of forest management has been the change in relative species composition in some types of forest (e.g., Mladenoff et al. 1993; Carleton 2000). Modeling such changes over time often indicates further declines in certain dominant tree species (e.g., Baker et al. 1996). Therefore, long-term, cumulative, incremental loss of particular habitats may occur, especially for current types of old-growth forest.

Although it is difficult to measure the biological fitness of an individual in a forest, various surrogate measures correlate with fitness. Thompson and Colgan (1987) provided an example for marten in Canadian boreal forests. In old primary forests, whether food was abundant or scarce, marten maintained smaller home ranges than in successional forests that had been logged 10–40 yr earlier. In that study, the size of individual home ranges correlated with habitat quality defined in terms of the abundance of the marten's prey. This has also been suggested more generally for animals in terms of other aspects of habitat quality (Harestad and Bunnell 1979). The latter Ontario study, as well as studies of marten in Maine (Chapin et al. 1998a) and Quebec (Potvin et al. 2000), have all shown that marten establish territories in postlogging successional forests. However, closer examination revealed that the home ranges of marten always included some uncut forest, were larger than in nearby unlogged areas, and were significantly larger than in old primary boreal forests in Ontario. Furthermore, mortality rates were generally much higher in young forests than was the case in old forests. This evidence suggests that marten can indeed live in successional forests, but at a cost in terms of individual fitness, indicating that young forests provide suboptimal habitats. This example suggests that many animal species, especially those that prefer old forests, may only be tolerating suboptimal habitats in managed forests, resulting in small populations that are susceptible to catastrophe in the event of environmental variation. Such suboptimal habitats would also occur in landscapes subject to natural disturbance, so managers must understand the acceptable range in the amount of high quality habitat and its distribution in space and time.

Forest harvesting may not always result in a poorer habitat (at comparable age) than under natural disturbance regimes, but much of the evidence to date suggests that it may, especially for species that prefer old forest (Bunnell and Huggard 1999). However, managers who observe animals living in suboptimal conditions and have no basis for defining high quality habitats begin to accept suboptimal habitats as the norm. A good example is provided by the grizzly bear (*Ursus arctos*), a species that now occupies only the fringes of a former range that once included much of the North American Great Plains (Craighead and Mitchell 1982). (In this example, logging is clearly not the only or even the primary factor for the decreased range.) Furthermore, temporal variation, at least at finer spatial scales, does not mean that a species' presence guarantees persistence. Under typical environmental conditions, a species can occupy a range of habitats that differ in quality, but the range of suitable habitats is likely to wax and wane as environmental conditions change.

Disturbance Emulation, Ecosystem Processes, and Conservation Ecology

Maintaining Ecosystem Processes and Functions

A relationship exists between biodiversity and forest productivity

Partly as a result of the number of species that have gone extinct in the past century and the current large number of species on endangered lists, there has been increased research into the roles that species play in the functioning of ecosystems. There are concerns over the capability of ecosystems to continue providing a range of goods and services with reduced numbers of species or with different species complements. Several recent literature reviews examine the re-lationship between species richness (often called "diversity" in these papers) and various ecosystem functions and processes (e.g., Chapin et al. 1998b; Schlapfer and Schmid 1999; Loreau 2000; Hector et al. 2001). The general consensus has been that there is still limited evidence on which to base general conclusions about the relationship between species richness and ecosystem function. However, almost all of the studies reviewed were conducted over short periods, and the effects of changes in biodiversity may occur over the long term and relate to broader changes than the simple short-term perturbations in state that are most often studied (Lawton and Brown 1993; Loreau and Mouquet 1999). It is important to remember that biodiversity losses across broad regions may be more consequential than losses at a local scale (Tilman 1999; Hector et al. 2001).

Another unknown is the functional role that rare species may play in buffering an ecosystem against extreme changes (Schulze and Mooney 1993). For example, after the massive 1998 ice storm in eastern North America, some rarer species (e.g., red spruce) might increase their growth and abundance in response to reduced competition for light or nutrients on the disturbed sites. The relationship between stability in systems and species diversity is still unknown, but most studies suggest that biodiversity plays a role in an ecosystem's resilience, resistance to disturbance, and variability in time (McNaughton 1977; Naeem and Li 1997; Tilman 1999; Diaz and Cabido 2001; Kennedy et al. 2002). However, most models of diversity have produced inconclusive results (May 1972; Loreau 2000). Doak et al. (1998) have suggested that the effects of one species may compensate for those of another species and thus may dampen the effect of environmental variation on ecosystem function. In other words, greater biodiversity provides a greater range of variability and thus an increased ability to withstand variable selective pressures. Unfortunately, none of the work has been scaled up from simple patch studies to whole landscapes, and the only definitive studies have been in structurally simple grasslands rather than in complex forests.

Changes in species composition could affect the long-term stability of an ecosystem

There appears to be widespread functional redundancy in forest communities (Diaz and Cabido 2001). That is, several species may fulfil the same roles (although possibly in different ways), and there are strong interrelationships among processes in forests (Perry and Amaranthus 1997). However, functional diversity appears to be important to the overall functioning of the ecosystem (Diaz and Cabido 2001). The evidence for this is good enough to suggest that simplification of forest ecosystems through the elimination of species could cause instability. Certainly, in the face of the projected broad-scale influences of climate change, continued human population growth, and invasive exotic species, it would be prudent to maintain diversity of ecosystems and processes as much as possible, and thus to possibly buffer the effects of excessive changes (Chapin et al. 1998b).

Systems may function at different levels, depending on the community structure, but multiple stable states probably exist for any system, and these states may vary in time and in space (Cardinale et al. 2000). Recovery of the forest ecosystem after a perturbation depends on the intensity of the disturbance and the availability of species to recolonize sites both locally and across regions. To ensure that ecosystems are used sustainably, the management regime may need to maintain landscape and stand levels of structural diversity that are similar in time and in space to those produced by natural disturbance, thus preserving niches for the regional suite of species. This means that managers take into account forest ages, forest types, and their spatial configuration and extent across a landscape (e.g., Hunter 1990)—in other words, they must try to emulate the variability caused by natural disturbance. Although species richness generally increases in response to disturbance (Petraitis et al. 1989), some species are primarily or only associated with older forests, so providing these habitats is important to the survival of these species (Thomas 1979; Bunnell 1995).

Disturbance can move an ecosystem into a different state

Forests are complex systems, with multiple processes and feedback loops. Forest ecosystems all result from disturbance, whether small or large and at various temporal scales, and disturbance maintains the forest's structures and processes over time. Forests are constantly changing, so the term *stability* is a relative concept, and a forest's structure, function, and composition are normally maintained within certain bounds by natural disturbance. However, Frelich (2002) presented evidence that unusually severe or frequent disturbances will move an ecosystem

into a different state. Holling (1992a) also advanced this concept more generally, and suggested that catastrophic disturbances could preclude recovery of the ecosystem, often resulting in an altered state with reduced or different biodiversity.

For example, if the long-term natural disturbance pattern in a type of forest is governed by small-gap dynamics, and the clearcutting silvicultural system is used to harvest the forest, many species will be unable to recover and the system will be invaded by species adapted to severe disturbance and large openings, resulting in a new forest type with different biodiversity (e.g., Carleton 2000; Lindenmayer et al. 2002). This lack of convergence between pre- and postharvest forests means that the ecosystem's resistance to change has been overcome, a threshold has been exceeded, and the resulting processes have changed. Examples of such changes are found in degraded forests in many regions of the world, where soil processes have often been altered to the extent that trees can no longer be supported or the species capable of occupying the site have changed (e.g., Friedman et al. 1989; Dahlgren and Driscoll 1994). Other examples are less clear, and only with time will we learn whether new and potentially less desirable states of forest ecosystems are stable; this kind of change has occurred during the invasion of logged black spruce sites by shrub communities in central boreal Canada or the invasion of salal (*Gaultheria shallon* Pursh) on logged low-elevation conifer sites in the western coastal areas of North America. Perry and Amaranthus (1997) also cite examples from Oregon, where the loss of soil function after poor harvesting and silvicultural practices has resulted in unsuccessful regeneration of tree species. In all these cases, the thresholds of change in the energy flows between plants and soils and in moisture-control feedbacks in the soil may have been greatly altered by harvesting, imposing severe new constraints on plant communities and possibly culminating in new stable states for these systems. Matching silviculture to natural patterns of disturbance may eliminate or mitigate some of these effects.

Legacies may help maintain ecosystem stability

Franklin et al. (2002) discussed the importance of maintaining structural attributes in stands and landscapes regenerating after harvesting because these structures relate directly to stand-level processes. This illustrates the concept of maintaining *legacies* in managed systems so as to retain sources of organisms that can repopulate surrounding areas as the forests regenerate (e.g., Dahlberg et al. 2001; Thysell and Carey 2001; Lindenmayer et al. 2002). The concept is particularly appropriate for the many species that have limited dispersal capability, including those involved in soil processes. For example, the presence of certain lichens common in older spruce forests can be predicted based on the proximity of sources of these lichens (Hilmo and Såstad 2001). In addition, retention of legacies can supply old forest structures that are needed by wildlife well into the subsequent sere. The lack of such legacies in managed forests has been a major difference between landscapes burned by wildfires and those that have been commercially logged, but is a factor that could be corrected under a management plan that emulates natural disturbance.

Landscape pattern can also affect ecosystem stability

As previously noted, past forest management has produced landscape patterns that differ from those that developed in response to natural disturbance. At larger scales, if forestry alters landscape patterns and this change affects the species assemblage and regional species richness, then some ecosystem functions and biodiversity will be lost or at least diminished. Landscape patterns can affect disturbance too; for example, the spread of fires (Turner and Romme 1994), the spread of herbivorous insects (Fleming et al. 2000), and the capacity of exotic species to invade regions (Hobbs and Huenneke 1992; Cadenasso and Pickett 2001) can all be affected by landscape patterns. Fire control and forest management, which lead to extensive areas of even-aged timber, have enhanced the extent of outbreaks of certain defoliating insects (e.g., Holling 1992b). Consequently, paying attention to the natural variability in landscape patterns may provide the best model for reducing the risk of severe changes caused by forest management.

The Problem of Emulating Temporal and Spatial Scales

Some forms of environmental variation occur over a very long time

Much of the above discussion of emulating natural disturbance pertains to spatial scaling, but it is also important to understand the temporal scales for natural disturbance. In other words, what is "natural" in a changing environment (Sprugel 1991; Suffling and Perera, chapter 4, this

volume)? In boreal forests, the problem may seem fairly easy to resolve, because the fire rotations are known for most forest types. However, fire frequency has changed several times over the past 500 yr (and longer periods), probably as a result of climatic variation (Johnson et al. 1998; Bergeron et al. 2001). Furthermore, how can we reconcile fire rotations with the ultimate longer-term, broader-scale impacts that climate has on a forest landscape based solely on data for the past few hundred years?

For example, Ontario's temperate deciduous forest occurred as far north as Timmins only 3000 yr ago, in an area now occupied by boreal forest (Liu 1990). Similarly, the present distribution of deciduous forests at the eastern edge of the Great Plains in the United States may have resulted from climatic changes that facilitated the successful invasion of trees into grasslands (Grimm 1984). Current climate patterns and predicted changes in fire frequencies over much of the boreal forests may once again alter forest types on a very broad scale (Weber and Flannigan 1997; Thompson et al. 1998). Insect infestations, although shorter-term in their action, may also disturb potential equilibria over time. Hence, forest landscapes may never be at equilibrium over long temporal scales (Sprugel 1991; Suffling 1995). Furthermore, the scale at which climate operates is broader than that of forest management units (Johnson and Wowchuk 1993). By attempting to emulate natural disturbances based on a time period of, say, 300 yr in boreal forests (e.g., Bergeron and Harvey 1997; Bergeron et al. 2001), managers are implicitly selecting current disturbance patterns as their model. This may not be inappropriate, but managers must recognize the existence of natural, long-term indeterminism in forest systems (i.e., current forests are only stable within the bounds of the current climate): what they plan today may not be congruent with stochastic environmental changes that occur over a time interval as brief as the next 100 yr. Therefore, maintaining variability (from genes to landscapes) in forests is an important management objective to ensure that species have a chance to adapt to anticipated environmental variations (Bergeron et al. 1998).

Even in temperate and tropical forests where gap-phase models of disturbance dynamics seem to be appropriate, longer-term climatic influences may override the effects of gap-level processes. For example, although the old forests of the coastal northwestern United States and southwestern Canada appear to have sustained gap-phase dynamics for many centuries, some of these forests are dominated by Douglas-fir (*Pseudotsuga menziesii* [Mirb.] Franco), which generally regenerates best after fire. Work by Hemstrom and Franklin (1982) and by Yamaguchi (1986) determined that a disturbance regime with a fire frequency of 450 yr probably established many of these trees, suggesting the existence of multiple cohorts after fire and blowdown in many of these temperate rainforests. In contrast, some sites in other forest types in the same region have not been disturbed by fire in 6000 yr (Lertzman et al. 2002). These results suggest a definite direction for managers: to create a multistoried canopy within the stands, and pay close attention to the distribution of individual species across these landscapes. This direction includes the goal of maintaining variability in space and time.

The Species-Area Relationship: Size Does Matter

Species richness is related to the amount of forest

Large areas of habitat maintain more species than small areas (MacArthur and Wilson 1967), and to preserve many individual species, managing based on a few large areas will be more effective than that based on a network of smaller patches (Connor et al. 2000). There is a well-documented relationship between species richness (S) and habitat area (A) of the form $S = cA^z$, where c is a constant, z is usually between 0.15 and 0.40 (Preston 1980; Hubbell 2001). The shape and asymptote of the curve depends to a large extent on the beta diversity (and size) of the landscape: a high beta diversity will result in high regional diversity, but low local richness (Cornell and Lawton 1992). Studies of truly fragmented landscapes (i.e., those in which neither continuous forest nor interpatch connectivity exists) show that small patches generally support fewer species than do large fragments (Lovejoy et al. 1986; Andren 1994; Laurance et al. 1997), and isolated protected areas cannot maintain their full species complements, particularly if the areas are small (Dudley and Stolton 1999; Gurd et al. 2001).

In continuous forests, fragmentation is ephemeral but patch size matters

Most forests in North America are managed under some form of a sustainable management regime that contrasts greatly with the fragmented pattern described earlier in this chapter. In these forests, fragmentation is a temporary phenomenon for many (but not necessarily all) species, because the hostility of the matrix (i.e., the likelihood of mortality in the intervening cut blocks) diminishes as the forest regrows (Harestad and Sutherland 2001). There is evidence that forest management alters landscape patterns compared with the natural patterns in all forest types examined (Mladenoff et al. 1993; Cissel et al. 1999; Perera and Baldwin 2000). Depending on the ecosystem, these changes result from harvesting practices that create abnormal openings and stand sizes. The abnormality may be in the mean size, the variation about the mean size, or both. Given that the effects of these changes on animals and plants are largely unknown, managing forests to produce a more natural structure is a reasonable and precautionary approach that should be adopted until there is a better understanding of how landscape patterns affect the long-term conservation of biological diversity (e.g., Hunter 1990).

The species-area relationship provides a theoretical basis to guide managers in determining the size of the forest habitat patches to be retained within the landscape after forest harvesting. For example, understanding the patch sizes of old-forest habitats and dispersal distances would help managers maintain biodiversity in managed forests (Bunnell 1995). Hence, the emulation of natural disturbance must not only retain large patches of old forest but must also ensure that disturbances are large enough to regenerate large patches of old-forest habitat in the future. Furthermore, the relationship has clear implications for the establishment of protected areas—especially in forests—that could be used as benchmarks against which to measure management objectives and success under an adaptive management system. Size is important both because of the species-area relationship and because the likelihood of including sufficient variation in habitats (based on site conditions and disturbance history) increases as the size of the landscape increases.

Conclusions

Emulating natural disturbance has been proposed as a coarse filter approach to forest management and as a major component of management planning, with the goal of conserving biological diversity and ecosystem function (e.g., Niemela 1999). The emulation of natural disturbance is in some ways an application of the precautionary approach (e.g., Restrepo et al. 1999) within active forest management; with this approach, we also advocate the inclusion of large natural areas that can serve as benchmarks and the careful selection of smaller protected areas that represent various local values. The approach is also precautionary because of current uncertainty over the capacity of ecosystems to replenish their structures and functions after anthropogenic disturbance and concerns over presenting species with environmental variation that surpasses their ability to adapt. Although there is a compelling theoretical basis for emulating natural disturbance, direct empirical evidence to support this approach is scant for many of the principal concerns related to effects of forest harvesting, beyond simple stand-level effects on a species or on species richness. Such evidence is not easily gathered for ecological processes. In addition, such evidence is often not gathered because some of the effects are likely to be expressed at spatial and temporal scales greater than those typically monitored for forestry practices. Even with efforts to collect such data, the study design will likely be challenged by high variation and factors such as climate change. Hence, a precautionary approach should inform the implementation of management plans that attempt to emulate natural disturbance.

The foundation for the concept of emulating natural disturbance rests on two ideas: a landscape's disturbance history provides a model for managing its ecosystems, and disturbance is integral to maintaining biological diversity (Landres et al. 1999). This concept is intuitively simple: logging and natural disturbance are similar at least to the extent that they open the forest canopy and thus promote succession. The concept assumes, at a broad spatial scale, that the preharvest (pre-European settlement) forest landscape in North America had a stable age structure and species composition within the bounds of variability around some mean value. It also assumes that this forest resulted from a "normal" natural disturbance regime, and thus that deviations from this regime may alter forest processes enough to produce a different forest. The basis for emulating natural disturbance rests in part on the theoretical principles of ecology and population genetics, and more recently, on the science of conservation biology. Gene flow, population size, and dispersal must therefore be taken into account. In addition to their magnitude, intensity, and variation, natural disturbances have multiple temporal and spatial scales. An important component of the emulation approach involves attempting to match management planning to these scales.

However, questions arise over our ability to emulate natural processes through forest management and the ultimate convergence (or lack thereof) of the resultant plant and animal communities with those that would follow natural disturbance. Furthermore, a time scale for disturbance much longer than that typically considered may result in ecological indeterminism. Important processes—from mutualism to dispersal, at scales from the site level to large landscapes—are all affected by disturbance and by the sum of these processes, few of which are well understood. This complexity may lead to a lack of predictability and may therefore compromise the utility of emulating natural disturbance as a management protocol.

There has been considerable debate about the ability of forest managers to truly emulate natural disturbance (e.g., Hunter 1993; DeLong and Tanner 1996; Andison and Marshall 1999; Landres et al. 1999; Reich et al. 2001). In practice, it is untenable to suggest that we can completely emulate natural disturbance when so many of the underlying processes differ so dramatically in their type (e.g., chemical versus physical), result (e.g., biomass removal, nutrient loss versus immediate nutrient cycling and accumulated dead wood), and altered landscape pattern (e.g., Christensen et al. 1996; McRae et al. 2001). Does this therefore mean that the paradigm is invalid? Not necessarily. At one level, emulating natural disturbance is about trying to make a series of harvests resemble a fire or a large blowdown. However, emulating natural disturbance does not mean precisely duplicating natural events, but rather creating conditions in managed stands and landscapes that let ecological processes perform similarly to those in naturally disturbed systems. Assessments of forestry practices that attempt to emulate natural disturbance should evaluate criteria central to these ecological processes and ecosystem functioning. These practices must be based on more than opinions or simple visual comparisons. Hence, we must advance our understanding of the ecological processes in areas undisturbed for long periods, in areas more recently affected by natural disturbance, and in those disturbed by forest management. It is important to learn what we can from the past and to improve practices in the future.

The debate should shift from whether we can perfectly replicate natural disturbances to a discussion of what forestry practices support the ecological processes that maintain ecosystems and their species assemblages, and how much deviation from these natural processes is ecologically or socially acceptable. Such debate must progress to well-designed studies in the framework of adaptive management. This will help provide the evidence needed to direct managers toward practices that maintain the broad range of forest values and the ecological processes that support these values. Without these studies, there will be continued tension between the social and ecological rationales for management. The social rationales will tend to pull forest management beyond the limits of natural variation, whereas the ecological rationales will direct managers to remain within these limits. Ecological and genetic theory suggest that although there may be limits, forest managers should not manage solely for these limits but rather for a range of conditions.

CHAPTER 4

Characterizing Natural Forest Disturbance Regimes
Concepts and Approaches

ROGER SUFFLING and AJITH H. PERERA

A clear understanding of natural disturbance is a prerequisite for emulating natural disturbance within the forest landscape. In addition to adopting an acceptable set of terms and concepts related to disturbance, we must understand how human and the forest's natural disturbance regimes function. In this chapter, we show that defining *natural* requires, of necessity, a value-based choice among various possibilities. We also describe the diversity of processes involved in natural disturbance, their complexity, and the intricacy of the interactions among them.

The array of physical, biotic, and chemical factors involved in the initiation, progress, and cessation of disturbance makes this process endlessly and sometimes unpredictably variable in its effects (Attiwill 1994). This variety is at once fascinating for the scientist and vexing for the manager. Disturbance regimes can be understood in the context of factors that drive disturbance, including the causal agents (e.g., fire, windstorm), the scale of the disturbance, how often the disturbance occurs, its intensity, and its severity (how much the ecosystem is perturbed). The characteristics of the whole population of disturbance events are also critical.

Some Basic Concepts in Disturbance Regimes

Definitions and Related Terms

White and Pickett (1985, p. 7) defined forest disturbance as "[a] relatively discrete event in time that disrupts ecosystems . . . and changes resources, substrate availability, or the physical environment." However, Dunster and Dunster (1996, p. 94) were more specific in seeing disturbance as "a discrete force that causes significant change in structure and/or composition through natural events such as fire, flood, wind or earthquake; mortality caused by insect or disease outbreaks, or by human-caused events such as the harvest of a forest."

Single kinds of disturbance, such as wildfire, and single disturbance events, such as an individual fire, should be seen as conceptually distinct from a natural disturbance regime that incorporates a suite of disturbance types. For instance, a comprehensive disturbance regime may incorporate fires, insect attacks, and windstorms. We were unable to find suitable definitions of such a disturbance regime in the literature, so we suggest the following:

A comprehensive forest disturbance regime in a forest landscape is characterized by all the kinds of natural and human-caused disturbance drivers that are present, their stochastic and regular spatial and temporal distributions, their intensities, and the severities of their effects on the landscape's component ecosystems. The disturbance regime also includes the synergies and antagonisms between different disturbance agents, such as fire and wind.

Several authors (e.g., White and Pickett 1985; Turner et al. 2001; Bergeron et al. 2002) have variously identified six components of disturbance regimes: frequency (return interval), rotation period, intensity, severity, residuals, and size of disturbance (table 4.1). The first two components characterize how often events occur and how long they take to affect an area equivalent to that of the entire study area, respectively. Intensity and severity describe the magnitude of an event, and the term *residuals* refers to the nature of the forest that remains after the disturbance. The last

TABLE 4.1. The Main Components of a Natural Forest Disturbance Regime

Disturbance Regime Component	Definition	Example
Frequency (return interval)	Number of events caused by a given disturbance agent per time period at a given point in the landscape. Return interval = 1/frequency.	Five blowdown events in 250 yr in a given stand
Rotation period	The time over which an area equal to that of the study area is disturbed. This component integrates the frequency and size of a disturbance event.	If 0.5% of the landscape is burned per year, the rotation period is 200 yr
Intensity	How much energy is released by a disturbance event per unit area per unit time.	4000 metric tonnes of earth move on average each year in landslides in a given valley
Severity	The impact of the disturbance on the organism, community, or ecosystem.	An average of 55% of the soil's O horizon burned in one fire versus 5% in another
Patch size	The sizes of individual disturbance patches as well as the size distribution of all patches.	In a management unit, the average wildfire size is 5000 ha and the modal size is 50 ha
Residual structure	The complex of physical and biological materials and conditions left after a disturbance event.	Shapes of patches, number of live trees per ha remaining, number of snags >50 cm diameter per ha, average silt depth deposited after a flood
Causal agent	The kinds of disturbance present in the study area, as well as their relative occurrence.	Fire, windstorm, and floods
Relative influence of agents	How often each agent occurs and the magnitude of its impact relative to those of other agents.	The average return interval of 5 yr for flooding in flood plains, affecting 7% of a region; a fire return period of 75 yr affecting 85% of a region
Interactions: synergism and antagonism between agents	How different agents influence each other.	Synergism: a windstorm increases fire likelihood; antagonism: a fire decreases the severity of a subsequent insect outbreak

Note: The table incorporates elements defined by White and Pickett (1985), but adds others that have been acknowledged since that time.

component, size, addresses the spatial extent of disturbance. To these we propose adding the diversity of disturbance, as discussed below.

The *rotation period* has been distinguished from the frequency (or return interval) because it measures how much time passes before an area equivalent to the entire study area has been disturbed (White and Pickett 1985; McKenzie et al. 2000). However, during the rotation period, some sites may have been disturbed several times, and others not at all. Baker and Ehle (2001) have argued that the rotation period and the mean return interval must be equivalent, and Harvey et al. (2002) and Bergeron et al. (2001) have explored the implications of the re-

lationship between the two for emulating natural disturbance.

Broadly, *intensity* refers to the amount of energy expended during the disturbance. For instance, fire intensity has been loosely defined as the rate of heat release from a fire front (Byram 1959). *Severity* is closely related, but not identical, to intensity; it measures the effects of the disturbance, which can vary among stands and sites for disturbances of equal intensity. Intensity has been most often used to describe fires (e.g., Flannigan 1993; Van Wagner 1998; Dahlberg et al. 2001; Henig-Sever et al. 2001; Kafka et al. 2001; Ryan 2002), but has also been studied in such contexts as flooding (Vuori et al. 1998) and hurricanes (Boose et al. 2001). Fire intensity has been interpreted in different ways that complicate its use as a descriptive term: for instance, Angelståm (1998) has associated fire intensity with fire frequency. Severity is the amount of ecological disruption caused by the disturbance (e.g., Schimmel and Granstrom 1996; Perez and Moreno 1998; Dahlberg et al. 2001); there is a complex relationship between intensity and severity that is determined by such variables as seasonality (Ryan 2002) and the location of the disturbance within the ecosystem (e.g., surface, crown, ground fires). Johnson and Miyanishi (1996) have discussed many of the factors responsible for fire behavior as they relate to prescribed burns used to emulate natural disturbance.

Pickett and White's (1985) last factor, *residuals,* is often referred to as *disturbance legacy,* and represents the aggregate of the physical, chemical, and biological conditions that remain after a disturbance event. We treat the patch-size characteristics of a disturbance as a landscape-scale residual effect. The form of the disturbance legacy governs the success of emulating natural disturbance, for it both represents the outcome of the previous disturbance and establishes the conditions that govern the next one. And between disturbance events, it governs the organisms (including humans) that use the forest.

The lists of disturbance components compiled by White and Pickett (1985) and Turner et al. (2001) could be applied to any single causal agent or kind of disturbance, but the context of these works makes it clear that their authors are acutely aware of multiple disturbance agents. Traditionally, a disturbance regime may be considered to be synonymous with a single dominant agent, such as fire, or various agents may be treated separately. This results from researchers' practical difficulties in dealing with all kinds of disturbances in a system at once. However, as we demonstrate, various agents interact strongly on a regular basis. Therefore, we have added three items to the customary list of six components of disturbance regimes to address the diversity of disturbance types in the regime. These are the causal agent, relative influence of causal agents, and interactions among causal agents (see table 4.1). Each is integrally concerned with the mix of disturbance agents present and their interactions. These interactions can be subdivided into the categories of synergism and antagonism. We define *synergism* as the interaction of two or more kinds of disturbance that acts to increase the magnitude of one disturbance (intensity or severity) without a commensurate decrease in the remaining disturbance types. Conversely, *antagonism* is the interaction between two or more kinds of disturbance that acts to decrease the magnitude of one or more disturbance types without a commensurate increase in the remaining ones. Theoretically, increased disturbance by one agent could exactly balance the decrease in another agent, but we have not encountered this in the literature or our own work.

Although the definitions of disturbance proposed by Pickett and White (1985) and Dunster and Dunster (1996) are widely cited, neither addresses how artificial disturbance approximates (or fails to approximate) natural disturbance. One cannot begin to solve this problem until the definition of *natural* has been addressed.

What Is the "Natural" State that Might Be Emulated during Forest Management?

Does *natural disturbance* imply that humans are absent or excluded from forest landscapes, or are humans an integral and traditional part of forest landscapes? The human-excluded definition provides a simple ecological reference point, but one that is often impractical because of the recent and historical pervasiveness of human influence. Conversely, if the definition includes human influences, our view will encompass socioeconomic and cultural biases in addition to ecological reasoning, thus complicating policy-making.

From the beginnings of the North American conservation movement, naturalist philosophers have argued for the preservation of "wilderness" (e.g., Leopold 1966; Thoreau 1988; Elbers 1991; Muir and Gifford 1996). Other observers reason that, in North America, such a totally natural system is a subjective social and political construction of European origin (e.g., Gardner 1978; Silvester 1986). If, however, natural forests are an objective reality, one must develop a relatively objective basis for deciding what constitutes a natural state. Debates about defining old-growth forests (e.g., U.S. Department of Agriculture Forest Service 1986; Cundiff 1989; Hunter 1989; Carleton and Gordon 1992; Peterken 1996; Larson et al. 1999) demonstrate how these relatively simple philosophical ideas tend to become intractable and technical in their application.

A system without human influence is seemingly natural, but even Earth's remotest systems have been altered by humans over at least several millennia; for example, humans have influenced the environment of caves (Culver 1986), the deep ocean floor (Thiel and Schriever 1990; Thiel et al. 2001), and Antarctica (International Union for the Conservation of Nature and Natural Resources 1991). Thus, debate over whether individual systems are natural really focuses on how and to what extent they have been modified. In North America, this passionate and sometimes acrimonious debate often leads to questions about whether indigenous (pre-European) cultures existed in benign equilibrium with their environment.

On the one hand, there is evidence that humans caused mass extinctions of Pleistocene megafauna (e.g., Martin 1967), although others (e.g., Krech 1999) reject this argument. Likewise, there is debate about how native people could have partaken so readily in the excesses of the fur trade (Kay 1985). Others argue that native peoples were, or are, an integral part of their ecosystem (e.g., Martin 1978; Steegman 1983; Orton 1995), and Myers (1992) has summarized similar arguments proposing indigenous harmony with the environment of the humid tropics. As Kay (1985) and various authors in Vale (2002) have argued, the relationships of indigenous people to their environment varied at the time of European contact, and were specific to particular times, peoples, and places. Several of the above authors have assumed that native land use influence was pervasive, but such impacts were often concentrated in well-defined regions (e.g., Suffling et al. 2003), and Krech (1999) has disputed whether native people had much impact over most of the continent. With such a variety of circumstances and interpretations, as well as a continuing stream of fresh information, it is extraordinarily difficult to make broad, valid generalizations. Moreover, the passionate spiritual and political emotions aroused by this debate make consensus unlikely. Thus, in the absence of a priori generalities, one must ask of any particular landscape: are and were the human inhabitants practicing sustainable land use?

Some human communities have evolved an enduring environmental harmony in spite of, or even because of, anthropogenic disturbance. North American native peoples and others regularly burned their ecosystems to drive animals, promote the growth of good forage, and facilitate crop gathering (e.g., Vale 2002). Martin (1978) has argued that native hunter-gatherer populations were regulated at relatively low densities by ecological and/or cultural constraints. However, shifting forest agriculture has also been

found, as with the Iroquois (Hasenstab 1990; Snow 1994) and the Hurons (Heidenreich 1970; Suffling et al. 2003). Such systems diverted only a small proportion of a region's natural resources to human use, but agriculture was a recent innovation in the northern part of the continent at the time of European contact, so one cannot assume that indigenous cultivators were in ecological equilibrium at that time.

Change in ecosocial systems is marked by both increases and decreases in ecological disturbance. When humans modify the disturbance regime (e.g., Romme and Despain 1989; Angelstäm 1998; Veblen et al. 2000), this can be deliberate, as is the case when people burn grasslands to encourage game animals (e.g., Vale 2002); it may also be inadvertent (e.g., Guyette et al. 2002), or a combination of these two alternatives (Weir et al. 2000; Suffling et al. 2003). The effects on the ecosystem have varied in intensity (Vale 2002). Europeans colonizing North America mostly decried the numbers and extent of fire events, whether natural or caused by native peoples or themselves. The prevailing pattern in North America has been an initial increase in the frequency of fire after European contact, followed by an eventual diminution in fires. By controlling fire in target communities, European colonizers eventually reduced its prevalence in such nontarget situations as remnant grasslands (e.g., Brown and Sieg 1999). Thus, human-influenced disturbance regimes in nontarget systems often mimic those of neighboring human-targeted systems (e.g., Suffling et al. 2003).

These observations beg the question of whether temporal changes in regional-scale disturbance occur in natural forests, as they do in those influenced by humans. Some landscape-scale studies in the boreal and mixedwood forests of Canada and the United States suggest regional stability in unlogged forests (e.g., Van Wagner 1978; Yarie 1981), whereas others (Suffling 1983) point to the opposite conclusion. Thus, there is no consensus on this issue.

Whether in equilibrium or not, traditional land management systems are frequently central to their cultures, and, like many natural systems, they are often threatened by sociopolitical and economic change. In this sense, traditional land uses can be regarded as "natural" and integral to ecosystem function (e.g., Naveh 1993).

In light of these difficulties, one strategy for emulating natural disturbance is to ask questions about the origin of the present forest. However, if the forest is, or has been, part of a traditional land-use system, one should first assess whether this management has been benign. A *traditional system* is one that broadly adheres to long-existing land use practices of a long-established culture, such as the outport dwellers of the island of Newfoundland or the Cree of northern Quebec. Conversely, the adoption of nonindigenous practices can broadly be regarded as nontraditional, as in the adoption of farming by some native groups in the Canadian prairie provinces. An existing system is not necessarily traditional: the prairie farm system is strongly culturally embedded, but is not one of great antiquity; nor is it necessarily highly sustainable in all situations. Equally, a traditional land use system is not, a priori, a benign one. A traditional forestry system can be considered benign if it is indefinitely sustainable in terms of the ecosystem's soils, hydrology, and biota, and if it will support the indigenous population in good health, while allowing their culture to survive and adapt. If they are benign, traditional practices, including disturbance, might be incorporated into the emulation of natural disturbance. (This could be termed "emulation of traditional forest disturbance"). In addition, some forests have long been managed successfully on a nontraditional technological basis, but using a traditional value system. Here, there are options for emulating natural disturbance while retaining the most benign elements of the traditional system (e.g., Pecore 1992; Hunter 1993).

The Osnaburgh House region in northwestern Ontario's naturally fire-prone boreal forest (Fritz et al. 1993) illustrates the above dilemmas. The 125-yr-long Hudson's Bay Company journal record provides a semi-quantitative record of lightning- and human-caused fires during the fur-trade era and before significant timber exploitation had occurred. The indigenous Ojibwa did not generally set fires deliberately, but did start them inadvertently, and did little to control fire. They were responsible for much of the burned area along trade routes, but we do not know whether this pattern predates the fur trade. Conversely, Osnaburgh House's European traders used fire as a tool and as an economic weapon, and actively tried to control it. Thus, humans significantly affected the fire regime even before forestry was introduced.

In emulating natural disturbance, should we take such cultural influences into account? In the particular case of Osnaburgh House, should a constructed disturbance regime emulate the era before the fur trade (about which we know little), the era of the fur trade, or an artificial past that ignores the human presence? Those whose cultural views will shape the management philosophy may reach radically different answers to these questions, depending on the value that each assigns to natural and human disturbances. Although the decisions on policy should be informed by science, they are ultimately best reached by means of societal consensus (Rapport et al. 1998).

Understanding Disturbance Regimes

Disturbance Drivers

The disturbance regime and its various components (table 4.1) are governed largely by disturbance drivers. These include relevant climatic factors (e.g., macroclimate, seasonality, cyclicity, secular trends, stochasticity), site factors (e.g., geology, soils, microclimate), and management (e.g., market demand for wood, laws and customs). Agents such as fire, windstorms, and logging are, in one sense, the cause of ecological disturbance, but they can also be seen as the medium through which disturbance drivers make their mark on the landscape. Management is driven by ethics, values and beliefs, perceived needs, traditions, and capabilities. As a result of the interactions among these drivers, the disturbance regime may be manifested as a series of discrete events, as in the case of fire (White and Pickett 1985), or as an ongoing and integrative process, as is the case for wind.

The disturbance regime also has legacy characteristics resulting from the underlying drivers, but mediated through the effects of the different disturbance agents (Foster et al. 1998). Legacy factors include residual live and dead organic matter (e.g., Lee et al. 1997), nutrients (Van Cleve et al. 1996), soil organic material and charcoal (e.g., Wardle et al. 1998), mycorrhizae (e.g., Dahlberg 2002), soil enzymes (e.g., Staddon et al. 1998), the residual aboveground ecosystem structure, species composition, and surviving propagules (e.g., Tellier et al. 1995; Lavoie and Sirois 1998; Skre et al. 1998; Irwin et al. 2000; Wilson and Carey 2000; Wimberly and Spies 2002), underground biological structures (Schmidt et al. 2001), storage and cycling of materials (e.g., Vose et al. 1999), altered hydrology (Bayley et al. 1992), altered energy flux, and size of disturbance.

These factors in turn influence forest processes, including the disturbance regime (Foster et al. 1998). Legacy is important not only because it defines how and to what extent humans can use or enjoy forest resources, but also because legacy factors are drivers for the next disturbance.

Causal Agents

In any region, foresters and ecologists are accustomed to working with a few common disturbance agents. In the boreal forest, the "usual suspects" include fire, defoliating insects, and windstorms. It is easy, however, to dismiss narrower-scale but still pervasive disturbance agents, such as beavers, who create ponds (and eventually meadows) and remove hardwoods from shorelines. One can also easily overlook events that occur only occasionally but with devastating results. The 1998 ice storm in eastern Canada and the northeastern United States was a recent reminder of this. The real list of agents is usually much longer than we know or care to admit. Thus, in recent decades, eastern Ontario—although not a coastal, high-energy mountainous, or volcanic zone—has endured fires, windstorms, defoliating insects, earthquake-related landslides, a massive ice storm, urbanization, flooding related to weather and beavers, and fungal tree diseases. In other regional environments, different factors appear, such as volcanism and tsunamis (e.g., Whittaker et al. 1999). The amazing variety of disturbance agents is illustrated by the following four examples:

- Extreme and rapid freeze-thaw events occur in temperate continental climates, especially near mountains affected by Chinook or *firn* winds. Whole regions can be severely disturbed by these events (Auclair et al. 1990; Hogg et al. 2002), but the impacts are poorly documented. These events are really extremes of ongoing temperature stress at varying levels.
- Natural air pollution that damages forests is a regular occurrence in volcanic regions (Cimino and Toscano 1998; Huebert et al. 1999; Delmelle et al. 2002), and there are records of extensive tree-killing pollution caused by mass nesting of the now extinct passenger pigeon (*Ectopistes migratorius*) (Schorger 1955; Blockstein 2002). This process can still be seen locally in heron and cormorant colonies.
- The 1908 Tunguska incident in Siberia (Baxter 1976) flattened and burned 500 km² of

forest (Svetzov 2002). Such events of extraterrestrial origin, although rare, cause extensive fires (Jones and Lim 2000; Svetsov 2002).
- Newly introduced forms of disturbance often add to or act synergistically with natural disturbance. The Fraser fir (*Abies fraseri* [Pursh] Poir.) of the Great Smoky Mountains in the United States is a typically dismal example. These high-altitude forests are being disrupted by a combination of anthropogenic acidic fog and the introduced woolly adelgid (*Adelges piceae* [Ratzeburg]), in addition to the natural regime of ice damage and blowdown (Busing and Pauley 1994).

Scale in Time and Space

One should not ignore the scale dependence of landscape functions and disturbance regimes (Delcourt et al. 1983; Reed et al. 1993; Jentsch et al. 2002). The array of forest disturbance agents can be viewed at broad and fine scales by relating their temporal and spatial dimensions (Delcourt et al. 1983; Urban et al. 1987). Thus, such forest disturbances as treefalls, windthrow (e.g., Ulanova 2000), forest fires, pathogens, and violent floods can be scaled in time (based on the return period or rotation) and in magnitude (based on the size distribution of events) as spatiotemporal domains. Scales range widely, from years to millennia and from a few square centimeters to millions of hectares. This simple message—that disturbance occurs at almost all spatial and temporal scales—poses an important dilemma in our attempts to emulate natural disturbance: at what scales should we characterize disturbances and emulate them? The answer is complicated. For example, forest fires can occur annually and burn only small areas, or they can occur at intervals of several centuries and burn millions of hectares. Most studies of forest disturbance have inevitably focused on finer scales (e.g., Lertzman and Fall 1998), but judicious extrapolation of accumulated knowledge lets us understand disturbance at broader and longer scales (e.g., Wells et al. 2002; Wimberly and Spies 2002). However, as McKenzie et al. (1996a) point out, this approach has inherent problems because it does not help us understand extreme events.

Very large and infrequent disturbances are being recognized as a distinct category that requires special attention (e.g., Alvarado et al. 1998; Romme et al. 1998; Turner and Dale 1998). Managers must decide whether such large events should be emulated. From the point of view of commercial harvesting, one might argue, for instance, that large clearcut areas emulate the effects of large fires. However, if large wildfires occur simultaneously in spite of fire management, then the total number of large disturbance patches will increase beyond natural levels under such a policy. More research is needed in this area.

Attempts to understand and characterize disturbance must thus emphasize temporal issues (e.g., the frequency of disturbance, the length of the disturbance period) and spatial issues (e.g., the area of disturbance, the total area affected) before determining the study's scale. That scale should specify the disturbance's spatial and temporal resolution, spatial extent, and time frame. Allen and Hoekstra (1992), among others, stated that ignoring scale in studies of ecological processes often leads to incorrect conclusions. At the other end of the scale of studies, one should not ignore Goldstein's (1999) argument that managing ecosystem processes alone, without reference to the ecology of component species, entails many conservation-related dangers (e.g., Thompson and Harestad, chapter 3, this volume).

Various authors working on emulation issues have emphasized the characteristics and shapes of patches. Baskent and Jordan (1995) suggested a hierarchical approach to characterizing patch shapes and sizes, and Lundquist and Beatty (2002) formulated a method for characterizing and mimicking the canopy gaps caused by different types of disturbance. Reed et al. (1996) used patch characteristics to discuss strategies for restoring landscape functions in the Rocky Mountains of Wyoming. Mladenoff et al. (1993) and Gluck and Rempel (1996) have investigated the similarities and differences between naturally and artificially disturbed landscapes. Clearly, these approaches are important, but they must be replicated and augmented in a wider variety of forest environments to build on the considerable theoretical literature in this area.

Interactions among Causal Agents

Because different kinds of disturbance agents can act together, the drivers and causes of individual events are often difficult to determine (e.g., Lundquist and Negron 2000). For instance, the cause of a fire may be the ignition source (e.g., lightning, a careless smoker), the volume of drying fuel engendered by a previous tree-killing insect outbreak, increased fuel connectiv-

ity in the wake of a tree-felling windstorm, or the weather before and during the fire. Interactions between these causal agents are critical in determining individual events and shaping the overall disturbance regime. However, much current research on disturbance concentrates on solitary disturbance events of a single kind (e.g., Rasmussen et al. 1996). Very few authors have attempted to integrate multiple factors, as Bailey and Whitman (2002) did in characterizing insect diversity in terms of the interactions among aspen, elk (*Cervus canadensis*), and fire.

The literature amply demonstrates that various casual agents interact synergistically or antagonistically. Some disturbance agents, such as defoliating insects, are present as ongoing stressors at varying levels, rather than as discrete events, and these agents should also be included in the description of the disturbance regime. Later in this chapter, we describe a synergistic interaction between forest fire and insect outbreaks (e.g., Romme and Despain 1989; Veblen et al. 1994). Insect outbreaks can both precede and follow other disturbance (e.g., McCullough et al. 1998), and Radeloff et al. (2000) have demonstrated that the sequence of overlapping disturbance agents in a given landscape patch significantly influences recolonization of that patch. Stocks (1987) documented the increased susceptibility to wildfire of balsam fir (*Abies balsamea* [L.] Mill.) stands attacked by eastern spruce budworm (*Choristoneura fumiferana* Clem.). Likewise, McCullough (2000) suggested that jack pine budworm (*C. pinus pinus* Free.) may kill or damage enough jack pine (*Pinus banksiana* Lamb.) to change moisture regimes beneath the canopy, thereby creating conditions suitable for sustaining the fires needed for promoting recolonization of a site by jack pine. Conversely, Bergeron and Leduc (1998) observed that changes in fire cycle could explain a large portion of the spatiotemporal variations in tree mortality caused by insect outbreaks in the southeastern Canadian boreal forest. However, Methven and Feunekes (1991) found no direct link between spruce budworm attacks on balsam fir and white spruce (*Picea glauca* [Moench] Voss) and increased risk of forest fire. In other instances, the absence of fire may promote insect and disease damage. Knight (1987) discussed interactions between forest fire and disturbance caused by pathogens. He cited many instances of the absence of forest fire for long periods promoting insect outbreaks, and the proliferation of parasites changing the tree species composition over large areas. Similarly, Veblen et al. (1994) found that after a fire or avalanche, Engelmann spruce (*Picea engelmanii* Parry) forests were less likely to support outbreaks of the spruce beetle (*Dendroctonus rufipennis* Kby.). They also found that lightning strikes that fail to ignite fires predispose the affected trees to attacks by pests and parasites and accelerate tree mortality. This mortality increases the number of fallen trees and standing snags, which in turn increase a stand's vulnerability to fire.

Large, drying fuel loads after windstorms increase the potential for fire. In the Boundary Waters–Quetico region on the border between Minnesota and Ontario, a 1999 summer cold front blew down nearly 1600 km² of forest, which dramatically increased fuel loads and threatens to engender large, stand-replacing fires (USDA Forest Service 2000). Heinselman (1996) had noted a similar potential for fire in this area after earlier large blowdowns. Windstorms also interact with other disturbance agents. Large blowdowns in Colorado triggered outbreaks of the spruce beetle and played a key role in the establishment of old-growth forest (Veblen et al. 1989). Moreover, wind-disturbed forests were more likely to sustain extensive beetle outbreaks (Eisenhart and Veblen 2000). Busing and Pauley (1994) demonstrated that the loss of Fraser fir after adelgid infestations increased the wind exposure of the remaining fir and red spruce (*Picea rubens* Sarg.) trees, which are then more likely to blow over. Conversely, forests damaged by other disturbance agents, such as eastern spruce budworm, are more prone to wind disturbance (Ruel 2000).

Interactions between disturbance agents can also be purely physical. Although earthquakes can directly damage forests (Allen at al. 1999), they are more commonly responsible for extensive landslides (Rodriguez et al. 1999; Vittoz et al. 2001). Around the world, ecologically productive thixotropic marine clays in coastal regions are prone to being shaken and liquefied in seismically active regions, resulting in slope failure and forest disturbance (e.g., Tuttle et al. 1990). Landslips, debris flows, and avalanches are also common in mountainous regions, especially those with high rainfall or seismic activity, as in the coastal rainforests of western North America. Nakamura et al. (2000) have described a cascade of interacting disturbance effects from the slopes through riverine forests and into stream systems. Thus, combinations of geomorphological fea-

tures can increase the risk of forest disturbance. There is considerable evidence for interactions between forest harvesting and landslides (e.g., Sidle 1991, 1992; Jakob 2000; Montgomery et al. 2000; Schmidt et al. 2001).

The causal agents behind disturbance can also interact antagonistically. For instance, fire can control the outbreak and spread of defoliating insects. Both the intensity and severity of western spruce budworm (*Choristoneura occidentalis* [Freeman]) outbreaks have increased with human exclusion of fire (Anderson et al. 1987), and Morin et al. (1993) concluded that the reduced fire frequency increased the likelihood of spruce budworm outbreaks. Veblen et al. (1994) found that stand-replacing fires limit the spread of spruce beetle outbreaks by influencing the age distribution of the postfire stands. A variety of studies have shown that grazing in forests limits the spread and intensity of fires (e.g., Liedloff et al. 2001; Kitzberger and Veblen 2002). Decreased forest flammability has been associated with flooding caused by beavers (e.g., Remillard et al. 1987). Suffling (1993) and Veblen et al. (1994) showed that avalanches and debris flow tracks inhibit the spread of fire in cold, mountainous regions, thereby governing fire return intervals and influencing vertical vegetation zonation.

Similarly, heterogeneous landscapes resulting from recurrent fires can limit the suitable foraging opportunities for defoliating insects, as well as increasing rates of parasitism on defoliators. This limits the risk for future large outbreaks in the homogeneous, old, fir-dominated landscapes that would otherwise develop without fire (Holling 1992b; Heinselman 1996; Cappuccino et al. 1998). Patchy, fire-disturbed landscapes may, however, improve the survival of forest tent caterpillars (*Malacosoma disstria* Hbn.) due to increased light penetration into the stands (Roland 1993).

The interaction between different causal agents is complex, not always intuitively obvious, and not yet well explored. Lertzman and Fall (1998) noted that, in general terms, forest disturbance types are hierarchically nested: broader-scale physical perturbations, such as forest fires, act as constraints and as context for fine-scale disturbance events, such as pest attack. Hierarchy theory (e.g., Allen and Starr 1982; O'Neill et al. 1986) may prove useful in unraveling the complex interactions among forest disturbance types, but the results must always be verified by means of field observations and experimentation. However, the immediate implication for emulating natural disturbance is that managing exclusively for a single cause of disturbance is likely to generate artificial and unintended results.

Characterizing Natural Disturbance Regimes

Variability in Space and Time

The size distribution of the events caused by a single causal agent, such as fire, often follows a Poisson distribution, with many small events and few huge ones. Large events usually occur at extended intervals, but in a stochastic series may also bunch together, as when bad fire years occur back to back. This contrasts strongly with the constant delivery over time that managers strive for in forest harvesting. The Delcourt et al. (1983) model, which displays the relationship between magnitude and return period for random events, is a tool for considering the likelihood of occurrence of events of particular sizes. For example, the average time between impacts on Earth's surface by very small extraterrestrial particles is on the order of minutes or days, whereas on average a Tunguska-size event occurs somewhere on Earth about every 1000 yr (Brown et al. 2002). Although such large events are rare, they can occur within any forest region. Even larger and more extreme events, such as the asteroid impact that occurred during the late Cretaceous near Chicxulub, Mexico, have an average return interval of about one hundred million yr (Grieve and Therriault 2000), but have devastating effects that are felt worldwide. An approach based on relative risk permits a prediction of the likelihood of events greater than a certain size within the management time horizon for a given region. And—critically in emulating natural disturbance—this approach lets managers compare the relationships between magnitude and frequency for various natural and anthropogenic disturbance processes.

Some disturbance events are partly or wholly cyclical rather than stochastic (e.g., Kitzberger and Veblen 2002; Parish and Antos 2002). These events range in frequency from waves lapping on lakeshores to recurrent ice ages. Many cycles are diurnal, as in the changes in fire behavior between daytime and nighttime atmospheric conditions. Likewise, annual rhythms are so familiar that we overlook them as cyclical events; for instance, fire seasons and the annual life cycles of many defoliating insects. Cycles with longer periodicity are partly governed by intrinsic population dynamics—which have been vigorously debated by theoretical and systems ecologists—as is the case for populations of some defoliating insects (e.g., Berryman 1988). Other multiyear cycles are driven extrinsically by climate, as is the case for the El Niño cycle and its effects on fire occurrence (Heyerdahl et al. 2002; Kitzberger and Veblen 2002).

Longer-term changes might be regarded, for practical purposes, as a one-way or imperceptible trend. A typical example is the change in the frequency of fire occurrence since the end of the Little Ice Age (Miyanishi and Johnson 2001). However, the same long-term changes are possibly cyclical (e.g., Suffling and Speller 1998). The warming of climate since the end of the most recent (Wisconsinan) ice age is an extreme example, as it represents the fourth in a series of such cyclic Pleistocene events. The introduction of exotic species that alter disturbance regimes typically results in long-term and often irreversible changes to natural disturbance regimes; for example, the introduction of chestnut blight (*Cryphonectria parasitica* [Murrill] M. Barr) to North American deciduous forests has dramatically changed the effects of all processes that depend on the existence of mature chestnut trees (*Castanea dentatum* [Marsh.] Borkh.) (Roane et al. 1986). Longer-term periodicity for cyclical disturbance, combined with long-term trends and even random variation, result in *drift* of the disturbance regime, in which the regime changes gradually over time (e.g., Masters 1990; Suffling and Speller 1998). This underscores the need for practitioners of disturbance emulation to collect information on disturbance regimes from as long a time period as possible, and to place specific data into a general temporal context of climate fluctuation and regional ecological trends.

As forest disturbance processes result from complex interactions among numerous physical and biological variables, they are only rarely deterministic at any scale. Because the occurrence of the causal factors follows both spatial and temporal trends, forest disturbances exhibit inherent variability. Just as some disturbance is predictable over time within specific scale limits, so some perturbations are spatially predictable, again within specific scales. Riverine and coastal flooding are highly geographically predictable at most scales (e.g., Suzuki et al. 2002). In contrast, windstorms are predictable at a continental scale, but their occurrence is wildly random within regions (Peterson 2000). It is possible, therefore, to predict the temporal and spatial variations in disturbance for some causal agents, but not for all. For some types of natural disturbance, such as flooding and avalanches, there is a long tradition of risk assessment. In the case of fire and insects, a new generation of models is beginning to allow the characterization of spatial risk in various parts of the landscape based on both location within the physical environment and a forest stand's history (e.g., Keane, Parsons, and Rollins, chapter 5, this volume; MacLean, chapter 6, this volume).

Many have studied, demonstrated, and discussed the temporal (e.g., Heinselman 1973) and spatial variability in forest disturbance (e.g., Baker 1993). Lertzman et al. (1998) provide an excellent, detailed discussion of these sources of variability in relation to forest fire. In addition, because the probability of occurrence of a given combination of causal factors depends on their individual probabilities and the interactions, disturbance may not always recur in an identical manner even in the same time and space. This third aspect of variability emerges as intrinsic randomness of disturbance (e.g., Armstrong 1999). This randomness is more a statistical attribute than an ecological property of forest disturbance (Dutilleul and Legendre 1993).

The confusing multidimensional variability of disturbance regimes is better understood using ecological scale and hierarchy relationships. Natural disturbance is controlled by different factors at different temporal and spatial scales. For example, climatic trends at regional scales, topographic and vegetational cover patterns at subregional scales, and fuel availability at stand-level scales may all partially control forest fires. Thus, characterizing the frequency, return interval, and size distribution of various disturbance agents is a critical part of any attempt to emulate natural disturbance.

To be useful, the characterization of disturbance must include all three aspects of variability:

- Spatial variability reveals the effect of the characteristics of a given landscape (e.g., terrain, geology, geometry, heterogeneity of land cover), reflected in variations in both the disturbances and their impacts on the forest cover (e.g., Johnson et al. 1998).
- Temporal variability explains the effects of the trends in the causal factors underlying disturbance (e.g., climate, life cycles of pests and pathogens).

- Stochastic variability, which can be understood only through replicated simulations of disturbance, explains the nondeterminism in disturbance as well as its impacts.

Methodological Approaches

Many attempts have been made to categorize the diverse methods available to characterize disturbance. Without duplicating the excellent information available in these reviews (e.g., Gardener et al. 1999; Keane, Parsons, and Rollins, chapter 5, this volume), we examine here only the broadest premises of the methodological approaches. Our obvious bias toward fire as a focal disturbance is an artifact of the available literature. The methods we address are those that describe the salient features of the disturbance regimes without explicitly attempting to forecast specific occurrence of a disturbance, and are thus analogous to weather forecasts. This differentiation is important, because the methods that characterize disturbance regimes are more useful in emulating natural disturbance than are those that forecast individual disturbance events. There appear to be two major approaches: historical and simulation.

Historical approaches to characterizing disturbance regimes are based on deductive analysis using historical evidence about the former states and disturbances of forest ecosystems. The sources of evidence include historical documents, such as survey records (e.g., Suffling and Wilson 1994; Delcourt and Delcourt 1996); cultural evidence, such as photographs and drawings (e.g., Hart and Laycock 1996); anecdotal writings; palynological evidence (e.g., McLachlan and Brubacker 1995; Lynch 1998); dendrochronology (e.g., Batek et al. 1999; Brown et al. 1999); ethnographic records of disturbance (e.g., Fritz et al. 1993); and palaeomagnetism records (Kletetschka and Banerjee 1995). See Egan and Howell (2001) for an extensive review of these methods.

Scenario simulation approaches include modeling methods that characterize and quantify disturbances based on inductive reasoning about what could happen. There are three major means of simulation modeling: statistical, mechanistic, and hybrid approaches. Statistical extrapolation models generally use empirical observations of past conditions to construct generalized statistical distributions for the prediction of future events. Examples include those reported by Van Wagner (1978) and Johnson and Van Wagner

TABLE 4.2. Comparison of the Historical (Observational) and Scenario Simulation Approaches to Defining Disturbance Regimes

Historical (Observational) Approach	Scenario Simulation Approach
Usually single or few events	Multiple events, whole regimes
Often no error terms included in the description of events	Variation is predicted or used in predictions
Concrete	Abstract
Usually depends on few assumptions	Tends to depend on numerous assumptions
Politically potent	Often viewed skeptically by the public

(1985). The strength of these models is that their empirical basis is relatively easily understood, and they relate readily to data and methods familiar to forest managers. However, they have remained largely nonspatial thus far, and rely upon many assumptions (table 4.2). Mechanistic models are rooted in physical and biological science, and most use parameters and information defined by experiments. The fire model FARSITE (Finney 1998) is a classic example of this group. These models tend to be spatially explicit, make relatively few assumptions, and are incrementally improvable. Most often, modelers combine these two methods, resulting in hybrid models based on both empirical evidence and mechanistic knowledge. Such models overcome limitations imposed by the computational demands of statistical models and the extensive knowledge requirements of purely mechanistic models. The hybrid models are spatially explicit, incrementally improvable, and most importantly, can be applied to large areas and long time frames. Examples include DISPATCH (Baker et al. 1991), Fire-BGC (Keane et al. 1996a), LANDIS (Mladenoff and He 1999; He et al., chapter 10, this volume), SEM-LAND (Li, chapter 8, this volume), and BFOLDS (Perera et al., chapter 9, this volume).

Although none of these approaches is incorrect, those who wish to characterize disturbance regimes must choose among them wisely. Many criteria must be addressed in selecting a method. Ideally, the methods selected should:

- Characterize disturbance regimes comprehensively in terms of their spatial (sufficiently large area) and temporal (sufficiently long period) extents, and in terms of their attributes (e.g., disturbance intensity, post-disturbance changes).

- Address potential synergism among different disturbances, so as to understand the synergistic (reinforcing) and antagonistic (damping) interactions among causal agents and disturbance drivers.

- Describe the stochastic nature of disturbance regimes and thus express the nondeterminism inherent in natural disturbance.

- Portray the spatial and temporal variability of disturbance and so capture the spatial biases and temporal fluctuations in the occurrence of disturbances and their outcomes.

- Explicitly present the estimation errors and their propagation associated with the methods, so as to estimate the level of confidence in the characterization of the disturbance regime.

- Present the premises and assumptions of the methods and so identify the limitations of the characterization and provide a basis for continued improvements of the methods.

Conclusions and Recommendations

Emulating natural forest disturbance demands more of land managers than ever before. The practice of explicitly mimicking nature will lead to new practical and theoretical insights, as well as to profound difficulties as we begin to appreciate how little we really understand about even relatively simple northern forests. Frustration will be minimized if foresters, ecologists, modelers, and others adhere to a common set of simple principles for documenting their work. The following guidelines should be applied in building a framework for the emulation of natural disturbance:

- To emulate the natural disturbance regime in a given area, managers must first characterize that regime. Predictions of disturbance regimes can usually only describe and characterize natural disturbance over time and space, rather than prophesying specific events. Disturbances can be described as individual events (e.g., the size of one fire or intensity of one ice storm) or as comprehensive regimes (e.g., a population of disturbance events). To characterize a disturbance regime requires explicitly recognizing the temporal and spatial patterns of natural disturbance as well as their interactions in time and in space.

- Characterizing a natural disturbance regime is more than a formality. Most systems are highly altered compared with their historical condition, and many have a traditional or indigenous human occupancy that can be validly considered as a factor in the disturbance regime. In characterizing any disturbance regime, one should explicitly state whether human disturbance has been considered as part of the regime. If anthropogenic disturbance is considered integral to the regime, then the historical period and kinds of human disturbance that are incorporated into the regime should be stated clearly. The decision to incorporate or omit the human component of disturbance, and what era and culture(s) to include, is inherently a cultural and political decision and should not be treated as a strictly technical issue.

- Disturbance regimes drift over time. They also change character in response to changes in climate, land use, and such factors as the introduction of exotic species; some of these factors go back a long time. In characterizing a disturbance regime, both the reference time period over which the regime is described and the rationale for picking that time period should be made explicit.

- It is often useful or necessary to characterize a large region through a detailed study of a subarea of the region, such as a watershed. When such extrapolations are made, there should be clear reference to the information base on which the management assumptions are based. Thus, characterizations of disturbance regimes should explicitly define the reference area.

- An objective comparison should be made between the natural regime and that imposed by management. Attention to these details will ensure that the disturbance caused by forest management relates to specific natural disturbance processes. For each disturbance agent, one should attempt to determine the frequency (return interval), intensity, severity, patch sizes, and legacies (residuals). In addition, one should understand the drivers of disturbance for each disturbance agent. Some disturbance agents are present as ongoing stressors rather than as discrete events, and these agents should also be included in the description of the disturbance regime.

- A disturbance regime may be described in such terms as the probability distribution for the size of discrete events, the spatial tendency of occurrence, or the variance in return intervals. In characterizing a disturbance regime, it is essential that descriptions of disturbance regimes include stochasticity (measures of central tendency, extremes, and probability distribution functions).

- The interactions of causal agents in a disturbance regime are of critical importance, but can also be counterintuitive and subtle. One should not view disturbance emulation in the context of a single disturbance agent (such as fire) unless it has already been established that this agent operates in isolation from other factors. Where there is more than one disturbance agent in a disturbance regime, one should attempt to determine the relative influence of each agent and the interactions among agents.

- The numerous means of characterizing a disturbance regime include observational, experimental, and at least three modeling methodologies. Each approach has specific strengths and weaknesses, and none is sufficient by itself. As the methodologies available for characterizing disturbance regimes are generally complementary, they should be employed in focused combinations.

- In practice, land managers must often work with incomplete or imprecise information that does not meet the standards suggested above. Therefore, land managers who intend to emulate natural disturbance should document all their assumptions and the gaps in their knowledge of the disturbance regime.

Acknowledgments

We thank our students and associates (especially Harry Doran, Francy Gertsch, Bonnie Hui, Brian Kutas, Rafael Muñoz-Marquez, Brett Woodman, and Xie Yan) whose challenging questions made starting this chapter imperative, and finishing it almost impossible.

CHAPTER 5

Predicting Fire Regimes at Multiple Scales

ROBERT E. KEANE, RUSSELL E. PARSONS, and MATHEW G. ROLLINS

Wildland fire is the primary form of disturbance in many North American ecosystems, where it recycles nutrients, regulates succession by selecting and regenerating plants, maintains diversity, reduces biomass, controls insect and disease populations, triggers and regulates interactions between vegetation and animals, and most importantly, maintains biological and biogeochemical processes (Johnson 1992; Agee 1993; Crutzen and Goldammer 1993; DeBano et al. 1998). Many ecosystems have been so highly influenced by fire that many biota developed adaptations that let them survive fires or disperse into burned areas (Wright and Bailey 1982; Pyne et al. 1996; DeBano et al. 1998). Historical landscapes have consisted of shifting mosaics of plant communities and seral stages created by the variable fire behaviors and their effects in time and space (Knight 1987; Swanson et al. 1997). These landscapes supported diverse flora and fauna, many of which depended on fire for their continued existence (Chang 1996). However, the successful and extensive government fire suppression campaign that started in the early 1930s in the United States and Canada—whose technology and resources have increased up to the present day—has altered fire regimes (frequency and severity of fires over time) in many landscapes (Keane et al. 2002c).

Recently, fire has been recognized as a valuable ecosystem process for conserving ecosystem and landscape biodiversity, and many land management agencies are now initiating major efforts to restore the role of fire to western landscapes by using prescribed fire and silviculture to mimic the effects of historical fires (e.g., Babbitt 1995).

If such efforts are to be successful, land managers will need comprehensive and accurate temporal and spatial descriptions of historical and current wildland fire processes (Brown 1995). For example, characterizations of native fire regimes can aid in the planning and design of silvicultural treatments, such as target stand structures and the shapes, sizes, and locations of harvest units (Morgan et al. 2001). These characterizations can also be used to prioritize and schedule future treatments based on the expected fire frequency and severity. Maps of historical fire regimes can be used to develop input parameters for landscape-scale models used to simulate effects of alternative management strategies and changing biophysical processes, such as climate change, on landscape dynamics (Mutch 1994; Keane et al. 1998; White et al. 1998; Li 2000a).

The prediction of fire regimes is a difficult, problematic, and costly task that can be discouraging to many land management agencies, in part because the task requires extensive expertise in fire ecology, fire history sampling, statistical analysis, and GIS technology (Morgan et al. 2001). This chapter describes the challenges of predicting fire regimes, then reviews three strategies (classification, statistical analysis, and simulation modeling) stratified by three approaches (stochastic, empirical, and biophysical) for describing, predicting, and mapping fire regimes using examples from the literature and modeling experiments. Challenges in the prediction of fire regimes are also discussed, such as field sampling techniques, spatial and temporal scales, and estimates of variability. Morgan et al. (2001) have published an excellent companion paper to this chapter.

Background

Fire regimes are often described in terms of fire frequency, size, pattern, seasonality, intensity, and severity (Heinselman 1981; Agee 1993). Fire frequency is defined in many ways, depending on the scale and objective of the exercise. Point-level measures, such as fire-return interval and fire probability, describe the number of fire events experienced over time at one point on the landscape (Skinner and Chang 1996; Baker and Ehle 2001). Spatial measures of fire rotation and fire cycle estimate the number of years it takes to burn an area the size of the defined landscape (Van Wagner 1978; Johnson and Gutsell 1994; Reed et al. 1998). Fire size is measured using standard surveying or mapping techniques. The frequency distribution of fire sizes within a landscape or region depends primarily on the size and number of the largest fire events and landscape complexity (Yarie 1981; Strauss et al. 1989; Bessie and Johnson 1995). Vegetation, topography, antecedent weather, and fuels often dictate the mosaic of burned patches within the landscape (Skinner and Chang 1996; Kushla and Ripple 1997). Fire intensity describes the heat output from a fire, whereas fire severity describes the subsequent damage to the biota (Whelan 1995). Fire intensity and severity are not always related; a low-intensity, smoldering duff fire can be quite severe, because high temperatures driven downward through the soil can kill roots, soil biota, and regenerating plant parts (DeBano et al. 1998).

In this chapter, fire regimes are primarily described by the frequency and severity of fires, because these two factors are most important in determining the effects of fire and are used in the majority of predictive studies and fire management planning activities. The point-based average fire-return interval (years) is used to describe fire frequency, and fire severity is described by classifying fires into three categories: non-lethal surface fires, mixed-severity fires, and stand-replacement fires. A *nonlethal surface fire* burns surface fuels at low intensities and kills few individuals in the overstory (<10%). *Stand-replacement fires* kill the majority (90%) of the dominant vegetation, often trees and shrubs (Brown 1995), and include both lethal surface fires and active crown fires (Agee 1993). *Mixed-severity fires* contain elements of both nonlethal surface fires and stand-replacement fires, but mixed in time and space. Passive crown fires, patchy stand-replacement fires, and understory fires are com-

mon in mixed-severity fires (DeBano et al. 1998). Typically, the term *mixed-severity fire* describes the patchy burn patterns created during a single fire. However, *mixed severity* can also be used to describe regimes in which fires vary in severity over time, such as when a nonlethal surface fire is followed by a stand-replacement fire (Shinneman and Baker 1997). There are other severity types, including ground fires (i.e., fire that burns extensive duff layers), but these are not as prevalent as the three categories we have defined here and use in this chapter (Agee 1993). We emphasize forest fire regimes because of the abundance of studies and field data, but the same concepts could be used for rangelands.

Fire regimes generally result from the cumulative interaction of fire, vegetation, climate, humans, and topography over time (Crutzen and Goldammer 1993). These interactions are spatially and temporally correlated: future fires are influenced in space by the adjacency of an area to burnable stands and fire-prone topographic features (e.g., upper slopes, mountain grasslands), and in time by the timing and severity of past climate (e.g., drought, wind patterns) and disturbance events (e.g., previous fires, insect outbreaks). A change in any of these factors will ultimately result in a change in the fire regime, and because all five factors are constantly changing, fire regimes are inherently dynamic. For example, climate change can affect the fire regime by altering fire ignition patterns (e.g., lightning), vegetation characteristics, and fuels (Flannigan and Van Wagner 1991; Balling et al. 1992; Swetnam 1993; Bachelet et al. 2000). Moreover, invasions by such exotic weeds as cheatgrass (*Bromus tectorum*) and spotted knapweed (*Centaurea biebersteinii*) have changed fire regimes in many semiarid ecosystems (Whisenant 1990).

The fire regime of a region is a spatial disturbance gradient that does not follow discrete mapping units, so it should not be viewed as an attribute or characteristic of an ecosystem or cover type. Attempts to predict fire regimes solely from fuels (Olsen 1981), vegetation (Frost 1998), or topography (Keane et al. 1996b) have only partially succeeded, because they have not recognized the pervasiveness of fire on the landscape and the interactions of the many factors that control fire dynamics.

Historically, models that predict fire regime have been mostly aspatial, and were usually developed by using complex statistical analyses of empirical data to create regression equations, logistic and discriminant functions, or classification keys (Morgan et al. 2001). However, land managers now need more than a single equation to predict fire regime characteristics, especially if a spatial context is desired to support management. Because fire is a spatial process, it is essential that the equations, keys, or simulation models used to create the models be fully integrated into GIS databases. The application of simple predictive models to create spatially explicit fire regime layers in GIS requires that the independent or predictor variables also be mapped across the landscape being analyzed. Unfortunately, the latter data rarely exist for many areas. In this chapter, we emphasize the development of fire regime models for creating GIS maps (i.e., spatial data layers) of fire regimes.

Many techniques have been used to predict fire regimes for stands, landscapes, and regions (table 5.1). In this chapter, we describe these techniques in terms of the approach and strategy used to estimate the fire regime. *Approaches* are general descriptions of the methods used to create models for predicting fire regimes, and include stochastic, empirical, and physical approaches (see reviews in Gardner et al. [1999] and Turner

et al. [2001]). *Stochastic* approaches use probabilities to quantify or describe the fire regime. *Empirical* approaches use field data to derive deterministic relationships that represent the characteristics of a fire regime. *Physical* approaches formulate the basic physical processes that drive ecosystems and landscapes and use the formulations to create fire regime descriptions.

These approaches are not mutually exclusive. In fact, the best predictive models often combine these approaches, thus making it difficult to categorize models solely by approach. For example, stochastic prediction models can be derived from empirical data describing a physical process (e.g., the probability of a lightning strike). Consequently, in this chapter, we have chosen to categorize models based on the strategy employed to create the predictive model as well as the approach, and have used these categories in our discussion of efforts to predict fire regime.

Strategies for predicting fire regimes are described by the set of tools and protocols used in developing the predictive models (table 5.1). We identified three broad strategies: classification, statistical analysis, and simulation modeling. The

TABLE 5.1. Examples of the Approaches and Strategies Employed for Predicting Fire Regimes

	Strategy		
Approach	Classification	Statistical Analysis	Simulation Modeling
Stochastic	Fire regimes are stochastically assigned to land type classifications.	Logistic regression is used to predict fire regime from topography and weather.	A simulation model that simulates fire ignition based on probability distributions is used to estimate fire regimes.
	RARE	COMMON	COMMON
Empirical	Regression trees or discriminant analysis are used to predict fire regimes from fire history and topographical data.	Regression analysis is used to predict fire frequency from topography, site, and vegetation data.	Deterministic models of fire and succession are used to describe fire regimes.
	COMMON	COMMON	RARE
Physical	Regression trees are used to predict fire regimes from simulated respiration, evapotranspiration, and productivity, along with topography.	Multivariate modeling techniques (regression, general linear models) are based on biophysical processes, such as decomposition, transpiration, and outflow.	A mechanistic process model is used to simulate fire regimes for complex landscapes.
	RARE	RARE	RARE

Notes: The frequency of use of a particular combination in the literature is indicated as either rare or common. Shaded cells represent combinations that were not evaluated in the model comparison discussed in this chapter.

predictive models presented here can take many forms, ranging from simple classifications and empirical equations to complex mechanistic simulation models, depending on the combination of strategy and approach.

Classification Strategy

The classification strategy involves assigning a fire regime to the most appropriate category in a given classification scheme, based on the vegetation, biophysical settings, soils, climate, or any other mapped ecological process responsible for the regime. For example, a regime with frequent low-severity fires could be assigned to the ponderosa pine (*Pinus ponderosa* Laws.) cover type. Gardner et al. (1999) refer to this strategy as a theoretical approach and Morgan et al. (2001) call it a rule-based approach. The classification strategy commonly uses rules or keys to assign fire regimes, and these keys can be constructed using expert opinions (stochastic approach), field data coupled with simple to complex statistical techniques, such as regression trees and neural networks (empirical approach), or measurements of ecosystem dynamics (physical approach).

Fire regimes are typically classified based on potential vegetation type, such as habitat types, plant associations, or habitat-type groups (Fischer and Bradley 1987). Barrett (1988) used an empirical approach to assign fire frequency and severity classes to the special groupings of the habitat types of Pfister et al. (1977). This classification was further refined by using the terrain characteristics of aspect, slope, elevation, and drainage pattern (Barrett and Arno 1991). Several fire ecology publications described fire regimes for groups of habitat types in the Rocky Mountains (Davis 1980; Crane and Fischer 1986; Fischer and Bradley 1987; Bradley et al. 1992a,b). At the fine scale, a fire regime map of the Selway-Bitterroot Wilderness, Idaho, was initially developed by Barrett and Arno (1991), then refined by Brown et al. (1994) by using potential vegetation, topographic, and landform classifications. At broad scales, Morgan et al. (1996) used an empirical approach to assign a fire-return interval (four categories) and a fire severity type (three categories) to 1-km pixels, based on classifications of cover type, habitat-type group, and ecosystem region derived from field data on fire history for the Columbia River Basin. Hardy et al. (2001) mapped historical fire regimes across the conterminous United States by using local experts to assign one of five categories (defined to integrate fire frequency and severity) to combinations of categories. These were based on a refined Küchler (1975) map of potential natural vegetation, a regional ecosystems map (Bailey 1995), and satellite-derived land type and forest density maps (Loveland et al. 1991; Zhu 1994). In addition, Frost (1998) created a fire-frequency layer for the lower 48 U.S. states by using regional vegetation classifications of structure and composition. More recently, Kasischke et al. (2002) used GIS to compare the spatial distribution of fires in Alaska with topographic, vegetation cover, and climatic features.

The main limitation of the classification strategy is that it does not account for spatial relationships across the landscape. For example, the adjacency of burnable and nonburnable stands within a landscape plays an important role in fire dynamics, because recent burns offer some protection to forests on the leeward side of the burn; such complex spatial processes are rarely included in a classification strategy (Heyerdahl et al. 2001). For another example, stands at higher elevations within a watershed are more likely to burn than are stands on the leeward sides of lakes or on talus slopes (Agee 1993). Other spatial relationships that would be difficult to include in a classification strategy include a site's prevailing wind pattern, topographic orientation, and vegetational mosaic.

Another drawback of the classification strategy is that fire regime categories do not always match categories based on vegetation or biophysical classification, because of the lack of spatial relationships between the two (caused by scale problems), and because of the factors that determine fire weather. Wind is a primary factor in the spread of large fires (Johnson 1992), yet wind is rarely integrated into landscape or vegetation classifications. Most land type classifications describe only the dominant vegetation or physical attributes, and inadequately represent underlying disturbance or ecophysiological processes; moreover, most land type categories are dynamic, which limits their usefulness for predicting fire regimes (McKelvey and Busse 1996). Fire regime classifications are only as good as the parent classification systems used to develop them. Errors in fire regime assignments can be compounded and propagated by errors in the cover type, topography, or potential vegetation type classifications and maps. This results in a misleading fire regime layer, which is why it is so important to validate both the fire regime and the parent classifications and maps with abundant, high quality field data.

The advantages of the classification strategy are its simplicity, ease of use, and ability to provide an adequate first approximation. Expert knowledge can be easily summarized by common vegetation or biophysical classifications to create useful fire regime maps when field data are scarce. In addition, many ancillary classifications can be integrated across scales to augment fire regime assignments to account for diverse physical and ecosystem processes that affect fire dynamics (Hardy et al. 2001). Fire regime maps can easily be refined and modified, once additional data become available, because the computer (GIS) commands and programs have already been developed. However, an accuracy assessment must be performed on the resultant fire regime map to quantify the spatial and classification errors to guide interpretation of the regime and implementation in a management plan.

Statistical Analysis Strategy

The statistical analysis strategy can make use of the entire suite of multivariate statistical techniques, such as regression, ordination, general additive models, and logistic regression, to create deterministic or stochastic fire regime predictive models. This strategy is easily the most popular, because of its simple, data-driven approach. A common statistical analysis correlates point samples of fire history with site characteristics. Long (1998) used field data on fire history and ancillary spatial data layers to stochastically predict the fire frequency and severity categories for stands within a large landscape in western Montana by using logistic regression. Kessell (1979) used statistical gradient modeling techniques to characterize fire regime components based on environmental gradients of climate, topography, and vegetation in Glacier National Park, Montana. Agee and Flewelling (1983) and D'Elia (1998) used data on lightning activity, climate, and topography to predict fire cycles for the Olympic Mountains in Washington and the Great Basin region of the United States, respectively. McKenzie et al. (2000) used complex statistical models to predict fire frequency for the Interior Columbia River Basin (United States), based on data from a network of fire history sites.

Another common and reliable spatial statistical analysis of fire regime involves the synthesis of fire atlases, which are maps that document the spatial extent and dates of past fires to provide a representation of fire frequency and severity (Morgan et al. 2001). This is usually accomplished by fitting the spatial data on fire history to a probability distribution curve (most often a two- or three-parameter Weibull function) stratified by topographic or vegetation types. Fire atlases can be created from digitized maps of past fires (McKelvey and Busse 1996; Rollins et al. 2001); from maps of stand age based on airphoto interpretation (Johnson and Larsen 1991; Camp et al. 1997; Agee and Krusemark 2001); from maps of fire occurrence based on spatially extrapolated fire scar data or other factors (Barrett et al. 1991; Barrett 1994; Niklasson and Grandström 2000); and from satellite imagery coupled with field data (Minnich and Chou 1997; Russell-Smith et al. 1998). Another variant of this method derives fire atlases from georeferenced point-level fire history information (typically, fire dates calculated from fire scars or stand age) by using spatial surface modeling or interpolation (Habeck 1994; Heyerdahl 1997). Spatially explicit fire dates can then be summarized using GIS "moving window" techniques and ancillary biophysical stratifications to estimate the parameters of the most appropriate probability distribution (Johnson and Gutsell 1994). Fire severity must be inferred from postburn vegetation characteristics when severity was not documented at the time of the fire (Johnson 1992; Agee 1993; Grissino-Meyer 1995; Swetnam and Baisan 1996; Reed 1997).

The statistical strategy is simple and efficient, and can be the most accurate of the three strategies, because it is ultimately based on real field data and thus, depicts the most realistic, but most recent, fire regimes. Predictive equations can be easily derived if advanced statistical expertise and extensive field data sets are available. This strategy can be accomplished using any of the three approaches and usually provides an estimate of variance. A stochastic approach can generate probability estimates for the occurrence of a particular fire regime using logistic regression, for example, whereas an empirical approach is used to predict categorical or continuous estimates of fire regime from standard multivariate models, such as regression equations, discriminant functions, and regression trees. The physical approach can be used when the ecosystem process variables that govern fire dynamics, such as climate and hydrology, are included as independent variables in the statistical analysis. The statistical strategy is most appropriate when extensive field data exist.

The limitations of the statistical strategy are well known. First, statistical models rarely imply cause and effect, so they should not be used to draw conclusions about landscape-scale fire dynamics. The application of statistically derived predictive models is also limited to the realm described by the field data (e.g., geographic location, ecosystem, period of record for the fire history). Data sets are rarely robust enough to extrapolate to other landscapes, because of differences in climate, topography, soils, vegetation, and lightning occurrence (Knight 1987; Baker 1989). This strategy is also data-intensive, requiring costly collection of fire history and vegetation data across long time spans and over large areas, as well as extensive expertise in spatial statistical analysis. It is also limited to a single spatial scale, because multiple scale factors are difficult to include in a single statistical model. For example, it is difficult to include wind interactions in the regression equations, because these broad-scale data are difficult to collect and summarize at the scale typically used for prediction. Finally, the statistical strategy is entirely limited to the time period covered by the fire history data, and this period may not include the full array of factors responsible for fire dynamics in the landscape.

Simulation Modeling Strategy

The simulation modeling strategy uses stand- or landscape-scale models to simulate fires and succession to generate some expression of fire regime. For example, Li (2000a) used the spatial landscape model SEM-LAND to simulate fire and successional dynamics over long time periods (e.g., 1000 yr) and generate output that summarized fire frequency and severity in a fire regime map. This strategy is somewhat new, because recent advancements in computer technology have allowed an independent spatial simulation of fire spread based on weather and topography (Finney 1998). There are usually four components in a landscape fire model: vegetational succession, fire ignition, fire spread, and fire effects (see Keane and Finney [2003] for a review of this topic). Succession is simulated using a variety of approaches, such as a stochastic Markov transition model (Usher 1981, 1992; Acevedo et al. 1995); a species-based vital attributes scheme (Roberts and Betz 1999); an empirical frame-based multiple pathway model (Keane et al. 1998, 2002b; Chew 1997; Wimberly, Spies, and Nonaka, chapter 12, this volume); a deterministic age-since-disturbance function (Baker 1994; Li et al. 1997); or a physical biogeochemical model (Keane et al. 1996b). Fire ignition is usually modeled with stochastic functions based on Weibull

probability distributions (He and Mladenoff 1999a; Keane et al. 2002b) that can be linked to indices of fire weather (Gardner et al. 1996; Li et al. 2000). Fire spread is simulated using cellular automata, percolation, or physical fire behavior models (Gardner et al. 1999; Turner et al. 2001). The effects of a fire are often modeled using a rule-based approach, probabilistic functions, or explicit simulations of fire damage (Keane and Finney 2003).

Our review of the literature found a few examples in which simulation modeling has been used to simulate fire regimes, although not all authors output fire regime maps from these simulations. The SEM-LAND model estimated fire regimes for Alberta (Li et al. 1997; Li 2000a), and Perera et al. (2003) used BFOLDS to simulate crown fire regimes in the boreal mixedwood forest (also see Perera et al., chapter 9, this volume). He and Mladenoff (1999a) used the LANDIS model (see He et al., chapter 10, this volume) to simulate fire regimes in northern Wisconsin, and Keane and Long (1998) simulated broad-scale fire regimes by using the CRBSUM model for the Interior Columbia River Basin (United States). A deterministic predator-prey fire regime model using differential equations and based on biomass components was developed by Casagrandi and Rinaldi (1999) for Mediterranean forests. The DISPATCH model was used to simulate historical and current fire regimes within a high-elevation landscape in Wyoming (Baker 1993, 1994). Roberts and Betz (1999) used the LANDSIM model to simulate fire regimes within a landscape in Bryce Canyon National Park (United States) and Gardner et al. (1996) simulated fire regimes in Yellowstone National Park, Wyoming, with the EMBYR model. Antonovski et al. (1992) characterized fire regimes in Siberia by using a spatial landscape-scale fire succession model. In each effort, simulations of thousands of years were summarized at the pixel level to create spatial predictions of the fire frequency and severity.

The major drawback of the simulation approach is that most models are oversimplifications of reality and thus, do not always accurately represent all landscape, ecosystem, climate, and fire dynamics and their interactions. Many succession modules in a model are so simple that detailed fire behavior and growth predictions are not appropriate (Finney 1998). Fire spread algorithms may not generate realistic fire patterns, because detailed hourly weather patterns and fine-scale fuel distributions are nearly impossible to obtain for large landscapes and long time spans. Hourly time steps for a simulation, combined with high-resolution landscapes, may overwhelm computer resources, resulting in excessively long and complex simulations. Fire ignitions may not match those that occur in nature, because of the lack of knowledge, paucity of data, and inherent complexity of the ignition processes; for example, storm tracks, seasonal variations, lightning strikes, and fuelbed receptivity can all influence ignition. Landscape models can be expensive to build, test, and apply because of this inherent complexity. Many management agencies lack the experience or computer resources to implement complex landscape models for their land area. Finally, the diversity of landscape models with respect to their simulation of ecosystem processes, application, scale of development, geography, and computer resources makes it difficult to pick the optimal model for a given landscape or application without detailed knowledge of all available models.

A major advantage of simulation modeling over the other methods is its ability to integrate diverse and cross-scale processes into a single application in an attempt to account for most factors that govern fire regimes. Climate can be modeled at a relatively broad scale and linked to stand-development and fire ignition models at a finer scale (Keane et al. 1996a). Simulation models also provide a means to integrate point-level data, such as fire frequency, with physical processes to generate spatially explicit computations of the fire regime. Traditional fire history chronologies can be used as inputs for simulation models to compute fire rotation (time required to burn an area the size of the landscape in question), cycle, and pattern within the landscape. Another advantage is that models can be used for purposes other than fire regime prediction, such as comparing management alternatives and understanding landscape dynamics. In addition, new fire regime maps can be easily derived as the models integrate new research findings. And finally, fire regimes can be described via simulations over time periods much longer than the recorded period of fire history.

Comparison Exercise

We have evaluated some of the approaches and strategies used for developing fire regime maps by expanding on research in gradient modeling and analysis (Keane et al. 2002a). Five combinations of the three approaches and strategies defined in the present chapter were used to cre-

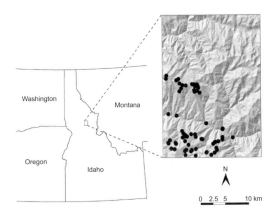

FIGURE 5.1. Geographic location of the lower Selway landscape in the northwest United States, used in the comparison of strategies and approaches. Dots on the enlarged section of the map represent the locations of sample points used to provide data for the analysis.

ate fire regime maps (fire frequency and severity) with identical classification categories, map resolutions, and field data sets (see table 5.1). A portion of the lower Selway watershed (51,761 ha) located on the western edge of the Selway-Bitterroot Wilderness in the mountains of central Idaho (United States) was simulated using a 30-m spatial grid (figure 5.1). Field data used in this comparison were taken from 64 plots located within this landscape in 1995 by Keane et al. (2002a).

We used two maps to describe the fire regime for each of five approach-strategy combinations. The fire frequency map contained three categories of fire-return interval: less than 40, 40–100, and greater than 100 yr. The three categories defined at the beginning of this chapter were also used to create the fire severity map: nonlethal surface fire, mixed-severity fire, and stand-replacement fire. One pair of fire regime maps was developed by using a classification-empirical combination from the rule sets used by Barrett and Arno (1991) for the Selway-Bitterroot Wilderness. Another pair of maps was created by employing statistical-empirical discriminant analysis and regression modeling on an extensive gradient-based set of field data on soils, topography, vegetation, biophysical site type, and fuels (Keane et al. 2002a). We also simulated more than twenty ecosystem process variables, including net primary productivity, evapotranspiration, and soil moisture, for each plot and used these new variables in the discriminant and regression analyses to achieve a statistical-physical combination. Logistic regression was

TABLE 5.2. Comparison of Fire-Return Interval Classes Generated from Combinations of Approaches and Strategies

Strategy-Approach Combination	Fire Interval Class (% landscape)			Map Agreement with Plot Data	
	Frequent (< 40 yr)	Moderate (40–100 yr)	Infrequent (>100 yr)	Overall Percentage Correct	Kappa[1]
Classification-empirical	46.2	36.4	17.4	33.0	0.15
Statistical-stochastic	14.0	11.0	75.0	41.7	0.05
Statistical-empirical	13.1	48.7	38.2	54.2	0.22
Statistical-physical	17.8	41.5	40.7	64.6	0.40
Simulation-stochastic	14.4	82.3	3.3	31.2	0.00
Percentage of field plots	12.5	50.0	37.5		

Notes: The fire interval classes are generated from the combinations presented in table 5.1. Numbers in the table represent the predicted percentage of the lower Selway landscape in each class. The last row presents the number of plots in each fire-return class.

[1]The Kappa statistic adjusts for unequal plot distribution across categories (Congalton and Green 1999).

also used on the same data set, with the process variables used to represent the statistical-stochastic combination. Finally, the probabilistic LANDSUM landscape-scale fire succession simulation model (Keane et al. 1997, 2002b) was used to generate fire regimes in a simulation-stochastic combination.

The results from this comparison showed that fire regime maps created by the five combinations of strategies and approaches differed significantly (tables 5.2, 5.3; color plates 1, 2). Interestingly, a broad-scale comparison of rule-based (classification), simulation, and statistical strategies performed by Morgan et al. (2001) for the Interior Columbia River Basin (United States) showed strikingly similar findings. For fire frequency, the classification-empirical combination resulted in most fires being classified as frequent (<40 yr return interval), the simulation-stochastic combination classified most fires as moderate (40–100 yr), and the statistical-stochastic combination suggested that more of the landscape would be classified in the infrequent fire-return interval (>100 yr) (table 5.2). For fire severity (table 5.3), the mixed-severity class was more commonly predicted by the classification-empirical, statistical-empirical, and statistical-physical analyses. However, the simulation-stochastic and statistical-stochastic combinations predicted that more of the landscape would fall in the stand-replacement severity class.

The fire frequency and severity results created by each analysis are spatially portrayed in the maps in color plates 1 and 2, respectively. Infre-

quent stand-replacement fires are generally predicted to occur at high elevations in nearly all analyses, whereas frequent, nonlethal surface fires generally occur at lower elevations. However, the predicted distribution of moderate-interval, mixed-severity fires differed significantly across strategy-approach combinations (color plates 1, 2). Curiously, no map was especially accurate when compared with the historical plot data, because of the inadequate sample size and biophysical representation (tables 5.2, 5.3).

The differences between the predicted fire regimes resulted from a variety of factors. Statistical strategies and empirical approaches ultimately rely on field data, and although the field data from this study were reasonably representative of the biophysical environment, they were limited by sample size (only 64 plots) and temporal depth (only 200–300 yr of fire history). Statistical strategies are the most accurate, because they best represent the data used to construct and validate the models (table 5.2). The value of physical variables in statistical analysis was surprising. Ecosystem process variables increased accuracies by 10% for both maps and also tended to better portray spatial relationships in the fire regimes. For example, statistical analysis based solely on topographical variables exploited inconsistencies in the digital elevation model at the pixel level, as witnessed by the cross-hatching in color plates 1d,e and 2d,e. The statistical-stochastic analysis tends to suggest that most of the landscape fell into the infrequent, stand-replacement fire regime, because the clas-

TABLE 5.3. Comparison of Fire Severity Classes Generated from Combinations of Approaches and Strategies

Strategy-Approach Combination	Fire Severity Class (% landscape)			Map Agreement with Plot Data	
	Nonlethal Surface	Mixed-Severity	Stand-Replacement	Overall Percentage Correct	Kappa[1]
Classification-empirical	14.6	66.3	20.1	50.0	0.19
Statistical-stochastic	2.4	8.0	89.6	35.4	0.00
Statistical-empirical	3.3	65.0	31.7	64.6	0.34
Statistical-physical	5.7	52.4	41.9	72.9	0.51
Simulation-stochastic	7.1	22.4	70.5	39.6	0.00
Percentage of field plots	8.3	56.3	35.4		

Notes: The fire severity classes are generated from the combinations presented in table 5.1. Numbers in the table represent the predicted percentage of the lower Selway landscape in each class. The last row presents the number of plots in each fire severity category.

[1]The Kappa statistic adjusts for unequal plot distribution across categories (Congalton and Green 1999).

sification of logistic regression is heavily influenced by the prior probabilities of the modeled categories (i.e., most plots were in the moderate frequency, mixed-severity fire regime).

Differing analysis scales were the main reason that the simulation and classification strategies produced such different results from the statistical strategies. Spatially explicit simulations and rule-based classifications based on biophysical setting and topography tend to produce fire regimes that implicitly represent longer periods over broader mapping units. Simulation results are summarized over 1000 yr and directly integrate the spatial context through the modeling of fire spread in time and space. Because the plot data only represent the past 200–300 yr, it is probably unrealistic to expect that the simulated results would ever agree precisely with the plot data, especially given that fire intervals and fire severity have been highly variable in the lower Selway landscape. The three statistical strategies (empirical, physical, and stochastic) are all strongly driven by data from the sample points, and thus had quite similar spatial patterns for both fire interval and fire severity. The spatial patterns generated by the simulation model are the most complex, as they represent an integration of random fire ignitions interacting with dynamic vegetation and topography over a long period. The pattern of fire regimes in the rule-based method is probably most similar to the maps developed by the simulation, because both incorporate maps of potential vegetation based on elevation, aspect, and other factors that tend to span scales of time and space.

Of the analytic combinations that we studied, we recommend simulation modeling as a fire regime prediction package for several reasons. First, simulation modeling integrates fire history and ecological data into a comprehensive application that spans the time and space scales needed to realistically describe fire regimes. Simulation also allows the integration of other ecosystem processes so as to evaluate changes in fire regime brought about by changes in other processes, such as climate, management, and invasions by exotic species or successional changes in vegetation. Diverse management alternatives can be evaluated with simulation modeling to determine how management can change fire regimes. Finally, simulation models provide a tool for understanding and exploring the interactions of fire with the vegetation communities, climate, and biophysical site.

Challenges

There are many challenges in accurately and comprehensively predicting fire regimes across many scales and for different purposes. Here we discuss the most important factors that influence the prediction and simulation of fire regimes.

Field Data

Field data on fire history are essential for many prediction tasks, because they provide the only facts that can be used for understanding, predicting, and interpreting fire dynamics (Morgan et al. 2001; Suffling and Perera, chapter 4, this volume). Field data are needed to describe fire regime characteristics and the resultant classifications. They are also necessary for the creation of predictive algorithms using statistical techniques, and the subsequent validation and accuracy assessment for those equations. Georeferenced (i.e., spatially located) fire history data are especially useful for the development of fire regime maps and assessment of their accuracy (Congalton and Green 1999). Field data sets that contain more than fire dates (e.g., elevation, aspect, soils, weather) can also provide insights into and context for the causal mechanisms that underlie fire regime dynamics. Finally, field data on fire history can be used to initialize, parameterize, and validate landscape-scale ecosystem models of fire dynamics (Li, chapter 8, this volume).

Fire histories (i.e., the dates of past fires) can be developed from many sources, including fire scars on trees; charcoal in lakes, bogs, or soil; and postfire tree establishment dates. However, each of these sources has important shortcomings. Fire scars are the most valuable and commonly sampled source of data, and offer annual to subannual temporal resolution and fine spatial resolution of fire chronologies, but they are also point samples that may be expensive to collect and analyze, and many such samples are needed to adequately describe spatial fire regimes (Heyerdahl et al. 1995). Moreover, the temporal depth of the fire scar record is often dictated by the longevity of the tree species that record the presence of fires and the length of time the scars are preserved after the tree's death. In addition, some fires are missed because they did not scar any trees (Baker and Ehle 2001). In contrast, lake sediments offer long temporal records (thousands of years), but their low precision (10–500 yr) and the difficulty of confirming the source area for a deposit limit their application to broad-scale phenomena such as climate-fire interactions (Clark 1988a; Millspaugh and Whitlock 1995). Tree establishment dates provide spatial descriptions of fire regimes, but are only useful for the mixed-severity and stand-replacement fire regimes, and are also limited to the longevity of the tree species (Johnson and Gutsell 1994; Finney 1995; Reed et al. 1998). Finally, soil charcoal provides a fire record with low temporal and spatial resolutions. For all these techniques, it is costly to collect and analyze the extensive data needed to determine the spatial extent or pattern of fire events. Empirical and physical strategies require prohibitively dense fire history samples before they can provide statistically significant predictive equations. However, even with these limitations, fire history records are invaluable, and their collection is essential for predictive modeling of fire regimes. Unfortunately, these records are rapidly being lost across many landscapes, because many forests have developed high densities as a result of (in the case of the United States) 70 yr of fire exclusion that have increased the potential for intense fires that can destroy fire-scarred individuals. In addition, erosion has removed soil carbon from many stands on steep sites. It is important for fire managers to sample, record, and store existing historical evidence of fire before it is lost forever (Schmoldt et al. 1999).

Scale

A primary challenge in predicting fire regimes involves rectifying (accurately matching) the spatial and temporal scales that govern fire and landscape dynamics (Simard 1991; Heyerdahl et al. 2001). Fire is a complex disturbance process that manifests itself at many time and space scales, yet many fire regime studies attempt to describe fire dynamics only at the stand level across relatively short time spans. For example, prevailing winds and dominant slope are considered to be the primary landscape factors responsible for fire pattern (Agee 1993; Johnson and Wowchuk 1993; Schmoldt et al. 1999), yet most empirical predictive modeling of fire regimes is done at the stand level without accounting for wind and slope effects. Drought is another important factor that dictates fire dynamics, and some fire regime studies integrate this factor over large areas in predictive models (Agee 1995). Lightning dynamics is yet another important and complex factor that is rarely integrated into predictive fire regime models (Knight 1987; Agee 1991).

Most predictive studies rely on data that describe historical fire characteristics (see the previous section of this chapter), but the scale of these data often fails to match the scale of the analysis. Many statistical models of fire regime are temporally limited to the tree ring or fire scar record, and this truncated period may not adequately represent the full historical range of fire regimes in many ecosystems. Fire chronologies are temporally limited to the recent past (i.e., 400–1000 yr BP), which may not provide an appropriate description of future fire regimes, because the historical data rarely include changing climates and land use (Weber and Flannigan 1997). This truncated fire record may not adequately describe fire dynamics in ecosystems with infrequent fires; even a 500-yr fire record has too few fire dates to accurately describe fire dynamics in landscapes where fires are rare (Baker and Ehle 2001). Broad-scale climate processes, such as the El Niño Southern Oscillation and the Pacific Decadal Oscillation, are rarely integrated into predictive modeling, but doing so would help modelers more accurately portray fire regime dynamics linked with climate across large areas and long time intervals (Swetnam and Betancourt 1990).

Other elements that characterize fire regimes, such as pattern and seasonality, are rarely included in predictive models because of scale problems. Fire frequency can be easily quantified from point- or stand-level data, but fire patterns (i.e., the size and shape of the burned patches) require a landscape-scale (broad-scale) spatial design (i.e., fire atlases or a systematic grid of fire scar samples). The prediction of fire seasonality requires not only long-term climate records but also an assessment of the phenology of the trees that sustain fire scars. Fire severity and frequency depend on stand-level fuel characteristics (Ryan and Noste 1985), the location of stands within the landscape matrix (Camp et al. 1997), and the dynamics of broad-scale wind patterns for stands (Swanson et al. 1997). Most variables used to predict fire regime describe attributes at only one temporal or spatial scale and would thus almost certainly exclude important processes that act across scales, such as climate, hydrology, and geomorphology (McKenzie 1998). Geostatistical techniques are needed to integrate spatial scale processes into truly comprehensive predictive models. Simulation modeling provides another, perhaps better, tool for integrating multiscale input data in models that predict fire regimes.

Variability

Inherent spatial and temporal variability within a fire regime is ultimately responsible for landscape structure and composition (Heinselman 1981; Agee 1993; Gill and McCarthy 1998; McKenzie 1998). Lertzman et al. (1998) describe three types of heterogeneity in fire regimes (temporal, spatial, and severity) that play a critical role in landscape dynamics and subsequent management. The variability in the fire-return interval is of utmost importance in assessing the degree of departure from historical conditions, as

well as in designing fire treatments, assessments, and schedules for landscape management (Landres et al. 1999). Diversity in fire pattern characteristics dictates the size and shape of future treatments that attempt to emulate natural disturbance by fire (Hunter 1993). In addition, the range of fire severities experienced within a fire regime will guide treatment design and implementation (Keane 2000). Yet despite recognizing these principles, few developers of fire regime models characterize the intrinsic variability of the factors used to predict fire frequency or severity.

The effects of fire regime variability can manifest at many scales. At the tree level, Keane et al. (1990) found that occasional extended fire-return intervals were needed to maintain ponderosa pine in xeric western Montana forests, because the pine needed the extra time to grow tall enough to escape lethal scorching. In the western United States, many upper-elevation glades, bare areas, and meadows in landscapes with a long fire-return interval were created by "double" burns, in which a second fire occurred within a decade or so of an initial stand-replacement fire (Agee 1993). Many Australian fauna depend on the shifting patchwork of fire severity inherent in that country's fire regimes (Gill 1998). For these reasons, a representation of the variability in each element of a fire regime is essential to support land management, and should be included in any predictive model.

Fire Regime Classification

It is a significant challenge to develop a fire regime classification that adequately portrays fire dynamics across diverse landscapes. Perhaps the most significant challenge lies in quantifying fire frequency. Most predictive models represent fire frequency as either a class (Morgan et al. 1996) or an average (Long 1998). However, it is the variability in fire-return intervals that affects stand and landscape structures and compositions, so each fire regime class also requires an explicit and suitable measure of variation. For example, a fire frequency class of 0–30 yr is meaningless if the field data used to develop the class have a variability of 25 yr. Despite this, most fire regime classifications do not address this fascinating aspect of fire dynamics. A possible solution would be to define each class by using a moving window based on the expected variability and error within the class.

The severity of fire regimes is not always static in time, even when the climate is somewhat stable. It is possible for a stand with a history of nonlethal surface fires (e.g., ponderosa pine) to experience a stand-replacement fire in an extremely droughty or windy year, or when shade-tolerant competitors have created a dense understory after a rare prolonged fire-free interval (Ferry et al. 1995). Some ecosystems, such as those dominated by lodgepole pine (*P. contorta* Dougl.), have a history of both nonlethal surface fires and stand-replacement fires (Brown 1973; Barrett 1994). This mix of severities is very important to fire management, yet the mix is seldom described in the fire history record or in classifications based on this record. Statistical analysis and simulation modeling using stochastic or empirical approaches can provide estimates of temporal variation in fire severity, and new classification categories should be developed to address this phenomenon.

The challenge, then, is to design useful fire regime classifications for diverse local, regional, and national applications. The range of fire-return intervals used to define a given frequency category may not be optimal across all landscapes in a region or across all spatial scales. For example, Hardy et al. (2001) defined their "frequent" fire category to encompass mean fire-return intervals between 0 and 35 yr, yet 35 yr might be far too long for some grasslands and too short for some forests. Broad ranges may not provide sufficient discrimination between important fire regimes, especially in dry, fire-prone landscapes. Thus, the interpretation of this somewhat arbitrary range might be misleading for many management applications. Some managers might use the midpoint of this class as a target fire-return interval without first evaluating field evidence collected from their local landscapes and comparing sampled interval variability within design categories. A possible solution is to create predictive equations that perform their calculations based on continuous fire-return intervals instead of discrete categories, so as to better depict the spatial and temporal variability across a landscape. These frequency models must be augmented by including measures of variability and error.

Future Research

Future fire regime prediction will require the identification of all factors that influence fire dynamics and their interactions. Predictive models will need to explicitly represent those ecosystem processes that contribute to fire ignition, spread, and severity. Two general types of model may

have potential: bottom-up and top-down models (Lertzman et al. 1998, Heyerdahl et al. 2001).

Bottom-up models explicitly account for all factors and interactions that govern fire regimes, including climate, flora, fauna, and topography, in predictive models. Such models are so complex that classification and empirical approaches are not sufficiently sophisticated to link all relevant factors into a comprehensive and cohesive model of fire regimes. Moreover, a field campaign to collect the data for the vast number of variables required for process-based classification and statistical analysis would be logistically prohibitive, costly, and complicated. For these reasons, developing a bottom-up model should use a simulation approach, in which each factor can be simulated and linked to all other factors in a single computer program. Unfortunately, many ecosystem processes have not been studied in the great detail needed for quantitative simulation. Therefore, future research must investigate both the causal mechanisms that underlie fire and landscape dynamics, so that comprehensive simulation models can be developed.

In contrast, top-down models retain those factors important to landscape and fire dynamics as the developer progresses from broad to fine temporal and spatial scales. This means that an explicit simulation of climate at broad temporal and spatial scales should yield weather variables that are important for estimating those mid-scale processes related to such factors as fuels, vegetation, drought, and lightning that govern fire regimes. These variables would then allow fine-scale simulations of fire spread, vegetation succession, and fire effects, which would in turn feed into broader scales to link with climate, thereby completing the cycle. Only the most important processes are retained as one progresses downward toward increasingly fine scales and then back upward in scale. Research must be conducted to identify and study these important processes at each scale that governs the fire regime. This research should be linked to simulation models that integrate the research results into comprehensive frameworks that explore ecosystem, fire, and climate interactions and, in turn, identify future research needs.

A significant dilemma with most efforts to predict fire regimes is that some models will become obsolete if predicted changes in climate become reality (Weber and Flannigan 1997). These changes will render most maps incorrect, unless the models used to describe fire regimes contain a dynamic link to models of climatic processes and thus incorporate the changing conditions into their predictions. It is essential that future predictive efforts integrate climatic variables into the underlying model's design, so that fire regime maps can be updated based on new or alternative weather records. This emphasizes the importance of the simulation strategy in creating future fire regime maps, because simulation models can be designed to explicitly link to such detailed climate drivers as temperature and precipitation (Waring and Running 1998).

The integration of predictive models into fire management activities will require thoughtful analysis and explicit recognition of the assumptions and limitations made during model development. Fire regime maps or equations reflect highly variable natural systems that are constantly changing because of human activities, exotic invasions, and climate. Therefore, fire treatments should never be scheduled at fixed intervals, because this inherent variability affects the ecosystems and thus, how they will sustain fires (Keane 2000). The inclusion of this inherent variability in fire planning requires novel approaches. For example, the fire-return interval for an area could be represented by two maps or models, one describing the twentieth percentile return interval and the other describing the eightieth percentile return interval. Alternatively, the two models could predict the average and standard deviation of the fire-return interval. Results generated by these predictive tools should never be taken as absolute truth but rather as a "best first guess" or as a factor used in relative comparisons.

We think that landscape simulation modeling is the most robust and flexible strategy for mapping fire regimes because (1) the models developed can be used for many other purposes, including comparison and prioritization of alternatives; (2) large regions can be mapped consistently and comprehensively; (3) models allow the integration of processes at multiple scales; and (4) models can extrapolate from relatively minimal field data across large domains. However, simulation models may not accurately portray important fire and vegetation dynamics that will influence the fire regime predictions, because any model must necessarily simplify reality. Landscape-scale fire and succession models can be overly complex, difficult to understand, and costly to execute, thereby precluding their application in land management planning. Moreover, the fire history parameters used as inputs in predictive

simulation modeling represent a relatively small slice of time and a small area in space, so interpretation of the results must account for these limitations in scale. Modelers and users of the models must recognize that fire regimes are not static and will change as patterns of climate, land management, and vegetation change, eventually rendering any predictive map outdated. Developing fire regime models based on the fundamental processes that govern fire dynamics, such as weather, fuels, and ignition patterns, can delay or eliminate this obsolescence.

The best predictive model or map in the world may be useless if there is no comprehensive and ecologically based strategy for integrating the model into land management planning. Ironically, building the predictive model is probably easier than implementing it, because many other factors are involved in the design of fire treatments, such as accessibility (roads, trails), resources (expertise, personnel, monies, time), and landscape characteristics (topography, complexity). Designing landscape-scale fire treatments requires the manager to determine the objective of land management, which could be the restoration of fire-prone ecosystems, the reduction of fuel loads in fire-excluded landscapes, or the creation of fire breaks to reduce fire danger at the urban-forest interface. This provides the context for defining what treatments will be considered, and when, where, and how they will be implemented. Emulating natural fire processes with silvicultural treatments, such as prescribed burning, demands a comprehensive prediction of fire regimes and their variability that will allow managers to determine the size, frequency, and location of harvesting, burning, and other silvicultural treatments. But it is also important to note that more research is needed to enable managers to design prescriptions that truly emulate natural processes.

Conclusions

We draw the following conclusions from our discussion of predictive modeling for fire regimes:

- The strategy and approach for development of fire regime predictive equations and maps should be matched with the available expertise, computing resources, field data, and management objectives.

- Simulation modeling provides the most flexible and robust means of characterizing the fire regime because it (1) allows summarization across broad spatial and temporal scales;

(2) characterizes the landscape composition and pattern; (3) allows comparisons of different management and climate scenarios; (4) extrapolates from relatively limited field data to broader spatial and temporal scales; and (5) enables exploration of the interactions between fire, climate, topography and management.

- Statistical strategies are more accurate because they are developed from actual field data. However, statistical models are limited in application to conditions similar to those described by the field data.

- Field data are essential for fire regime prediction and modeling.

- Temporally and spatially deep field data provide a source for building, testing, validating, and assessing the accuracy of models and parameterization of the simulation model.

- Fire history data have many limitations that should be acknowledged during the modeling of fire regimes, so that their consequences can be mitigated.

- Every effort should be made to collect fire history data in landscapes where these data are being lost through logging, fire, decomposition, or other processes.

- An estimate of fire frequency and its variability is critical for assessing fire dynamics across landscapes.

- Fire frequency should be reported as a continuous variable, rather than as discrete intervals, to permit the documentation of spatial variability.

- Measures of error and variability are critical for understanding the effects of fire across multiple scales.

- Fire treatments should not be scheduled at fixed intervals; they should be implemented when conditions require or facilitate a fire treatment.

- A process-based approach for predicting fire regimes is essential to account for changes in climates and ecosystems.

- Climate and weather variables should be included in a model's design.

- A simulation strategy that incorporates climatic factors is most appropriate for predicting future fire regimes.

- Enlightened fire management requires that predictive models be implemented as part of a comprehensive strategy that includes other sources of information and other tools.

- Maps of vegetation and fire occurrence are beneficial for designing the type, location, timing, and severity of a proposed treatment intended to emulate natural disturbance by fire.

- Models of succession provide a useful temporal context for investigating fire and vegetation dynamics.

CHAPTER 6

Predicting Forest Insect Disturbance Regimes for Use in Emulating Natural Disturbance

DAVID A. MACLEAN

Disturbance is both a major source of temporal and spatial heterogeneity in the structure and dynamics of natural communities and an agent of natural selection in the evolution of species (Sousa 1984). The concept of using natural disturbance regimes as a guide for appropriate forest management has received increasing attention over the past two decades (e.g., Pickett and White 1985; Attiwill 1994; Kohm and Franklin 1996; Hunter 1999; Y. Bergeron et al. 1999; Harvey et al. 2002). Information on the historic variability in the magnitude and pattern of natural disturbance can help forest managers to maintain structural and compositional patterns and regional mosaics of forests that are consistent with the natural disturbance regime (Mladenoff et al. 1993; Landres et al. 1999). Such information can also help managers to select stand-level silvicultural treatments that are appropriate, given the natural disturbance regime, silvics of the species, and stand dynamics (Franklin 1993; MacLean et al. 1999; Seymour and Hunter 1999; Bergeron 2000; Harvey et al. 2002). Maintaining the regional diversity of forest types and applying appropriate stand-level treatments reflect the essence of the "coarse filter" approach to maintaining biodiversity (Woodley and Forbes 1997; Thompson and Harestad, chapter 3, this volume).

Several factors have resulted in an increasing need for landscape-scale models of forests that include natural disturbance. First, various forest certification systems demand information on the spatial pattern and distribution of natural forest communities; their level of fragmentation and connectivity; the maintenance of habitats and landscape diversity; or the management of size, shape, and placement of harvests. Second, the move toward ecosystem management includes assessing conditions at a much-expanded (broader) scale and over longer time frames and considering the full ecological, economic, legal, and social ramifications of management actions (Thomas 1999). Third, natural disturbance largely determines the spatial and temporal patterns of forests. Hence, modeling the effects of fire, insects, and wind is increasingly being used to guide the design of appropriate forest management.

Fire is the most commonly cited natural forest disturbance (e.g., Keane, Parsons, and Rollins, chapter 5, this volume; Suffling and Perera, chapter 4, this volume). However, insect outbreaks are another major natural disturbance in North America and other regions. Schowalter et al. (1986) reviewed the role of herbivory in forested ecosystems, the factors that influence herbivory, and its consequences at the tree and ecosystem levels of resolution. Timber supply losses caused by insects and diseases have been estimated at between 81 and 107 million $m^3 \cdot yr^{-1}$ in Canada, an amount equal to roughly half of annual harvest levels (Sterner and Davidson 1982; MacLean 1990; Power 1991). Somewhat surprisingly, pest-caused losses in Canada are about three to four times as large as those caused by wildfires, in part due to the greater feasibility of salvaging trees after a fire, which affects discrete areas rather than individual trees dispersed through a larger stand. In the most recent (1996) U.S. forest inventory, mortality (the volume of sound wood in trees that died from natural causes during a specified year) was 179 million $m^3 \cdot yr^{-1}$, in comparison with removals (harvesting, silvicultural operations, or clearing of timber land for nonforestry use) of 453 million m³·yr⁻¹ (U.S. Department of Agriculture Forest Service 2001).

Major insect outbreaks result in stand-replacing disturbances, whereas less severe insect damage often causes fine-scale disturbances (gaps). In the latter cases, single trees or clumps of trees are killed, creating gaps that permit subsequent regeneration of trees or other vegetation. Categorizing and analyzing insect-caused disturbance regimes can help to inform an approach to forest management that is consistent with natural disturbance. It is somewhat questionable whether we can truly emulate insect-caused disturbance—there will always be differences between harvesting and natural processes—but such analysis clearly can help managers to design (I prefer the term *inspire*) harvesting systems that are consistent with both stand- and landscape-scale disturbance. In this chapter, I review the effects and prediction of insect disturbance regimes for use in emulating natural disturbance.

Major Insect Disturbance Regimes in North American Forests

To understand or emulate insect-caused natural disturbance, we need to understand the temporal and spatial patterns and the scale of insect outbreaks, as well as the relationships that link outbreak characteristics (defoliation, or population levels for nondefoliators) with the impacts of the disturbance on trees and stands. Insect-created disturbance regimes differ from fire disturbance regimes in that they are host-species dependent. That is, only the host species of a given insect are directly affected. Therefore, the range of an insect's host species provides an initial estimate of the potential spatial extent of outbreaks. Overlaid on this spatial pattern are

the outbreak dynamics of insect populations, which include such considerations as whether outbreaks begin at epicenters and spread or whether they erupt synchronously over large areas as a result of climatic or biotic influences. For example, Fleming et al. (2000) explored four alternative theories of outbreak dynamics for the eastern spruce budworm (*Choristoneura fumiferana* Clem.). Historical data provide the starting point for interpreting outbreak patterns and scales. Many insects can disperse over long distances via flights, and can effectively locate host trees within primarily nonhost stands. However, the spread of a few insects, such as the gypsy moth (*Lymantria dispar* [L.]), is limited by their lack of dispersal ability (in this case, because female gypsy moths cannot fly).

From 1975 to 2000 in Canada, forest insects caused a staggering 709.2 million ha of moderate-to-severe defoliation, plus areas of beetle-killed trees (figure 6.1a) (Canadian Council of Forest Ministers 2002). Moderate and severe defoliation are defined as affecting 30–70% and 71–100% of the current-year foliage, respectively, in New Brunswick, Canada (MacLean and MacKinnon 1996), with similar class definitions used in other regions. To put this estimated 709.2 million ha in perspective, Hebert (2002) reported that of the 234 million ha of commercial timber in Canada, about 0.4% are harvested annually and 0.5% are depleted by fire and insects annually. The latter estimate equates to 11.7 million ha·yr⁻¹ of commercial timber. In comparison, from 1980 to 2000 in the United States, the six major insect pests affected 167.1 million ha (figure 6.1b) (U.S. Department of Agriculture Forest Service 2002); note that this total does not include data for "other insects," and that the spruce beetle (*Den-*

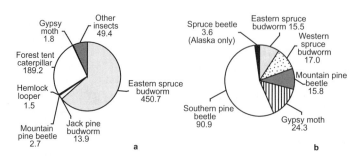

FIGURE 6.1. Distribution, by insect type, of the area of moderate-to-severe defoliation caused by forest insects and the area of beetle-killed trees. (a) The total of 709.2 million ha of damage that occurred from 1975 to 2000 in Canada; (b) the total of 167.1 million ha of damage that occurred from 1980 to 2000 in the United States. Data for (a) are from the Canadian Council of Forest Ministers (2002); data for (b) are from U.S. Department of Agriculture Forest Service (2002).

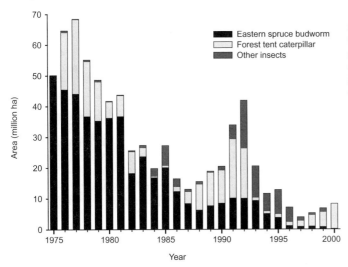

FIGURE 6.2. Area in which moderate-to-severe defoliation caused by forest insects and beetle-killed trees occurred in Canada from 1975 to 2000. Data are from the Canadian Council of Forest Ministers (2002).

droctonus rufipennis Kby.) estimate for Alaska is for the 1980–1999 period only. However, the area of moderate-to-severe defoliation in Canada has varied widely over time, ranging from almost 70 million ha in 1977 to less than 4 million ha in 1997 (figure 6.2). This temporal variation reflects outbreak periods for major insect pests, especially eastern spruce budworm and forest tent caterpillar (*Malacosoma disstria* Hbn.), which caused 64 and 27% of the defoliation, respectively (figure 6.1a).

There are several caveats that apply to the interpretation of insect survey data for large areas (Canadian Council of Forest Ministers 2002). More than one insect may defoliate any forested area, and thus, there can be considerable overlap in the reported figures. The area in which moderate-to-severe defoliation occurs can also include roads, cultivated areas, small lakes, or burned areas. Areas reported as defoliated may include patches in different severity classes. In addition, some areas of defoliation may be missed in the overall surveys. For beetles, the areas reported are those in which trees have been killed; areas classified as having sustained beetle damage do not usually include defoliation. However, all of these concerns are minor relative to the overall effect of defoliation, and the only available data for interpretation over large areas come from aerial surveys.

Six insects have caused the majority of damage in Canada (figure 6.1a), and five of these are native species:

- Eastern spruce budworm is the most destructive pest of balsam fir (*Abies balsamea* [L.]

Mill.) and spruce (*Picea* spp.) forests. It occurs throughout the native range of balsam fir and has been reported on 25 tree species, including other firs (*Abies* spp.), spruces, pines (*Pinus* spp.), hemlocks (*Tsuga* spp.), larches (*Larix* spp.), and one species of juniper (*Juniperus* sp.).

- Jack pine budworm (*Choristoneura pinus pinus* Freeman), a close relative of the eastern spruce budworm, is a solitary defoliator whose range coincides almost exactly with that of its preferred host, jack pine (*Pinus banksiana* Lamb.). In Canada, it is found from New Brunswick to British Columbia, but the major outbreaks have been reported in Ontario, Manitoba, and Saskatchewan, and have covered about 14 million ha from 1975 to 2000.

- Mountain pine beetle (*Dendroctonus ponderosae* Hopk.) is a serious pest of mature pines in western Canada. Major infestations totaling 2.7 million ha have occurred in western forests with a significant lodgepole pine (*Pinus contorta* Dougl.) component, especially in British Columbia. Outbreaks last 8–10 yr and can cause extensive mortality in pine stands.

- Eastern hemlock looper (*Lambdina fiscellaria fiscellaria* [Gn.]) is a defoliator that occurs from the Atlantic coast westward to Alberta. Its preferred hosts are balsam fir and hemlock (*Tsuga canadensis* [L.] Carr.). The insect initiates sudden outbreaks, and can kill trees in 1–2 yr, especially in Newfoundland, where outbreaks are frequent and serious. Mountain pine beetle and eastern hemlock looper differ from the other insects in figure 6.1a in that the area infested often approximates the area of mortality, whereas the other insects require repeated annual defoliation over several years to cause mortality.

- Forest tent caterpillar is a pest of hardwoods (mainly trembling aspen, *Populus tremuloides* Michx.), and is widely distributed from coast to coast. During outbreaks, it will also attack other hardwood species.

- Gypsy moth is a nonnative defoliating insect introduced into North America in 1869. This exotic has become a serious pest of hardwoods and has expanded its range to include the Maritimes, Quebec, Ontario, and British Columbia, as well as extensive areas in the eastern United States (e.g., Muzika and Liebhold 2001; Sharov et al. 2002).

A total of 36 other insect species have caused 49 million ha of moderate-to-severe defoliation (7% of the total) in Canada from 1975 to 2000 (table 6.1). These include several other budworm species, plus a variety of aspen defoliators, beetles, and sawflies.

Six insects also caused the majority of the damage in the United States, including three pests that are prominent in Canada (eastern spruce budworm, mountain pine beetle, and gypsy moth). The other major insects are the southern pine beetle (*Dendroctonus frontalis* Zimmerman), western spruce budworm (*Choristoneura occidentalis* [Freeman]), and spruce beetle. As is the case in Canada, the area of the United States affected

TABLE 6.1. Total Area in Canada of Moderate-to-Severe Defoliation Caused by Other[1] Insects or Total Area of Beetle-Killed Trees from 1990 to 2000

Insect	Area (ha)	Percentage
Blackheaded budworm (*Acleris variana* [Fern.] and *A. glovernana* [Wlshm.])	34,650	0.07
Balsam fir sawfly (*Neodiprion abietis* [Harr.])	111,853	0.23
Swaine jack pine sawfly (*N. swainei* Midd.)	3055	0.01
Bruce spanworm (*Operophtera bruceata* [Hulst])	827,116	1.67
Western spruce budworm	1,059,041	2.14
Large aspen tortrix (*Choristoneura conflictana* [Wlk.])	2,048,887	4.15
Two-year-cycle spruce budworm (*C. biennis* Free.)	894,195	1.81
Spruce beetle	245,301	0.50
Aspen twoleaf tier (*Enargia decolor* [Wlk.])	9,451,110	19.13
All remaining species in the category *other*[1,2]	34,715,251	70.29
Total insects	49,390,459	100.00

Source: Data are from Canadian Council of Forest Ministers (2002).

[1]See figure 6.1a for a definition of *other* insects.

[2]By individual provinces, the remaining insects include:

Newfoundland, New Brunswick, Nova Scotia: white pine weevil (*Pissodes strobi* [Peck]), yellowheaded spruce sawfly (*Pikonema alaskensis* [Roh.]), eastern larch beetle (*Dendroctonus simplex* Lec.), birch casebearer (*Coleophora serratella* [L.]), larch casebearer (*Coleophora laricella* [Hbn.]), birch skeletonizer (*Bucculatrix canadensisella* Cham.), larch sawfly (*Pristiphora erichsonii* [Htg.]), spruce budmoth (*Zeiraphera canadensis* Mut. & Free.), oak leafroller (*Archips semiferana* [Wlk.]), spruce beetle, balsam woolly adelgid (*Adelges piceae* [Ratz.]), eastern larch beetle (*Dendroctonus simplex* Lec.), white-marked tussock moth (*Orygia leucostigma* [J. E. Smith]), variable oakleaf caterpillar (*Lochmaeus manted* Dbly.), satin moth (*Leucoma salicis* [L.]), cedar leafminers (*Argyresthia canadensis* Free.), birch sawfly (*Arge pectoralis* [Leach]).

Ontario: birch skeletonizer, early aspen leafcurler (*Pseudexentera oregonana* [Wlsm.]), aspen serpentine leafminer (*Phyllocnistis populiella* [Cham.]), greenstriped mapleworm (*Dryocampa rubicunda rubicunda* [F.]), pine false webworm (*Acantholyda erythrocephala* [L.]), oak leafshredder (*Croesia semipurpurana* [Kft.]), oak leafroller.

Saskatchewan: large aspen tortrix, aspen leafroller (*Pseudexentera oregonana* [Wlsm.]).

Alberta: large aspen tortrix, aspen leafroller, spruce gall midge (*Mayetiola piceae* [Felt]).

British Columbia: balsam fir bark beetle (*Pissodes dubius* Rand.), large aspen tortrix, spruce beetle, Douglas-fir beetle (*Dendroctonus pseudotsugae* Hopk.), gray spruce looper (*Caripeta divisata* Wlk.), blackheaded budworm, greenstriped forest looper (*Melanolophia imitata* [Wlk.]), satin moth.

by forest insects has varied widely over time, with maximum areas of defoliation reaching about 10.7 million ha in 1988 for southern pine beetle, 5.3 million ha in 1981 for gypsy moth, 3.1 million ha in 1978 for eastern spruce budworm, 2.4 million ha in 1986 for western spruce budworm, 1.9 million ha in 1981 for mountain pine beetle, and 0.5 million ha in 1996 for spruce beetle (U.S. Department of Agriculture Forest Service 2002). The U.S. Department of Agriculture Forest Service (2002) reported on the condition of 45 native species and 16 nonnative species, but the areas affected by these insects were not summarized in many cases. The majority of the damage was caused by the abovementioned six species.

However, natural disturbance is not equivalent to the area of moderate-to-severe defoliation, but rather to the effects of that defoliation on trees. Although it is clear that the biological effects of natural disturbance include growth reduction and changes in nutrient cycling, the strongest indicator of a major natural disturbance is the area that sustains substantial mortality caused by defoliation or other insect damage. Estimating this area requires the ability to predict tree and stand responses to defoliation, which are discussed later in this chapter. However, some data are available from the annual losses caused by pests and fire in Canada (also termed *depletion estimates;* table 6.2) and from forest inventory reports on natural mortality in the United States (U.S. Department of Agriculture Forest Service 2001). Major eastern spruce budworm outbreaks occurred in the 1970s and early 1980s; during this period, the estimated wood volume lost to pests in Canada reached 107 million $m^3 \cdot yr^{-1}$ (table 6.2). This compares with an estimated volume lost to fire of only 25 million $m^3 \cdot yr^{-1}$ (considerably less than the insect losses, because it is more feasible to salvage trees affected by fire) and, for example, a harvest volume

of 180 million $m^3 \cdot yr^{-1}$ in 1988. The sustainable harvest level for Canada's forests, estimated in terms of the annual volume production at about 240 million $m^3 \cdot yr^{-1}$, is only a little more than double the annual losses to pests (table 6.2). The U.S. Department of Agriculture Forest Service (2001) reported natural mortality of 179 million $m^3 \cdot yr^{-1}$ in 1996, in comparison with total removals by harvesting, silviculture, and clearing of timber land equaling 453 million $m^3 \cdot yr^{-1}$.

Which pests cause the biggest losses? Sterner and Davidson (1982) provided the following breakdown for Canada for the 1977–1981 period: eastern and western spruce budworms 41%, wood decay organisms 23%, cankers (*Hypoxylon* spp.) 10%, aspen defoliators 7%, mountain pine beetle 5%, dwarf mistletoe (*Arceuthobium* spp.) 3%, and spruce beetle 3%, with miscellaneous defoliators, bark beetles, and diseases accounting for the remaining 8%. Major eastern spruce budworm outbreaks in North America began in about 1910, 1940, and 1970 (Blais 1983).

Characterizing Temporal Patterns of Insect-Caused Disturbance Regimes

It is clear that information about the temporal and spatial patterns of insect outbreaks is needed at finer scales than are provided by country- or continent-wide statistics. To examine this problem, I explore outbreaks for four insect groups in Canada, and then consider eastern spruce budworm at even finer scales in eastern Canada as an example of spatial and temporal analyses. Figure 6.3 presents the temporal distribution of defoliation by eastern spruce budworm, forest tent caterpillar, jack pine budworm, and other insects in four regions of Canada. Most of the defoliation by eastern spruce budworm occurred in Ontario and Quebec, whereas defoliation caused by forest tent caterpillar and jack pine budworm mainly occurred in Ontario and the western provinces. Atlantic Canada sustained major eastern spruce

TABLE 6.2. Annual Losses Caused by Pests and Fire Compared with Annual Harvest Levels and the Annual Allowable Cut (AAC) for Canadian Forests

Disturbance	Volume ($m^3 \cdot yr^{-1}$)	Years	Source
Pests	107,000,000	1977–1981	Sterner and Davidson (1982)
Pests	81,265,000	1982–1987	Power (1991)
Fire	24,600,000	1977–1981	Bickerstaff et al. (1981)
Harvest	180,000,000	1988	Rotherham (1991)
AAC	240,000,000	1980–1988	Rotherham (1991)

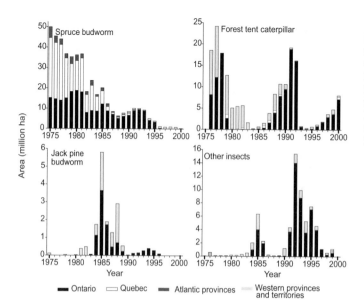

FIGURE 6.3. Area within which moderate-to-severe defoliation occurred (including the area of beetle-killed trees), by geographic region in Canada, from 1975 to 1999. From the Canadian Council of Forest Ministers (2002); data sources are the Canadian Forest Service and the National Forest Database.

budworm outbreaks, but occupies only a small area relative to the other regions.

Figure 6.3 gives some indication of the broad outbreak patterns for several insects, especially forest tent caterpillar and jack pine budworm, but finer-scale data show a pattern of more discrete outbreaks. Data on moderate-to-severe defoliation in New Brunswick indicate the occurrence of two eastern spruce budworm outbreaks since 1950 (figure 6.4a). The total area of New Brunswick is 7.4 million ha, and about 4.7 million ha contains some budworm-susceptible spruce and fir. At the peak of the outbreaks in the 1950s and in the 1970–1980s period, 4.1 million and 3.6 million ha, respectively, sustained moderate-to-severe defoliation (figure 6.4a). However, even the province-level depiction in the figure gives a somewhat misleading indication of the outbreak pattern, as can be seen by focusing on a 190,000-ha district in northern New Brunswick (figure 6.4b). This finer-scale view reveals a 13-yr period of negligible or light defoliation between two discrete periods of moderate-to-severe defoliation. Defoliation began and declined earlier in some parts of the province than in others. As this example shows, scale is important in such analyses. Royama (1984) provides more details on budworm outbreak dynamics.

Eastern spruce budworm outbreak patterns have been characterized for specific zones in Ontario (Candau et al. 1998, 2000; Fleming et al. 2000) and Quebec (Gray et al. 1998), thereby permitting further refinement of outbreak predictions. Candau et al. (1998) and Gray et al. (1998)

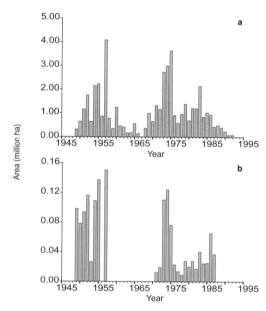

FIGURE 6.4. Moderate-to-severe spruce budworm defoliation from 1945 to 1995 in (a) New Brunswick, Canada; (b) the 190,000-ha Black Brook District in northern New Brunswick.

noted differences in both the onset of defoliation (beginning of the outbreak) and the duration of the defoliation (length of outbreak) among different spatial zones. Outbreak patterns can also be altered by the use of insecticides, which were used to treat 3.6 million ha in Canada from 1990 to 2000 (table 6.3). Of this area, 72% was treated using the biological insecticide *Bacillus thurin-*

TABLE 6.3. Application of Forest Insecticides in Canada, by Product, from 1990 to 1999

Year	Area treated (ha)				
	Bacillus thuringiensis	Fenitrothion	Tebufenozide	Other	Total
1990	855,535	457,323	0	57	1,312,915
1991	370,614	224,270	0	1355	596,239
1992	176,679	178,506	0	11,199	366,384
1993	162,254	72,987	0	7597	242,838
1994	95,445	4709	400	147	100,701
1995	243,850	0	2762	225	246,837
1996	216,038	0	592	419	217,049
1997	0	0	16,557	0	16,557
1998	200,013	0	0	20,957	220,970
1999	212,637	0	0	0	212,637
Total	2,533,065	937,795	20,311	41,956	3,533,127
Percentage of total	72	26	1	1	100

Source: Data are from Canadian Council of Forest Ministers (2002).

giensis (B.t.), primarily against eastern spruce budworm. Fenitrothion was used extensively against eastern spruce budworm (26% of the area sprayed) in the early 1990s and tebufenozide and other materials were used on only 2% of the total area sprayed. Insecticide use varied widely among regions, depending on forest policy, with most spraying in the early 1990s occurring in New Brunswick and Quebec, and spraying in Saskatchewan, Manitoba, and Alberta occurring later in the decade. Studies have found little impact of insecticide spraying on large-scale outbreak patterns (e.g., Fleming et al. 1984; Royama 1984; Lysyk 1990). However, insecticide use does reduce defoliation locally and thus alters local patterns of defoliation. Porter, Hemens, and Mac-Lean (chapter 11, this volume) used spray efficacy data to estimate defoliation levels in the absence of spraying for the 1970s and 1980s budworm outbreak for the land base presented in figure 6.4b, and found a strong effect of insecticide spraying, which reduced defoliation in the Black Brook District.

It is worth noting that some authors believe that insect outbreaks function to stabilize forest production in some systems rather than acting as a destabilizing force. This hypothesis is supported by a large body of literature that quantifies the impacts of insect outbreaks on tree growth, mortality, succession, and forest structure. Matt-

son and Addy (1975) discussed the role of phytophagous insects as regulators of primary production and concluded that the presence of many defoliating insects tends to ensure consistent and optimal output of plant production for a particular site. Insect actions appear to vary inversely with the vigor or productivity of the system, probably as a consequence of the long history of coevolution between plants and their consumers (Mattson and Addy 1975). Romme et al. (1986) considered the hypothesis that mountain pine beetle outbreaks in the Rocky Mountains function as a regulator of primary productivity in lodgepole pine, through their selective killing of dominant trees and the subsequent redistribution of resources among the survivors. Although surviving trees did grow significantly faster, wood production was redistributed across stands, and annual wood production per hectare returned to preattack levels in 10–15 yr, analyses showed that the beetles introduced more variability in wood production than would have existed in their absence. Romme et al. (1986) concluded that the mountain pine beetle–lodgepole pine system showed great resilience.

Lowman and Heatwole (1992) studied spatial and temporal variability in the defoliation of *Eucalyptus* spp. Some eucalyptus species sustained much greater defoliation than others, and this was related to their susceptibility to dieback. Two species (*E. nova-anglica* and *E. stellulata*) lost up to 300 and 274%, respectively, of their initial leaf surface area in 1 yr and sustained severe dieback, whereas other species lost as little as 20 or 8%. Defoliation can result in enhancement of the dietary quality of the remaining leaves of *E. blakelyi* (Landsberg 1990), leading to self-perpetuating chronic herbivory.

Insect Outbreak Patterns over Time

It is possible that current and future insect outbreak patterns may differ substantially from historic outbreak patterns as a result of changes in host density, forest spatial patterns, climate change, and other factors. For example, fire suppression has resulted in changes in species composition and forest structure in the western United States, and this has increased the area of host species for western spruce budworm. Similarly, Blais (1983) hypothesized that the increasing area of eastern spruce budworm outbreaks in eastern North America during the twentieth century resulted from an increased area of balsam fir, the most susceptible host species, in response to a policy of fire suppression. In addition, climate change may alter host species distributions and insect outbreak dynamics (e.g., Volney 1996). This possibility needs to be further researched.

Clearly, we cannot simply assume that historic outbreak patterns will reoccur in the future. Understanding the mechanisms of insect population dynamics, the interaction of these dynamics with climatic and forest factors, and the causes of outbreak initiation and decline is the ultimate goal, but this understanding will be difficult to attain. Retrospective analyses of historical trends provide useful information in support of this goal.

Understanding the silvics of individual species and stand dynamics can also help to provide directions for emulating natural forest disturbance. For example, thinning or a return to a more natural fire regime (especially to reduce populations of invasive, fire-intolerant species) could help return forest structures in the western United States to more historically typical conditions that would be less prone to insect outbreaks. Precommercial and commercial thinning to remove balsam fir (Blum and MacLean 1984), planting nonsusceptible or slightly susceptible species, and mixedwood management (Needham et al. 1999) all have a role to play in reducing eastern spruce budworm damage.

Quantitative Assessment of the Impacts of Forest Insect Disturbances

To determine the effects of insect outbreaks, we need methods to quantify and model stand responses to defoliation or other insect damage (e.g., mountain pine beetle is not a defoliator). This requires quantification of the relationships of growth and mortality to defoliation or other measures of insect occurrence or damage. Erdle and MacLean (1999) proposed a method for quantitatively assessing the impacts of forest pests based on forecasting pest populations, defining the nature and extent of the damage inflicted on trees by those populations, and translating the impact across scales, from tree to stand to forest. Central to this process were stand development forecasts that included tree-level impacts and provided input to forest-level models. Calibration of stand growth models in an implementation for eastern spruce budworm involved using 20 yr of permanent sample plot data to define the relationships of tree survival and growth to defoliation (Erdle and MacLean 1999).

The first requirement is to define pest incidence over time by predicting where, when, and how much defoliation will occur. In many cases, it is difficult to accurately predict the onset of defoliation or the beginning of an outbreak. Analyses of historical temporal and spatial patterns in outbreaks and their relation to climatic and other factors (e.g., Candau et al. 1998; Gray et al. 1998; Fleming et al. 2000) and to insect and disease survey data provide a strong basis for predictions. In addition, scenarios based on historic or generalized outbreak patterns can be used to determine the "what if?" effects of outbreaks. MacLean et al. (2001) defined such general scenarios for eastern spruce budworm outbreak patterns in Atlantic Canada. Alternatively, actual historical defoliation data, such as those in figure 6.4, can also be used.

The incremental effects of pest damage are generally a function of the severity of the damage, which can be measured (e.g., as percentage defoliation by defoliating insects), and the vulnerability of the trees being affected, which may result in different responses to a given level of damage (Erdle and MacLean 1999). For example, a given level of defoliation will more strongly decrease survival in a mature balsam fir stand than in an immature spruce stand (MacLean 1980). Erdle and MacLean (1999) determined that the relationships between growth and budworm defoliation were similar among balsam fir, white spruce (*Picea glauca* [Moench] Voss), and the red spruce–black spruce complex (*Picea rubens* Sarg. and *Picea mariana* [Mill.] B.S.P.; here, these species are lumped together because they hybridize), but mortality for a given level of defoliation decreased in the order balsam fir > white spruce > red spruce–black spruce. When such relationships are used to calibrate a defoliation-based stand growth model, as in Erdle and MacLean (1999), the model can become the basis for assessing and modeling natural disturbance.

There is a long history of modeling eastern spruce budworm dynamics and their effects on forests (e.g., Baskerville 1976; Holling 1978; Baskerville and Kleinschmidt 1981; Stedinger 1984; Steinman and MacLean 1994). The Spruce Budworm Decision Support System (DSS) (MacLean et al. 2000, 2001) is a recent result of these efforts and uses stand and timber-supply models combined with GIS to predict the volume reduction that will result, stand by stand, for large areas of forest. The Spruce Budworm DSS is useful in estimating the natural disturbance effects caused by eastern spruce budworm, because mortality causes the majority of the stand-level changes associated with these outbreaks. The Spruce Budworm DSS has been implemented for all forests in New Brunswick (MacLean et al. 2002) and for test land bases in Quebec, Ontario, Saskatchewan, and Alberta. Models that predict stand-level growth reduction and mortality are the critical linchpin in DSS for pest management (MacLean 1996a). Because of the large amount of research and quantitative sample plot data available for eastern spruce budworm's effects on forests, mortality caused by this budworm is largely predictable if data on defoliation levels are available.

Other empirical models of stand dynamics that include pest effects include modifications of the Stand Prognosis model (Stage 1973; Wykoff et al. 1982) for western spruce budworm (Colbert et al. 1981; Stage et al. 1986; Crookston 1991), Douglas-fir tussock moth (*Orgyia pseudotsugata* McD.) (Monserud and Crookston 1982), and various root diseases (Stage et al. 1990; Shaw et al. 1991) and the yield correction factor methodology of Thomson and Alfaro (1990). The dynamic model of Baskerville and Kleinschmidt (1981) differs from these models by being one of the few process-based stand models that includes insect effects. Considerable research has been conducted to quantify and model gypsy moth impacts (Muzika and Liebhold 1999, 2001; Liebhold et al. 2000; Sharov et al. 2002), including the use of GIS software for forecasting gypsy moth defoliation (Liebhold et al. 1998).

Prediction versus Scenario Planning

It is currently difficult (and perhaps impossible) to accurately predict the onset of an eastern spruce budworm outbreak, although pheromone traps do give some early warning of population increases and are thus the primary sampling tool for low-level populations. Increases in moth catches in pheromone traps are generally evident several years before defoliation becomes noticeable. Pheromone traps, population estimation using branch samples, and aerial defoliation surveys are all established survey methods.

With knowledge of population levels, we can predict the effect of a given defoliation sequence via scenario planning, which involves "what if?" analyses and offers one of the most powerful methods for conveying the consequences of alternative management actions or natural disturbance regimes (MacLean 1998). Scenario planning is a rules-based method for imagining possible futures that has been applied to a wide range of issues (Schoemaker 1995). Among the many tools a manager can use for strategic planning, scenario planning is unique in its ability to capture a whole range of possibilities, thereby allowing managers to see a wider range of possible futures (Schoemaker 1995). This approach is particularly effective in evaluating public concerns or issues related to the preparation of forest management plans, because it provides objectivity concerning such contentious issues as clearcutting, the establishment of conservation areas, sustainability, and wildlife habitat preservation, and communicates the impacts of insects on forests or natural disturbance (e.g., MacLean et al. 2001).

Normal and *severe* scenarios for eastern spruce budworm outbreaks, based on historical defoliation patterns for New Brunswick and Nova Scotia, have been used extensively in the implementation of the Spruce Budworm DSS (MacLean 1998; MacLean et al. 2001, 2002). The predicted volume loss for a severe eastern spruce budworm outbreak affecting more than 3 million ha in northern New Brunswick is presented in color plate 3. Volume loss offers a good surrogate for mortality; the higher the volume reduction, the greater the mortality and relative intensity of the natural disturbance. This simulation produces a landscape-scale spatial pattern that primarily reflects the distribution of the host tree species.

Scenario planning was also used to develop a consensus-based, multistakeholder management planning process for a 114,000-ha land base in New Brunswick (MacLean et al. 1999). Public consultation and stakeholder input were used to define 25 scenarios for determining the effects of alternative means of managing riparian strips, building roads, controlling competing vegetation and insects, harvesting, maintaining biodiversity, and establishing plantations. The Woodstock forest modeling software (Remsoft 1996) was used to determine the effects of each scenario on timber supply, forest structure, various measures of biodiversity and ecological integrity, the area of mature forest, and wildlife habitat. This general approach is applicable to evaluating and communicating the magnitude and effects of natural disturbance and various disturbance emulation scenarios to forest managers and the public.

Natural Disturbance as the Basis for Designing Management Approaches

Fire and insects are natural shapers of forests. Therefore, understanding the natural disturbance patterns they create provides a basis for defining forest-level objectives and appropriate silvicultural treatments that are consistent with the natural disturbance and the silvics of the affected trees (Seymour et al. 2002). The three primary measures of landscape-scale effects are species composition (i.e., host density), age-class distribution, and patch sizes, and the more that forest management differs from the natural range of variability of these measures, the greater the cause for concern.

Seymour et al. (2002) evaluated natural disturbance regimes in northeastern North America in terms of graphs of the "contiguous area disturbed and regenerated" versus the "interval between disturbances (at the same point on the landscape)." This is a useful evaluation measure, but Seymour et al. (2002) focused exclusively on northern hardwood and mixed-coniferous forest types disturbed only by natural canopy gaps (with gap sizes between 4 and 1135 m^2 and a return interval of 50–200 yr) and severe wind and fire disturbances (affecting from 2 ha to <80,000 ha, with a return interval of 800–14,000 yr). Seymour et al. downplayed eastern spruce budworm outbreak dynamics, but these dynamics are clearly a natural disturbance in this region that would extend the boundaries of variation in

any studies of natural disturbance, as depicted in their analysis. In cases of partial mortality, the budworm creates contiguous disturbed patches of about 0.01–1 ha, with a return interval between disturbances of about 30–70 yr. Severe mortality caused by stand-replacing budworm outbreaks results in contiguous disturbed patches of about 2–1000 ha (and potentially larger), with a return interval between disturbances of roughly 60–120 yr. These disturbance intervals assume that a stand-replacing outbreak results in regeneration to mixedwood or balsam fir stands, which will still be immature at the time of the next outbreak (in 30–60 yr), and will thus sustain only partial mortality (creating patches). The analytic technique proposed by Seymour et al. (2002) is elegant and useful, but the analysis deserves to be extended to include the extensive eastern spruce budworm outbreaks that occur in the region.

In terms of emulating natural disturbance, there are clear roles for both even-aged and uneven-aged management. Stand-replacing disturbance can be approximated by clearcutting, whereas patch-type disturbance can be approximated by partial cutting. Thus, insights into natural disturbance can provide useful guidance for designing harvesting and silvicultural treatments. A key factor in these designs is the amount of mortality within stands that will result from an outbreak. A good rule of thumb that has been supported by multiple studies is that mortality levels will average 85% in mature (>50-yr-old) balsam fir stands, 42% in immature (<50-yr-old) balsam fir stands, 36% in mature spruce stands, and 13% in immature spruce stands (MacLean 1980). The presence of hardwoods in coniferous stands has been demonstrated to reduce eastern spruce budworm defoliation of balsam fir (Su et al. 1996) and has been implicated in reducing fir mortality (MacLean 1980; Bergeron et al. 1995; Needham et al. 1999). The spatial pattern of mortality within stands is also of interest in prescribing the spatial harvesting pattern in a stand. Relatively few data are available on this aspect of outbreak dynamics, but both Baskerville and MacLean (1979) and MacLean and Piene (1995) demonstrated "contagious" or "clumped" distributions of mortality within stands, which may suggest that the presence of nonhost species can interfere with the movement of eastern spruce budworm between host trees. Thus, stand-level mortality can be used as a guide for silvicultural treatment.

Insect outbreaks influence forest structure and composition at both the stand and landscape level. At the stand level, harvest prescriptions can be based on removing a number of trees equivalent to what would be killed by insects. At the landscape level, data about insect effects can help managers set targets for species composition, age-class distribution, patch-size distribution, and connectivity. For example, the proportion of a stand killed by insects could be approximated by the amount of forest harvested and regenerated to a younger age class; changes in species composition created by harvesting could be based on insect-caused differential mortality; patch size and connectivity patterns could be implemented using spatially based harvest planning. Note that such a disturbance-based analysis must be tailored to individual stand types and regions; the analysis is not transferable to other stand types nor even necessarily to the same stand type in other regions.

Realistically, no jurisdiction or forest company would blindly "emulate" insect-caused mortality over large regions. However, in some cases, stand-level harvesting that approximates insect effects may make good ecological sense, such as where it results in mixed-species stands or uneven-aged stands that would otherwise be rendered homogeneous (even-aged) by clear-cutting. In particular, analyzing insect effects at the landscape level (e.g., color plate 3) can be instructive in a coarse-filter approach to conserving biodiversity. Insect-caused mortality influences species composition, age-class distribution, and patch sizes. Some DSS are designed to assist in strategic planning to account for insect effects on the annual allowable cut (AAC) and harvest scheduling, and in operational planning of insecticide application or restructuring forests to reduce losses. These can also be used to determine the spatial patterns resulting from insect disturbance; for example, the Spruce Budworm DSS (MacLean et al. 2000, 2001, 2002) is being used in this way in New Brunswick in an effort to move from a finer scale (trees and stands) to a broader scale (landscapes).

Interactions between Insect Outbreaks and Fire

It is relevant to consider the interrelationships between fire and insect outbreaks, the two major natural disturbances in North America. At first glance, there would seem to be a direct relationship, because insect outbreaks often result in dead trees, broken crowns, and an accumulation of dead fuel over large areas. The insect-wildfire

hypothesis, which states that the dead and dying forests remaining after insect outbreaks are more flammable or more susceptible to wildfire than undamaged forests (Furyaev et al. 1983), has been gradually accepted by laypersons, scientists, and forest managers. Several authors in the early literature suggested an increased fire hazard in balsam fir stands killed by eastern spruce budworm (e.g., Graham 1923; Swaine and Craighead 1924; Graham and Orr 1940; Burgar 1963; Flieger 1970; Prebble 1975). Historical information indicates that severe forest fires may follow budworm outbreaks, such as the 1825 Miramichi fire in New Brunswick, the 1936 Isle Royale fire, and the 1948 Chapleau-Mississagi fire in Ontario (Flieger 1970; Stocks 1987). Interactions between fire and insects have also been studied in several other systems (e.g., Anderson et al. 1987; McCullough et al. 1998), but in this chapter, I concentrate on the relevant literature for the interaction between eastern spruce budworm and fire.

During an eastern spruce budworm outbreak, trees usually start to die after 4–5 yr of severe defoliation, and when the outbreak collapses (usually after 10–15 yr), mortality may have claimed 70–100% of the trees in balsam fir stands (MacLean 1980). These dead trees provide drier aerial fuels than do live trees, and faster drying of surface fuels occurs because the canopies of dead stands allow more sun and wind to reach the fuel. In addition, the accumulation of surface fuels increases as trees and their crowns break and fall to the forest floor. The mortality that results from a budworm outbreak can thus increase fire potential, but not to the extent that it overwhelms the effects of topography, weather, fuels, and the ignition source. Van Wagner (1983) noted that the three requirements for a spreading fire are sufficient fuel of an appropriate size and spatial arrangement, sufficient dryness of the fuel to support a spreading combustion reaction, and an ignition source.

Stocks (1987) determined that experimental fires conducted in standing budworm-killed forest in Ontario varied widely in severity. Spring fires were described as "explosive" when conditions were right (e.g., low relative humidity, high temperature, high wind). Stocks concluded that no agency could control fires of the intensity levels reached by spring fires in the budworm outbreak areas. The fires also caused *spotting* (the development of new spot fires by transport of burning embers via the fire's convection column) and quickly reached the crowns of the trees (*crowning*). Fallen dead branches, tops, and trees that had not yet decomposed and standing dead trees with their associated fallen crowns were the primary influences in increasing fuel loads and facilitating crowning.

In contrast, trial fires set in the summer depended on the amount of surface fuel available, because more fuel was required to support a fire sufficiently hot to consume the moist green understory vegetation that proliferates several years after defoliation. Hence, fire hazard was very seasonal, being most severe before vegetation green-up occurs in late spring and less severe thereafter (Stocks 1987). Fleming et al. (2002) determined that the window of opportunity for fire to follow eastern spruce budworm outbreaks opens as available fuel accumulates, but closes as the fuels begin to decompose and understory vegetation proliferates. They also demonstrated regional differences in this window that they attributed largely to climatic differences.

Péch (1993) evaluated changes in fire hazard in budworm-killed forests on Cape Breton Island, Nova Scotia. He expected that over time, the surface fuel loading would increase as mortality and blowdown increased, similar to Stocks's (1987) results. However, Péch found that the incorporation of fallen fuels into the ground and surrounding vegetation was quick enough to preclude the fallen fuels from contributing to the surface fuel layers. Openings left by the fallen trees provided light to understory vegetation, which rapidly increased the amount of ground cover (MacLean 1988) and further reduced levels of surface combustibles. Although no prescribed burning was attempted in Péch's study, two wildfires spread into the budworm-killed stands in late spring. Neither of these fires spread out of control—both were prevented from spreading by a lack of combustible surface fuels. Péch attributed the differences between the Cape Breton and Ontario studies to differences in local climate; colder weather and higher relative humidity in Nova Scotia led to quick disintegration of surface fuels through decomposition and shrub overgrowth.

Thus, the risk of fire posed by the increased amount of available fuel may increase after an insect outbreak, but if the fire weather is unsuitable or an ignition source is lacking, the fire frequency may not increase in insect-killed balsam fir forests. It is likely that regional differences in postoutbreak fire risk will arise due to differences in climate and in the responses of the understory vegetation.

Bergeron and Leduc (1998) discussed the interrelationships between fire cycle, forest composition, and eastern spruce budworm outbreaks for the southern boreal forest. At later stages in the successional sequence, longer fire cycles tend to increase the proportion of stands that are dominated by balsam fir and spruce. Hence, low fire frequency tends to create a forest landscape that is more vulnerable to eastern spruce budworm outbreaks, whereas high fire frequency tends to remove balsam fir, which cannot survive even low-intensity fires (Furyaev et al. 1983). It is clear that fire removes balsam fir, whereas fire exclusion promotes this species. Fir species generally have thin bark and thus do not survive fire; moreover, they do not create seed banks in the forest floor or soil and do not disseminate seeds over long distances; because they are shade-tolerant, seedlings and small trees of these species are often found in the understory of stands (Furyaev et al. 1983).

Outbreaks of eastern spruce budworm in the southern boreal forest have the broad-scale (regional) effect of reducing the forest's softwood component and increasing the mixedwood component (Bergeron and Dansereau 1993; Harvey et al. 2002). These mixedwood stands eventually develop into balsam fir-spruce-white cedar (*Thuja occidentalis* L.) stands, with a hardwood component maintained in local gaps by relatively long-lived birch (*Betula* spp.) (Kneeshaw and Bergeron 1998). These stands are very vulnerable to eastern spruce budworm outbreaks, which usually result in heavy tree mortality (MacLean 1980; Bergeron et al. 1995). In the absence of fire, fir-spruce stands with heavy budworm-caused mortality can develop either a mixedwood composition (Harvey et al. 2002) or succeed to softwood-dominated stands (Baskerville 1975; MacLean 1988). In some cases, the interactions between fire and insects can cause major changes in forest composition (Payette et al. 2000).

Veblen et al. (1994) conducted an interesting analysis of disturbance regimes and the interactions among avalanches, fire, and spruce beetle outbreaks. Their results suggested that there were predictable relationships in space and time among disturbances, and that these relationships helped to explain broad-scale (landscape-scale) vegetation patterns. Large and frequent snow avalanches created areas of reduced forest biomass that served as fire breaks. Avalanches combined with fire created young spruce stands that were too small to support spruce beetle outbreaks even after 70 yr of growth. Therefore, the spatial distribution of recent avalanches and fires did not overlap with that of recent spruce beetle outbreaks (Veblen et al. 1994). Interactions among disturbance agents undoubtedly occur in other ecosystems as well.

Test Cases: Implementation of Harvesting Regimes Based on Natural Disturbance

In 1998, a group of ecologists in the Maritime Provinces of Canada designed a categorization system for the natural disturbance regimes that occurred in their region (Clowater et al. 1999). The major disturbance agents included fire, eastern spruce budworm, windstorms, and ongoing tree mortality (gap-level replacement). Each disturbance regime was categorized in terms of:

- Agent;
- Disturbance cycle (return interval): continuous/ongoing, short (5–30 yr), intermediate (31–60 yr), long (61–100 yr), and very long (101–1000 yr);
- Extent: regional, multistand (101–500 ha), single stand (5–100 ha), and tree (within-stand); and
- Postdisturbance stand condition: stand-replacing (0–25% of the stand remaining), partial stand-replacing (26–50% remaining), patch-replacing (51–95% remaining), and single-tree replacing (gap creating, with 95–100% remaining).

The categories included six fire regimes, ranging from very infrequent, severe fires to frequent, mild fires. They also included three windstorm regimes, as well as the three budworm and two gap-creation regimes shown in table 6.4.

The five disturbance regimes in table 6.4 are being implemented experimentally by my colleagues and me in 2600 ha of forest owned by J. D. Irving, Ltd. in northern New Brunswick. Disturbance categories B1, B2, and B3 are based on the level of mortality caused by eastern spruce budworm, and vary based on the amount of balsam fir and spruce in the stand; emulating these disturbance regimes will involve harvesting an average of about 75–100%, 51–75%, and 5–50%, respectively, of the fir-spruce content (midpoints of these classes were set equal to 87.5, 62.5, and 27.5%, respectively). These regimes are equivalent to stand-replacing, partial stand-replacing, and patch-level replacement disturbances. The actual proportion of the stands harvested will vary depending upon each plot's species composition, with 20 plots per disturbance category; harvesting prescriptions will be based on average mortality levels for each species (e.g., removing eight to nine out of every ten mature balsam fir, four out of every ten mature white spruce or red spruce, and two out of every ten black spruce).

Disturbance regimes G1 and G2 apply to shade-tolerant hardwood stands (sugar maple [*Acer saccharum* Marsh.] and yellow birch [*Betula lutea* Michx. f.]) and are essentially group and

TABLE 6.4. Percentage of the Stand Remaining after Disturbance as a Guide to Target Removal Levels during Harvesting

Disturbance Category[1]	Severity	Stand Remaining (%)		Target Removal Intensity (%)
		Range	Midpoint	
B1: Budworm—stand killing	Stand-replacing	0–25	12.5	87.5 (75–100)
B2: Budworm—partial mortality	Partial stand-replacing	26–50	37.5	62.5 (51–75)
B3: Budworm—minor effects	Patch-replacing	51–95	72.5	27.5 (5–50)
G1: Gap creation—patch creation	Patch-replacing	80–95	87.5	12.5 (5–20)
G2: Gap creation—natural stand dynamics	Gap replacement	96–100	97.5	2.5 (0–5)

Source: Based on Clowater et al. (1999).

[1]Because eastern spruce budworm only influences host species (balsam fir, white spruce, red spruce, and black spruce), the harvest removals for categories B1–B3 only apply to the host species component of the stand.

single-tree selection treatments. These approximate the creation of gaps by the death of small clumps of trees or single trees, caused by blowdown, root rot, insects other than the budworm, natural mortality due to aging, and other factors. A set of 240 permanent sample plots is being used to monitor a suite of specific ecological indicators and assess the biotic integrity of stands in untreated reserves, stands subject to the five natural disturbance-inspired harvesting treatments, and those subject to current operational forestry treatments in the working forest. This will permit not only testing and implementation of the natural disturbance-based harvesting treatments (table 6.4), but also assessment of the pre- and posttreatment responses and between-treatment comparisons of stand structure.

In Maine, using the TRIAD approach to forest management, Seymour and Hunter (1992) proposed the creation of reserves and plantations within a landscape managed by means of alternative silvicultural systems (Franklin 1989; Gillis 1990). Because silvicultural systems should be patterned after local natural disturbance regimes in this approach (Seymour and Hunter 1992), the following types of areas should coexist within the landscape: intensively managed stands (e.g., plantations, thinned stands), extensively managed stands (e.g., naturally regenerated, perhaps longer rotations), and protected areas (unmanaged areas meant to sustain values that could otherwise be lost). The management challenge in this approach is to determine how much of each type of area should be created, in what locations, and at what times to produce a given set of values. In principle, focusing intensive management on a portion of the land base would concentrate silvicultural investments, permit enhanced protection against insects and fire, and permit high yields, while reducing the pressure on extensively managed forest. The prudent desire to take appropriate precautions mandated the inclusion of protected areas: because we do not yet know all the consequences of our silvicultural actions, unmanaged (benchmark) areas should be retained for scientific study and "to err on the side of caution." In this philosophy, there is clearly a role for emulating natural disturbance.

Conclusions

Natural disturbance is an important part of ecosystem management. Understanding its mechanisms and consequences provides important inputs for the design of forest management systems and helps managers to identify ecologically appropriate harvesting systems.

Insect outbreaks and fires are the major natural disturbances in North America. In certain areas, insect effects predominate; for example, eastern spruce-fir forests are strongly affected by eastern spruce budworm, whereas aspen forests are affected by forest tent caterpillar. In other areas, insects share the role of natural disturbance with fire; for example, southern pine beetle and fire coexist in the southeastern United States, and jack pine budworm and fire coexist in boreal jack pine stands. In other areas, fire dominates. There are definite interactions between insects and fire, and these complicate all attempts to implement forest management that emulates natural disturbance.

Although it is difficult to precisely predict the onset of an insect outbreak, sampling of current population trends using pheromone traps and other survey methodologies can provide data—before defoliation is observed—on an insect's current status and rate of change. In addition, analysis of the patterns, timing, and duration of past disturbances in terms of percentage defoliation of the current foliage for each year during an outbreak provides the basis for planning analyses of the effects of the insects on stands and landscapes. The linchpin in these analyses is a modeling approach that relates defoliation to the impacts on trees, stands, and landscapes (e.g., growth reduction, increased mortality). With stand-level models, impact scenarios can be used as templates to help managers emulate natural forest disturbance and can also provide useful input into forest management planning.

It does not make sense to blindly follow a prescription for emulating natural forest disturbance without considering the prescription's effects on other values. For example, forest management actions should be gauged in terms of their effect on a forest's future susceptibility to insect outbreaks (risk of an outbreak occurring) and subsequent vulnerability (level of damage once an outbreak occurs). We do not want to perpetuate or accentuate insect outbreaks in areas with a primary focus on timber production. Nevertheless, information on natural disturbances can definitely assist forest managers in the rational selection and justification of objectives related to forest composition (e.g., species composition, age-class distribution, range of patch sizes) and of silvicultural and harvesting treatments that are consistent with natural disturbance regimes.

PART II

APPLICATIONS:
UNDERSTANDING FOREST DISTURBANCES

Empirical Approaches to Modeling Wildland Fire in the Pacific Northwest Region of the United States
Methods and Applications to Landscape Simulation

DONALD MCKENZIE, SUSAN PRICHARD, AMY E. HESSL, and DAVID L. PETERSON

Vegetation dynamics, disturbance (especially fire), and climatic variability are key ingredients in simulations of the future condition of heterogeneous landscapes (Lenihan et al. 1998; Keane et al. 1999; Schmoldt et al. 1999; Dale et al. 2001; He et al. 2002). Spatially explicit models in particular require large amounts of empirical data as inputs, but existing data are rarely adequate. The extent and resolution of the available empirical data are important considerations when simulating vegetation dynamics, because different types of data in a database are often collected at different spatial and temporal scales (McKenzie et al. 1996a; McKenzie 1998). In addition, the spatial pattern of sample data may not reflect the spatial pattern of variability in the landscape being modeled. For example, when viewed at broad spatial scales, data points will often be clustered because of the relatively small spatial extent of most data collection. In such cases, it is easy to underestimate the intrinsic variability of the data and difficult to discern the autocorrelation structure of the data (Rossi et al. 1992); it therefore becomes difficult to aggregate these data to create the continuous coverage that is necessary for landscape-scale simulations.

Empirical models (Keane, Parsons, and Rollins, chapter 5, this volume) are an important source of both input data and model parameters, particularly if careful attention is paid to questions of scaling and aggregation in their development. The empirical data that form the basis for models of fire and vegetation are of three types: climatic data; fire history reconstructions; and data on vegetation, fuels, and topography. Each type of data raises key questions regarding extent, resolution, and spatial pattern, and presents key problems that affect the quality of the data and its usefulness for modeling (table 7.1). For example, empirical climatic data come from weather stations that are generally not optimally distributed to support spatial modeling (e.g., they are too widely spaced in steep, complex topography). To model landscapes, gridded data sets must be created via interpolation and extrapolation from these weather station records (Daly et al. 1994; Thornton et al. 1997). In addition, these data typically extend back to only around 1900. Thus, climate reconstructions are necessary before we can investigate associations between climate and historical fire regimes (Swetnam and Betancourt 1990; Swetnam 1993; Grissino-Meyer and Swetnam 2000; Veblen et al. 2000).

In studies of local fire regimes, most fire history data have been collected with spatially intensive rather than spatially extensive sampling. Continuous, extensive spatial coverage is generally needed for landscape modeling, but at regional scales, fire history sites are often clustered, with large gaps between sites (Heyerdahl et al. 1995). Models are needed to transform fire history records into data structures that are suitable for spatially explicit simulations.

Vegetation data are available in a variety of classification schemes, including historical and current cover types (Quigley et al. 1996; Sierra Nevada Ecosystem Project 1996), potential vegetation (Küchler 1964), and structural stage (Schmidt et al. 2002). These data are available in a form suitable for landscape modeling, but the links between vegetation and fire regimes are not always spatially well defined, although Schmidt et al. (2002) discuss exceptions. In addition, most vegetation coverage information can no longer

TABLE 7.1. Key Limitations of Raw Empirical Data (Scale and Pattern) that Affect the Success of Large-Scale Fire Modeling

Climatic Data	Fire History Data	Data on Vegetation, Fuels, and Topography
Instrumental data have limited temporal extent and resolution.	Fire history sites do not span the range of environmental variability of the modeled sites.	There may be unequal taxonomic resolution, the data are often qualitative, and biophysical factors are not incorporated.
Interpolation and extrapolation are difficult in complex terrain.	Data points are clustered, making interpolation difficult.	The heterogeneity is not captured at scales critical for fire modeling.

be considered raw data, having passed through considerable qualitative and quantitative transformation and aggregation before attaining its final form.

Perhaps the most useful inputs to landscape fire models—and the most difficult to acquire—are models of the relationships among climate, fire, and vegetation that are robust at the spatial and temporal scales of the applications. We suggest that the ideal input data or empirical model would have the following attributes:

- Data are spatially explicit, capture heterogeneity at a range of scales, and represent multiple environmental gradients;
- Time series data have sufficient extent to capture the historical range of variability, but sufficient resolution to capture short-term variability;
- Empirical models are applicable at the temporal and spatial scales (resolution) of the landscape simulation models; and
- Model output is robust over the historical range of variability being simulated.

In this chapter, we illustrate methods for linking empirical research to landscape modeling, using as examples four existing studies conducted at different spatial and temporal scales (McKenzie et al. 1996b, 2000; Hessl et al. in press; Prichard 2003). These studies have produced data layers or time series that capture the spatial and temporal variation in the parameters that define the relationships among climate, fire regimes, and fuels and vegetation. We have categorized these empirical studies based on the types of landscape simulation models for which they provide suitable input, and on the spatial and temporal scales at which they are appropriate (table 7.2). We give only a brief description of

each study; details on their methods and results can be found in the associated citations. We focus on key characteristics of each study that determine its value for the appropriate class of landscape simulation model. The first two studies were initiated and completed specifically for the purpose of providing inputs and parameters for broad-scale landscape-level modeling of fires, whereas the last two have these and other objectives and are ongoing research projects.

Model Categories and Examples

Qualitative or Rule-Based Models

Qualitative or rule-based models are based on expert opinions gathered during discussions or workshops, decision structures that rely on either verbal rules or numerical thresholds, specific algorithms designed for qualitative modeling (Puccia and Levins 1985), or knowledge-based systems (Schmoldt and Rauscher 1995; Reynolds et al. 1996). They are appropriate not only when quantitative data are insufficient, but also when logical inference is believed to be superior to statistical inference (Reynolds et al. 1996).

Subcontinental-scale qualitative model of vegetation transitions

The model discussed here is described in more detail in McKenzie et al. (1996b).

Description. Fire is the principal form of natural disturbance in western North America. As such, it is expected to constrain vegetation development, even though this development is principally controlled by climate (Woodward 1987; Woodward and McKee 1991). In general, fire maintains earlier successional stages than would exist in an undisturbed landscape, and changes in fire frequency are eventually reflected in a quasi-equilibrium represented by successional

TABLE 7.2. Scales, Model Types, and Target Applications for Four Example Empirical Studies

Study	Spatial Scale	Temporal Scale	Model Categories	Target Application
McKenzie et al. (1996b)	Subcontinental	Century	Qualitative; equilibrium	Rule-based models
McKenzie et al. (2000)	Regional	Annual to century	Semiqualitative; quantitative; statistical; equilibrium	Process-based models
Hessl et al. (2003)	Subregional; multiple scales	Annual to century; multiple scales	Quantitative; statistical; time-series; dynamic	Process-based or stochastic models
Prichard (2003)	Local	Millennial	Semiqualitative; time-series; dynamic	Process-based models with broad temporal scale

stages. "Potential vegetation," which is assumed to represent the vegetation that would exist in an undisturbed landscape, was aggregated for the study region based on Küchler's (1964) system into 44 vegetation types, based on a qualitative assessment of the observed or inferred similarities in fire regimes. A set of one-step vegetation transition rules was developed, based on an exhaustive literature review of the effects of fire on dominant vegetation. The transition rules are summarized for the western United States by means of a flowchart (figure 7.1). When applied to the landscape in the region being modeled, the transition rules suggest shifts in the proportional cover and spatial patterns of the dominant vegetation (color plate 4).

Target application. The model of McKenzie et al. (1996b) was designed specifically to inform a qualitative fire-disturbance module for the biogeographical Mapped Atmosphere Plant Soil System (MAPSS) model (Neilson 1995). A cross-tabulation was established between the physiognomically based vegetation simulated by MAPSS and the species-oriented vegetation types in color plate 4 (McKenzie et al. 1996b), so that the MAPSS output could be interpreted at the species level.

Semiqualitative Statistical Models

Semiqualitative statistical models rely on both qualitative and quantitative procedures to estimate the model's parameters.

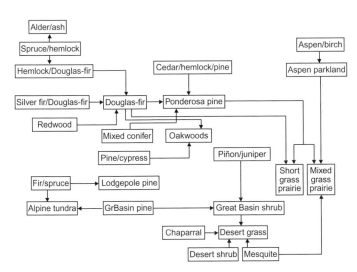

FIGURE 7.1. Summary of the one-step transition rules for vegetation types in the western United States as a function of increased fire frequency. The vegetation type names represent the "aggregated types" of McKenzie et al. (1996b) (see also chapter text), not always species names. Adapted from McKenzie et al. (1996b).

McKenzie et al. (2000) describe the model discussed here in more detail.

Description. The Pacific Northwest region of the United States exemplifies the difficulties of adapting empirical data for use in landscape-level fire models. Steep gradients in elevation, precipitation, and temperature exist across multiple scales. The diversity of climatic conditions, topography, and elevations supports a variety of ecosystem types, including coastal temperate rainforest, subalpine parkland and alpine meadows, drier mixed coniferous forests, and semi-arid shrublands and grasslands (Daubenmire 1978; Lassoie et al. 1985). A variety of fire regimes occur in the Pacific Northwest (Agee 1993), including large, stand-replacing fires (Agee and Smith 1984; Huff 1984; Henderson et al. 1989); mixed-severity, medium-frequency fires (Morrison and Swanson 1990; Taylor and Halpern 1991); and low-severity, variable-frequency fires (Bork 1985; Kertis 1986). Severe fires, particularly in moist, high-elevation forests, are usually associated with synoptic weather patterns (Agee 1993; Ferguson 1997; Schmoldt et al. 1999). In drier ecosystems on or east of the crest of the Cascade Range in the Interior Columbia River Basin, altered fire regimes in the past century and the potential effects of global climate change are of particular interest to modelers and managers.

Qualitative component. A hierarchical model was developed to rank the dominant vegetation cover types in the Interior Columbia River Basin, based on relative fire frequency and severity, and the results were displayed as a dendrogram (figure 7.2). The hierarchical structure reflects the similarity in fire frequency and severity between vegetation types that are adjacent in the dendrogram, and an aggregation algorithm whereby vegetation types can be clustered into groups, depending on the required resolution, in a "taxonomy" specifically associated with fire regimes.

Quantitative/statistical component. Multiple-regression models were used to predict broad-scale (1-km) patterns of fire frequency for forested areas in the Interior Columbia River Basin (figure 7.3) using fire-return intervals from a fire history database as the response variable (Heyerdahl et al. 1995) and GIS data layers for vegetation (Quigley et al. 1996), precipitation (Daly et al. 1994), and elevation as predictors. The vegetation variable was assigned a rank based on the hierarchical model and incorporated into the multiple regressions. Predictions from the models were then mapped to the landscape of the Interior Columbia River Basin at 1-km resolution. Because of the varying quality of the fire history database, two models were fit to the data: one that used the full database, and another that used only the highest-quality sites—those with fully cross-dated fire records (McKenzie et al. 2000).

The models predicted fire-return intervals of between 2 and 375 yr in the Interior Columbia River Basin (color plate 5). Elevation, summer precipitation, vegetation type, and latitude were all significantly associated with fire frequency,

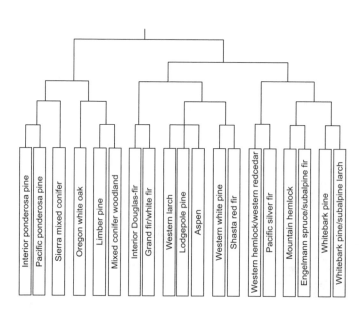

FIGURE 7.2. Dendrogram for the dominant cover types of the Interior Columbia River Basin of the western United States. Moving from the top to the bottom of the dendrogram, the fire-free interval increases. Adapted from McKenzie et al. (2000).

Interior ponderosa pine
Pacific ponderosa pine
Sierra mixed conifer
Oregon white oak
Limber pine
Mixed conifer woodland
Interior Douglas-fir
Grand fir/white fir
Western larch
Lodgepole pine
Aspen
Western white pine
Shasta red fir
Western hemlock/western redcedar
Pacific silver fir
Mountain hemlock
Engelmann spruce/subalpine fir
Whitebark pine
Whitebark pine/subalpine larch

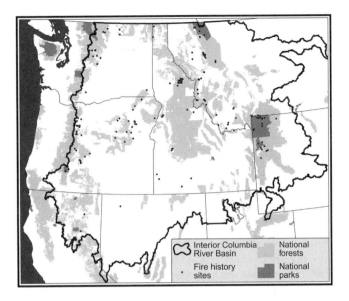

FIGURE 7.3. Fire history sites in the Interior Columbia River Basin of the western United States. Adapted from McKenzie et al. (2000).

Legend:
- Interior Columbia River Basin
- National forests
- Fire history sites
- National parks

although the vegetation type was only barely significant, and of all the predictors, explained the least amount of the variance. Unlike all previous mapped estimates of fire frequency for the Interior Columbia River Basin, which predicted broad ranges based on vegetation alone, the statistical models produced by this study found environmental variables to be better predictors than vegetation, and estimated a mean and a variance for each 1-km cell instead of a range of values.

Target applications. The models were designed to assist broad-scale simulations that use fire frequency as a basic input variable (e.g., Keane et al. 1996c). The limits imposed by the coarse resolution mean that the models will be less useful for fine-scale than for broad-scale applications, although local managers may be able to integrate model predictions with local qualitative data and knowledge about systems similar to theirs to estimate historical conditions as a basis for emulating natural disturbance (Morgan et al. 1994; Landres et al. 1999; Swetnam et al. 1999). The statistical approach allows modelers to incorporate means and variances in applications in which fire starts are simulated stochastically. For example, in mechanistic models (e.g., Keane et al. 1996a), predicted fire-return intervals for each pixel in the model could be used directly or used to represent the mean value in a candidate probability distribution such as the Weibull (Grissino-Meyer 1999). From this distribution, input fire-return intervals could be chosen randomly. Broad-scale modeling will probably need

to incorporate semiqualitative elements for the foreseeable future, because of the unavailability of sufficient high-quality empirical data (Keane and Long 1998; McKenzie 1998). Our results suggest that heuristic, knowledge-based methods (the hierarchical vegetation model) and rigorous statistical methods can be successfully combined.

Quantitative Nonequilibrium Models

Quantitative nonequilibrium models use statistical techniques, but simulate temporal dynamics rather than a static view. Typically, time-series analysis is involved, either explicitly if temporal autocorrelation is to be incorporated, or as a preprocessor to remove autocorrelation, so that other statistical methods can be used (e.g., Cook and Kairiukstis 1990).

Fire and climate in the inland Pacific Northwest region of the United States

The model described here is from the study by Hessl et al. (in press).

Description. The geographic extent of this study spans two national forests in the Interior Columbia River Basin: the Okanogan-Wenatchee and the Colville, which extend across the Northern Cascades and Okanogan Highlands physiographic provinces (Franklin and Dyrness 1988). In the Northern Cascades, the topography is extremely rugged, with deep and steep-sided valleys and eastward- and westward-flowing streams. Further east, the Okanogan Highlands present moderate slopes and broad, rounded summits. A variety of soil types appear in both

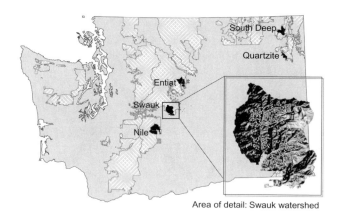

FIGURE 7.4. Five focal watersheds in eastern Washington, for which spatially explicit fire records are available. Hatched areas represent national forests. The inset shows the detailed topography and the spatial pattern of fire-scarred trees in the Swauk watershed. Adapted from Hessl et al. (in press).

Area of detail: Swauk watershed

provinces, reflecting the influence of Pleistocene glaciers, with glacial soils predominant on valley bottoms and residual soils on hillslopes and ridgetops. The climate is intermediate between the maritime climate west of the Cascade Crest and the continental climate east of the Rocky Mountains, and is characterized by summer drought.

Climate reconstructions (Stahle et al. 1998; Cook et al. 1999; Gedalof and Smith 2001) were compared with data on fire extent and occurrence documented in a spatially explicit fire history data set from five watersheds in the Okanogan-Wenatchee and Colville National Forests of Washington State (Everett et al. 2000) (figure 7.4). Correlation analysis, superposed-epoch analysis (Grissino-Meyer 1995), and cross-spectral analysis (Bloomfield 1976) were used to identify significant relationships between fire occurrence (as measured by the percentage of trees that recorded a fire in a particular year) and three climate indices: the Palmer Drought Severity Index (Alley 1984), for which lower values indicate drier conditions, the magnitude of the El Niño Southern Oscillation, and the magnitude of the Pacific Decadal Oscillation (Mantua and Hare 2002). Relationships between these climatic indices and fire were established at both the watershed and regional scales. In addition, mean fire-free intervals and Weibull median probability intervals (Grissino-Meyer 1999), the latter a robust estimator of central tendency in fire-interval distributions, were estimated on a gradient of spatial scales from point to watershed. Composite fire intervals at spatial extents from point-level to the entire watershed were fit to a two-parameter Weibull distribution to estimate the *hazard function* (Johnson and Gutsell 1994), which represents the risk of fire in a given

year. If the slope of this function is zero, fires are considered equally likely in any year following a fire, whereas a positive slope indicates increasing hazard over time, possibly associated with the buildup of fuels (Johnson and Gutsell 1994).

Fire occurrence was negatively correlated with the Palmer Drought Severity Index and thus was positively correlated with drought over the period of record (1700–1975), but the correlation was stronger before 1900 (figure 7.5). No significant correlation was identified with either the El Niño Southern Oscillation or the Pacific Decadal Oscillation. Superposed-epoch analysis also identified a significant relationship between fire occurrence and the Palmer Drought Severity Index (figure 7.6); in each case, the mean index was lower in or around the year of the fire (lag year 0), confirming the study's observation that fires were associated with increasing drought. Cross-spectral analysis identified a significant association between fire occurrence (aggregated to 5-yr sums) and the Pacific Decadal Oscillation, at a period (1/spectral frequency) of approximately 47 yr—roughly the length of a complete cycle of the Pacific Decadal Oscillation—with a lag of approximately 5 yr. This last result is promising, in that it and the superposed-epoch analysis suggest that the causes of temporal variability in fire regimes may be identified at multiple scales. These results are preliminary, however, and must be carefully validated.

Comparison of composite fire intervals for watersheds as a whole with the fire-free intervals for individual points (table 7.3) while accounting for the different areas represented by the different watersheds suggests the possibility of modeling the relationship between increasing sample area and decreasing estimates of fire-free intervals. Initial simulations indicated that a

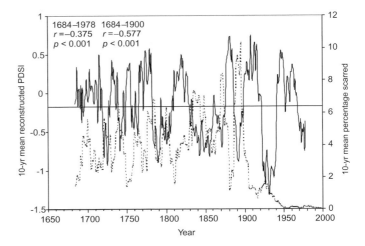

FIGURE 7.5. Ten-year average of the Palmer Drought Severity Index (PDSI) reconstructed from tree-rings for the 1684–1978 period (Cook et al. 1999) (solid line), and the 10-yr average of the percentage of trees scarred by fire over time (dotted line). Adapted from Hessl et al. (in press).

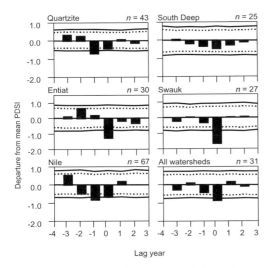

FIGURE 7.6. Superposed-epoch analysis for each watershed in the area studied by Hessl et al. (in press), showing departures from the mean annual Palmer Drought Severity Index (PDSI) during fires that affected about 10% of the trees used to reconstruct the fire history. *Lag year* represents the time before (negative values) or after (positive values) the fire; the fire itself occurs in lag year 0. PDSI is shown during, prior to, and following the fire year. The horizontal dashed and solid lines represent 95% and 99% confidence intervals, respectively. Adapted from Hessl et al. (in press).

strong nonlinear inverse relationship exists between these two variables in *neutral landscapes* (landscapes simulated to have no specific constraints on fire regimes) (Gardner and O'Neill 1991) and that models of this type should be pursued further.

Target applications. If modelers have access to empirical fire history data at appropriate scales,

some difficult problems associated with data aggregation, such as the scaling of fire frequency estimates from point to stand to landscape, are mitigated (McKenzie et al. 1996a, 2000). Additionally, because constraints on fire regimes change across scales, appropriate parameters can be more easily applied in mechanistic models if they have been identified at multiple scales. For example, fine-scale models might take advantage of fire probabilities based on Weibull distributions to simulate fire starts (e.g., Keane et al. 1996a), whereas broad-scale models might use empirical data on fire distributions based on climatological inputs. These spatially explicit, extensive fire history data can be used to identify fire-climate interactions and the key spatial scales of variability in fire frequency (table 7.4). These scales and interactions are then used to estimate key mechanisms at different scales (table 7.5) and predict the characteristics of fire regimes in unsampled watersheds throughout the study area, so as to provide the extensive coverage needed for broad-scale simulations.

Spatial and temporal dynamics of fire and forest succession in a montane ecosystem

The model described here can be found in more detail in Prichard (2003).

Description. This model focused on Thunder Creek, a large (30,000-ha) watershed in the heart of North Cascades National Park, Washington. The watershed lies in the rain shadow of several large peaks; thus, local climate is considerably drier than in watersheds west of these mountains. An existing fire history data set (Agee et al. 1990) provided a record of the mixed-severity fire regime that typifies transitional climatic zones

TABLE 7.3. Fire Interval Statistics from Eastern Washington for the Pre-European Settlement Period and the Modern Period

Watershed	Point WMPI (yr)	Composite WMPI (yr)	Point MFI (yr)	Composite MFI (yr)
Pre-European settlement period (1700–1900)				
Entiat	13	5	15	6
Nile	16	6	17	6
Swauk	20	7	22	8
South Deep	32	7	35	10
Quartzite	15	4	16	5
Modern Period (1901–1990)				
Entiat	58	—	59	—
Nile	30	—	41	—
Swauk	20	—	29	—
South Deep	36	10	45	22
Quartzite	36	—	48	—

Source: Data are from Hessl et al. (in press).

Notes: Statistics are for years in which about 10% of trees (with a minimum of two trees) recorded fire scars. MFI, mean fire-free interval; WMPI, Weibull median probability interval; —, composite estimate could not be computed, because of the paucity of fire-free intervals. Intervals are rounded to the nearest year.

of the North Cascades, but it is temporally limited to the age of the existing forests.

Records of lake sediment charcoal can provide an extensive temporal record of fire; in the North Cascades, this record extends throughout the Holocene (<10,000 YBP to the present) (Cwynar 1987). In the same sediment record, macrofossils and pollen can be used to evaluate changes in species composition and abundance associated with historical fires (Dunwiddie 1986; Gavin and Brubaker 1999). Although lake sediments have excellent temporal resolution and extent, they

TABLE 7.4. Characteristics of a Multiscale Analysis of the Influences on Fire Occurrence

	Fine Scale (20–75 ha)	Medium Scale (200–15,000 ha)	Broad Scale (0.5–1.0 million ha)
Application	Watershed-scale modeling of fuels and fire (sub-district level)	Fuel succession modeling (district to national forest level)	Fire event prediction using fire-climate relationships (national forest to regional level)
Unit of analysis	Trees or points	Watersheds	National forests
Climatic variables	Reconstructions of climatic variables from watershed-level chronologies	Climatic reconstructions from broad-scale climatic data	Regional climatic variables (Palmer Drought Severity Index, El Niño Southern Oscillation, Pacific Decadal Oscillation)
Biophysical variables	Aspect, slope, elevation, solar radiation (30-m resolution)	Potential vegetation, solar radiation (1-km resolution), slope, elevation, aspect (summarized by watershed)	Geographic gradients of medium-scale variables

TABLE 7.5. Candidate Mechanisms Associated with Temporal and Spatial Patterns of Fire

Fire Occurrence	Fire Pattern	Probable Driving Factor
Dispersed	Fires are less likely soon after a fire.	Fuel buildup is the primary constraint.
Random	Fires are equally likely at any time.	The dominant factor is stochastic (e.g., local extreme weather conditions).
Aggregated	Fires are more likely soon after a fire.	Periods of favorable climate (weather) for fire, potentially corresponding to cycles of productivity and drought.
Synchronous	Multiple fires occur across polygons with similar aspect, or across watersheds or regions in the same year.	Local to regional-scale climate patterns, depending on the geographic extent of the synchrony.
Asynchronous	No relationship between the timing of fires across the landscape.	Climate is not an important driver; topography, fuels, or human influence may be dominant.

represent an indefinite area surrounding lake basins and do not record spatial variability in fire history and vegetation (Dunwiddie 1987; Whitlock and Millspaugh 1996).

In this study, the spatial and temporal dynamics of fire and vegetation were investigated in a 4-km^2 area of the Thunder Creek watershed. Forest succession after fires was reconstructed by using age-structure analysis based on tree ages (increment cores) and tree-size distributions along a precipitation gradient parallel to Thunder Creek and along an elevation gradient up the steep northeastern slope of the watershed. Holocene fire and vegetation histories were reconstructed from lake sediment records sampled from a small (0.4-ha) lake in a montane forest zone. Charcoal records were used to reconstruct a continuous record of local (1- to 3-ha) fires over the past 10,000 yr, and macrofossils of conifer needles, twigs, and seeds were used to assess fluctuations in forest assemblages in association with changes in fire frequency. Charcoal accumulation rates and total macrofossil accumulation rates were compared by means of superposed-epoch analysis (see the previous section) to evaluate whether charcoal peaks represented local fires.

The successional dynamics after fire varied dramatically across the 4-km^2 study area. Even at similar elevations, current forest assemblages are strikingly different at the dry end of the precipitation gradient parallel to Thunder Creek, where lodgepole pine (*Pinus contorta* Dougl.) and Douglas-fir (*Pseudotsuga menziesii* [Mirb.] Franco) codominate the stands, and at the moist end of the precipitation gradient, where western hemlock (*Tsuga heterophylla* [Raf.] Sarg.), western red cedar (*Thuja plicata* Donn) and Douglas-fir codominate (figure 7.7). Throughout the more than 10,000-yr lake sediment core, rapid sedimentation rates yielded a high-resolution (average, 10.5 yr cm^{-1}) record of fires (figure 7.8). Superposed-epoch analysis of the charcoal accumulation rates and total macrofossil accumulation rates demonstrated a statistically significant ($p = 0.001$) decrease in macrofossil accumulation rates following peaks in charcoal accumulation rates. Because macrofossils represent the local vegetation that typically lies within 30 m of a lake (Dunwiddie 1987), this association of macrofossils with peaks in charcoal accumulation rates suggests that the charcoal record represents local fires. In the Thunder Creek watershed, fire regimes appear to have varied during the broad climatic changes of the past 10,000 yr. The mean fire-return interval for the Holocene period is 227 yr, with more frequent fires in the early Holocene (an interval of 158 yr) than in the mid-Holocene (214 yr) and late Holocene (308 yr).

Target applications. This study provides a unique reconstruction of the historical range of variability in mixed-severity fire regimes by combining the spatial dynamics of vegetation relatively soon after a fire (<150 yr) with the temporal dynamics of fire and vegetation throughout the Holocene period. Mixed-severity fire regimes, common in transitional climates such as that of the Thunder Creek watershed, are often more complex in terms of spatial patchiness and temporal variability than are low- or high-severity regimes (Agee 1998), and are consequently more difficult to simulate. Furthermore, in spite of how

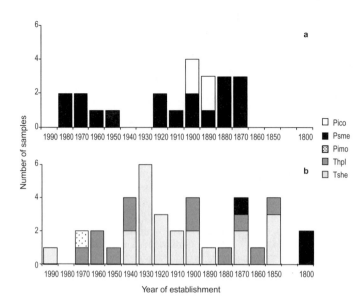

FIGURE 7.7. Age-class frequency distributions for (a) dry forest and (b) moist forest in the Thunder Creek watershed. Pico, lodgepole pine; Pimo, western white pine (*Pinus monticola* Dougl.); Psme, Douglas-fir; Thpl, western red cedar; Tshe, western hemlock. Adapted from Prichard (2003).

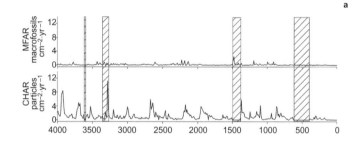

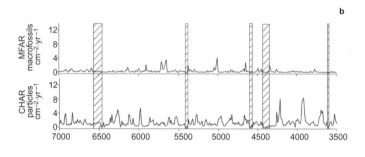

FIGURE 7.8. Macrofossil (MFAR) and charcoal (CHAR) accumulation rates in the (a) Late Holocene, (b) Mid-Holocene, and (c) Early Holocene periods. Crosshatched sections indicate areas of contamination or missing samples between cores. Adapted from Prichard (2003).

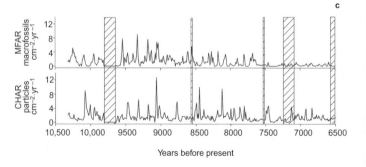

common mixed-severity fire regimes are in mountain forest ecosystems, they have received little attention in fire history studies or modeling (Agee 1993). Although the Holocene results lie outside the typical temporal range of landscape modeling, this ongoing study will provide both a baseline and a historical record that can guide long-term simulations and future monitoring of fire and forest dynamics in the Cascade Range. For example, these research results are currently being used to validate the application of a process-based model of fires and succession for mixed-severity fire regimes (Keane et al. 1999; Fagre et al. 2003).

Linking Empirical Studies to Landscape Modeling

We suggest that modelers should address several considerations when designing empirical research to support landscape-level modeling.

Determine the Extent and Resolution of the Modeling

Equilibrium models may adequately support simulations based on current conditions, whereas simulations that incorporate climate change may require methods based on dynamic time series. If extrapolation from existing data is part of the simulation exercise, empirical models should be robust enough that they are not just valid for the conditions under which they were built. Qualitative models should provide for the inclusion of categories not represented in the initial data. For example, the hierarchical model of fire regimes (see figure 7.2) allowed the prediction of fire frequency for vegetation types not represented in the fire history database.

Determine the Robustness of Existing Data Sources

What is the predictive ability of models based on available data? For example, the transition-rule model (see figure 7.1) was qualitative, because insufficient quantitative data were available at the appropriate spatial scale. Similarly, the Interior Columbia River Basin fire frequency model could not incorporate spatial autocorrelation, even though fire is known to be contagious, because the spatial pattern in the raw data was clustered. In contrast, the fire history data from eastern Washington permits spatial modeling within watersheds, but not among them. However, the qualitative change in fire regimes that seems to have occurred around 1900 prevents extrapolation of quantitative fire-climate relationships from an earlier period (1700–1900) to the present day. The Thunder Creek study records broad-scale temporal variation in fire regimes over more than 10,000 yr, but extrapolation to regions outside the North Cascade Range would be unwise.

Tune the Complexity of the Empirical Model to the Structure of the Simulation

Most simulation models use explicit time steps, between which values of each variable change as a mechanistic function of the variable's current state and that of other variables. An autoregressive model built from empirical data for one of these variables would be superfluous, unless it could be combined with mechanistic functions—a process that entails considerable additional complexity, with little theory to support it. Similarly, some mechanistic models use biogeochemical cycling to compute litterfall, mortality, decomposition, and the resulting load of dead fuel (e.g., Keane et al. 1999). An empirical model of fuel succession, in which a statistical relationship between fuel abundance and the elapsed time since disturbance is modeled, would be superfluous, because its function is superseded by routines within the simulation model. In both these examples, the empirical models could be of considerable scientific interest, but not of particular value to the simulation.

Report the Uncertainties and Limitations of the Empirical Model

Stochastic simulation models can explicitly incorporate the error structure of any empirical relationships. Deterministic models can at least use knowledge of this error structure to design sensitivity analyses efficiently. The statistical error structure of a quantitative empirical model (e.g., variances, confidence intervals, patterns of residuals) is the easiest type to report, but other errors based on dubious assumptions, extrapolations, or inadequate data collection should also be evaluated. For example, in the Interior Columbia River Basin fire frequency study, substantial uncertainties in the fire history data were suspected, because of the dearth of data points for many of the reconstructions. A sensitivity analysis was therefore conducted of the regional-scale prediction errors associated with inadequate cross-dating of fire scars (Madany et al. 1982; McKenzie et al. 2000).

Conclusions

Emulating historical fire disturbance in the landscape requires a quantitative understanding of fire regimes at multiple scales (Keane, Parsons, and Rollins, chapter 5, this volume). In this chapter, we illustrated some empirical approaches for generating statistical models and new data layers for landscape-level fire simulations. Many other empirical studies, with different spatial, temporal, and taxonomic resolution, would inform or help parameterize landscape simulations. Some examples include:

- A fuelbed classification with sufficient links between vegetation and fuel load that it can be applied to existing vegetation coverage and to initialize fire behavior modules (Sandberg et al. 2001);
- Statistical models of fire-climate interactions in the American Southwest (Swetnam and Betancourt 1990), the southern Rocky Mountains in Colorado (Veblen et al. 2000), and the Canadian boreal forest (Skinner et al. 1999, 2002); and
- Other studies that use paleoecological records (Long et al. 1998, Gavin and Brubaker 1999, Millspaugh et al. 2000).

The limitations of each of our examples suggest future data needs and research directions. The models of fire frequency would be improved by more extensive collection of fire history data, especially in ecosystems with long fire-return intervals. Even so, there are limits to the interpretation of fire history data collected at different spatial scales; fire history models from eastern Washington indicate that both fire frequency and the constraints on fire occurrence are scale-dependent. To be most useful to modelers, we need to identify scaling laws that can translate local fire history information to the broad scales at which models of landscape disturbance are applied. Because we may never have enough empirical data to characterize every forested landscape adequately for modeling, we must instead develop robust methods for making inferences across spatial and temporal scales and varying levels of taxonomic (e.g., species versus life-form) resolution.

Landscape-level modeling of fire is one method of simulating natural forest disturbance. We have focused on applications of empirical models to landscape simulation, but empirical models can also directly inform the emulation of natural disturbance on real landscapes to help achieve management objectives. Our work suggests that for management applications and the support of policy development, we need to be able to inte-

grate empirical models from multiple scales, not only because of the hierarchical nature of administrative agencies but also because of the varying scales that we identified for landscape patterns and processes, in part, by empirical modeling.

For example, in the dry forest ecosystems of the Pacific Northwest of the United States, the region that the majority of our research represents, the principal means of emulating natural disturbance are prescribed fire under controlled conditions and mechanical treatments designed to emulate the disturbance processes that gave rise to historical landscapes. Our analysis of historical low-severity fire regimes can be directly applied to temporal patterns of prescribed fire by identifying the extent and frequency of the fires, and indirectly applied to silvicultural prescriptions by predicting the density and age structure of trees in the posttreatment stands. These prescriptions for emulating natural disturbance are best applied at the (local) scale of national-forest districts.

The fire frequency models for the Interior Columbia River Basin are best applied to broad-scale (regional) management, in which an array of individual strategies may be necessary to emulate natural disturbance in different landscapes. Particularly in ecosystems characterized by high-severity, low-frequency fires, emulation will be very difficult, because unlike in much of the Canadian boreal forest, fire-return intervals (200 to >300 yr) are up to four times as long as even the longest silvicultural rotations, and prescribed crown fire is not currently a management option.

The fire-succession study in the Thunder Creek watershed suggests that ecosystem response to moderate-severity fires is complex, and simple attempts to emulate natural disturbance (e.g., clearcutting followed by planting economically desirable species) are unlikely to reproduce the species composition and spatial pattern expected after a fire. At a minimum, a careful selection of the residual trees that will serve as a seed source and an attempt to recapture prefire species composition through replanting are necessary.

Finally, our empirical studies exemplify the variety of methods and data sources necessary for both a multiscale understanding of natural disturbance and identification of the range of opportunities and limitations for emulating them via management. At local scales, detailed quantitative knowledge of ecosystems is the key to successfully emulating and possibly restoring natural disturbance regimes. At the policy level, understanding the variability in natural disturbance regimes across geographic regions will inform more flexible, adaptive, and efficient strategies for managing ecosystems.

Acknowledgments

We thank Robert Keane and Ajith Perera for encouraging us to develop the ideas in this chapter and for providing the initial venues for their presentation. Alynne Bayard, Lara Kellogg, and Robert Norheim produced the maps. Ze'ev Gedalof and Charles Halpern provided helpful comments on an earlier draft. Research was funded by the U.S. Department of Agriculture Forest Service's Pacific Northwest Research Station, the Canon National Park Science Scholars Program, and the Global Change Research Program of the U.S. Geological Survey.

CHAPTER 8

Simulating Forest Fire Regimes in the Foothills of the Canadian Rocky Mountains

CHAO LI

Forest resources and a healthy level of biodiversity should be sustainable through management practices that emulate natural disturbance (primarily fire). Ontario's *Crown Forest Sustainability Act* (Statutes of Ontario 1995) and such research programs as Ecosystem Management by Emulating Natural Disturbance that link harvesting methods with regeneration procedures to promote holistic, ecologically sensitive silviculture (Spence and Volney 1999) reflect this belief. Such a philosophy requires a better understanding of natural fire patterns, and challenges forest managers and researchers because of the relative lack of empirical data on natural disturbance (i.e., disturbance without human intervention). Figure 8.1 shows the conceptual linkages between natural and observed fire regimes and the corresponding forest dynamics.

Whereas natural fire regimes are associated with natural forest dynamics, observed fire regimes are associated with observed forest dynamics. The observed fire regimes result from the direct and indirect effects of human activities, such as fire suppression, increased sources of fire ignition, pest management, harvesting, and changes in land use, all of which have significantly influenced natural fire regimes. Changes in climatic variables might also have influenced natural fire regimes. Only factors directly related to fire appear in figure 8.1 and are discussed in this chapter. Natural fire and forest dynamics result from long-term interactions among natural ignitions, the landscape structure (including the vegetation or cover composition, plus patch and age patterns), topography, and weather variables. In boreal forests in Canada, forest dynamics and fire regimes are closely associated and

should therefore be discussed together. Empirical data on fire regimes have often been collected for several decades only (i.e., under various levels of fire suppression); consequently, they may not truly reflect long-term natural fire patterns. The challenge for managers and researchers is how to isolate the influence of human activities based on available observations so as to discern natural fire patterns.

The intensity of human activities determines the extent of the difference between natural and observed fire regimes (see Suffling and Perera, chapter 4, this volume). For example, human activities generally increase the number of fires. In Alberta, human-caused fires accounted for 47% of the total number of fires from 1961 to 1995 (table 8.1). However, this increase does not necessarily increase the area burned annually proportionally, possibly because of fire suppression. In Alberta, human-caused fires burned about 16% of the total burned area. An analogous situation holds for Ontario, where longer fire cycles have been attributed to the province's fire suppression policy (Ward and Tithecott 1993; Ward et al. 2001). Similar anthropogenic changes in fire frequency and fire cycle were found in Alberta's Banff, Jasper, and Wood Buffalo National Parks (Tande 1979; White 1985; Larsen 1997), Prince Albert National Park in Saskatchewan (Weir et al. 2000), and the Boundary Waters Canoe Area in Minnesota (Baker 1992). Consequently, fire size distributions in these areas must reflect a significant increase in the number of small fires and a reduced number of intermediate and larger fires. One challenge in emulating natural fire patterns is, therefore, to determine fire size distributions under natural conditions.

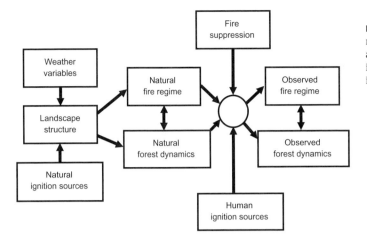

FIGURE 8.1. Relationships between natural and observed fire regimes and the associated factors that influence these dynamics. Arrows indicate the direction of influence.

TABLE 8.1. Statistics for Lightning- and Human-Caused Fires in Alberta

Fire Cause	Proportion of Total (%)		
	Number of Fires	Area Burned	Mean Fire Size (ha)
Lightning	53.1	84.2	75.7
Human	46.9	15.8	16.1

In this chapter, I present an Alberta, Canada, case study that demonstrates why existing empirical fire data may be an unsuitable basis for emulating natural fire patterns and suggest how ecological modeling could help managers partially compensate for the lack of suitable empirical data. Existing empirical data, however, are still important for providing background for the development of an ecological model, simulation of fire regimes, and validation of the results. The case study focuses on the information currently required for emulating natural fire patterns, such as fire size distributions and the construction of fire probability maps.

This chapter comprises:

- An empirical analysis of fire data to characterize the patterns observed under fire suppression;
- A rationale and methodology for simulating fire regimes that describes Li's (2000a) Spatially Explicit Model for LANdscape Dynamics (SEM-LAND) model for emulating natural fire patterns; and

- A presentation of simulated natural and observed fire patterns for an area in the foothills of the Canadian Rocky Mountains to demonstrate the model's ability to emulate natural fire patterns.

Analysis of Fire Regimes Observed in Alberta

Alberta covers an area of 66.1 million ha and lies immediately east of the Rocky Mountain Cordillera of North America. The province has 38.2 million ha of forested land (Canadian Council of Forest Ministers 1997). Historical fire records from 1961 to 1995 were provided by the Forest Protection Division of Alberta Sustainable Resource Development for this study. Data on the locations and causes of fire ignition, fire types, and final fire sizes were used in the analysis.

Fire Size Distribution

Various systems can be used to classify fire sizes. In Canada's national wildfire statistics, for example, the post-1986 system differed from the pre-1986 system (Ramsey and Higgins 1991). The breakpoints for pre-1986 categories were 0.1, 4, 40, and 200 ha. After 1986, the number of categories increased and a logarithmic scale was used, with breakpoints of 0.1, 1, 10, 100, 1000, 10,000, and 100,000 ha. The uneven size classes in these two systems give greater weights to smaller fires than to larger fires. The uneven classes are convenient for mathematical treatment, and a negative-exponential distribution of fire sizes (i.e., a straight line in a log-log plot of frequency versus size class) can be relatively easy to demonstrate, as Cumming (2001) did for 2898 lightning fires from 1980 to 1993 in part of northeastern Alberta.

TABLE 8.2. Fire Size Distribution in Alberta
(1961–1995)

Fire Size Class (ha)	Frequency (% of total number of fires)
<0.1	32.67
0.11–1.0	29.12
1.1–10.0	25.26
10.1–20.0	3.19
20.1–50.0	3.52
50.1–100.0	1.90
100.1–200.0	1.42
200.1–500.0	1.27
500.1–1000.0	0.38
1000.1–10,000.0	0.90
10,000.1–100,000.0	0.29
>100,000.0	0.08

Source: Data are from the Forest Protection Division of Alberta Sustainable Resource Development.

Note: The distribution is based on the size classes proposed by the author.

For emulating natural patterns, however, such systems might not help managers to choose appropriate sizes for harvesting blocks. Adding more categories to describe larger fires could improve the quality of the information. One such system would have breakpoints of 0.1, 1, 10, 20, 50, 100, 200, 500, 1000, 10,000, and 100,000 ha. Table 8.2 shows the resulting fire size distribution for this system, based on Alberta's historical fire records from 1961 to 1995.

Table 8.2 shows that more than 95% of all fires burned less than 50 ha, whereas slightly more than 1% of all fires burned more than 1000 ha. These results are consistent with the conclusion that the 2–3% of fires that grow larger than 200 ha account for 97–98% of the total area burned in Canada (Stocks 1991).

One big advantage of using uneven size classes is that the resulting fire size distribution can be manipulated to fit a negative-exponential distribution; data for the classes can then be conveniently used as inputs for a fire scenario model that investigates the consequences of known fire regimes in terms of fire cycle and size distribution. When such a theoretical distribution was applied in one scenario model, however, the simulated fire size distributions differed from the empirical observations (He and Mladenoff 1999a).

The different effects that result from the various types of fires that occur in a forest landscape can be emulated by appropriate harvesting methods. For example, clearcutting removes all trees from a site in a manner similar to the effect of a stand-replacing fire (e.g., a crown fire), which kills most trees in the burned area. In contrast, a partial cut removes some percentage of trees in a cut block and thus resembles a partial stand-replacing fire (e.g., a surface fire). As these examples reveal, understanding natural fire regimes that include crown and surface fires is vital for emulating natural fire patterns. However, fires and harvesting have different effects on forest landscapes (e.g., McRae et al. 2001), although these differences are beyond the scope of this chapter. My analysis of observed fire regimes in Alberta documents the fire size distributions for different types of fires.

Alberta's historical fire data record the fire type for each fire from 1983 to 1995, thereby permitting compilation of the fire size distribution for each fire type. Figure 8.2 shows that discontinuous fire size distributions can be identified for ground, surface, and crown fires, based on a system of even-interval size classes.

Discontinuous patterns are commonly reported when using even-interval size classes (e.g., White 1985; Li et al. 1996; Li and Perera 1997). This type of classification system can provide more detailed fire frequencies across different size classes. Because the vertical axes in figure 8.2 are logarithmic, a straight line across size classes would indicate that the fire size distribution is a negative exponential, a distribution that is often assumed to exist. No such line is evident in this figure; therefore, this assumption does not appear to hold for the observed fire regime in Alberta.

The observed discontinuous fire size distributions have three possible primary causes. First, the empirical data cover a relatively short period, and the total number of fires may not have been sufficiently large to capture the long-term reality. If this is true, then a continuous fire size distribution might emerge if more data were available. Simulating a negative-exponential probability distribution can force agreement with this expectation, but the empirical fire data might not support the simulation. In my study, records for 12,345 fires were included in the data for the 1983–1995 period and 28,328 fire records from 1961 to 1995; yet these large numbers failed to produce a continuous fire size distribution, which suggests that the observed fire sizes may not follow mathematically convenient probability distributions

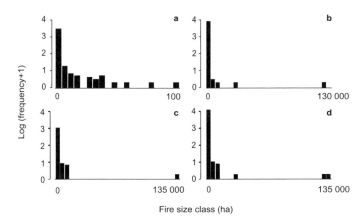

FIGURE 8.2. Observed fire size distributions in Alberta (1983–1995) using a 5000-ha even-interval classification system for surface and crown fires and a 5-ha even-interval system for ground fires. (a) Ground fires; (b) surface fires; (c) crown fires; and (d) all classes of fire combined.

suitable for modeling. This result is consistent with the experience of Reed and McKelvey (2002), who demonstrated simple power law or negative-exponential behavior for fire size distribution by using a constant fire-extinguishment rate, but the results of their model did not appear to conform with the six empirical data sets used in their study. Thus, mathematical models of fire size distribution based on a simple power law or a negative exponential do not adequately simulate the observed fire regimes, and alternative approaches are necessary.

A second possibility is that fire suppression has reduced the ultimate size of some fires, thereby increasing the frequency of smaller fires. If this is true, one might expect to observe fewer discontinuous fire size distributions under natural conditions, and gaps in the intermediate fire size classes would be a function of the effectiveness of fire suppression. This effectiveness would, in turn, be a function of the investment made in suppression and the fire intensity, rate of spread, and size when suppression resources arrived at the scene. Stocks (1991) reported fire size distributions in northwestern Ontario for fires that prompted and those that failed to prompt suppression actions. He demonstrated that "in the absence of active suppression activities, fire-size distribution is somewhat normal, with large to very large fires common. Actioned fires, however, exhibit a negative exponential distribution, reflecting successful fire suppression activity" (p. 201). Stocks's reasonable inference can be explored using simulation models such as SEM-LAND (Li 2000a), as I discuss later in this chapter.

The third possibility is that human activities can result in climate change that could also alter fire regimes; however, these alterations depend on the nature of climate change. If the climate becomes warmer, for instance, a shortened fire cycle with more intermediate and large fires can be expected (e.g., Clark 1988b; Swetnam 1993; Stocks et al. 1998). If the climate becomes cooler, drought may become less frequent, as has been observed in the Duparquet region of Quebec's southern boreal forest; as a result, a longer fire-return interval and more small fires can be expected (e.g., Bergeron 1991; Flannigan et al. 1998). It has been predicted that climate change will increase the average Fire Weather Index (FWI) in western Canada but decrease the index in eastern Canada (Bergeron and Flannigan 1995). Changed FWI values were also predicted for large regions of the northern hemisphere for the twenty-first century (Flannigan et al. 1998). Although there is no evidence for a cooling climate in west-central Alberta during the nineteenth century, warming was predicted for the region in a scenario with double the current levels of carbon dioxide using a first generation coupled global climate model (Flannigan et al. 1998). Therefore, it is unlikely that the observed discontinuities in fire size distributions have been caused by climate change.

Mean Fire Size and Fire Cycle

By definition, the concept of a fire cycle (FC) focuses on the total area burned over the long term for a large region (Li 2002), and the mean fire size is calculated from the total area burned divided by the total number of fires. In the present chapter, fire records from 1983 to 1995 were used to calculate fire cycles in Alberta for crown, surface, and ground fires, and for all fires combined, using equation 8.1:

$$FC = A \times Y / \sum_{i=1}^{n} S_i, \qquad (8.1)$$

TABLE 8.3. Observed Fire Regime Statistics for Alberta

Fire Type	Proportion of Total (%)			
	Area Burned	Number of Fires	Mean Fire Size (ha)	Fire Cycle (yr)
Ground	0.3	25.6	0.5	314,081
Surface	43.4	65.7	31.6	1943
Crown	56.3	8.7	309.9	1498

Note: The data are based on Alberta's 38.3 million ha of forested land.

where A is the total area, Y is the total number of years for the investigation, S_i is the size of the i-th fire, and n is the total number of fires.

The statistics that characterize various types of fires are quite different. Ground fires accounted for 25.6% of total fire ignitions (table 8.3); however, ground fires only burned 0.3% of the total burned area, and their mean size was less than 1 ha. Surface fires accounted for 65.7% of all fires, and burned 43.4% of the total burned area, with a mean fire size of about 32 ha. Crown fires usually cause the greatest tree mortality and stand replacement, and even though they accounted for only 8.7% of the total number of fires, they accounted for more than half of the total burned area (56.3%). The fire cycle estimated from this data set is about 844 yr for all categories of fire combined.

These results suggest that surface and crown fires accounted for more than 99% of the total burned area, with crown fires responsible for stand replacement in more than half of this area. Severe surface fires also caused stand replacement, but the data used for this analysis did not distinguish between these types of fires. Nevertheless, these data support the assumption that most fires had a stand-replacing effect in the simulation case study presented here.

The mean fire size was calculated as 47.8 ha. However, 75% of reported fires were smaller than 10 ha (see table 8.2), probably as a result of fire suppression. Excluding these small fires provides a more reasonable estimate of the impact of significant fires, which averaged 359.5 ha. The number of large fires was relatively small, with only 246 fires that burned more than 1000 ha (about 1.3% of the total fires).

Table 8.1 shows that the mean size of fires caused by lightning (75.7 ha) was larger than that of fires caused by humans (16.1 ha), and 84.2% of the total burned area resulted from lightning fires (53.1% of the total number of fires). The results are consistent with those of Stocks (1991), who reported that lightning was responsible for 85% of the total area burned in Canada and that "lightning fires generally grow larger, as detection and subsequent initial attack is often delayed" (p. 201). Thus, a lightning-caused fire is more likely than a human-caused fire to escape from fire suppression.

The fire cycle is not a static property of a fire regime; instead, it is temporally dynamic. It has been commonly reported that the fire cycle has changed during the past several decades or since the beginning of the previous century (e.g., Larsen 1997; Li 2000a; Weir et al. 2000). This change may have resulted from changes in climate, the intrinsic cyclic pattern of disturbance, or land use patterns. The distribution of times since the last fire for the case study area in this chapter indicates that the fire cycle changed significantly around 1900, with an estimated fire cycle of about 105 yr before 1900 and about 632 yr thereafter.

Spatial Patterns of Fire Ignition

Figure 8.3 shows the fire ignition patterns for lightning- and human-caused fires from 1983 to 1995 (excluding national parks). Note that lightning-caused fires are more randomly distributed in forests across the province than are human-caused fires, which seem more likely to ignite near rivers and lakes (perhaps associated with recreation sites) and at the boundaries between forests and agricultural lands, where human activities are common. These factors are unlikely to affect the distribution of lightning fires.

The Cumulated Fire Map

A cumulated fire map comprises overlays of annual fire maps for the period under investigation. The cumulated fire map for Alberta shows the areas burned during the 1961–1995 period

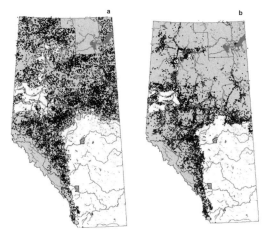

FIGURE 8.3. Patterns of fire ignition sources in Alberta (1983–1995) for (a) lightning-caused fires and (b) human-caused fires. White indicates agricultural areas, and gray indicates the forested area of Alberta. No vegetation data were included for national parks. Black dots indicate the locations of fire ignition.

(figure 8.4). This map is similar to the cumulated fire map for Ontario from 1951 to 1995 (Perera and Baldwin 2000), in the sense that very little overlapping of areas burned by two or more different fires can be identified. This result could be due to the relatively short period covered by the data; that is, an increasingly meaningful cumulated fire map might develop as more annual fire maps become available. The observation could

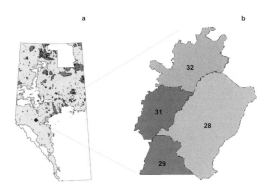

FIGURE 8.4. (a) Cumulated map of historical burns in Alberta from 1961 to 1995. White indicates agricultural areas and national parks, for which no fire data were included in the data set. Dark areas burned during the period. Gray indicates forested lands. (b) The location of the study area (Athabasca Working Circle compartments 28, 29, 31, and 32 of Weldwood of Canada Limited's Hinton Division forest management area in west-central Alberta).

also result from a reduction in the area burned as a consequence of intensive fire suppression during this period. The small degree of overlap in the areas burned means that the relative fire probability for forests is either 0 (if the site had never been burned) or 1 (if the site had been burned once). Consequently, the relative fire probability map might not be a very useful tool for managers intending to locate harvest blocks so as to emulate the natural fire regime.

In summary, this analysis of the empirical data suggests that the observed fire patterns might not be a suitable template for emulating natural fire patterns. For example, the observed fire size distribution consists of a large number of very small fires and a few very large fires, but this may represent the effect of fire suppression rather than the natural fire size distribution. The observed fire size distributions also provide insufficient information for forest managers to use them as the basis for defining harvest block sizes and numbers. The cumulated fire map for Alberta indicates that few burned areas overlapped, and thus provides insufficient information for resource managers to use as the basis for deciding where to allocate harvest blocks.

Rationale and Methodology for Simulating Fire Regimes

Ecological models have taken advantage of the increasing computational power made available over the past few decades. These models now represent useful tools for synthesizing the available knowledge and answering various types of ecological questions (Rykiel 1996). Modeling can help identify the key information required to analyze questions and thus can identify gaps in our knowledge that require more attention in future research. All fire models developed so far are abstractions of reality based on assumptions required to simplify the real world and so permit simulation. The simulations that result from such models thus necessarily represent possibilities or the logical consequences of the model's assumptions and quantitative relationships. The SEM-LAND model was developed with the aid of systems analysis to investigate the long-term dynamics of forest landscapes influenced by fires that are often irregular in location, frequency, size, and intensity. In developing this model, I emphasized simulating the interactions among fire, weather, and the forest landscape, because long-term fire dynamics cannot be understood without understanding the underlying forest dynamics.

The design of SEM-LAND's components began with an analysis of forest succession theory. Two of the model's major components (forest growth and regeneration) represent two functions in the classical theory of forest succession (Clements 1916): conservation and exploitation. Conservation, which emphasizes slow accumulation and storage of energy and material, was described using growth equations for major tree species (Alberta Forest Service 1984). Exploitation, which emphasizes the rapid colonization of recently disturbed areas, was characterized by simulating regeneration after stand collapse in the wake of a fire. For example, after stand-replacing fires, regeneration was assumed to begin immediately, and this was simulated by resetting stand age to zero in the SEM-LAND model. The third major component of SEM-LAND describes fire disturbances using the release function in the so-called "4-box" model (Holling 1986, 1992b). In this model, release (also called "creative destruction") describes changes in which tightly bound accumulations of biomass and nutrients are suddenly released by disturbances such as fire and insect infestations.

A fire regime can be characterized by descriptors that each reflect one aspect of the regime. Empirical fire data are not ideal for evaluating possible linkages among the different aspects of a fire regime, because not all aspects have been measured or recorded. Conceptually, it appears that different descriptors of a fire regime should be either directly or indirectly related, because of the interactions among ecosystem components. Therefore, an effort was made to explore possible correlations during the research that supported the development of SEM-LAND, such as the relationships between fire frequency and fire size distribution (Li et al. 1999), the pattern of fire ignition sources and the dynamics of the fire regime (Li 2000b), the forest's age-class distribution and fire disturbance (Li and Barclay 2001), and fire frequency versus the fire cycle (Li 2002). The premise of such research is that if different descriptors of a fire regime are correlated, then different aspects of the regime can be estimated by an ecological model calibrated using a reduced number of attributes that encapsulate other factors.

These research results can be useful in the reconstruction of natural fire regimes. Complex correlations can be expected to arise from the interactions among ecosystem components, such as from self-organizing behavior, in which fire disturbance functions as an organizing force that alters forest landscape structure; the resulting landscape structure may in turn influence the fire process in subsequent years. Consequently, such theoretical considerations and the resulting modeling rationale can improve confidence in the simulation's results.

From an ecological perspective, the theory of self-organization (Prigogine 1980; Holling et al. 1996) can be used to support the belief that forest ecosystems and landscapes are naturally sustainable (Bradbury et al. 2000). This theory is relevant because it provides forest managers and researchers with a theoretical basis for understanding the long-term dynamics of the forest landscape. The linkage of research on natural fire regimes to the theory of self-organization reflects an effort to seek support from the contemporary knowledge base on ecosystem dynamics and theoretical ecology. Although at first glance such theory seems irrelevant to short-term fire and landscape dynamics, it may be important in guiding research on short-term dynamics from the perspective of long-term ecosystem dynamics: any short-term fire dynamics are obviously elements of the long-term fire dynamics in a given period.

Fire regimes, as organizing forces for the dynamics of ecosystems and landscapes, should thus be viewed as self-organized (Malamud et al. 1998). The theory of self-organization addresses the question of how intrinsic properties and interactions act to create order within complex systems (Green 2000). The theory also provides the insight that ecosystem dynamics do not necessarily display either a stable equilibrium or an unstable collapse after severe disturbances; instead, ecosystem dynamics may display different behaviors over time. Because ecosystem dynamics are thus moving targets, their management should not always aim for a fixed result. Therefore, an investigation of whether the model's behavior displays the self-organization of forest and fire dynamics over a long period would provide additional data with which to evaluate how well harvest planning can emulate natural fire patterns.

Simulation of Fire Regimes Using SEM-LAND

Simulations of fire disturbance using the raster-based SEM-LAND model have already been published (Li 2000b). In that analysis, the default spatial resolution was set to 1 ha. The structure of a revised SEM-LAND model that includes a simulation of fire suppression can be described by equations 8.2, 8.3 and 8.4:

$$P_{Initiation} = P_{Baseinitiate} \times F_{Weather} \times F_{Fuel} \qquad (8.2)$$

$$P_{Spread} = \begin{cases} 0 & (R \geq R_{Crit}) \\ P_{Basespread} \times F_{Weather} \times F_{Fuel} \times F_{Slope} \\ \quad \times (1\text{-FSE}) & (R < R_{Crit}, S < S_{Crit}) \\ \\ P_{Basespread} \times F_{Weather} \times F_{Fuel} \\ \quad \times F_{Slope} & (R < R_{Crit}, S \geq S_{Crit}), \end{cases} \qquad (8.3)$$

where $P_{Initiation}$ is the probability of fire initiation, $P_{Baseinitiate}$ and $P_{Basespread}$ are the baseline fire probabilities for the initiation and spread stages, respectively, and the two latter probabilities are characterized by a logistic equation:

$$P_{Base} = k/(1 + \exp(a - b \times Age)). \qquad (8.4)$$

In these equations, $F_{Weather}$ and F_{Fuel} are scale factors calculated according to the Canadian Forest Fire Behaviour Prediction System (Forestry Canada Fire Danger Group 1992) that represent the influence of fuel type and weather conditions; F_{Slope} is the scale factor due to slope. Detailed formulas for calculating these scale factors can be found in Li (2000a). FSE is the fire suppression efficiency; R represents the daily precipitation; R_{Crit} is the critical value of daily precipitation, beyond which any precipitation stops a fire; S represents fire size; S_{Crit} is the critical value of fire size, beyond which any fire escapes fire suppression; a, b, and k are regression parameters; and Age is the stand age (defined as the time since the last stand-replacing fire). This logistic equation describes the possible patterns of fuel accumulation since the last burn, and assumes that the quantity of fuel is a function of forest age.

The usefulness of the revised SEM-LAND model was improved by replacing the use of a random wind direction with a frequency distribution of wind direction based on historical data (Wolfe and Ponomarenko 2001). The model was further validated by comparing the simulated and observed fire patterns, with a simulation of fire suppression included.

Distribution of Wind Direction

Wind direction governs the final shape of a fire and its spatial location, and hence plays an important role in forest dynamics. Wolfe and Ponomarenko (2001) summarized the wind regimes of 34 weather stations in Alberta, Saskatchewan, Manitoba, and the Northwest Territories using hourly Environment Canada wind data classed by wind speed and direction. For example, figure 8.5 shows the frequency distribution for wind direction measured at three weather stations.

The distribution of wind directions was incorporated into SEM-LAND so that wind directions could be probabilistically assigned to fire events. In the current study, the frequency distribution of wind directions at Whitecourt (Alberta) was used, because this weather station was nearest to the study area.

Simulation of Fire Suppression

The development of lightning-caused fire under a regime of fire suppression was simulated in three stages. Stage one begins with fire ignition, and lasts until the fire has been reported and fire crews have arrived at the scene. During this stage, the fire spreads freely under natural conditions. The increasing capability of wildfire detection and advances in transportation equipment, such as helicopters, have shortened the duration of this stage over the past century. The second stage begins with the initial attack on the fire, and

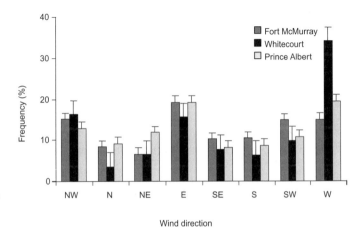

FIGURE 8.5. Frequency distribution for wind direction recorded at three weather stations: Fort McMurray and Whitecourt (Alberta), and Prince Albert (Saskatchewan). Data from Wolfe and Ponomarenko (2001).

lasts until the fire has been suppressed or has escaped the fire suppression effort. The value of P_{Spread} is reduced by fire suppression efforts during this stage. The third stage begins when a fire escapes from suppression efforts and ends when the fire finally stops. During this stage, fire spread is unconstrained, and P_{Spread} reverts to the probability under natural conditions. This three-stage description of fire suppression is not intended to capture every detail of suppression operations; instead, it captures a broader picture that can be implemented in simulations of fire regime to determine the possible long-term consequences of fire suppression on the dynamics of fire regimes and forest ecosystems.

The three-stage fire suppression scenario was simulated by SEM-LAND through the combined effect of *FSE* and S_{Crit}. The parameter *FSE* indicates the level of protection from fire suppression agencies, which is determined by the resource investment in the operation and the equipment and facilities used. For example, an increased budget for fire suppression could allocate more fire fighters, helicopters, air tankers, and other resources to a given fire. Thus, the rate of construction of fire control lines could be increased, and the rate of fire growth reduced. Consequently, the probability of a fire escaping from suppression would decrease. The parameter *FSE* is expressed as a reduced probability of fire spread. As noted earlier, S_{Crit} indicates the threshold fire size, beyond which any fire escapes from fire suppression. This parameter was considered to be independent of *FSE*.

Simulated Fire Regime for a Study Area in the Canadian Rocky Mountain Foothills

Study Area

The study area consists of the Athabasca Working Circle (compartments 28, 29, 31, and 32; see figure 8.4b) of Weldwood of Canada Limited's (Hinton Division) forest management area in west-central Alberta. It is located in zone 11 of the Universal Transverse Mercator grid system from easting 473357 to 494457 and from northing 5958944 to 5984544, and encompasses an area of 31,444 ha. Most of the area (98.5%) is covered by forest, with only 470 ha of nonforested area. Of the forested area, lodgepole pine (*Pinus contorta* Dougl.) occupies 50.4% (15,610 ha) of the total, black spruce (*Picea mariana* [Mill.] B.S.P.) occupies 13.7% (4252 ha), white spruce (*Picea glauca* [Moench] Voss) occupies 10.4%

(3226 ha), and hardwoods (mainly trembling aspen, *Populus tremuloides* Michx.) occupy 13.0% (4022 ha). The rest of the area (3864 ha) is nonproductive land.

Color plate 6a–c shows the spatial input data for the simulation. The current mosaic of forest ages from the province's forest inventory data was used to approximate the time since the last fire, based on the assumption that all stands in the study area originated from stand-replacing fires. This assumption appears reasonable, because the youngest stand recorded in the Alberta Vegetation Inventory (Alberta Forest Service 1984) for this area was older than 80 yr at the time of the inventory (around 1980), and there were no records of large-scale timber harvesting for this area during that period. Twenty-four fires were reported in the study area from 1961 to 1995. Only one of these fires was caused by human activities, with the rest caused by lightning. The largest fire, which burned 30 ha, was caused by a lightning strike in 1980, and Can$28,273 was spent on fire suppression. The second-largest fire was also a lightning fire (in 1976) that burned 12 ha, with a fire suppression cost of Can$20,052. One fire burned 2 ha, with a Can$2945 fire-suppression cost, and two fires burned 1 ha each, with suppression costs of Can$8271 and Can$2823, respectively. Nineteen other fires each burned less than 0.5 ha, and their suppression costs ranged from nonexistent to Can$8278. Without fire suppression, these fires could have burned larger areas.

Simulation Experiment

For each year in the simulation, the sources of fire ignition were sampled from a normal probability distribution with a mean probability expressed per unit area across Alberta. The mean value equaled 0.26 per 10,000 ha·yr^{-1}, with a variance of 1. The value of $P_{Baseinitiate}$ was set at 0.25 to approximate the proportion of lightning fires larger than 1 ha, based on historical fire data for Alberta. The parameter $P_{Basespread}$ was assumed to be greater than $P_{Baseinitiate}$, because a fire can more easily spread to adjacent stands from a burning stand than from a lightning strike. In reality, S_{Crit} can vary with individual fires, depending upon such factors as the investment in fire suppression, the location of the fire ignition, and weather conditions. In the current simulation, it was set to 50 ha to represent an average situation. The parameter *FSE* is a complex function of several factors, and was set at 0.5; that is, $P_{Basespread}$ was modified by a factor of 0.5.

Ten replications were carried out for scenarios with and without fire suppression.

The model was written in FORTRAN 77, and the model experiment was carried out on a SunBlade 1000 UNIX workstation running the Solaris operating system. The average CPU time for each simulation was about 2.5 h.

Reconstructed Natural Fire Regimes

Fire Cycle

The mean simulated fire cycle under the fire suppression scenario was 675.63 yr, with a range of 427.67 to 1523.42 yr. The observed fire cycle (estimated to be 844 yr under a regime of fire suppression based on historical fire data) and the fire cycle of 632 yr after 1900 (estimated from the distribution of times since the last fire for the study area) fell well within the simulated range. The simulated mean fire cycle without fire suppression was 104.61 yr, with a range of 96.14 to 107.95 yr. The fire cycle of 105 yr before 1900 (based on historical data) is also within the simulated range.

Fire Size Distribution

Simulated fire size distributions with and without fire suppression are shown in figure 8.6. The observed discontinuous patterns of fire-size distribution (see figure 8.2) also appeared in the simulation of fire regimes under a fire suppression scenario. Without fire suppression, the distribution of fire sizes displayed fewer discontinuities (figure 8.6). The shape of the fire size distribution in the natural scenario (without fire suppression) was consistent with that reported by Li and Corns (1998); that is, it had a negative-exponential shape, but could not be adequately characterized by a negative-exponential probability function. Cumming (1997) reported that changes in fire sizes can be expected under different levels of fire suppression, and this assumption has been applied in studies of fire regimes and forest dynamics in Alberta (e.g., Armstrong 1999). It can thus be hypothesized that the extent of the discontinuities in intermediate size classes are a function of fire suppression efficiency, which can in turn be a func-

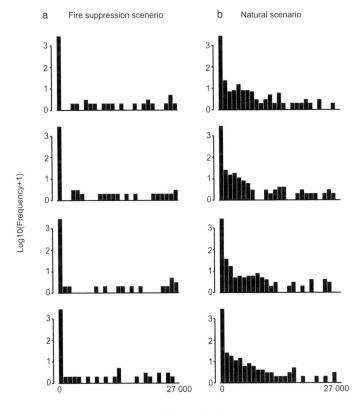

FIGURE 8.6. Examples of simulated fire size distributions (a) under a fire-suppression scenario and (b) under natural conditions for the study area.

a   Fire suppression scenerio   b   Natural scenario

Log10(Frequency+1)

Fire size class (ha)

tion of the investment in suppression actions, of fire intensity, of the rate of fire spread, and of the fire's size when the resources arrive at the scene. Simulation results using different *FSE* values confirmed this expectation.

Fire Probability Map

A map of fire probability under the natural scenario (without fire suppression) was constructed (color plate 6d) based on long-term interactions among fire events, landscape structure, forest conditions, and weather and fuel conditions. This map appeared to be relatively stable from simulation to simulation, and for different values of $P_{Baseinitiate}$, $P_{Basespread}$, *FSE*, and S_{Crit}. Thus, the simulated map produced for the fire suppression scenario is very similar. This is reasonable, because fire suppression is less likely to alter the relative fire probability for any specific forest stand. The stability of the map of relative fire probability indicates that the map is likely to be useful for the development of forestry policy and for long-term resource-management planning.

Discussion

Emulation of Natural Fire Patterns

It is important to emulate natural fire patterns in forest management based on the fire patterns under natural conditions, without human intervention. This may be difficult in practice because of a relative lack of empirical data. Most available data have been collected for recent decades or centuries, but human impacts began before the period covered by this data. Therefore, managers must understand that the observed patterns might not truly resemble the natural patterns. Figure 8.1 separates natural and observed fire regimes and forest dynamics, because the human influence on fire regimes might not be complementary to that of natural phenomena.

Miyanishi and Johnson (2001) suggested that there was little good data to support the assumption that fire suppression alters natural fire regimes and ecosystem dynamics; in their view, changes in observed fire frequency are attributable to climate change. Given that fire suppression became efficient only after the 1950s, the reported changes in fire frequency and fire cycle (e.g., Tande 1979; White 1985; Baker 1992; Larsen 1997; Weir et al. 2000) that occurred around 1900 must have been caused by more than just fire suppression. Therefore, the mechanisms and factors capable of reducing fire size must be investigated, and the simulation of fire suppression

efficiency in this chapter represents only one possible mechanism behind the change in fire size distributions. In the absence of evidence for cooler climates in west-central Alberta during the past century, exploration of the effects of fire suppression remains a meaningful way to assess the impact of human activities. The algorithm used in the revised SEM-LAND model is only one possibility. Reed and McKelvey (2002) pointed out the difficulties in estimating the effects of fire suppression, and suggested that the role of fire suppression may be similar to that of rain; both may reduce the spread of fires, leading to smaller fires. A similar idea was tested during the development of SEM-LAND, but no significant influence of rain was found on the fire cycle over long periods (Li 2000a).

The emulation of natural fire patterns is not targeted at any specific fire, but rather at the overall fire regime that characterizes long-term dynamic patterns of fire for a given area (Merrill and Alexander 1987). Fire is a semirandom process that operates at the landscape scale. The final shape and size of a fire are determined by the interactions among environmental factors and the landscape, so the ability to predict fires will always be limited by incomplete knowledge, unavailability of data with sufficient resolution and accuracy, and various uncertainties (Costanza and Jorgensen 2002). Slight changes in weather, for example, could significantly change a fire's final shape and size, even when all other conditions remain constant. Therefore, more attention should be given to the long-term dynamics of fire regimes, so as to provide meaningful templates for emulating natural fire patterns.

As reviewed by Keane, Parsons, and Rollins (chapter 5, this volume), different approaches can be used to reconstruct natural fire regimes (e.g., Li and Perera 1997; Li 2000a,c). In traditional studies of fire history, the researchers' capacity to complete the reconstruction was essentially limited by the lack of resources and time (e.g., Larsen 1997), and some technical limitations could result in biased estimations (e.g., Baker and Ehle 2001). Reconstruction of stand-origin maps for a region (e.g., Weir et al. 2000) allows researchers to detect changes in fire frequency; nevertheless, considerable resources are required, and these may be unavailable, limiting the application of such approaches to forest management and policy development. With growing computational power available to them, simulation models are increasingly capable of helping researchers reconstruct natural fire regimes (Li

2000a). Such models simulate fire size distributions and generate fire probability maps without fire suppression so as to emulate natural conditions.

Ecological Modeling to Reconstruct Natural Fire Regimes

Ecological modeling to reconstruct natural fire regimes must be performed within an appropriate conceptual framework supported by contemporary ecological theories of ecosystem dynamics. The complex, adaptive, and hierarchical properties of ecosystem dynamics revealed by ecologists (Costanza and Jorgensen 2002) explain why accurate forecasting of short-term fire and forest landscape dynamics is so difficult. The emergent properties of ecosystem dynamics also mean that the dynamics of forest landscapes may not simply equal the sum of the dynamics of their parts (Costanza and Jorgensen 2002). The situation is further complicated by gaps in our knowledge and the uncertainties involved in describing ecosystem dynamics (Holling 1986, 1998). Self-organizing behavior (Prigogine 1980) provides guidance for studying complex ecosystem dynamics. This has been an interesting topic in ecology for many years (e.g., Kay and Schneider 1994). The theory suggests that interactions among system components or the exchange of materials and information in open systems induces order from seeming disorder. This insight could have a profound influence on research into forest landscape dynamics, including natural fire regimes.

The perspectives of forest and fire managers on fire may differ. Fire management focuses on predicting the behavior of individual fires, such as the speed of fire front movements, possibility of ignition, and fire danger rating. In contrast, forest management may emphasize the fire dynamics over years or decades and the impacts of such dynamics on forest resources. To implement or evaluate natural fire patterns within harvesting plans, for example, managers might need to know (1) the natural fire-size classes to design harvest block sizes; (2) the natural fire-size distribution to determine the number of blocks in each size class; (3) the likelihood of natural fires across landscapes to locate harvest blocks within the landscape; and (4) the nature and impact of human disturbance to evaluate the consequences of different management options. The first two factors require an understanding of the fire-size distribution under natural conditions for the area under management, whereas the third requires estimates of relative fire probabilities for different sites. These three factors are discussed in Li (2000a,c). The fourth factor is necessary to avoid potentially irreversible negative impacts of management.

These different perspectives suggest that it will be difficult to obtain all the answers from any single model, and that different kinds of models may be needed—for example, fire *event* simulators versus fire *regime* simulators. Fire event simulators may not be appropriate for addressing long-term fire dynamics, whereas fire regime simulators may fail to address short-term fire behavior. It may be inappropriate to determine the reasonableness of any fire regime simulator by its ability to reproduce every detail of individual fires at the time scales of interest in simulations.

The appropriate modeling approach is determined by the purposes and objectives behind model development. Fire researchers usually develop fire growth (mechanistic) models that use physical relationships obtained from fire behavior research to simulate fire propagation. Landscape ecologists usually develop theory-based probabilistic models to simulate fire spread. Scenario-based models simulate fires using predetermined fire regimes, based on fire-size distributions and fire frequencies. The approach used to develop SEM-LAND combined mechanistic and probabilistic approaches.

Traditional models simulate fire behavior using weather variables and assume that accurate input of sufficient weather data will allow such models to simulate fire initiation and spread with great accuracy. In reality, gaps in existing knowledge and other uncertainties undermine this premise. The algorithms used in fire growth models may differ, but the underlying premise does not change.

Fire event simulators originally attempted to simulate every detail of a single fire to provide information that would guide operational fire suppression. Therefore, they simulate fires using short time steps (e.g., daily or hourly) in a fire season. In this context, insufficient attention is given to temporal changes in fuel conditions; for example, fuel maps do not change over time in these models. Although the models could theoretically be run for long periods to simulate the dynamics of fire regimes, the underlying assumption of a deterministic fire process and the lack of high-resolution temporal and spatial weather data make such simulations difficult.

Possible improvements include addressing temporal changes in fuel conditions across a landscape, the suitability and availability of weather data or predictions, the rules that govern the extinction of fires, the existence of remnants within burned areas, and irregular fire shapes. SEM-LAND was improved by incorporating probabilistic components into the mechanistic physical relationships from fire event simulators. The probability of fire in a given cell was estimated from the forest type (e.g., softwoods burn more easily than hardwoods), landscape structure (e.g., fragmentation of fuel continuity), topography (e.g., direction of fire approach, uphill or downhill spread, adjacency to fire-prone stands), level of the ignition source (e.g., probability of the presence of a fire ignition source), and weather conditions. Most of SEM-LAND's parameters were obtained from the Canadian Forest Fire Weather Index and Forest Fire Behaviour Prediction systems; these established relationships were used to reduce the number of parameters required to calibrate the model, thereby reducing the level of uncertainty.

The model displayed the expected significant influence of weather on simulated fire and forest landscape dynamics under current conditions and in response to climate change (Li et al. 2000), and the results were consistent with the summary by Weber and Stocks (1998). Although further improvement is possible, SEM-LAND already provides a possible solution to the above-mentioned problems, and its behavior is conceptually and qualitatively consistent with our understanding of fire regimes (Li 2000a, 2002; Li and Barclay 2001).

Due to a lack of empirical fire data and the assumptions used to simplify the simulation sufficiently to permit computation, fire modeling research should not be expected to emulate reality in detail. This is true of both fire event and fire regime simulators, and is reflected in the shortage of published validations of models against actual fire records. Our inability to fully predict future fires and the semirandom behavior of fires make it more important to validate qualitative models than quantitative models (Rykiel 1996). Such validations could be based on the experience of fire managers and researchers and on observations of actual fire disturbances. Validation examines the reasonableness of the model's assumptions, statistics for simulated fire regimes, fire size distributions, irregular fire shape, remnant unburned stands within burned areas, the formation and dissolution of spatial mosaics, and other considerations (Li 2002). SEM-LAND's assumptions have been carefully examined, and

the model is consistent with the expected characteristics of real fire regimes (Li et al. 2000; Li 2000a, 2002; Li and Barclay 2001). This increases confidence in the model's simulation of fire-size distribution and other predictions.

Practical Considerations

It may not always be possible to reconcile the sustainability of natural systems with human expectations. Co-evolutionary processes govern the outcome of long-term interactions in natural systems. Human influences add a complex and challenging dimension. One purpose of emulating natural disturbance in forest resource management is to align human expectations with the sustainability of natural systems, thereby sustaining forest resources for future generations. Irregular large fires can create severe problems for local industries and communities, and human society cannot wholly eliminate or exclude these fires. Recognizing that such large fires are natural does not necessarily mean that large cut blocks must be implemented, nor that alternatives should be dismissed. The reconstruction of natural fire regimes appears to be the first step in implementing or evaluating a policy of emulating natural disturbance patterns, and offers a logical means of evaluating the benefits and costs of management options. An improved understanding is needed of how often and where large fires occur (e.g., Li et al. 1997), and of their impacts on forest resources (e.g., Li et al. 2000). Simulation results provide possible scenarios for forest landscape dynamics, with and without fire suppression, and these scenarios provide part of the understanding required to support forest management decisionmaking. The use of this information involves complex considerations of human needs from social, economic, and environmental perspectives.

There are different ways to emulate natural fire patterns. For example, Alberta-Pacific Forest Industries plans harvest rates in their Forest Management Agreement area in northeastern Alberta on the assumption that the average annual harvest over long periods should not exceed the size of the areas prevented from burning by fire suppression (Armstrong 1999). Furthermore, a viable strategy for offsetting any negative impacts on such a harvest quota could involve salvaging timber killed by blowdown, fire, or insects and diseases. Thus, the total harvested and burned area will not exceed the level of natural mortality. Simulated fire regimes provide a basis for estimating the mean area burned annually and the associated fire-size distribution. Because different fire cycles appear for different forest ecosystems (Turner and Romme 1994), the harvest rate should be based on the conditions found in each region.

The map of relative fire probability generated by the simulations can guide the choice of location of harvest blocks. There are many ways to generate such maps, some of which are intuitively obvious from the model structure. For example, long-term simulations based on a fixed fuel-condition map would produce a map of fire probability that corresponds to the fuel map if enough fires are simulated. In SEM-LAND, different situations can be seen by, for example, comparing panels d and b in color plate 6. Increasing the number of fires simulated results in better maps of fire probability. A similar spatial pattern in maps of fire probability could be iden-tified among all the simulations, thereby providing a more reliable average map. Such a map would obviously facilitate decisionmaking.

Acknowledgments

The Alberta Vegetation Inventory data were provided by Weldwood of Canada Limited (Hinton Division). I thank Harinder Hans for his GIS assistance, and Brenda Laishley for her critical reading of an earlier version of this manuscript. The helpful comments and suggestions from the editors of this volume and two anonymous reviewers also improved the quality of an earlier version of this manuscript. This research was partially supported by the Climate Change Impact on Energy Sector (CCIES) program of the Federal Panel on Energy Research and Development (PERD).

CHAPTER 9

Spatial Simulation of Broad-Scale Fire Regimes as a Tool for Emulating Natural Forest Landscape Disturbance

AJITH H. PERERA, DENNIS YEMSHANOV, FRANK SCHNEKENBURGER, DAVID J. B. BALDWIN, DEN BOYCHUK, and KEVIN WEAVER

To emulate natural forest disturbance, forest managers must select appropriate disturbance regimes to emulate, based on the relative significance of such disturbance agents as fire, wind, and insect epidemics in the landscape. This decision also depends on the spatiotemporal scale of the proposed management activities, such as the extent of the unit being managed, the spatial resolution, the planning horizon, and the time interval between planning steps. Next, the forest manager must understand the nature of the disturbance regimes and be able to quantify the regimes to set unambiguous management goals. These steps presume the availability of reliable knowledge about the disturbance regimes of interest. In other words, explicit information about natural disturbance regimes is a prerequisite to moving from the concept of emulating natural disturbance (e.g., Perera and Buse, chapter 1, this volume) to the practice.

Although there are many possible approaches to understanding natural disturbance (Suffling and Perera, chapter 4, this volume; Keane, Parsons, and Rollins, chapter 5, this volume), the approaches can be broadly grouped into historical reconstruction and simulation modeling. The first group includes various methods for obtaining evidence of past disturbances (e.g., Heinselman 1973; Bergeron et al. 2001). The second group includes theoretical predictions based on empirical information (e.g., Johnson and Van Wagner 1985), mechanistic processes (e.g., Finney 1999), or a hybrid of both (e.g., Keane et al. 1996a). Further descriptions of these approaches are available in Egan and Howell (2001) for the historical methods and Gardner et al. (1999) for the modeling methods.

Simulation modeling can be further categorized based on the nature of the variation in the predicted outcome (deterministic versus stochastic) and the explicitness of the outcome (spatial versus nonspatial). Deterministic models use functions based on nonprobabilistic processes. In contrast, stochastic models of disturbance account for inherent variation in processes (e.g., fire ignition, spread, extinction) and inputs (e.g., weather, fuel), and thereby predict a probabilistic outcome. Spatially explicit models use geo-referenced data for inputs (e.g., forest cover, climate, terrain), encapsulate spatial interactions within the model's functions, and produce geo-referenced outputs (e.g., age, species). They also explicitly scale the model's functions (e.g., fire spread in 100-m steps determined by weather that changes at 1-h intervals) and account for spatial processes (e.g., propagule dispersal).

Our objective in this chapter is to describe a spatially explicit stochastic simulation modeling method and demonstrate its use for quantifying broad-scale natural fire disturbance regimes in the boreal forest. We believe it will provide useful information to support forest management based on emulating natural forest disturbance. Specifically, we simulate a series of spatial and aspatial characteristics of a *null fire regime*—one that starts with the existing forest cover and present-day climate—as if fires were allowed to ignite and spread without human interference. This does not assume a "natural" or pre-European settlement fire regime. We ask readers to keep in mind that this chapter only presents a case study meant to illustrate the approach, not a detailed ecological examination of the fire regime and its causal factors in the study area in the strictest sense.

Simulating the Fire Regime of a Region in North-Central Ontario

Model Description

The Boreal Forest Landscape Dynamics Simulator (BFOLDS), a grid-based, spatially explicit model, was used in this case study. This model contains a simulation module for crown-fire regimes (FSM) and a vegetation transition module (VTM). BFOLDS simulates the fire regime and fire-induced forest cover dynamics at broad spatial and temporal scales (>10 million ha and ≤300 yr), but uses a relatively fine spatial scale for some processes (1-ha spatial resolution). The VTM operates on a fixed time step (1 yr), and the FSM operates on a variable time step (the time for fire fronts to propagate to the centers of adjacent 1-ha cells). If a 1-ha cell burns with sufficiently high intensity in the FSM, the model assumes complete canopy destruction, and the cell is subsequently occupied by an appropriate tree species determined by the VTM. If a pixel does not burn, its forest cover may or may not be replaced, based on the probability of succession in the VTM, with these probabilities based on a combination of age and site type. In both cases, the elapsed times since burning and since a species transition occurred are tracked separately in each year.

Figure 9.1 provides a conceptual overview of this model. BFOLDS uses several raster GIS data layers that include forest cover composition and stand age, terrain, fuel type, time since last disturbance, and a suite of geoclimatic variables (table 9.1). Details of the model, including its architecture, simulation functions, input data structure, output information, and operating instructions, are provided in Perera et al. (2004) and Yemshanov and Perera (2002). Here we provide only a brief overview of the model, based on figure 9.1.

The FSM is process-based, and contains submodules for fire ignition, fire spread, and fire extinguishment. It represents the current state of our knowledge of fire processes (e.g., influence of fuel, terrain, and weather on fire spread) rather than empirical data (e.g., mean fire size) or assumptions about the fire regime (e.g., size class distribution of fires, mean annual proportion burned). We used Ontario fire and weather history databases for the past four decades (1963–2001), which contain the locations of lightning-caused fire ignitions, to "seed" fire ignitions in the model. The number of ignitions was selected randomly from a Poisson distribution (after Cunningham and Martell 1973), and the spatial positioning of these potential ignitions in the study area was also random. The actual ignition

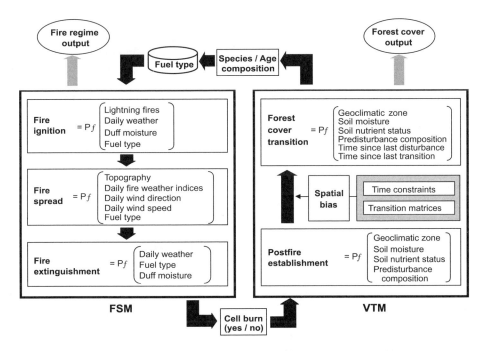

FIGURE 9.1. Conceptual overview of the Boreal Forest Landscape Dynamics Simulator (BFOLDS), including the Fire Simulation Module (FSM) and the Vegetation Transition Module (VTM). P_f denotes that the probability of the outcome is a function of the stated variables.

TABLE 9.1. Spatial Data Layers Used as Input in the Boreal Forest Landscape Dynamics Simulator (BFOLDS) to Simulate a 200-yr Fire Regime

Data Type	Source
Slope and aspect	Digital elevation model (Centre for Topographic Information 2000) corrected for watershed and stream networks
Time since last fire	Ontario forest fire history database (Perera et al. 1998)
Geoclimatic zone	Forest ecoregion map of Ontario (Hills 1959; Rowe 1972)
Soil nutrient status	Soil texture database (Ontario Ministry of Natural Resources 1977), corrected by using a flow accumulation index (ESRI 1998) for lowlands and 30-m LANDSAT-TM land-cover classification (Spectranalysis 1999) for outcrops and treed bogs
Soil moisture	Soil moisture database (Agriculture and Agri-Food Canada 1996, Ontario Ministry of Natural Resources 1977), corrected by using a topographic wetness index (Wolock and McCabe 1995)
Forest stand age	Ontario Forest Resource Inventory database (Ontario Ministry of Natural Resources 1996b)
Forest cover composition	Ontario Forest Resource Inventory database (Ontario Ministry of Natural Resources 1996b)
1962–2002 daily occurrence of fire ignitions[1]	
1962–2002 daily fire weather indices (FFMC, DMC, DC)[1]	Fire Science and Technology Unit, Aviation Forest Fire Management Branch, Ontario Ministry of Natural Resources
1962–2002 daily wind speed and direction data[1]	

[1]Point-source data. All others are raster data, with 100-m resolution.

and subsequent spread of these fires resulted from the local weather patterns (simulated by using daily fire weather data for the 1963–2001 period) and their interactions with the spatial patterns of fuel and terrain. The FSM uses cellular automata to simulate the processes of fire ignition, spread, and extinguishment, based on the indices of the Canadian Forest Fire Weather Index System (Van Wagner 1987); these include the Fine Fuel Moisture Code, Duff Moisture Code, and Buildup Index, as well as wind speed, wind direction, and terrain characteristics. The model predicts fire intensity and rate of spread on an hourly basis when any fires are burning by using the Canadian Forest Fire Behaviour Prediction System (Hirsch 1996).

The VTM contains two steps: postdisturbance recruitment of tree species, which is based on knowledge of the regeneration ecology of boreal tree species, and postdisturbance succession of the forest cover. The latter is simulated as a semi-Markovian transition process that predicts succession of the forest cover, based on the probability of the persistence of individual tree species in the canopy, the length of the postdisturbance period, and geoclimatic conditions, with discrete states corresponding to the dominant tree species. The probabilities of discrete state transitions are stratified spatially, based on geoclimate, soil moisture, and edaphic gradients. BFOLDS also contains a submodule that applies a spatial bias to the transition probabilities used to predict the change in forest cover; to do so, the submodule uses local spatial autocorrelation of environmental site conditions (Weaver and Perera, in press) to reduce the artificial spatial randomness introduced by Markov models.

Study Area

The area selected for this simulation exercise occupied 2.15 million ha in north-central Ontario, Canada (figure 9.2). It covers portions of the Northern Clay Belt (in the northern third of the study area) and the Missinabi-Cabonga sections of Rowe's (1972) classification of the boreal forest region of Canada. The major land use in this area since the early twentieth century has been commercial forestry, with widespread timber harvesting (Perera and Baldwin 2000). In addition to forest harvesting, this area has been subjected to

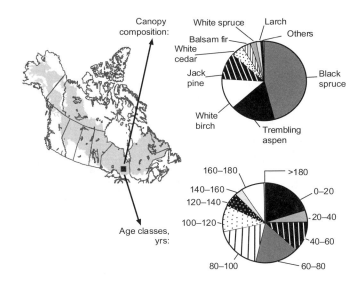

FIGURE 9.2. Location of the boreal forest landscape used in the simulation study, and its forest cover and age composition. (White spruce = *Picea glauca* [Moench] Voss, balsam fir = *Abies balsamea* [L.] Mill., white cedar = *Thuja occidentalis* L., and larch = *Larix* spp.)

periodic forest fires (Perera et al. 1998) and insect epidemics (Candau et al. 1998) during the past century. The present forest cover in the area has therefore resulted from extensive anthropogenic and natural disturbances. Spatial data are readily available for this area from databases on geo-climate (Ontario Ministry of Natural Resources 2000), soils (Ontario Ministry of Natural Resources 1977; Agriculture and Agri-Food Canada 1996), terrain (Centre for Topographic Information 2000), and forest inventory (Ontario Ministry of Natural Resources 1996b). Based on the forest inventory data, roughly 50% of the forested area is composed of stands 40–100 yr old. Around 20% of the forest is younger than 20 yr, and around 13% is older than 120 yr. Forest cover is primarily boreal and is dominated by black spruce (*Picea mariana* [Mill.] B.S.P., 45% by area), followed by trembling aspen (*Populus tremuloides* Michx., 17%), birch (*Betula* spp., 13%), and jack pine (*Pinus banksiana* Lamb., 11%), as shown in figure 9.2.

Simulation Runs and Study Assumptions

The fire regime in the study area (2.15 million ha) was simulated at 1-ha resolution for 200 yr, with annual reporting of the number of fires, area burned by each fire, spatial geometry of the burns, succession of the forest cover, composition of the forest cover, and age composition of the forest. Forest age was tracked with respect to the times elapsed since the last disturbance ("site age") and since the last change in forest cover ("canopy age"). We replicated the entire 200-yr simulation 20 times, always starting the simula-

tion runs with the currently existing forest cover and age composition. Therefore, the simulation results provide insight into the uncontrolled natural fire regime of the study area starting with the present forest cover (representing the legacy of more than a century of human intervention) rather than with "pristine" forest conditions.

This approach was chosen because no methods were available to generate the potential natural forest cover to use as input in a reliable, spatially explicit manner. Similarly, we assumed that climatic conditions would remain stable during the 200-yr simulation period, because spatially explicit models of climate change are not yet available for Ontario. Because spatially detailed fire weather data have only been available since the 1960s, we could not simulate fire weather scenarios that lie beyond the extremes that have occurred during the past 40 yr. We address the significance of these assumptions in the discussion on applying the study results. The simulation parameters used here (a 200-yr period, 2.15 million ha extent, and 20 replications) were arbitrary, and have been selected solely to illustrate the use of the method. In consequence, they lack real ecological significance.

Results of the Simulation

Once a simulation was complete, we imposed an internal 5-km-wide buffer (0.250 million ha) to eliminate edge effects caused by fires originating outside the area being simulated and spreading into the study area. This decision reduced the extent of the study area to 1.902 million ha. A further 238,000 ha (including water

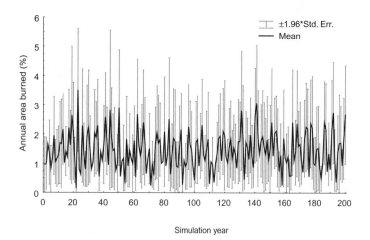

FIGURE 9.3. The mean percentage of the total area burned annually (annual burn rate) during the 200-yr simulation period, and standard error bars ($n = 20$).

and unvegetated areas, such as settlements and bedrock) was deemed unburnable under the Canadian system of fuel classification (Forestry Canada Fire Danger Group 1992). Consequently, the simulations applied to only 1.664 million ha of the study area, and this value was used in reporting the results of the simulations.

Annual Burn Rate

For the 200-yr simulation period, 1730 ± 38 (mean $\pm$ standard error) fires occurred within the burnable extent of the study area. For the 4000 yr of simulations (200 yr × 20 replicates), at least one fire occurred in 73.3% of the years. The overall mean annual burn rate for the 200-yr period (i.e., the proportion of the total area burned) was $1.45 \pm 0.03\%$, with values during the simulation period that ranged from less than 0.2% to more than 3.5% (figure 9.3). The corresponding mean fire cycle, commonly estimated as the reciprocal of the annual burn rate (Johnson and Gutsell 1994), was 69 yr.

Size Class Distribution of the Fires

The mean fire size in the simulated fire regime was 2814.7 ± 42.6 ha, with the largest individual fire exceeding 145,000 ha (8.7% of the burnable area). The number of fires per year and the total annual area burned were significantly correlated, as would be expected. The size class distribution for fires during the simulation period showed a negative-exponential trend, with numerous small fires (figure 9.4). Based on the total number of fires and the total area burned during the simulation period, we constructed probability distributions for the number of fires and for the area burned for each size class (figure 9.5). The probability of a fire being larger than 1000 ha was $41.0 \pm 1.0\%$, versus $16.6 \pm 0.4\%$ for fires larger than 5000 ha, $8.2 \pm 0.3\%$ for fires larger than 10,000 ha, and $3.2 \pm 0.2\%$ for fires over 20,000 ha. In other words, most fires (nearly 60%) were smaller than 1000 ha. Only a very small proportion (1.6%) of the burned area was caused by these

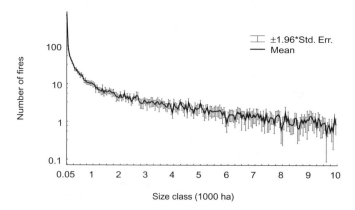

FIGURE 9.4. The mean size class distribution for fires during the 200-yr simulation period, and standard errors of the values ($n = 20$).

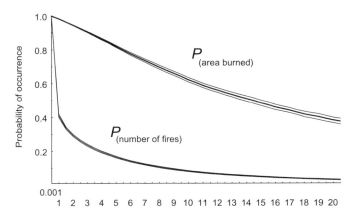

FIGURE 9.5. The cumulative probability of occurrence of fires of different sizes ($P_{\text{number of fires}}$) and the cumulative probability of the study area being burned by fires of different sizes ($P_{\text{area burned}}$) during the 200-yr simulation period, plus the 95% confidence interval for these probabilities ($n = 20$).

small fires, whereas $98.4 \pm 0.08\%$ of the burned area was caused by fires larger than 1000 ha; of the latter percentage, $82.0 \pm 0.6\%$ was caused by fires larger than 5000 ha, $62.2 \pm 1.3\%$ by fires larger than 10,000 ha, and $38.3 \pm 1.6\%$ by fires larger than 20,000 ha.

Mean Interval between Fires

The overall mean interval between fires for the study area, which is equivalent to the "point fire frequency" (Lertzman et al. 1998), was 85.06 ± 1.56 yr. This was estimated as the average period between two consecutive fires for each 1-ha pixel in the burnable portion of the study area during the 200-yr simulation, averaged over 20 replicates. The mean interval between fires for most of the area (52%) was 40–80 yr, with another 29% of the area having a mean interval of 80–120 yr. Only 15% of the area had an interval of 120–180 yr. The spatial distribution of these intervals was clustered, which suggests the existence of spatial biases in repeatedly burned areas (color plate 7).

Fire Probability

We defined the probability of a given 1-ha pixel burning as the number of times that pixel burned during the 200-yr period, expressed as a proportion of the maximum possible number of opportunities the model allowed for it to burn. A very large proportion of the burnable area (nearly 77%) has an 8–20% probability of burning during a given 200-yr period. The spatial distribution of these probabilities was also clustered (color plate 8), because of the spatial biases in repeated burns, as discussed in the previous section.

In addition, we examined the spatial probability of occurrence of fires in different size classes. These values were estimated as the number of times a given pixel was burned by a fire of a given size class, expressed as a proportion of the maximum possible number of opportunities for it to burn allowed by the model. The spatial probability of occurrence of very small fires (<250 ha) was low: more than 60% of the area had a probability of less than 0.2% of being burned by a very small fire. This is because most of the very small fires (although highly frequent) are spatially random, without any apparent biases (color plate 9a). With the next-largest size class, small fires (250–1000 ha), the probability of occurrence increased (>60% of the area had a probability >0.3%), and the spatial patterning was still dispersed (color plate 9b). As size class increased from medium (1000–5000 ha; color plate 9c) to large (5000–10,000 ha; color plate 9d), the spatial probability of occurrence also increased (>60% of the area with a probability >1.2% for medium fires, and >2.0% for large fires), and the spatial patterns became more clustered. With very large fires (>10,000 ha; color plate 9e), the spatial probability of occurrence was even higher (>60% of the area with a probability >8.0%), and the spatial patterns were even more strongly clustered. This is because larger fires are more likely to overlap (hence repeat) in space.

Age Composition

The age composition of the forest cover changed considerably during the simulation period (figure 9.6). The age composition shifted from an initial multimodal distribution (with peaks at <20, 50–100, and 140–160 yr) toward a negative-

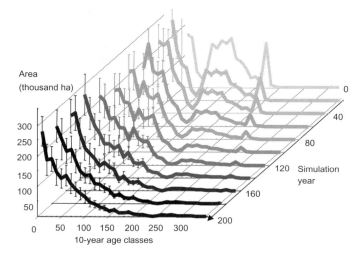

FIGURE 9.6. The initial age class distribution of the study area (year 0) and changes in the mean age class distribution over the course of the 200-yr simulation period (at 20-yr intervals), plus the standard error of the values ($n = 20$).

exponential curve by year 200. The maximum-age cohort also increased from 250 yr to more than 350 yr, because some parts of the study area did not burn during the simulation period. In addition, we tracked the maximum age that every 1-ha pixel in the study area attained during the simulation period. The spatial pattern for the mean maximum site age that was attained (over 20 replicates) is illustrated in color plate 10.

Variability in the Simulated Fire Regime

The inherent variability in a regional fire regime can be manifested in three dimensions: temporal, spatial, and stochastic (Lertzman et al. 1998). For example, in any given year of the simulation, the number of fires that occurred varied from 0 to 75, resulting in annual burn rates as low as 0% and as high as nearly 20%; these values correspond to fire cycles of more than 1000 yr to as little as 5 yr. Every aspect of the fire regime (e.g., size class distribution, location of fires) varied between years. This *temporal* variability results from annual variations in weather and long-term variation in forest cover, whether these changes occur randomly or as a result of autocorrelation. *Spatial* variability in the fire regime results from the spatial heterogeneity of the geoclimate, terrain, and forest cover in the study area. For example, the overall 85-yr mean interval between fires for the study area can be expressed in a spatially explicit manner so as to describe the spatial variability and biases in the study area, as shown in color plate 7. Another example is the burn probability with very large fires. The mean spatial probability of occurrence (9.2%) can be more effectively expressed using a choropleth map

(such as color plate 9e) that captures the spatial variability.

The *stochastic* variability approximates the non-deterministic nature of the many processes that may be involved in any natural fire regime, random fluctuations in these processes, and the ensuing changes in forest landscapes. This variability arises in simulations as the result of the product of many stochastic model functions (e.g., fire ignition, spread, extinction), succession in the forest cover, and intrinsic variability in such input data as weather and forest cover. The stochastic variability is captured as the among-replicate variation of the simulations, and can be illustrated as error bars and confidence intervals (e.g., figures 9.3–9.6), or as spatially explicit "surface roughness" (the differences in elevation in color plate 11).

Applying Information on Simulated Fire Regimes to the Management of the Forest Landscape

In approaches to forest management that emulate natural disturbance, knowledge generated by the simulation of fire regimes can be useful for defining the emulation *criteria* (see Perera and Buse, chapter 1, this volume); these criteria represent the characteristics of a disturbance that can be used to quantitatively describe the disturbance regime. Thus, they can be used as a guide in developing strategies and practices for emulating natural forest disturbance, on the assumption that fire is the most prevalent broad-scale disturbance in the management area.

The *degree* to which a fire regime can be emulated based on these criteria may vary, depending on various socioeconomic and ecological

TABLE 9.2. Criteria for Emulating Natural Disturbance and Examples of Their Use in Forest Landscape Management

Emulation Criteria	Fire Regime Characteristic (Null Values)	Forest Management Characteristic
Overall rate of disturbance for a planning region	Mean annual burn rate Regional fire cycle Temporal and stochastic variability in annual burn rate	Mean annual area harvested Regional harvest rotation Bounds of variation for variability in annual area harvested
Spatial and temporal frequency of disturbance	Spatial probability of burning Spatial variability of mean interval between fires Stochastic variability in probability of burning and mean interval between fires	Selection of harvest regions Local harvest rotation Bounds of variation for harvest regions and local rotations
Sizes and patterns of disturbance	Fire size class distribution Temporal and stochastic variability in fire sizes Spatial probability of fire size classes	Harvest patch size class distribution Bounds of variation for harvest patch sizes Spatial distribution of harvest patches
Potential for landscape aging	Distribution of site age and canopy age class Stochastic variability of age class distribution Spatial probability of mean and maximum of site age and canopy age Stochastic variability in probability of site age and canopy age	Age cohort composition, including old-growth extent (by species) Bounds of variation for age-cohort composition Spatial demarcation of old-growth sites Bounds of variation for spatial demarcation of old-growth sites
Forest landscape composition and patterns	Forest cover composition and spatial probabilities Spatial geometry of forest cover Spatial, temporal, and stochastic variability in forest cover and spatial geometry	Future composition and spatial patterns of forest cover "Desired" landscape patch characteristics Bounds of variation for forest cover and patch characteristics

considerations. Several characteristics of a fire regime can be used as the emulation criteria, and the simulated values used as the basis for harvesting regimes (table 9.2). For example, in the broadest (aspatial) sense, annual rates of fire disturbance (annual burn rate) in the region (as illustrated in figure 9.3) provide insight into potential forest harvest rates (Hunter 1993; Armstrong 1999). The concept of a regional fire cycle is heavily debated (e.g., Lertzman et al. 1998; Armstrong 1999; Li 2002), but can nonetheless serve as the basis for "regional woodshed planning" that determines the long-term timber supply and hence the harvest cycles. The spatial variation in rates of disturbance (e.g., spatial patterns for the mean interval between fires in color plate 7 and the probability of burning in color plate 8) provides a spatially explicit template for the region's heterogeneity.

These values can guide the spatiotemporal allocation of rotation length and the size of harvest blocks. The simulated probability distribution of fire size classes (see figure 9.4) presents one possible template for planning the size class distribution of harvest patches, which is required by some forest policies (e.g., Ontario Ministry of Natural Resources 2002a; McNicol and Baker, chapter 21, this volume). Moreover, the spatial probabilities of occurrence for different fire size classes may be used to guide the placement of harvest patches in a landscape. For example, smaller harvest patches may be placed anywhere in the landscape (as per color plate 9a), whereas very large patches may be placed only in certain parts of the landscape (as per color plate 9e). Similarly, the age class distribution resulting from the simulated regime may provide a logical basis on which to plan future forest-age cohorts, something that has been advocated, based on empirically derived fire regimes (e.g., Y. Bergeron et al. 1999). Information on the age patterns that result from simulated fire regimes can also help answer planning questions for old-growth forest (Johnson et al. 1995). Specifically, the temporal and spatial patterns of a simulated age class distribution (see figure 9.6) and of simulated maximum age (color plate 10) can guide strategic decisions about the extent and locations of future old growth (Perera et al. 2003). Although we did not present the forest cover patterns resulting from the simulations described in this chapter, these postfire patterns provide an indication of potential patterns of forest cover in terms of stand composition, spatial tendencies, and the spatial geometry of patches.

In their attempts to emulate natural disturbance, managers recognize that natural disturbance regimes vary by using terms such as *bounds of natural variation* (e.g., Ontario Ministry of Natural Resources 2002a); when resorting to natural history techniques, they may use *historical range of variability* instead (e.g., Morgan et al. 1994). As discussed earlier, simulation can separate variability into three distinct sources (spatial, temporal, and stochastic), each of which has different significance in understanding disturbance regimes. Thus, forest managers can distinguish the spatial variation of a fire regime criterion within their planning area (e.g., the mean interval between fires in color plate 7); the temporal variation during the planning period (e.g., the age class distribution in figure 9.6); and most importantly, the stochastic variation of a spatial or temporal variable (e.g., the spatial probability of fire in color plate 11 and the annual area burned in figure 9.3).

Estimation of the magnitude of these sources of variability, especially the spatial and stochastic sources, is almost impossible in empirical methods of predicting fire regimes. In fact, as Lertzman et al. (1998, p. 4) wrote, the "historical dynamics of any real landscape are [only] one realization of a stochastic process." Furthermore, the historical dynamics are analogous to a single run of a simulation model, with no estimation of the potential for random fluctuations. This is a principal issue in validating these types of long-term models of landscape dynamics. A comparison of what can be reconstructed from the historical record with the results produced by simulation models provides unsatisfactory support for the validity of the model. The inherent spatial and temporal variability of fires is quite high, and the temporal extent of the historical data is limited, because records of past fires have been obliterated by more recent fires. Further complicating this situation is the large

spatial extent of most simulations, which consequently encompass high spatial heterogeneity.

The picture that can be reconstructed from the historical record is, as indicated above, only one instantiation of a highly stochastic process, so it can readily fit within the simulated range of variability from landscape models, some of which might not resemble known landscape processes. Comparison with historical data is only one method for validation (e.g., Sargent 2000). In this chapter, our approach was to construct a model using available, accepted components. Validation of the model is also consistent with the approach recommended by Kleindorfer et al. (1998), in which validation is provided by dialog with the model's stakeholders to take advantage of their judgment about what is understood and accepted. The most significant aspect of this approach is that forest managers are not required to predict the fire regime's characteristics in advance. Empirical knowledge of either individual values (e.g., mean fire interval, frequency, or cycle) or their probability distributions (e.g., a negative-exponential fire size distribution) is not decided a priori. Instead, forest managers discover the regional fire regime to be a function of patterns in geoclimate and forest cover in their region, and thereby avoid the circular reasoning inherent in many empirical methods of predicting fire regimes.

Nonetheless, our approach also requires some assumptions (albeit not about the outcome) that may influence the use of the predicted fire regimes and their emulation by forest managers. First, for reasons explained above, we assumed that regional climate remains stable during the 200-yr simulation period and that extremes in the weather patterns will remain within the limits experienced during the past 40 yr. Although this is more a limitation of the input data than a true assumption, it may nonetheless limit the incidence of large and infrequent disturbances, such as those described by Turner and Dale (1998). In addition, the temporal and spatial variability of the simulated fire regimes, and therefore the variability in the values of the emulation criteria, may be underestimated. As spatial databases expand with time and spatially explicit models of climate change become available, this assumption is likely to become unnecessary.

The second assumption is more of a premise: we simulated the fire regime starting from the existing forest cover rather than beginning with a pre-European settlement or pristine forest cover unaffected by humans. The values of the emulation criteria generated using this approach may not represent any historical fire regime (patterns in cover, climate, and fire weather in the absence of humans) or, more importantly, may fail to represent the natural fire regime. These simulations may more realistically portray disturbance regimes, because we must manage the existing forests. However, the length of our simulation period (200 yr) captures only a short period of temporal variability in the fire regime and in the dynamics of the forest cover, and thus, may not adequately reflect longer-term trends that extend beyond the life spans of the tree species in the region.

Conclusions

Scientists, forest managers, and other stakeholders regularly engage in debate about what constitutes a natural fire regime. The debate may be futile, because the high degree of temporal, spatial, and stochastic variability evident in disturbance regimes suggests that *natural* spans a large range of possibilities. Simulation modeling offers a sound alternative, because the modeled scenarios represent potential outcomes rather than an attempt to define a singular natural outcome.

In this chapter, our goal was to demonstrate the utility of a spatially explicit stochastic simulation model in generating a null fire disturbance regime that could be helpful in planning and management of the forest landscape. The results of our simulations illustrated the usefulness of this approach for emulating natural disturbance. Because this method provides insight into several aspects of disturbance regimes that cannot otherwise be estimated, it complements such empirical methods as natural history surveys and nonspatial empirical models. It also offers several advantages.

First, all characteristics of the simulated disturbance regime are a direct product of scientific knowledge and the mechanistic logic encapsulated in the model's functions. Users discover the potential fire regime based on the knowledge embedded in the model, as well as on geoclimatic and forest cover data. Therefore, the approach avoids the circular reasoning intrinsic in approaches that are based on a priori empirical assumptions about characteristics of the fire regime, such as the annual burn rate, fire cycle, and statistical distribution of the fire size classes. Moreover, because the simulated fire regime is an emergent property of the model's functions and input data, simulations may discover probable but infrequent and spatially rare fire events that are missing from historical empirical information.

Second, the simulation's results are spatially explicit, and go beyond traditional descriptions of fire regimes that represent large regions aspatially by single values, such as the length of the fire cycle or the fire frequency. Patterns in these and other descriptors of the fire regime can be explicitly portrayed as spatial probabilities, thereby allowing forest managers to understand the spatial patterns in the fire disturbance regime of an area and factors that bias these patterns. These patterns may guide the positioning of forest management activities in a planning area.

Third, this simulation approach provides the opportunity to estimate the variability of predictions of the fire regime from three points of view. Year-to-year temporal variation during the simulation period and the spatial variation of the fire regime's characteristics in the study area can be directly estimated to elucidate the spatio-temporal dynamics of the disturbances. In addition, the stochastic nature of the model's several functions lets modelers predict the magnitude of the variation by replicating the simulation. Together, these estimates of variability add robustness to the predictions and give forest managers confidence that they are truly emulating natural disturbance.

In general, there are few obstacles to broadening the use of spatially oriented models of fire regimes. The potential for error propagation due to inaccuracies in spatial databases and the validity of the model's assumptions are the most prominent technical issues, although ongoing advances in the development of spatial databases and in fire science may quickly remedy these shortcomings. In practice, the relative lack of trust in simulation models in comparison with field observations may interfere with adoption of these models. This problem is compounded by the constant calls for testing and validation of models, which itself leads to a complex philosophical debate. The general paucity of adequate (i.e., long-term, accurate, replicated) empirical observations, which are required to validate the simulation models, makes this task difficult. However, modeling methods based on spatial simulation improve incrementally as science provides better logic, data, and assumptions. These methods also provide opportunities to explore "what if?" scenarios by explicitly manipulating input data (e.g., climate change) and linking the simulations with models that simulate other types of disturbance (e.g., insect epidemics).

Acknowledgments

We thank the Ontario Ministry of Natural Resources (OMNR) for funding this research, Jim Caputo of the Fire Science and Technology Section of the OMNR for providing fire weather information, and four anonymous reviewers for their constructive comments on the draft manuscript.

CHAPTER 10

Simulating the Effects of Forest Fire and Timber Harvesting on the Hardwood Species of Central Missouri

HONG S. HE, STEPHEN R. SHIFLEY, WILLIAM DIJAK, and ERIC J. GUSTAFSON

Most forest managers have a good understanding of the effects of silviculture and disturbance on stand development (e.g., Oliver and Larson 1990; Nyland 2001; Johnson et al. 2002), but understanding the cumulative landscape-scale effects of individual management decisions and of natural disturbance is challenging, due to the broad spatial scales and long time frames involved (Foster et al. 1997; Dale et al. 2001). Nevertheless, individual forest management activities and disturbance events interact with and alter the landscape-scale processes. For example, mean fire-return intervals for Missouri's Ozark Highlands ranged from 3 to 16 yr during the 1800s and early 1900s (Guyette et al. 2002), but are now on the order of 300–400 yr as a result of successful fire suppression efforts (Westin 1992). The open forest understories that were previously maintained by frequent fires and open-range grazing now have abundant understory vegetation. Following the wave of pine (*Pinus* spp.) harvesting in the Ozarks during the early 1900s, a mix of oaks (*Quercus* spp.) and hickories (*Carya* spp.) frequently replaced the shortleaf pine (*P. echinata* Mill.) that were formerly abundant in the overstory. The black oak (*Q. velutina* Lam.) and scarlet oak (*Q. coccinea* Muenchh.) that regenerated are now reaching maturity, and many stands—particularly those on droughty sites— are experiencing high levels of mortality with concomitant increases in the site's fuel loading and fire susceptibility (Dey et al. 1996; Bruhn et al. 2000; Shumway et al. 2001). Management activities can mitigate these conditions, but cannot do so simultaneously throughout the millions of affected acres.

In this chapter, we investigate how forest fire and timber harvesting interact with forest succession in a large, heterogeneous landscape in the Ozark Highlands of Missouri, United States. More specifically, we examine the following questions:

1. Has fire suppression altered the fire regime, has an altered fire regime led to large, stand-replacing fires, and if so, what would be the effects on the hardwood ecosystems of the central United States?

2. How do forest fires interact with natural succession in this region, and what are the expected recovery paths of the tree species after disturbance?

3. What are the effects of forest harvesting, and will the cumulative harvesting activities performed at the stand scale alter species composition and spatial patterns at the landscape scale?

4. Will forest ecosystems under current harvesting and fire regimes eventually diverge from the central Missouri hardwood ecosystems under a regime of frequent fires?

In many cases, questions involving such large-scale ecological processes as fire disturbance and timber harvesting can be addressed by using landscape-scale simulation models (Mladenoff and Baker 1999). Compared with traditional plot-based ecological models, landscape models track fewer details across broader spatial scales, and the condition and juxtaposition of sites is explicitly taken into account to simulate such ecological processes as disturbance (e.g., wind, fire, harvesting) and seed dispersal (see Keane, Parsons, and Rollins, chapter 5, this volume). Thus, landscape models are ideally suited to studying species composition and the spatial patterning of ecosystems in response to natural and anthropogenic disturbance (see Suffling and Perera, chapter 4, this volume).

In this chapter, we use a stochastic, spatially explicit landscape model, LANDIS (Mladenoff et al. 1996; He et al. 1999; Mladenoff and He 1999; Gustafson et al. 2000), to explore these questions. LANDIS has been tested with different species and environmental settings (e.g., He and Mladenoff 1999a,b; Mladenoff and He 1999; Shifley et al. 1999, 2000; Gustafson et al. 2000, in press; Franklin et al. 2001; Hao et al. 2001; He et al. 2002; Pennanen and Kuuluvainen 2002). The simulation model uses 10-yr time increments and derives the simulated results at decade *t* starting from the results for decade *t* – 1. Thus, if the model starts at year 0 with realistic parameters at the level of sites or cells (species composition and age cohorts) and at the landscape level (species distribution), the simulated results at time *t* will be realistic. Due to the stochastic nature of the such processes as fire, windthrow, and seedling dispersal simulated by the model, LANDIS is not designed to predict the specific time or place of individual disturbance events. Rather, it is a cause-response model that simulates landscape patterns over time in response to the combined and interactive outcomes of succession and disturbance in realistic landscapes. The model thus guides managers toward management practices that can mitigate current or anticipated problems and provide a better understanding of the long-term, cumulative effects of the combination of natural disturbance and management practices. In this chapter, we used LANDIS in combination with a set of simulation scenarios to answer the four questions listed above.

Materials and Methods

Study Area

Our study area includes portions of the Mark Twain National Forest in the Eleven Point and Current River watersheds of the Missouri Ozark Highlands (figure 10.1). The study area covers approximately 400,000 ha, and is largely forested, with white oak (*Q. alba* L.), post oak (*Q. stellata* Wangenh.), black oak, and scarlet oak as the dominant species. The abundance of shortleaf pine is much lower than it was prior to the mas-

FIGURE 10.1. The approximate location of the study site in the Ozark Highlands of southeastern Missouri, United States.

sive logging that occurred in the late 1800s and early 1900s. Sugar maple (*Acer saccharum* Marsh.) and red maple (*A. rubrum* L.) have increased in abundance, due to their shade tolerance and decades of fire suppression. The forest's age structure is relatively simple, as a result of historical harvesting patterns and more than a century of human impacts. Topographic variation is high, with elevations ranging from 140 to 410 m above sea level, and many slopes of 30° or greater (Bellchamber et al. 2002). The climate is continental, with mean high temperatures ranging from 6°C (January) to 32°C (July) and lows ranging from –7°C (January) to 18°C (July). Mean annual precipitation is 107 cm, but most soils are cherty and many are excessively drained. These environmental variations are largely captured by means of ecological land types, which are available in the form of a GIS data layer for the study area (Nigh et al. 2000).

The LANDIS Model

LANDIS is a spatially explicit model that was developed to simulate forest change over large, heterogeneous landscapes and long time scales (Mladenoff et al. 1996; Mladenoff and He 1999; He and Mladenoff 1999a; He et al. 1999; Gustafson et al. 2000). The modeling approach integrates site-level ecological processes (e.g., competition, succession, establishment) with spatially explicit landscape processes (e.g., fire, windthrow, harvesting). LANDIS performs the following functions:

1. It simulates large landscapes that are heterogeneous in terms of site conditions (e.g., ecoregions or other land classes) and initial vegetation (species and age classes).

2. It simulates spatial processes, including fire, windthrow, and forest harvesting, as well as their interactions.

3. It includes species-level forest dynamics using realistic mechanisms and with modest requirements for the input parameters.

4. It accommodates a range of study scales and a variety of resolutions for the input map data.

5. It allows parameterization using empirical data and calibrations that reflect historical distributions (He and Mladenoff 1999a).

LANDIS is a raster-based model that grids the study area into cells ranging in size from 100 m² to more than 10,000 m² (figure 10.2). Each raster unit or cell is a spatial object that tracks the following parameters through time:

- The presence or absence of age cohorts for each species or species group;

- Fuel levels based on fuel accumulation and decomposition characteristics;

- Fire and windthrow probabilities based on mean fire- and windstorm-return intervals;

- The time since the last fire, windstorm, or harvest disturbance; and

- The probabilities of species establishment for site-specific environmental conditions.

The birth, growth, death, regeneration, mortality, and vegetative reproduction of a species are simulated by using vital attributes of the species (table 10.1) over 10-yr time steps for each cell. At the landscape scale, seed dispersal, windstorm or fire disturbances, and forest harvesting are also simulated at each time step (figure 10.2).

In heterogeneous landscapes, ecoregions derived from landforms and other environmental data are used to stratify the landscape. Within each ecoregion, environmental factors are assumed to be relatively homogeneous, as are some other factors, such as mean fire-return intervals, fuel decomposition rates, and the probabilities of species establishment (He et al. 1999; Mladenoff and He 1999). These assumptions have been

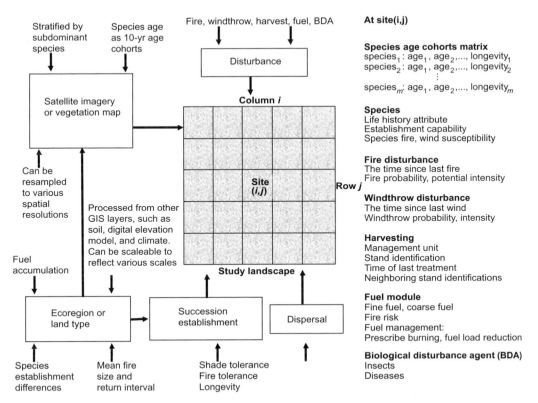

FIGURE 10.2. A conceptual flowchart for LANDIS. LANDIS is a spatially explicit model of succession and disturbance based on rasterization to grid the study area into cells that represent individual forest sites. LANDIS also simulates the interactions between spatial and nonspatial processes.

TABLE 10.1. Life History Attributes Parameterized for the Ozark Highlands of Missouri for Use in the LANDIS Model

Species Group	Longevity (yr)	Minimum Age for Seed Production (yr)	Shade Tolerance (Class)	Fire Tolerance (Class)	Effective Seeding Distance (m)	Maximum Seeding Distance (m)	Vegetative Reproduction (Sprouting Probability)	Minimum Age of Sprouting (yr)
Sugar maple	200	20	5	1	100	200	0.3	20
Shortleaf pine	200	20	3	4	40	80	0.5	20
White oak	250	20	3	4	30	800	0.5	50
Black oak	150	20	3	3	30	800	0.8	50

validated in numerous empirical and experimental studies (e.g., Brown et al. 1982; Kauffman et al. 1988). The initial species-age cohort information required by LANDIS can be derived from classified satellite imagery, forest inventory data, or both (see the section of this chapter titled "Model Parameterization" for details).

Currently, LANDIS simulates four spatial processes: fire, windthrow, seed dispersal, and harvesting (He and Mladenoff 1999a; Mladenoff and He 1999; Gustafson et al. 2000). Fire and windthrow disturbances are stochastic processes that are parameterized based on the historical size distributions and mean disturbance-return intervals for each ecoregion (He and Mladenoff 1999a). The former follows a log-normal distribution, whereas the latter is a negative-exponential distribution modified from the equation presented by Johnson (1992). Fire ignition in LANDIS involves randomly selecting locations in the modeled area and determining whether ignition occurs using a negative-exponential probability function (He et al. 1999). The model accounts for the observations that fires are more frequent in certain ecoregions than in others and that small, low-intensity fires occur more frequently than large, catastrophic fires. LANDIS simulates five levels of fire intensity, ranging from ground fires (intensity class 1) to crown fires (intensity class 5). The intensity is determined by the time since the last fire at each site, which acts as a surrogate for the amount of fuel accumulation, but is modified by using fuel accumulation and decomposition rates defined for each land type (He and Mladenoff 1999a). A more sophisticated LANDIS fuel module is being developed to track fine fuel, based on the live species present within a cell, and coarse fuel, based on the time available for fuel accumulation and decomposition. Tree species are grouped into five fire-tolerance classes, and fire severity (how many species or age cohorts are killed) represents the result of the interactions between the susceptibility of a species (based on age classes), its fire tolerance, and fire intensity (He and Mladenoff 1999a). Fire intensity can vary among cells, depending on fuel accumulation and the fire history for a given cell. In general, small, young trees are more susceptible to damage than large, older trees.

LANDIS simulates seed dispersal in a spatially explicit manner. The probability of seed dispersal is modeled by using an exponential distribution defined for each species, based on the effective and maximum seed-dispersal distances (He and Mladenoff 1999b). When seed is present on a site, species establishment is simulated by using a species-establishment coefficient (a number between 0 and 1) that quantifies how the environmental conditions favor or inhibit the establishment of the species (Mladenoff and He 1999). Species with high establishment coefficients have higher probabilities of establishment, and the establishment coefficient for a given species may vary from one ecological land type to another. These coefficients are derived by using the simulation results from a gap model (He et al. 1999), such as LINKAGES (Pastor and Post 1985; Post and Pastor 1996), or using estimates based on existing experimental or empirical studies (Shifley et al. 2000; Kabrick et al., 2002; Shifley and Kabrick, 2002).

An expanded forest harvesting module is a recent addition to LANDIS (Gustafson et al. 2000). This module has spatial and temporal components, and can remove age cohorts to simulate removal by harvesting. The spatial component controls how harvesting is affected by stand and management unit boundaries and by adjacency constraints, whereas the temporal component allows the simulation of iterative harvesting rotations and multiple-entry silvicultural treatments (e.g., shelterwood or periodic group-selection harvests within stands). The species-age cohort component allows specification of the species and age cohorts that will be removed by each harvesting activity (e.g., clearcutting, selection cutting, shelterwood cutting). Most harvesting prescriptions can be simulated by various combinations of these three components (Gustafson et al. 2000).

Model Parameterization

LANDIS was initially developed and applied in the northern Lake States of the United States. It was subsequently parameterized and applied in the Ozark Highlands (Shifley et al. 1997, 2000). Parameterization for a new ecoregion is a significant effort that involves compilation of existing site- and ecosystem-level data to derive patterns of wind and fire disturbance and to create map layers that include the forest cover type, age classes, species composition, ecological land types, and management units. Simulated patterns of succession were derived and evaluated by using inventory data from relatively undisturbed stands and in consultation with local experts.

We classified overstory species into four groups: white oak, black oak-red oak species assemblies (henceforth referred to simply as the *black oak group*), shortleaf pine, and sugar maple. Life-history attributes for each species group (see table 10.1) were parameterized in previous studies (Shifley et al. 1997, 1999, 2000). Initial landscape conditions were derived from spatial and tabular data maintained by the Mark Twain National Forest. Each $30 \times 30\text{-m}^2$ cell in the landscape was associated with a mapped stand (i.e., a management unit) and an ecological land type. These land types had been previously derived, based on information on soils, geology, landforms, and other environmental factors (Miller 1981). Land types were aggregated into seven categories: southwest slopes, northeast slopes, ridges or flat uplands, upland drainages, mesic coves or bottoms, sites with surficial limestone, and savannas and glades.

Stand age and forest type were known from a prior inventory. The combination of stand age and forest cover type was used to assign species and age classes to each site, based on the probabilities of occurrence observed from nearby sites for which species-level inventories had been previously conducted (Hansen et al. 1992; Shifley and Brookshire 2000; Miles 2001).

Species-establishment coefficients that quantify relative reproductive success by species and ecological land type (Nigh 1997; Nigh et al. 2000) were determined iteratively, based on the relative species abundance in undisturbed old-growth stands and relatively undisturbed mature second-growth stands. Coefficients were iteratively adjusted so that when the landscape started with an arbitrary species composition, the model's long-term species composition in the absence of disturbance would reach an equilibrium characterized by the approximate proportions of old-growth remnant stands and mature, undisturbed second-growth forests observed in the region.

Current fire regimes in the presence of fire suppression differ for each land type, with mean fire-return intervals varying from 300 to 415 yr; the exception is savanna, which has a much shorter mean fire-return interval (Westin 1992; Shifley et al. 1997, 1999). Records of the natural fire regime of this area are numerous, and studies of the fire history prior to European settlement show a very short mean fire-return interval (<10 yr) and low-intensity fires (Guyette 1995; Guyette and Dey 1997). These mean fire-return intervals are short because they were strongly influenced by the frequent burning conducted by Native Americans. Mean fire-return intervals in this region also have high temporal variation and variation among land types (Guyette 1995).

In this study, we used a maximum value of 110 yr for the mean fire-return interval for most land types in the simulations. This estimate is probably higher than the actual reported values. However, because many historical fires represented frequent, low-intensity fires that mainly affected grass or shrub species that are not simulated in the LANDIS model, this number is nonetheless comparable with the maximum mean fire-return interval used in this and other central Missouri hardwood areas (Shifley et al. 1997).

Simulation Scenarios and Data Analysis

We examined four simulation scenarios that contrasted two levels of timber harvesting—none (N) and uneven-aged harvesting by means of group selection (H)—and two levels of fire disturbance—suppressed (S) and frequent (F). This resulted in four simulated combinations (NS, NF, HS, and HF, in order of increasing disturbance) that illustrate and compare the outcome of widely differing disturbance regimes. The two intermediate disturbance regimes (NF and HS) represent disturbed landscapes dominated by fire and timber harvesting, respectively. The other disturbance regimes simulate the minimum (NS) and maximum (HF) levels of landscape disturbance. Thus, a comparison of the NS and NF scenarios reveals the effects of fire in the absence of harvesting (question 2 above). Comparison of the HS and NS scenarios reveals the effects of harvesting while minimizing the influence of fire (question 3). The NF and HS scenarios approximate the effects of natural fire disturbance and current anthropogenic influences (harvesting combined with fire suppression) on this landscape, respectively. Thus, comparing these scenarios reveals the effects of fire suppression in this landscape (question 1) and whether the landscape will diverge from its "natural" successional paths under current harvesting and fire regimes (question 4).

All scenarios were simulated for 300 yr within a 70,892-ha subset of the Mark Twain National Forest in the Ozark Highlands of southeastern Missouri; simulation values are provided in table 10.2. In each scenario, the model was run 10 times. A 300-yr time span was chosen as a compromise that allowed us to incorporate the effects of both fire and harvesting. Simulating fire requires a long time span (centuries) to capture repetition of the fire regimes, whereas short-term (decades) harvesting plans and decisions may become less meaningful over long time spans, even though harvesting may have long-

TABLE 10.2. Forest Harvesting Regime Used in Scenarios Involving Harvesting

Category	Value
Harvesting method	Group selection
Minimum stand age (yr)	0
Stand ranking method	Oldest first
Initial harvesting decade (yr)	10
Harvesting interval (yr)	10
Final harvesting decade (yr)	300
Harvest target (proportion of stands, %[1])	100
Stand proportion (%)	10
Mean group size (cells) [2]	2
Standard deviation (cells)	0.33

[1]Proportion of forest stands in the study area subject to harvest.

[2]Cells are 30×30 m^2 (0.09 ha).

term effects on species composition and spatial patterns (He et al. 2002). The landscape was represented by 30×30-m^2 cells (i.e., was rasterized). For each decade of a simulation, the attributes of each landscape cell were used to create maps of stand age and species composition and derive landscape statistics. We derived species trajectories (changes in the percentage of the landscape occupied by a species) over the 300-yr simulation and calculated an aggregation index (He et al. 2000) to quantify the spatial pattern of the species distribution. This aggregation index represents a class-specific landscape metric that allows a comparison of the degree of aggregation among different classes. The index has a value ranging between 0 and 1, with 0 indicating the lowest and 1 indicating the highest aggregation of a class (He et al. 2000).

Results

The disturbance processes of fire and timber harvesting alter the age structure and species composition of the forest landscape over time. The following sections compare the simulation results attributable to different fire and harvesting regimes and their combinations.

Size, Frequency, and Intensity of Fire Disturbance

The simulation results showed distinct differences between harvesting with fire suppression and the absence of harvesting but with frequent fires. With fire suppression, the simulated mean

fire size was 4.2 ha, and individual fires ranged from 0.09 to 81.2 ha in extent. The mean number of fires that occurred each year was low (7.2). With more frequent fires in the absence of fire suppression, the simulated mean fire size was 3.3 ha (smaller than in the fire-suppression regime), but the simulated maximum fire size reached 169.9 ha, more than twice the extent of the largest fire under the fire-suppression regime. The regime with frequent fires averaged more than 90 fires per year. This introduced greater variability into the simulated landscape than was the case with fire suppression. Under both fire regimes, small fires occurred more frequently than large fires. The cumulative areas burned for each decade illustrated that fires in the simulated regime with frequent fires burned significantly more area per decade than under the fire-suppression regime (figure 10.3). The mean fire-return intervals in the simulations were within 11% of the targeted mean fire-return intervals (110 yr for the regime with frequent fires and 460 yr for the fire-suppression regime). The discrepancy between simulated and target return intervals is acceptable, given the stochastic nature of fire (He and Mladenoff 1999a).

Under the fire-suppression regime, the total areas burned remained relatively constant throughout 300 yr of simulation (figure 10.3). Fire suppression did not lead to large catastrophic fires, at least for the duration we simulated. This fire regime is unlike those in northern hardwood forests, where forest fuel is expected to eventually build up sufficiently to support large and catastrophic fires (He and Mladenoff 1999a). However, in contrast with the scenario with no fire suppression, in which most simulated fires had low to moderate intensities (classes 1–3), fires simulated in the presence of fire suppression had predominantly high intensities (class 5). The total area burned per decade under the regime with frequent fires stabilized after about four decades (figure 10.3).

Fires in LANDIS are simulated in five intensity classes, based on the estimated fuel load at each site and on the site's fire history. For a xeric land type, the fuel decomposition rate is low and simulated fires tend to be more frequent than for more mesic land types. The interaction between fire and species can be examined in a spatially explicit manner, and the simulated results reflect the complex interactions among fire, species, the vegetation's age structure (significant because young trees are more likely to be killed by fire), and the environment (land type boundaries and

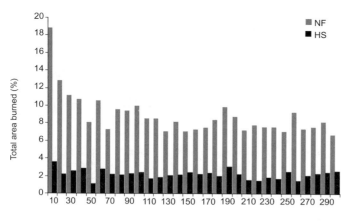

FIGURE 10.3. Cumulative fire disturbance (percentage of total area burned) over time under disturbance regimes with fire suppression (HS) and with frequent fires (NF).

conditions). An example of the simulation results at year 80 under the regime with frequent fires (color plate 12) illustrates the patterning that results from the stochastic nature of fire as it interacts with the species, age classes, and times since the last fire for each cell. In this example, only four different fire severity classes were present in the simulation results, due to the different fuel loads and fire histories for the individual sites. These fires interact explicitly with the tree species on the affected sites. At year 70 of the simulation, a small portion of the black oak group present at the beginning of the simulation has survived the frequent but low-intensity simulated fires that occurred in earlier decades and is now more than 140 yr old. However, high-intensity (class 4) fires in simulation year 80 removed many old cohorts of the black oak group, and the mean age class of this group across the landscape decreased substantially (illustrated by the color changes in the mapped landscape in

color plate 12). Many older cohorts in the black oak group were replaced by newly established cohorts in the 0–10-yr age class. Similar fire-related interactions occurred for all other species.

Effects of Fire Regime on Species Dynamics

To examine the effects of fire disturbance on species dynamics, we compared a scenario with fire suppression to one with frequent fires. The proportion of the landscape occupied by stands in the white oak group increased from approximately 40% to 70% under both regimes, but the increase was more rapid with frequent fires (figure 10.4a). In the absence of disturbance, the natural successional dynamic in this region is for the white oak group to persist and gradually dominate forests when the faster-growing but shorter-lived trees in the black oak group reach maturity and die. However, fire disturbance also favors trees in the white oak group, because trees in the black oak group are more susceptible to

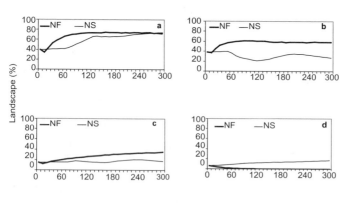

FIGURE 10.4. Effect of fire on the species dynamics of the white oak group (a), the black oak group (b), shortleaf pine (c), and the sugar maple group (d) in scenarios with no harvesting in the presence of fire suppression (NS) and no harvesting with frequent fires (NF).

fire-caused mortality. The proportion of the total area occupied by the white oak group in the fire suppression scenario remains relatively low (40%) for the first 60 yr and gradually increases to about 70% at year 130. As shown in figure 10.4b, the black oak group decreases substantially in abundance in the absence of fire. With frequent fires, cover by the white oak group increased from 40% of the landscape at year 0 to about 60% by year 60. It took about 100 yr for the white oak group to cover 70% of the landscape and become relatively stable for the remaining 200 yr of the simulation (figure 10.4a).

Fire is also a critical component in maintaining shortleaf pine in this landscape. Shortleaf pine increases in abundance gradually in a scenario with frequent fires and no timber harvesting. In contrast, frequent fires nearly eliminate the sugar maple group from this landscape within 80 yr. Fire suppression favors the establishment and survival of shade-tolerant maples in the understory, allowing them to occupy up to 20% of the sites by the end of the simulation period (figure 10.4d).

Effects of Forest Harvesting on Species Dynamics

We compared scenarios with and without timber harvesting to examine the differential effects of harvesting on species composition over time. A fire-suppression regime was used in conjunction with both scenarios to minimize any fire effects that might overshadow the effects of harvesting. Forest harvesting was simulated as group selection, which created small openings (one to a few pixels in size) covering approximately 10% of the area of each stand each decade. This harvesting regime was thus equivalent to a 100-yr harvesting rotation.

Simulated harvesting disturbance had the biggest impact on the proportions of the landscape occupied by the black oak group (figure 10.5b). Whereas the area occupied by this group decreased in the scenario with fire suppression but no harvesting, it increased from 40% of the landscape at year 0 to about 60% of the landscape at year 300 in the presence of harvesting and fire suppression. Shortleaf pine also increased in area by about 8% by year 300 in the latter scenario (figure 10.5c). In the absence of harvesting, a large proportion of the trees in the black oak group survive until their age of senescence and are replaced by longer-lived trees in the white oak or shade-tolerant maple groups. Under a regime with substantially more fire disturbance, the simulated increase in the sugar maple group (figure 10.5d) would be reduced or reversed. For the white oak group—the most common and dominant species—harvesting showed the least effect. Although small differences in the proportion of the white oak group occurred during the early years of the simulation, by year 300 (figure 10.5a), the size of the white oak group did not differ between scenarios. In general, our results suggest that forest harvesting influences species abundance in the same fashion as fire, but to a lesser degree.

Successional Dynamics in the Scenarios with Frequent Fires and with Fire Suppression

As shown in the last two sections, neither fire nor harvesting has a significant effect on the abundance of the white oak group within this landscape. In the scenario with harvesting and fire suppression, which approximates the current fire and anthropogenic disturbance regime, the white oak group also increased in abundance, but more slowly than in simulations with frequent fires.

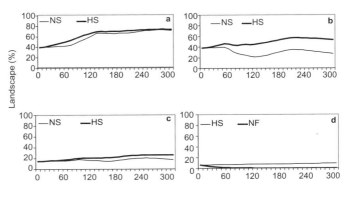

FIGURE 10.5. Effect of harvesting on the species dynamics of the white oak group (a), the black oak group (b), shortleaf pine (c), and the sugar maple group (d) in scenarios with fire suppression and no harvesting (NS) and with harvesting plus fire suppression (HS).

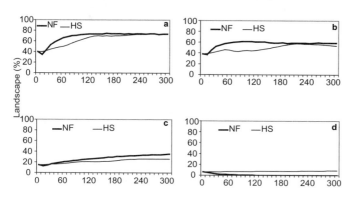

FIGURE 10.6. Dynamics of the white oak group (a), the black oak group (b), shortleaf pine (c), and the sugar maple group (d) under two moderate disturbance regimes: one with no harvesting but with frequent fires (NF) and one with harvesting combined with fire suppression (HS).

After 200 yr of simulation, the proportion of sites occupied by the white oak group was nearly identical in both scenarios (figure 10.6a).

As shown previously, both fire and harvesting favor the black oak group. The simulated effects of fire suppression and harvesting tend to cancel each other out. Thus, the proportion of the landscape occupied by the black oak group converges to approximately the same level in both scenarios after about 200 yr of simulation (figure 10.6b). With frequent fire but no timber harvesting, the black oak group responds in a similar fashion to that of the white oak group by increasing its presence from 40% of the landscape to more than 50% by year 40 and to about 60% by year 70, after which it remains relatively stable at 55–60% for the remainder of the simulation (figure 10.6b). With timber harvesting, the simulated increase in the black oak group in the landscape is slower. Cover by this group does not reach 50% of the landscape until after 160 yr of simulation.

Shortleaf pine was limited by the relatively few seed sources available in the landscape. The current area of shortleaf pine is much smaller than its historical level. In both scenarios, shortleaf pine showed a steady and substantial recovery throughout the 300-yr simulation. This increase was slightly faster in the scenario with frequent fires than in the scenario with timber harvesting, and the pine's presence reached a higher level too (about 40% and 30%, respectively).

The sugar maple group, being fire intolerant, initially occupied about 8% of the landscape. After five decades in the scenario with frequent fires, this group occupied only about 2% of the landscape. Under the fire-suppression regime with harvesting, the group remained steady at 8–10% of the landscape (figure 10.6d).

Dynamics of Age Structures and Spatial Patterns in the Scenarios with Frequent Fires and Fire Suppression

In this section, we further examine whether forest ecosystems under the current harvesting and fire regime eventually diverge from the central Missouri hardwood ecosystems under the regime with frequent fires. We compare the results from two scenarios: frequent fires with no harvesting, and harvesting with fire suppression.

The spatial and temporal distribution of the white oak group's age classes differed substantially over time when the primary disturbance was timber harvesting or frequent fires. At year 0 of the simulation, the landscape was generally young, with about 20% of the sites occupied by the white oak group less than 20 yr old, 30% in the 20–30-yr age class, and 50% in the 60–70-yr age class (color plate 13). By year 100, some of the white oak group under the harvesting disturbance regime had reached the 100–140-yr and 170–180-yr age groups, but large areas had also regenerated and were less than 40 yr old. In contrast, frequent fires with no harvesting created some large patches of young stands in the white oak group, and minimally disturbed areas grew into the older age classes (color plate 13).

By year 200, trees in the white oak group were much younger in the harvested landscape than in the landscape produced by frequent fires. With harvesting combined with fire suppression, 40% of the white oak group was less than 40 yr old, 30% was between 50 and 90 yr old, and 30% was older than 90 yr. With frequent fires, however, more than 70% of the white oak group was older than 100 yr (color plate 13). By year 300, frequent fires gave rise to consistently older trees in the white oak group than in the scenario with har-

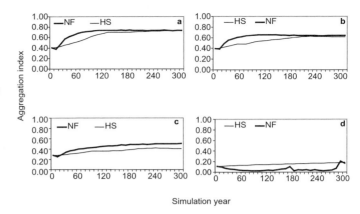

FIGURE 10.7. The aggregation index for the white oak group (a), the black oak group (b), shortleaf pine (c), and the sugar maple group (d) under two moderate disturbance regimes: one with no harvesting and frequent fires (NF) and one with harvesting and fire suppression (HS).

Simulation year

vesting. The spatial pattern of age structures in the white oak group was more homogeneous with frequent fires than with harvesting.

Measurement of the aggregation index for all four species groups indicated an increase in spatial aggregation over time for the white oak and black oak groups, as well as for shortleaf pine, in both scenarios (figure 10.7a–c). Group-selection harvesting tended to slow the rate of aggregation for oaks over time. The reverse was true for shortleaf pine and the sugar maple group, the two species groups that were initially much less abundant within the landscape. A lack of fire decreased aggregation for the sugar maple group (figure 10.7d), whose overall abundance within the landscape also decreased (figure 10.6). In general, the aggregation index was strongly correlated with the species' percentage of the total area. Under both disturbance regimes, the most abundant species (the white oak group) showed a high degree of aggregation, followed by the black oak group and shortleaf pine.

Discussion

Implications of Fire Suppression and Species Responses in Different Scenarios

The simulated results of fire suppression demonstrated the effects of an altered fire regime. The modern regime (which combines fire suppression with harvesting) tends to have fewer but higher-intensity fires than does the historical regime, with frequent fires. However, large, stand-replacing fires, which could profoundly alter landscape patterns and succession trajectories, were not simulated for this landscape in the fire suppression scenario. This probably resulted from the relatively fast fuel decomposition in the mesic land types of the central Missouri hardwood ecosystem.

Comparison of the results in the fire suppression and harvesting scenario with those of the scenario with frequent fires but no harvesting does not suggest that the simulated landscape will diverge from its natural successional trajectories. This result may be due to the following reasons:

- The lack of stand-replacing fires in the scenario with harvesting and fire suppression;
- Other disturbance agents (e.g., insects, diseases) were not simulated, and various disturbances can lead to more severe or more frequent fires; and
- The regime with frequent fires only reflects a moderate disturbance scenario (with a 110-yr mean fire-return interval), and this may not capture the catastrophic effects of natural fires.

Although species composition and distribution did not differ substantially in the two scenarios, the scenario with harvesting and fire suppression did produce a younger and less aggregated landscape than that produced in the scenario with frequent fires and no harvesting.

The results from the four simulation scenarios suggest that species respond independently to fire and harvesting within the LANDIS model. In all scenarios, the white oak group increased to cover approximately 80% of the landscape by the end of the simulation (year 300), although the temporal pattern of this increase varied among disturbance scenarios. In contrast, the response of the black oak and sugar maple groups depended much more on the particular disturbance scenario. Fire and harvesting increased the abundance of the black oak group at the expense of the sugar maple group, but the impact of frequent fire was greater than that of timber harvesting. Lack of disturbance favored the shade-tolerant, fire-intolerant sugar maple group at the expense of all other species.

Moderately disturbed landscapes, whether the disturbance was caused by fire or timber harvesting, maintained a greater proportion of the black oak group (figure 10.6b) than in landscapes with neither disturbance (figures 10.4 and 10.5). For the scenario without any disturbance, the proportion of the black oak group in the landscape decreased. With periodic disturbance by either fire or timber harvesting, the proportion of sites occupied by the black oak group reached an equilibrium at approximately 60% of the landscape after 180 yr. The combination of timber harvesting and fire suppression increased the relative proportion of the sugar maple group. Although this group never occupied more than 10% of the simulated landscape, it was far more abundant when fires were suppressed than when fires were frequent. The absence of fire overshadowed the effect of timber harvesting, which generally tended to reduce the abundance of the sugar maple group (figure 10.5).

The results also showed that the recovery of species from initially low abundance is limited by the available seed sources. Shortleaf pine never reached an equilibrium during the 300-yr simulation in any scenario. This was due to the comparatively low number of seed sources for this species in the simulated landscape. Although the real landscape historically (prior to 1890) had more shortleaf pine, and the simulated landscape could support more shortleaf pine, the simulated rate of increase for shortleaf pine was limited by its relatively low initial abundance in the landscape and by the correspondingly limited potential for seed dispersal. Similar results were found for the black oak group, whose trajectory in the scenario with harvesting did not catch up with the trajectory in the scenario with frequent

fires until year 200. It appears that once the species distribution reaches a certain threshold (e.g., 35% for shortleaf pine and 50% for the black oak group), the limitation imposed by seed sources is alleviated. This result is consistent with our findings for eastern hemlock (*Tsuga canadensis* [L.] Carr.) and other coniferous species in northern Wisconsin (He et al. 1999).

Implications for Landscape-Scale Forest Management

Although such moderate disturbance regimes as timber harvesting with fire suppression or frequent fires in the absence of timber harvesting can result in similar aggregate long-term species dynamics in terms of the percentage of the total area occupied by each species (see, e.g., figure 10.6), examination of landscape patterns over time reveals differences in patch sizes and age-class aggregation. These differences can only be discerned by using spatially explicit models, and may be important for some objectives, such as management for certain wildlife species or for maintaining species diversity. Simulation results for the uneven-aged harvesting provide a different spatial diversity of age classes across the landscape than would be expected in a regime with frequent fires. Natural fires vary greatly in terms of size and frequency. Differences in age structure may play a role in the value of timber harvested as well as in the value of the habitats required for a specific wildlife species. Multiple harvesting regimes used in combination can be simulated by using this methodology, and such simulations could produce greater variations in both harvest size and in the age-cohort distribution for the species than were simulated in the present study. Spatially explicit modeling systems such as LANDIS provide an opportunity to readily explore the consequences of a large number of alternatives.

CHAPTER 11

Using Insect-Caused Patterns of Disturbance in Northern New Brunswick to Inform Forest Management

KEVIN B. PORTER, BRENDAN HEMENS, and DAVID A. MACLEAN

The move toward ecosystem management that developed in the 1980s and 1990s incorporated a strong focus on the conservation of biodiversity (Grumbine 1994; Salwasser 1994). The *natural disturbance paradigm,* another developing idea, can be summarized as follows: if disturbances shape the composition and dynamics of vegetation communities (Pickett and White 1985), and the character of vegetation communities defines biodiversity (Huston 1994), then forest management strategies guided by an understanding of natural disturbance processes will maintain forests with attributes that conserve biodiversity (Attiwill 1994). The basic premise behind this paradigm is that the large structural species that dominate the environment shape the habitat of the smaller interstitial species (Huston 1994). Disturbances, including management actions, control structural species, such as trees, which in turn influence the habitat of interstitial species, such as understory plants, epiphytes, insects, birds, fish, and mammals. Therefore, if the changes in forest structure created by management are guided by natural disturbance (in terms of the patterns and distributions of the components influenced by disturbance), then the conditions necessary to sustain biodiversity will be provided (see Thompson and Harestad, chapter 3, this volume). The creation of important structural features with the expectation of maintaining those species that rely on them is known as the *coarse filter* approach to the maintenance of biodiversity (Noss 1987; Hunter et al. 1988).

Eastern spruce budworm (*Choristoneura fumiferana* [Clem.]) outbreaks are a major recurring disturbance in eastern North America (Baskerville 1975; Blais 1983; Royama 1984; MacLean et

al. 2002). As such, they are believed to have an important influence on the evolved adaptations of native species. However, this budworm competes with humanity for a necessary timber resource, and is thus typically suppressed by the application of insecticides. Consequently, there is an incentive to find ways to reproduce the forest characteristics created by eastern spruce budworm outbreaks through altered harvesting methods while continuing traditional budworm control practices. MacLean (chapter 6, this volume) suggests the use of three harvest classes that would remove roughly 33, 67, and 85% of the stand, respectively, in stand types that would sustain patch replacement, partial stand-replacing, and stand-replacing mortality levels. These differing mortality levels occur in stands with differing species compositions and age classes (MacLean 1980). Establishing the proportion of the three harvest strategies to assign across the forest requires a landscape-level assessment of the effects of uncontrolled budworm outbreaks.

We are working in cooperation with J. D. Irving, Ltd., a major forestry company in New Brunswick, Canada, to develop techniques and tools to facilitate the reconciliation of biodiversity management with timber production based on a coarse-filter approach. The general objective of the case study presented in this chapter is to predict the development of various attributes of stand structure and the abundance of stand classes across a forest landscape under the influence of eastern spruce budworm outbreaks, using the Black Brook District in northern New Brunswick as a test area. The specific objectives of the case study are to (1) determine the natural budworm-caused defoliation regime for a 190,000-ha area of northern New Brunswick, (2) model the development of the forest under budworm outbreak scenarios in the absence of human intervention, and (3) characterize the resulting forest landscape and compare it with a managed landscape as a basis for guiding future forest management.

Eastern Spruce Budworm Outbreaks and Their Effects on Forests

Populations of the eastern spruce budworm reach epidemic levels in New Brunswick at intervals of about 30–40 yr (Royama 1984), producing a repetitive disturbance regime (Blais 1983). Budworm larvae feed on the new foliage of balsam fir (*Abies balsamea* [L.] Mill.) and spruce (*Picea* spp.) trees, damaging their photosynthetic factories and causing growth loss, and (after 4 or 5 yr) causing mortality in some trees (MacLean 1980). An uncontrolled budworm outbreak that began in 1910 resulted in approximately 45 million m³ of dead spruce and balsam fir, covering an area of more than 4 million ha (Tothill 1921). The death of a portion or all of the trees in a stand alters the stand's structure and its potential development; the result is a very different stand than if the outbreak had not occurred. This change affects the flow of social and ecological values derived from the stand and the forest. Outbreaks of eastern spruce budworm are thus a dominant recurring disturbance in New Brunswick, affecting both stand structure and successional development (Baskerville 1975; MacLean 1988).

Given the influence of eastern spruce budworm in the forests of New Brunswick, it is likely that native tree species have evolved adaptations to the damage caused by this insect, as well as to the quantitative and qualitative aspects of structures that budworm outbreaks create. In the absence of complete information about the habitat requirements of every native species, we have assumed that providing these same structures through management would promote the survival of native species adapted to these types of disturbance. However, previous efforts to control budworm infestations have made the identification and characterization of these structures and their abundance difficult. Society has expended tremendous resources to control the budworm; an average of 20% of the forested area of New Brunswick has been sprayed annually with insecticides between 1952 and 1978 (Eidt and Weaver 1986). Spray programs have affected stand development by removing the influence of budworm on immature stands (thinning), and

by prolonging the existence of mature stands of spruce and balsam fir (Baskerville 1975). Moreover, harvesting has removed mature stands that might otherwise have been defoliated by the budworm. By altering the availability of certain habitat types in the landscape, spraying and harvesting have obscured any reference condition for a coarse-filter approach that we could deduce from current landscape conditions.

The effect of a budworm outbreak on a particular forest results from the integration of the outbreak's effects on all the constituent stands. Budworm-caused defoliation and mortality levels vary (1) among balsam fir and spruce and among mature and immature stands (MacLean 1980), (2) as a function of the quantity of hardwoods present in the stand (Su et al. 1996), and (3) as a function of site conditions (MacLean and MacKinnon 1997). Erdle and MacLean (1999) quantified the relationship between past defoliation and the resulting growth loss and mortality, including the influence of differences in species composition and age. They incorporated these relationships into a stand-table projection model (STAMAN) that can be used to forecast the development of stands enduring different sequences of annual defoliation. In the model, annual defoliation values for successive years are converted into a cumulative defoliation value that relates to the age structure of the foliage of spruce and balsam fir (MacLean et al. 2001).

STAMAN is a core component of the Spruce Budworm Decision Support System (DSS), which quantifies the marginal benefits for timber supply as a result of spraying insecticides and scheduling harvests so as to reduce budworm-caused losses (MacLean et al. 2001). Implementation of the Spruce Budworm DSS for all forested land in New Brunswick predicted total potential timber supply losses of 83 million and 195 million m³, respectively, of spruce-balsam fir for "normal" and "severe" budworm outbreak scenarios, defined based on past outbreaks in the Canadian provinces of New Brunswick and Nova Scotia (MacLean et al. 2002).

The link between stand-level forecasts and forest development lies in the pattern of defoliation across the landscape during an outbreak. The location, extent, and severity of defoliation can vary considerably from year to year (Fleming et al. 2000), resulting in different combinations of initial stand structure and defoliation sustained, and thus the potential for many different stand-level outcomes across the forest landscape. By determining the natural defoliation pattern in the absence of human intervention, we could forecast natural development of the forest and evaluate the characteristics that differentiate the natural landscape from a managed landscape. Emphasis should be placed on those features that are important in a natural landscape, but are lacking or threatened under management, to adjust future management practices accordingly.

Methods

Study Area

The study area was the 190,000-ha Black Brook District in northwestern New Brunswick, a forest management unit owned by J. D. Irving, Ltd. (figure 11.1). Outbreaks of eastern spruce budworm, and to a lesser extent, forest fires and windthrow events, are the dominant natural disturbances in this region. J.D. Irving, Ltd., purchased the land in 1943, and has carried out timber harvesting, planting, and silvicultural operations widely since then (color plate 14a). A substantial portion of the land base has been planted after clearcutting, and many plantations have undergone a first commercial thinning. The species planted are primarily white spruce (*Picea glauca* [Moench] Voss), black spruce (*Picea mariana* [Mill.] B.S.P.), and to a lesser extent, jack pine (*Pinus banksiana* Lamb.) and Norway spruce (*Picea abies* [L.] Karst.). Partial cuts, single-tree and group selection cuts, and thinned stands also make up a substantial portion of the area. The area's extensive road network is less apparent, due to the map scale. Relatively few large stands of balsam fir or spruce exist today, although natural stands containing a mixture of spruce and balsam fir are common along streams. Color plate 14a clearly reflects the managed con-

dition of this forest. The next step in the evolution of Black Brook's forest management is to evaluate and (as appropriate) incorporate the possibility of emulating the region's natural disturbance regimes within long-term planning.

Current and Historical Forest Conditions

The current forest inventory for the Black Brook District, based on GIS technology, provides a spatially explicit database of the forest's condition in 2001. A historical benchmark was derived from aerial photographs of New Brunswick taken by the Royal Canadian Air Force in 1944–1945. We acquired all photographs of the Black Brook District at that time, and photo-interpreted and digitized them to provide a reference forest condition (color plate 14b) that includes the spatial boundary, disturbance (if any), species composition, and maturity class for each stand in 1945. The interpretation and digitizing procedures were the same as those used in the modern process. Hardwood species were grouped as tolerant or intolerant, and identification of spruce was not always possible beyond the genus level. Field data to corroborate the interpretation were not obtainable, but qualitative comparisons of the 1945 inventory with the current inventory for the same pieces of land suggested that the stand typing for 1945 had been reasonably successful.

We classified the stands in the 1945 inventory based on their attributes by using the eastern spruce budworm impact classes defined in MacLean et al. (2001). This classification indicates stand vulnerability to budworm damage, and was based on the spruce-balsam fir content of the stands, as well as on their overall species composition, stand age, and silvicultural treatment. The New Brunswick Department of Natural Resources and Energy maintains a database of approximately 25,000 temporary forest inventory sample plots as part of their Forest Development Survey Program (White 2001). The stand tables (number of trees per species and per diameter class) in this database were classified by using the budworm impact classes and were matched with stands in the 1945 forest inventory based on spruce-balsam fir content, species composition, and age class.

Defoliation History

Annual aerial survey maps of eastern spruce budworm defoliation from 1945 to the present were digitized and analyzed to characterize past disturbances and provide a spatial chronology of outbreaks that would serve as inputs for a model

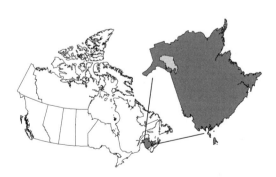

FIGURE 11.1. Location of the Black Brook District (light gray) in the province of New Brunswick (dark gray), Canada.

of stand growth. These defoliation assessments were made in mid-summer, when defoliation is most visible from an aircraft (MacLean and MacKinnon 1996). Assessments were categorical, and generally used three classes of defoliation severity: light (10–30% of the current foliage), moderate (31–70%), and severe (>70%). Because insecticide spraying to reduce defoliation is conducted earlier in the summer, defoliation assessments based on aerial surveys include the effects of the insecticide spraying, which was performed in several years in the 1950s, 1970s, and 1980s. To use the defoliation survey information for a simulation of natural disturbance, we estimated the effects of spraying on defoliation patterns. Maps showing the areas sprayed each year were digitized and digitally overlaid on the defoliation maps. This intersection of the historical records allowed us to adjust the observed defoliation for the effects of spray protection, based on annually published records of spray efficacy (e.g., Carter and Lavigne 1994; Kettela 1995). Generally, this correction moved sprayed areas into the next higher category of defoliation severity.

Because the forests of the study area have been intensively managed, it is possible that historical defoliation patterns were influenced by these management actions. The greatest change to the forest in terms of budworm susceptibility was the replacement of natural spruce-balsam fir or mixedwood stands with less vulnerable spruce plantations. However, there is no evidence that changing the species composition of a stand alters the underlying budworm population dynamics; white spruce and particularly black spruce sustain less mortality than balsam fir for a given level of defoliation (Erdle and MacLean 1999). In addition, 47% of the current plantations was established after 1987, the last year in which significant defoliation was recorded. A further 28% of the plantations was planted from 1979 to 1987, at least in part to replace stands salvaged after sustaining severe budworm damage.

The adjusted defoliation maps were digitally overlaid with the 1945 forest inventory to determine the individual annual defoliation history from 1945 to 2001 for each stand. Defoliation estimates were expressed as the percentage loss of current-year foliage. However, tree growth and mortality are more closely linked to *cumulative* defoliation, which integrates the sequence of annual foliage losses (Ostaff and MacLean 1995). Average cumulative defoliation for the period was calculated for each stand, based on weighting by foliage age (MacLean et al. 2001). Cumu-

lative defoliation values were calculated for each of the two successive budworm outbreaks (in the 1950s and in the 1970s–1980s).

Simulation of Stand Development and Budworm Impacts

The STAMAN model is designed to project stand tables, and has been calibrated for use in New Brunswick (Vanguard Forest Management Services 1993). These stand tables (expressed per hectare) serve as inputs for the model, which then applies diameter growth and mortality functions to these data to forecast stand development at 5-yr intervals. The observed relationships of defoliation by eastern spruce budworm to growth and mortality have been incorporated into the model, allowing stand forecasts to be made for user-defined defoliation sequences (Erdle and MacLean 1999).

The stands in the 1945 inventory were grouped, based on 530 observed combinations of 44 budworm impact classes (MacLean et al. 2001) and 43 different cumulative defoliation sequences in the 1950s outbreak. The budworm impact classification was also applied to 657 stand tables generated from Forest Development Survey Program data. For each of the 530 observed combinations of budworm impact class and defoliation sequence, the associated stand tables and cumulative defoliation sequences were used as inputs for STAMAN, which forecast new stand tables representing the postoutbreak stand conditions at the end of the 1950s and at the end of the 1970s–1980s outbreak. An average of 16 different initial stand tables (with a range of 4–46) was used to simulate the development of each class of stands, and then the results were averaged to obtain class-level outcomes.

At the end of the 1950s simulations, simulated stand characteristics were averaged across budworm impact classes, and then examined to determine the residual basal area and species composition for each class. Provincial data from permanent sample plots (Porter et al. 2001) were used to determine the average regeneration that would result from similar combinations of basal area and species composition, and this regeneration (number of trees per species and diameter class) was added to the simulated stand tables to represent ingrowth. The budworm impact classification was then applied to the results to obtain the likely stand composition resulting from the 1950s budworm outbreak. STAMAN simulations were then re-initiated for the second outbreak to simulate stand development until 2001.

The attributes of the resulting stand tables were examined for changes in species composition and stand age, then reclassified according to the same budworm impact classes used at the beginning of the simulation. Having compiled our results in a database, we then compared the 1945 stand conditions with the simulated 2001 conditions, and predicted the condition of the 1945 stands in 2001, based on the simulation results. From the results of this comparison, we prepared a new map layer in GIS to represent the 2001 landscape. Adjacent stands in this new map layer were merged if they belonged to the same age and budworm impact classes, so that stand boundaries represented areas of forest with homogeneous conditions in the simulated 2001 forest landscape.

The effect of protecting the forest by spraying insecticides was also simulated to contrast the resulting conditions with those in a simulation that represented natural disturbance by eastern spruce budworm. We used the target defoliation limit of 40% specified for balsam fir by the New Brunswick Department of Natural Resources and Energy (Carter and Lavigne 1994). The simulations for the budworm outbreak periods during the 1950s and the 1970s–1980s outbreaks were repeated, but with defoliation values capped at a maximum of 40%. Five defoliation scenarios were modeled: undefoliated, the defoliation sequence in the 1950s budworm outbreak with and without insecticide spraying, and the defoliation sequence in the 1970s–1980s budworm outbreak with and without spraying.

We also ran several simulations of a generalized "severe" budworm outbreak scenario defined by MacLean et al. (2001), based on an uncontrolled budworm outbreak in Nova Scotia during the 1970s and 1980s. The descriptions of stand structure produced by these STAMAN simulations were imported into the Stand Visualization System (SVS) software (McGaughey 1997) to create images of stand structures in the presence and absence of a severe budworm outbreak. For these visualizations, a simulation length of 15 yr was used, capturing a representative sequence of 10 yr of moderate-to-severe defoliation, followed by 5 yr of nil-to-light defoliation (MacLean et al. 2001). Using local volume tables, yield curves for spruce-balsam fir volumes in the five scenarios were also generated from STAMAN simulations.

To display details normally lost at smaller map scales, the results were compiled for a 5 × 6-km² subset of the Black Brook District, in addition to the results for the whole simulated forest. Results for the subset were tabulated and displayed as maps.

Results

Eastern Spruce Budworm Outbreak Patterns

The annual area of budworm defoliation for the Black Brook District from 1945 to 1995 is presented in figure 11.2a, using the following defoliation classes: light (10–30% of the current foliage), moderate (31–70%), and severe (>70%). Two discrete outbreaks occurred during this time period: from 1949 to 1958 and from 1971 to 1987.

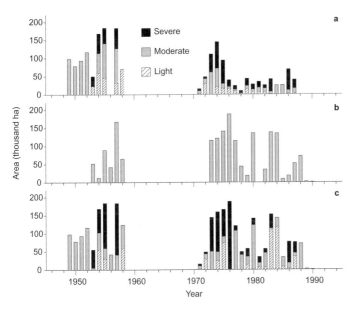

FIGURE 11.2. (a) Historical area of light (10–30%), moderate (31–70%), and severe (>70%) defoliation by eastern spruce budworm in the Black Brook District from 1945 to 1995. (b) The area protected from defoliation by insecticide spraying during the same time period. (c) Estimated defoliation levels in the absence of insecticide spraying.

Up to 97% of the 190,000-ha land base was defoliated in 1955 and 1957. The area sprayed with insecticide for the same period is shown in figure 11.2b. The aggressive insecticide protection programs in the 1970s and 1980s substantially lowered observed levels of defoliation during these decades. In the eight heaviest years of defoliation during these outbreaks, more than half the land base was protected, and nearly the entire forest was protected in 1976.

The estimated area of defoliation, adjusted to represent the absence of insecticide spraying, is shown in figure 11.2c. This figure represents the results of the natural budworm outbreak regime, which corresponds to the condition of the forests in the absence of human influence on the outbreak. The difference between figures 11.2a and c shows the effectiveness of insecticide spraying in reducing defoliation in the Black Brook District.

Effects of Eastern Spruce Budworm Outbreaks on Stand Structure

Changes in the simulated stand structure in response to a severe budworm outbreak are shown for three representative stand types in the Black Brook District in figure 11.3. The annual defoliation sequence was based on observed budworm population cycles in stands that showed severe mortality (MacLean et al. 2001). These simulations show a 92% average reduction in the numbers of balsam fir in most diameter classes in a mature balsam fir stand (figure 11.3a,b); a 92% reduction in balsam fir and a 74% reduction in spruce in a mature spruce-balsam fir stand (figure 11.3c,d); and an 80% reduction in balsam fir in a mature balsam fir-hardwood stand (figure 11.3e,f).

The resulting postoutbreak stands have very different species and diameter compositions than

they would have had in the absence of the budworm outbreak (figure 11.3). The mature balsam fir stand would be composed of 69% balsam fir and 31% hardwood, but a severe budworm outbreak changed this to 15% balsam fir and 85% hardwood. The mature spruce-balsam fir stand would consist of 23% balsam fir, 52% spruce, and 19% hardwood, versus 5% balsam fir, 34% spruce and 47% hardwood following the outbreak. The mature balsam fir-hardwood stand was made up of 59% balsam fir and 41% hardwoods, versus 22% balsam fir and 78% hardwood following the outbreak.

Figure 11.3 clearly demonstrates the magnitude of the changes in stand structure caused by budworm outbreaks. However, in many cases, considerable balsam fir advanced regeneration exists in the stands, and these trees survive budworm outbreaks and develop into the succeeding stand (Baskerville 1975; MacLean 1988). Because of its small diameter, this regeneration would not appear in figure 11.3. It seems likely, given the presence of young balsam fir, that the increased importance of hardwoods may be a temporary phase of stand development. This is particularly true given the high level of shade tolerance of this species in its immature stages. Therefore, stand species composition may well vary temporally after budworm damage.

An alternative visualization of the effects of a budworm outbreak in a mature balsam fir stand, created using the SVS software (McGaughey 1997), is shown in figure 11.4. Figure 11.4a corresponds to the mature balsam fir stand in figure 11.3a, whereas figure 11.4b depicts the condition after the budworm outbreak (figure 11.3b). Mortality of balsam fir yielded a stand dominated by hardwoods after the outbreak, with substan-

FIGURE 11.3. Diameter distributions simulated using STAMAN, with and without eastern spruce budworm outbreaks, for balsam fir, spruce-balsam fir, and balsam fir-hardwood stands. The three stand types were simulated for 15 yr (a, c, e) with no budworm outbreak, and (b, d, f) in the presence of a severe budworm outbreak. The latter scenario consisted of 10 yr of moderate-to-severe defoliation followed by 5 yr of nil-to-light defoliation. From MacLean et al. (2001).

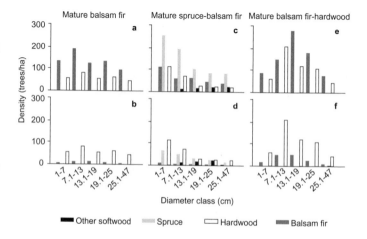

a   b

FIGURE 11.4. Visualization of a simulated mature balsam fir stand using the SVS software (McGaughey 1997) (a) without an eastern spruce budworm outbreak, and (b) 15 yr after a severe budworm outbreak. The coniferous tree images represent balsam fir; the deciduous trees are red maple (*Acer rubrum* L.).

tial coarse woody debris (budworm-caused mortality) still evident on the forest floor (figure 11.4b). Overall stand density declined from 1787 stems·ha^{-1} to 359 stems·ha^{-1} following the budworm outbreak, and hardwood content increased from 20 to 85%. Depending upon the spatial distribution of the budworm-killed trees, single-tree gaps or patches (holes) are created in the stand. Baskerville and MacLean (1979) noted a contagious distribution of budworm-caused tree mortality, in which the presence of a dead tree increased the likelihood of observing mortality of adjacent trees. The quasirealistic data visualization produced by the SVS software complements the graphs of density and size class in figure 11.3 by providing a more instantly recognizable view of the simulation results.

Simulated volume development of mature spruce, balsam fir, balsam fir-hardwood, and spruce-hardwood stands is presented in figure 11.5 for five defoliation scenarios. As expected, volume loss increased from the shorter, less severe 1950s outbreak to the more severe outbreak in the 1970s–1980s, and from the scenarios with insecticide spraying to the unprotected scenarios. The budworm outbreak during the 1970s–1980s resulted in a 30–45% reduction in the volumes of spruce-balsam fir stands (figure 11.5). Studies of volume loss in response to severe defoliation have observed larger volume loss in pure stands of balsam fir (MacLean 1980). However, the defoliation sequences that occurred in the Black Brook District during the simulation period were generally less severe, resulting in a lesser decline. For example, the defoliation scenarios in the absence of spraying in the 1950s and in the 1970s–1980s that were used in figure 11.5 resulted in cumulative defoliation of 30–50% in years 1–5 and of 70% in years 6–10; this contrasts with the

respective values of 67 and 85% for the severe outbreak scenario of MacLean et al. (2001) that was used in figures 11.3 and 11.4.

In simulations for spruce stands older than 100 yr (figure 11.5a), the volume was 24% less in unprotected stands during the 1970s–1980s budworm outbreak than in undefoliated stands 15 yr after the start of the outbreak; however, the stand structure remained largely intact, with substantial volume remaining for tree ages of up to 200 yr. Balsam fir stands older than 80 yr (figure 11.5b) that sustained the budworm outbreak in the absence of spraying in the 1970s and 1980s had 34% less volume than undefoliated stands 15 yr after the start of the outbreak, but experienced a precipitous decline in volume regardless of defoliation level, because of the normal stand breakup that occurs in stands of the relatively short-lived balsam fir. The decline was similar in balsam fir-hardwood stands (figure 11.5c), with lesser volumes throughout, because of lower balsam fir content, whereas the pattern in spruce-hardwood stands (figure 11.5d) resembled that in the spruce stands (figure 11.5a).

Effects on Landscape Composition

The 1945 forest landscape was dominated by stands of mixed spruce-balsam fir, balsam fir, and

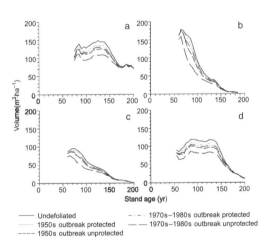

FIGURE 11.5. Comparisons of predicted spruce-balsam fir volumes in four stand types: (a) spruce older than 100 yr, (b) balsam fir older than 80 yr, (c) balsam-fir hardwood older than 80 yr, and (d) spruce-hardwood older than 100 yr. Five defoliation conditions were considered: undefoliated, protected by spraying and unprotected through the 1950s eastern spruce budworm outbreak, and protected by spraying and unprotected through the budworm outbreak in the 1970s–1980s.

spruce (color plate 15). In the 3400-ha area shown, there were 19 ha of pure balsam fir stands, 14 ha of pure spruce stands, 886 ha of mixed spruce-balsam fir stands, 1065 ha of mixedwood stands dominated by spruce and balsam fir, and 532 ha of hardwood-softwood mixedwood or hardwood stands. The current forest (color plate 15) includes many spruce plantations, partially cut or thinned stands (largely commercially thinned plantations), and large areas of hardwoods. In color plate 15b, there are 1040 ha of spruce plantations, 690 ha of commercially thinned spruce plantations, 10 ha of other plantations, 375 ha of stands harvested by single-tree selection (hardwoods) and shelterwood cuts, 310 ha of natural mixed spruce-balsam fir, and 760 ha of hardwoods. We separated out commercially thinned plantations, because observations suggest that important differences in species composition and biodiversity arise from thinning. As the stands open up with a first thinning, and especially with a second thinning, substantial ingrowth of hardwoods and natural softwoods has been observed.

The simulated 2001 forest, based on a simulation starting with the initial 1945 forest and including the budworm defoliation scenario in figure 11.2c, virtually eliminated stands dominated by balsam fir and spruce (color plate 15) as a result of heavy budworm-caused mortality of those species. Color plate 15c shows the simulated conditions in 2001, about 12 yr after the end of the 1970s–1980s budworm outbreak, and represents the results of natural disturbance only (i.e., no insecticide spraying). At this time, the forest is dominated by spruce-balsam fir mixedwoods (47% of the area) and hardwoods (39%), with some mixed hardwood-softwood stands, but as discussed earlier, this temporal stage will probably succeed to stands with higher proportions of spruce and balsam fir.

Table 11.1 summarizes landscape-level impacts on spruce-balsam fir volumes caused by simulated budworm outbreaks in the 1950s and in the 1970s–1980s (with and without insecticide spraying) for the 5 × 6-km² study area (color plate 15). The effect of species composition and the proportion of stand types on volume losses is ap-

TABLE 11.1. Simulated Impacts on Spruce-Balsam Fir Volumes Caused by Eastern Spruce Budworm Defoliation in the 1950s and in the 1970s–1980s

Susceptible Stand Type	Area (ha)	Undefoliated Volume ($m^3 \cdot ha^{-1}$)	Undefoliated Volume (m^3)	Protected Volume[1] (m^3)	Volume Reduction (m^3) 1950s Outbreak	Volume Reduction (m^3) 1970s–1980s Outbreak	Volume Reduction (%) 1950s Outbreak	Volume Reduction (%) 1970s–1980s Outbreak
BF, >80 yr	44	160	7040	6116	1584	3168	23	45
BF-SP, >80 yr	644	133	85,652	75,348	13,524	30,268	16	35
SP-BF, >100 yr	532	101	53,732	50,008	6384	17,556	12	33
SP, >100 yr	13	132	1716	1547	221	221	13	13
BF-HW, >80 yr	304	91	27,664	24,016	4864	10,336	18	37
SP-HW, >100 yr	1181	109	128,729	112,195	20,077	37,792	16	29
HW, 40–80 yr	141	20	2820	2538	564	987	20	35
HW-SF, 40–80 yr	422	37	15,614	14,348	1688	5064	11	32
HW-SF, >80 yr	117	40	4680	4212	468	1404	10	30
Total	3422		327,647	290,328	49,374	106,796	15[2]	33

Notes: Defoliation data are from figure 11.2c and cover the years from 1945 to 2001. The study area is the projected stand development of a 5 × 6-km² area of the Black Brook District (color plate 15). Abbreviations: BF, balsam fir; SP, spruce; HW, hardwood; SF, spruce-balsam fir.

[1]Protected volumes were calculated as the undefoliated volume minus the volume reductions resulting from the outbreak in the 1970s–1980s if protection had been applied to the entire land base.

[2]Total losses (%) were obtained by comparing the total loss to the total undefoliated volume rather than by summing up the volume reduction columns.

parent. The longer budworm outbreak in the 1970s–1980s, with more severe defoliation, resulted in more than double the losses of spruce-balsam fir volume (107,000 m^3 out of a total spruce-balsam fir volume of 327,000 m^3) that occurred during the 1950s outbreak (table 11.1). Even with the simulated application of insecticides during the budworm outbreak, the forest lost about 11% of its spruce-balsam fir volume. However, without insecticide, the volume loss was much higher, amounting to 33% of the standing spruce-balsam fir volume during the budworm outbreak of the 1970s–1980s. This type of loss is likely to have a substantial ecological impact on stand structure, species composition, and succession, as well as economic effects on the timber supply.

Color plate 15 presents only 1.7% of the 190,000-ha Black Brook District, primarily to permit the display of details that would be lost at smaller map scales. Species compositions, age classes, and patch size distributions in 1945, 2001, and the simulated 2001 forest (based on the initial 1945 forest and budworm disturbances only) for the entire Black Brook District are summarized in figure 11.6. In 1945, the forest was composed of 9% balsam fir, 31% spruce, 28% spruce-balsam fir-hardwood mixedwoods, and 32% hardwood-softwood mixedwoods and hardwood stands. The 2001 forest simulated with natural disturbance only included fewer stand types, and was composed of 47% spruce-balsam fir-hardwood mixedwoods and 53% hardwood-softwood mixedwoods and hardwood stands (figure 11.6). Less than 800 ha remained of pure balsam fir and pure spruce stands. Heavy mortality of mature balsam fir was the most notable change caused by the eastern spruce budworm disturbance, but the conversion of pure spruce stands to spruce-dominated mixedwoods was also considerable.

There was less change in the age-class (maturity) distributions between 1945 and the 2001 forest simulated with natural disturbance only, and both forests were dominated by mature stands. However, although forests dominated by spruce-balsam fir mixedwoods made up 83% of the mature component in the 1945 forest, 45% of mature stands in the simulated 2001 forest were pure stands of tolerant hardwoods and 43% were spruce-dominated mixedwoods. In contrast, the actual 2001 forest clearly has a more even age-class distribution, with more stands (primarily plantations) in the regeneration and young age classes.

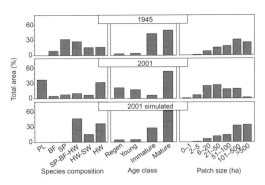

FIGURE 11.6. Species composition, age class (maturity of the dominant species in the stand, assigned during airphoto interpretation), and patch size in 1945, 2001, and simulated 2001. The simulation only considered eastern spruce budworm outbreak disturbances for the 190,000-ha Black Brook District. Abbreviations: BF, balsam fir; HW, hardwood; HW-SW, mix of hardwood and softwood; PL, plantation; SP, spruce; SP-BF-HW, mix of spruce, balsam fir and hardwood.

Comparisons of patch sizes show somewhat more large patches (>500 ha) in the simulated 2001 forest than in the 1945 forest (34% versus 26%), but the size distributions were generally similar (figure 11.6). However, the actual 2001 forest had far more small patches (especially those <50 ha) and far fewer large patches (>500 ha). The median patch size in the actual 2001 forest was 3 ha, compared with 18 and 19 ha for the 1945 and simulated 2001 forests, respectively. Average patch size increased from 46 ha in the 1945 forest to 63 ha in the simulated 2001 forest. These sizes were much larger than in the actual 2001 forest, which had an average patch size of 11 ha.

Discussion

In keeping with our goal of developing a coarse-filter strategy for the maintenance of biodiversity in the Black Brook District, we have focused our analysis on the examination of landscape-level effects of a budworm regime—changes in species composition, stand ages, stand size, and the degree of stand replacement. These characteristics have a large effect on the in-stand environment, and thus provide us with benchmarks for emulating patterns of natural disturbance that will sustain those components of biodiversity that depend on the influence of trees.

Given the historical preponderance of vulnerable balsam fir and spruce stands (color plate 14b), protection against eastern spruce budworm is clearly necessary to maintain timber production.

Replacement of natural balsam fir and spruce stands with spruce plantations (color plate 14a) is currently the primary long-term management strategy to reduce reliance on insecticides and reduce future budworm impacts (Blum and Mac-Lean 1984; MacLean 1996b). This approach has been a major element of the management of the Black Brook District by J. D. Irving, Ltd.

Simulation of the 1945 forest in response to successive budworm outbreaks in the 1950s and 1970s–1980s predicted substantial reductions in the volumes of balsam fir and spruce stands in the simulated 2001 land base. Mature balsam fir stands were largely killed and regenerated, whereas spruce stands were converted into spruce-hardwood mixedwoods. However, the reduction of overstory balsam fir after the budworm disturbances likely only persists for the length of time it takes the regenerating understory balsam fir to reach the overstory. That such periodic shortages of mature balsam fir are to be expected in a natural landscape may suggest that emulating this condition is relatively unimportant in terms of the persistence of most species. If plantations are to provide some of the characteristics and values of a budworm-influenced landscape, special attention must be paid to their patch-size distribution, and especially to the provision of larger (>500-ha) patches.

The opportunity to reconcile age-class distributions may be more elusive. Our results suggest that a clumped age distribution would result from a natural budworm disturbance regime. However, industrial users of timber demand a relatively even flow of wood, which is best supplied by a uniform distribution of stand ages. Resolving these conflicting goals depends on the scale at which a clumped age distribution is important; by zoning the forest, it may be possible to achieve the clumped distribution in individual areas while achieving a relatively uniform overall distribution. The key here may well lie in ensuring that sufficient older or multicohort stands (which result from natural disturbance in specific stand types) still occur within the landscape. In this sense, our analysis based on the broad maturity classes in figure 11.6 is overly simplistic and needs further refinement, since the mature class encompasses too broad a range of ages (e.g., see figure 11.5).

Our simulation converted mature balsam fir-dominated stands into immature stands via mortality and regeneration, a result that reflects field observations in previous studies (Baskerville 1975; MacLean 1988). However, mortality in spruce stands was often incomplete, owing largely to the reduced impact of defoliation on spruce mortality, as well as to the variable extent and severity of defoliation in subsequent years. This indicates that degrees of partial cutting are likely to be more appropriate than clearcutting in a large proportion of spruce-dominated stands, as has been proposed by MacLean (chapter 6, this volume). Clearcutting in mature balsam fir stands while protecting advanced regeneration would be consistent with our simulation results.

Our simulation results also suggest that pure and mixed stands with a hardwood component are an important feature of the natural landscape, at least intermittently. These stand types also warrant more study, because in many cases, they may be transitional stages in which budworm removes the softwood overstory and converts a mixedwood stand into a hardwood stand, followed by advanced softwood regeneration that eventually grows into the overstory. There are also cases of budworm having converted spruce stands into spruce-hardwood mixedwoods. Historically, precommercial thinning programs have favored the softwood component, converting mixedwood stands into predominantly softwood stands. To better emulate the effects of an eastern spruce budworm outbreak, a portion of future precommercial thinning efforts should aim to produce a mixedwood stand after harvesting. An associated benefit may well be reduced defoliation of balsam fir in the mixedwood stands (Su et al. 1996) and reduced timber losses. Needham et al. (1999) determined that when balsam fir stands undergo severe defoliation (85% of the current foliage was removed in their study) with no insecticide use, the balsam fir yield is maximized at a hardwood content of about 50%, whereas the optimum hardwood content shifted to 20% at moderate defoliation levels.

Note that other disturbances, albeit less important for this forest, do operate in the Black Brook District. Eastern spruce budworm outbreaks create a large volume of fine and coarse dead material that has been shown in other areas to affect the fire regime (Fleming et al. 2002), suggesting that this interaction warrants investigation for the study area. Future efforts in the larger context of our work in the Black Brook District will focus on determining the critical characteristics of other important disturbance regimes and synthesizing the interaction of these disturbances to understand their overall landscape effects.

The creation of progressively smaller patches via harvesting, combined with pressure to use ever-decreasing harvest block sizes, is a major inconsistency between current forest management and the results of our simulations. Our results used a measure of patch size based on species composition as an indicator of change. Clearly, the perception of what constitutes a patch depends on the species, and thus there are many potentially interesting definitions of patch size and landscape connectivity. An associated project is working to define the habitat needs of all vertebrate species native to the Black Brook region's forests, with the goal of using these definitions to compare the 1945 forest with the actual and simulated 2001 forests in terms of their relative abilities to support these species.

This case study is really only a starting point in the evaluation of emulating natural eastern spruce budworm disturbances. The budworm makes an ideal case study, because it is one of the few insects for which we have information and models that permit the simulation of natural disturbance at the stand and landscape levels (Erdle and MacLean 1999; MacLean et al. 2001; MacLean, chapter 6, this volume). However, important connections with other disturbances must still be studied. Future work will include more in-depth analyses of forest patterns and testing of new harvesting methods inspired by natural disturbance (eastern spruce budworm and gap replacement), including their application to the harvesting of 2600 ha of forest in the Black Brook District. We will use an extensive network of permanent sample plots to monitor a suite of specific indicators of forest structure and biodiversity before and after harvesting (MacLean, chapter 6, this volume). In addition to determining landscape-level patterns, part of the challenge is to develop harvesting prescriptions that emulate the level of mortality caused by the budworm. Once developed, implementation of these prescriptions on the ground will require information on in-stand spatial patterns of mortality, which have been shown to follow a clumped or contagious distribution (Baskerville and MacLean 1979; MacLean and Piene 1995).

Conclusions

It is noteworthy that only two of the chapters in this book (this one and MacLean, chapter 6) emphasize insect outbreaks as natural disturbances. Yet periodic eastern spruce budworm outbreaks are the most important natural disturbance in many areas of eastern Canada. Other insects, from mountain pine beetle (*Dendroctonus ponderosae*

Hopk.) in British Columbia to forest tent caterpillar (*Malacosoma disstria* Hbn.) in the Canadian Prairie provinces, have similarly important impacts on forest structure and development.

Our case study provides a methodology that links stand-level structures with forest-level defoliation patterns to forecast the types of stand structures that result from an eastern spruce budworm disturbance regime and to predict the size, location, and abundance of those structures across the landscape. If we are correct in our assumption about the dominant influence of trees on the quality of habitat for smaller interstitial organisms, then using harvesting and silviculture to create stand structures with sizes and abundances similar to those that develop after natural disturbances should help to sustain the populations of native species.

Our results suggest that both partial cutting and clearcutting of spruce and balsam fir stands have a place in harvesting methods designed to produce stand structures and forest composition consistent with those produced by budworm disturbance. Mixedwood management is clearly also a valid goal, and managers should consider avoiding the conversion of mixedwoods to pure softwood stands. Designing harvest blocks to achieve larger contiguous areas with homogeneous stand conditions also appears to be consistent with the natural disturbance regime. Overall, data on the effects of natural disturbance will provide a good basis for selecting and justifying forest composition objectives and appropriate silvicultural and harvesting treatments.

Acknowledgments

The analysis that forms the basis for this chapter was carried out within a project supported by J. D. Irving, Ltd., and by the Network Centre of Excellence Sustainable Forest Management Network. J. D. Irving, Ltd., funded the airphoto interpretation and digitizing of the 1945 aerial photographs for the entire land base specifically for this study, and the company's staff (notably Gaetan Pelletier, Charles Neveu, Joe Pelham, and Walter Emrich) provided data and much other assistance. A National Science and Engineering Research Council postgraduate scholarship supported one of us (BH). We appreciate the valuable assistance of John Henderson, Kathy Beaton, and Wayne MacKinnon in the preparation of this chapter. We are also indebted to the editors and two anonymous reviewers for valuable commentary that we used to improve the chapter.

CHAPTER 12

Using Criteria Based on the Natural Fire Regime to Evaluate Forest Management in the Oregon Coast Range of the United States

MICHAEL C. WIMBERLY, THOMAS A. SPIES, and ETSUKO NONAKA

Forest landscapes in the Oregon Coast Range have changed considerably since settlers first arrived in the mid-nineteenth century. Clearing of forests for agriculture and development, ignition of extensive forest fires by settlers and loggers, and conversion of natural forests to managed plantations have all contributed to the fragmentation of late-successional forests (Ripple et al. 2000; Wimberly et al. 2000). These rapid and widespread changes have led to concern for populations of native species, particularly those associated with old-growth forests and aquatic habitats. Some threatened or endangered species such as the northern spotted owl (*Strix occidentalis caurina* [Merriam]) have been the subject of intensive research, resulting in the development of detailed conservation plans (Thomas et al. 1990). There are many other species, however, for which comprehensive information on habitat requirements and demography is not available. Because of limited knowledge and resources, developing and implementing individual management plans for every species is not feasible. Instead, the *coarse filter* approach has been proposed as an alternative for conserving native biodiversity at the landscape scale (Noss 1987). This method entails preserving ecological diversity at the community level based on the assumption that a representative array of habitats will be sufficient to meet the needs of most species (see Thompson and Harestad, this volume, chapter 3).

Studies of landscape dynamics under natural disturbance regimes can play a critical role in the development of coarse-filter conservation strategies (Landres et al. 1999; Swetnam et al. 1999). Disturbance-based ecological assessments are grounded on the assumption that historical landscapes were subjected to both natural and human disturbances for millennia prior to the arrival of Europeans and other nonindigenous settlers. Despite these perturbations, pre-European settlement landscapes sustained the native species that we currently wish to preserve. Comparisons between present-day and pre-European settlement disturbance regimes can therefore serve as indicators of the potential for conserving species and sustaining ecosystem processes in the modern landscape. Analyses of natural disturbance regimes and the forest patterns they generate can also provide targets for future landscape restoration efforts and can suggest approaches to maintaining habitat diversity in dynamic landscapes. Because landscapes influenced by natural disturbance regimes seldom approach a steady state equilibrium, historical patterns must be characterized as a distribution of possible system states rather than as an arbitrary "snapshot" taken at a single point in time (Sprugel 1991).

Fire was a ubiquitous landscape-scale disturbance in the Pacific Northwest for thousands of years prior to the arrival of settlers (Agee 1993; see also Hessburg et al., this volume, chapter 13). Because fire has been almost entirely suppressed in the Oregon Coast Range for the past 50 yr, our ability to observe and study such fundamental processes as fire ignition, fire spread, and fuel consumption is limited. However, paleoecological, dendroecological, and historical research has yielded considerable information about the frequencies, sizes, and severities of past fires (Teensma et al. 1991; Impara 1997; Long et al. 1998; Weisberg and Swanson 2003). Therefore, we have adopted a primarily historical and empirical approach to characterizing and modeling fire regimes in this chapter. For the purposes of our study, we assumed that the "natural" fire regime is equivalent to the historical fire regime that existed prior to settlement. Fires initiated by lightning, as well as those ignited by Native Americans, are all considered to be part of this natural fire regime (Suffling and Perera, this volume, chapter 4).

The goal of this research was to reexamine current forest management practices and landscape patterns in the Oregon Coast Range by considering them in the context of the natural fire regime. Specific objectives were to (1) contrast the frequencies, sizes, and effects of historical fires with those of forest management disturbances; (2) examine differences in the abundance and pattern of seral stages between pre-European settlement and present-day forest landscapes; and (3) assess the potential for applying management based on pre-European settlement fire regimes in the Oregon Coast Range within the framework of present-day forest conditions and socioeconomic constraints.

Study Area

The Oregon Coast Range encompasses more than 23,000 km² in western Oregon, bounded by the Pacific Ocean to the west, the Coquille River to the south, the Willamette Valley to the east, and the Columbia River to the north (figure 12.1). Elevations range from sea level to more than 1000 m at the highest peaks. The physiography is characterized by highly dissected terrain with steep slopes and high stream densities. Soils are predominantly well-drained Andisols and Inceptisols (based on the Soil Science Society of America Classification System). Parent materials are mostly marine sandstones and shales, along with some basaltic volcanics and related intrusives. The climate is generally wet and mild, with most precipitation falling between October and March. Precipitation is highest and summer temperatures are lowest near the coast, resulting in low moisture stress during the growing season and high forest productivity. Decreasing precipitation and increasing temperature with increasing distance from the coast create a predominantly west-to-east gradient of increasing moisture stress (color plate 16).

Major coniferous species include Douglas-fir (*Pseudotsuga menziesii* [Mirb.] Franco) and western hemlock (*Tsuga heterophylla* [Raf.] Sarg.), with Sitka spruce (*Picea sitchensis* [Bong.] Carr.) prevalent along a narrow coastal strip. Hardwoods, including red alder (*Alnus rubra* Bong.) and bigleaf maple (*Acer macrophyllum* Pursh), are often found in mixed stands with young conifers and dominate many riparian areas. Long-lived conifers and a favorable climate combine to produce some of the largest accumulations of live-tree biomass in the world (Waring and Franklin 1979). The presence of large live and dead trees (often >100 cm in diameter at breast height [dbh]), along with a diverse multilayered canopy and spatial heterogeneity created by canopy gaps, are characteristics of Pacific Northwest old-growth forests (Franklin et al. 1981). These features typically require 200 yr or more to develop

FIGURE 12.1. Maps of the Oregon Coast Range study area. (a) Boundary of the study area, with major historical fires of the past 200 yr. (b) Location of the study area within the Pacific Northwest region. (c) Location of the study area within North America.

following a stand-replacing disturbance, although old-growth structure may develop more quickly following partial disturbances that leave a cohort of live remnant trees.

Major groups of forest landowners in the Oregon Coast Range include private industry, private nonindustrial owners, the state of Oregon, and the federal government (the U.S. Department of Agriculture Forest Service and the Bureau of Land Management). Landowner goals, forest management strategies, and regulatory constraints vary considerably among these ownership classes. Private industrial ownerships comprise 38% of the study area, concentrated in blocks in the northern, central, and southern portions of the Coast Range (color plate 16b). Private industrial lands are managed primarily for wood production within the regulatory constraints imposed by the Oregon State Forest Practices Act (Oregon Department of Forestry 2002). The predominant management practice on private industrial land is clearcutting, with rotations of 30 to 50 yr. Private nonindustrial ownerships cover 24% of the study area, primarily along the Willamette Valley margin and in the large river valleys. Private nonindustrial owners operate within the same regulatory framework as private industry, but have a broader range of goals and use a wider variety of management practices.

Federal lands managed by the U.S. Department of Agriculture Forest Service and the Bureau of Land Management also occupy substantial areas of the Coast Range (11 and 14% of the study area, respectively). A significant portion of the land controlled by the Bureau of Land Management is interspersed with private industrial land in a checkerboard pattern (color plate 16b). The federal lands are presently managed under the Northwest Forest Plan for primarily ecological goals related to the preservation and restoration of old-growth forests, conservation of threatened and endangered species, and protection of aquatic ecosystems (Forest Ecosystem Management and Assessment Team 1993). Timber harvests are restricted in a large network of late-successional and riparian reserves, and harvests in the remaining matrix lands are required to leave residual live trees, snags, and downed trees. State forest lands cover 12% of the study area. Current plans for the state forests propose active management for timber production, protection of the forest's health, and species conservation by combining retention of large trees, snags, logs, and other habitat elements following timber harvests with long rotations to create a range of forest structures across the landscape (Oregon Department of Forestry 2001).

Methods

Comparison of Disturbance Regimes

Historical fire regimes in the Coast Range were characterized through a literature review and an analysis of published data. Paleoecological research based on high-resolution charcoal analyses provided long-term fire history information for the watersheds surrounding Little Lake (Long et al. 1998) in the central Coast Range and Taylor Lake (Long and Whitlock 2002) in the northwestern Coast Range. Fire frequency data from these studies were summarized for different time periods as mean fire-return intervals, defined as the mean number of years between fires at a particular location (Agee 1993). Dendroecological research based on tree-ring and fire-scar data from a 1375-km^2 study area in the central Coast Range provided a more detailed picture of rates and patterns of burning over the past 500 yr (Impara 1997). Using published data from Impara's study, we computed the natural fire rotation to serve as a metric of fire frequency for the coastal-interior and Willamette Valley margin portions of the study area. The natural fire rotation represents the total number of years required to burn a particular landscape (Heinselman 1973), and is computed as:

$$NFR = \frac{YEARS}{\sum_{i=1}^{NFIRES} FSIZE_i \, / \, TAREA}, \qquad (12.1)$$

where NFR is the natural fire rotation, $YEARS$ is the length of the fire history record, $NFIRES$ is the total number of fires in the record, $FSIZE_i$ is the size of the ith fire in the record, and $TAREA$ is the total area of the landscape.

We chose this metric to measure fire frequency because it does not depend on a particular statistical model of fire frequency and therefore does not assume that the fire regime remains constant over time and space (Fall 1998). However, the estimated natural fire rotation will be biased upward if evidence of the oldest fire events is overwritten by more recent fires. We therefore used an alternative method, adapted from Morrison and Swanson (1990), to correct for our reduced ability to sample the oldest fires:

$$NFR = \frac{YEARS}{\sum_{i=1}^{NFIRES} (FSIZE_i \, / \, RAREA_i)}, \qquad (12.2)$$

where $RAREA_i$ is the area available to record each fire, which is computed as the total area that did not experience a stand-replacing fire more recently than the occurrence of the recorded fire.

Fire-size distributions were computed using estimates of fire size from historical maps of forest vegetation in 1850, 1890, 1920, and 1940 (Teensma et al. 1991). Most of the fires represented in these maps occurred after settlement of the study area had begun, but before the establishment of an effective fire suppression program in the latter half of the twentieth century. Human influences and a warm climate both contributed to widespread fires between 1850 and 1940 (Weisberg and Swanson 2003), and we assumed that the frequency distribution of fire sizes from this period was generally representative of the pre-European settlement fire regime. These historical fire regimes were contrasted with modern disturbance regimes measured in a remote-sensing study of landscape change from 1972 to 1995 (Cohen et al. 2002). A map of the locations and sizes of clearcuts that occurred during this time was used to determine the size distribution of the disturbances caused by timber harvesting and to calculate a harvest rotation using the same approach as used in equation 12.1. Because the time period of the study of landscape change (23 yr) was less than the harvest rotations commonly used in the Coast Range (>30 yr), we assumed that no clearcuts had been overwritten by subsequent disturbances.

Simulation of Pre-European Settlement Landscape Dynamics

Pre-European settlement landscape dynamics were simulated using the Landscape Age-Class Dynamics Simulator (LADS), a spatial simulation model that predicts the initiation, spread, and effects of fires (Wimberly 2002). Multiple spatial data layers delineated simulation area boundaries, topography, and major climatic zones as raster data layers with a 9-ha resolution (300×300-m^2 cells). The model used a statistical approach to simulate fire regimes, and modeled the frequencies, sizes, and severities of individual fires as probability distributions derived from empirical data (Gardner et al. 1999). Different probability distributions were derived for the coastal-interior and Willamette Valley margin climatic zones to reflect broad-scale spatial trends in the study area's fire history (see figure 12.1a). Fire ignition and spread were modeled using a stochastic cellular-automata algorithm that allowed fire to occur more frequently on more susceptible landforms and in stands with high levels of fuel accumulation. Stand dynamics were simulated using an age-driven model of forest cohorts, which assumed that structural diversity recovered more rapidly after moderate-severity fires than after stand-replacing fires. Forest vegetation was mapped using five structure classes (table 12.1): early successional, young, mature, old growth (early transition), and old growth (late transition—shifting mosaic). For the purposes of our analysis and to permit comparison with the present-day landscape, the two old-growth classes were combined into a single structure class. Additional details on the development, parameterization, and sensitivity analysis of the LADS model are provided by Wimberly (2002).

Previous research has demonstrated that the disturbance regimes and landscape dynamics of the Coast Range are scale dependent. Pre-European settlement landscape variability must be studied over relatively large areas (>100,000 ha) and long time intervals (>500 yr) because the occurrence of occasional large fires makes landscape dynamics unpredictable at finer scales (Wimberly et al. 2000). Our simulations of landscape dynamics therefore encompassed the spatial extent of the Coast Range (approximately 23,000 km^2) and were carried out based on the fire regime that characterized the 1000-yr period prior to settlement. This period was deemed most relevant to present-day management questions and was found in a previous study to represent an era in which climate, fire regimes, and species composition were generally similar to those of the present day (Wimberly 2002). The model was run for a total of 50,000 simulation years following a 1000-yr initialization period. Landscape patterns were output as digital maps at 200-yr intervals, generating 250 sample landscapes. Running the model for 50,000 simulation years does not imply that the simulated fire regime was representative of the actual 50,000 yr prior to settlement. Instead, the long simulation period was needed to generate a large number of independent landscapes that represented the range of variability that could have occurred over the 1000-yr time frame of the model.

The spatial pattern of each sample landscape was summarized by computing the total area and largest patch size for each structure class. In addition, we computed a spatial index that measured the relative isolation of each structure class from the other classes. In equation 12.3, this isolation index (I_s) was calculated for a given

TABLE 12.1. Classification of Forest Patches into Structure Classes based on AGE (Time Since Stand Initiation) and TFIRE (Time Since Last Fire)

AGE (yr)	TFIRE (yr)	Structure Class	Description
Any	<30	Early successional	Open canopy to dense, closed canopy with high tree densities. May have residual live trees, snags, and downed trees from the prefire stand.
>30	30–80	Young	Closed canopy with low levels of understory tree regeneration, shrubs, and herbs. May have residual live trees, snags, and downed trees from the prefire stand.
80–200	80–200	Mature	Canopy gaps allow light to reach the forest floor, facilitating reestablishment of the understory layer. Typically has low volumes of dead wood compared with older and younger age classes.
>200	80–500	Old growth (early transition)	Characterized by large living trees (>100 cm dbh); accumulations of large snags and downed trees; a diverse, multilayered canopy dominated by shade-tolerant trees in the mid-canopy layers; and a heterogeneous spatial pattern of canopy gaps and forest patches in various age classes.
>500	>80	Old growth (late transition– shifting mosaic)	Similar to the old growth (early transition) phase, but with decreasing numbers of living remnant trees from the establishment phase and increasing dominance of canopy-gap dynamics.

Notes: AGE = TFIRE for even-aged stands, but AGE > TFIRE for stands that have had one or more fires of moderate severity. See Wimberly et al. (2000) for a detailed explanation of this classification scheme. Descriptions of stand structure classes are from Spies and Franklin (1991) and Spies (1997).

structure class (*s*) by first computing the distances (d_{ij}) in km from every cell (*i*) in the raster map to its nearest neighbor (*j*) belonging to class *s*. The isolation index for structure class *s* was then computed as:

$$I_s = \frac{\sum_{i=1}^{N} \sigma_i \cdot d_{ij}}{\sum_{i=1}^{N} \sigma_i}, \qquad (12.3)$$

where *N* is the total number of cells in the simulated landscape, $\sigma_i = 0$ for cells belonging to structure class *s*, and $\sigma_i = 1$ for cells that do not belong to structure class *s*. This index was similar to the *GISfrag* index proposed by Ripple et al. (1991), except that distance values of 0 were not included in our computation. The isolation index computed for a particular structure class reflected the mean distance from each cell that was not a member of that structure class to its nearest neighboring cell that was a member. For example, computing a low isolation index for the old-growth class would indicate that forests in other classes tended to be located in close proximity to old-growth patches, suggesting a high potential for dispersal of organisms from old-growth refu-

gia into the younger forests. In contrast, a high isolation index computed for old-growth forest would suggest a limited potential for organisms to disperse from old-growth refugia into the surrounding landscape.

Simulations of the historical landscape were contrasted with a 1996 map of forest vegetation patterns derived from forest inventory plots, Landsat TM imagery, and GIS layers describing ownership, topography, and climate (Ohmann and Gregory 2002). This map provided a set of compositional and structural measurements for each forested patch, including an estimate of stand age. Patches in the 1996 map were reclassified into the same age-based structure classes used in the simulation model. The 1996 vegetation map initially had a 25-m spatial resolution, and was rescaled to a 300-m resolution to match the resolution of the simulation model's output. The total area, largest patch size, and isolation index were computed for each of the four structure classes based on the 1996 map. These values were compared with the probability distributions of the same indices derived from simulation of the pre-European settlement fire regimes.

Results

Comparison of Disturbance Regimes

The paleoecological records showed that fires have burned in the central Coast Range for at least the past 9000 yr (Long et al. 1998). However, mean fire-return intervals varied considerably as climate changed over this epoch. From 9000 to 6850 yr before the present (YBP), the climate was warmer and drier than the present climate and the mean fire-return interval at Little Lake averaged 110 yr. The interval at Little Lake lengthened over time as climate became cooler and more humid, averaging 160 yr between 6850 and 2750 YBP and 230 yr between 2750 YBP and the present day (Long et al. 1998). Values of the mean fire-return interval at Taylor Lake showed similar trends, averaging 140 yr between 4600 and 2700 YBP and increasing to 240 yr between 2700 YBP and the present day (Long and Whitlock 2002).

Using equation 12.1, Impara (1997) reported a natural fire rotation of 452 yr for the central Coast Range over the 367-yr period prior to settlement in 1845. Wimberly (2002) used the same data to compute natural fire rotation values of approximately 200 yr for the coastal-interior portion of the study area and 100 yr for the Willamette Valley margin portion of the study area. This calculation included only fires that were severe enough to initiate tree regeneration, and used equation 12.2 to correct for bias introduced by the erasure of evidence of older fires by subsequent burns. We used these values in our simulations of historical landscape dynamics because the 200-yr natural fire rotation computed for the coastal-interior region was reasonably close to the mean fire-return interval values reported for the past 1000 yr (Long et al. 1998, Long and Whitlock 2002). Running the simulations with natural fire rotation values at the shorter end of the range of estimates provided a conservative estimate of the amount of older forests in the pre-European settlement landscape.

Natural fire rotation decreased to 78 yr during the settlement period from 1845 to 1910, probably because of the warming climate combined with increased fire ignitions by settlers (Impara 1997). This increase in the rate of burning reflected a regional trend of increased fire frequency during the settlement era (Weisberg and Swanson 2003). In contrast, natural fire rotation increased considerably in the twentieth century as fire suppression became more effective, averaging 335 yr between 1910 and 1994 (Impara 1997).

Stand-replacing fires burned only 0.06% of the Coast Range between 1972 and 1995 (Cohen et al. 2002). If this rate of burning is extrapolated over long time periods, it is equivalent to a natural fire rotation of greater than 1600 yr.

Most of the fires reconstructed by Impara (1997) conformed to one of two distinctive patterns. Widespread fires were large burns that encompassed both the coastal-interior and Willamette Valley margin climate zones and were predominantly stand-replacing disturbances. Historically documented examples of such events (Loy et al. 1976) include the Nestucca fire of 1848 (120,000 ha); the Siletz fire of 1849 (325,000 ha); the Yaquina fire of 1853 (195,000 ha); the Coos Bay fire of 1868 (118,000 ha); and the Tillamook fires of 1933 (96,000 ha), 1939 (76,000 ha), and 1945 (72,000 ha). In contrast, Willamette Valley margin fires were smaller, predominantly moderate- and low-severity burns that left significant cohorts of remnant live trees (Impara 1997). Fire-size data from reconstructed forest age maps for the Oregon Coast Range in 1850, 1890, 1920, and 1940 (Teensma et al. 1991) also suggest a pattern of a few large, widespread fires combined with smaller, more numerous fires along the Willamette Valley margin. Mean fire size in the coastal-interior climate zone was 12,309 ha, whereas the mean fire size in the Willamette Valley margin climate zone was only 2576 ha (Wimberly 2002). These fire-size distributions were heavily skewed, with fires smaller than 10,000 ha accounting for 89% of the total number of fires. However, fires larger than 10,000 ha accounted for 84% of the total area burned (figure 12.2). The five largest fires, all larger than 50,000 ha, accounted for 71% of the total burned area.

Between 1972 and 1995, 27% of the Coast Range was clearcut, equivalent to a harvest rotation of 85 yr (Cohen et al. 2002). In contrast with fires, most clearcuts retain few live trees, snags, or large downed trees within their boundaries. Densities of large trees and snags (>50 cm dbh) can be three to five times higher in postfire stands than in logged stands of similar age (Hansen et al. 1991). The mean size of clearcuts was less than 200 ha, with no units larger than 5000 ha (Cohen et al. 2002). Spatial variability in harvest rotations and the sizes of timber harvests is presently controlled by the management regimes used in different classes of forest ownership (figure 12.3). Between 1972 and 1995, the harvest rotation on private industrial lands was 51 yr, compared with 100 yr on private nonindustrial lands, 189 yr on state-owned lands, and an average of

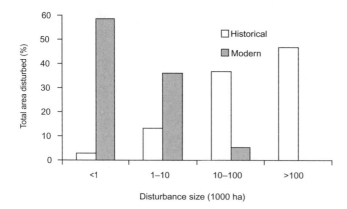

FIGURE 12.2. Size-class distribution for historical wildfires (1850–1940) and modern wildfire and timber harvest disturbances (1972–1995), expressed as percentages of the total area disturbed. Historical fire data are from Teensma et al. (1991), and timber harvest data are from Cohen et al. (2002).

160 yr on federal lands managed by the Bureau of Land Management and the U.S. Department of Agriculture Forest Service. Similarly, the mean size of clearcuts was more than twice as large on private industrial lands as in any other ownership class.

The harvest rotations presented in figure 12.3 must be interpreted with the caveat that they are based on only 23 yr of data. On state lands, little harvesting occurred from 1972 to 1995, because much of the forest was below harvestable age. The calculated harvest rotation on state lands will decrease in the future, as stands reach merchantable age and new management plans are implemented. On federal lands, the reported harvest rotations mostly reflect the period prior to the implementation of the Northwest Forest Plan. Under current policies, future rates of clear-cutting on federal lands will be lower than those reported here.

Simulation of Pre-European Settlement Landscape Dynamics

Old growth was the most common structure class in the simulated pre-European settlement landscape (color plate 17); in our simulations (figure 12.4), it occupied a median of 42% of the Coast Range, with values ranging from 29% (fifth percentile) to 52% (ninety-fifth percentile). In contrast, median abundances were 21% for the young class, 17% for the early-successional class, and 16% for the mature class. In the present-day Coast Range, the percentages of mature and old-growth forests are smaller than would be expected under the pre-European settlement fire regime, and the percentages of early-successional and young forests are greater than expected. In the simulated pre-European settlement landscapes, there was at least one large (>100,000-ha) patch of old-growth forest present at all times, but the largest old-growth patch in the present-day landscape occupies only 650 ha (figure 12.5). In contrast, a mosaic of early-successional and young forests dominates the present-day landscape, with the largest patch of early-successional forest larger than 500,000 ha and the largest patch of young forest larger than 300,000 ha. In the simulated pre-European settlement forests, most early-successional, young, and mature forests were within 1 km of an old-growth forest, as shown by

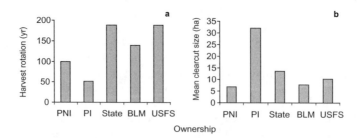

FIGURE 12.3. Distribution of (a) timber harvest rotations and (b) clearcut sizes in the Oregon Coast Range from 1972 to 1995. PNI = private nonindustrial, PI = private industrial, State = state of Oregon, BLM = Bureau of Land Management, USFS = U.S. Department of Agriculture Forest Service. Based on data from Cohen et al. (2002).

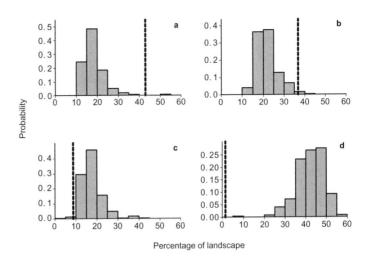

FIGURE 12.4. Simulated historical (pre-European settlement) ranges of variability for the percentages of the landscape covered by (a) early-successional, (b) young, (c) mature, and (d) old-growth forests in the Oregon Coast Range. Dashed lines represent the percentage of the landscape covered by each class in the present-day landscape.

Percentage of landscape

the isolation index values for old-growth forest, which were mostly less than 1 (figure 12.6). The spatial isolation of old-growth patches has increased considerably in the present-day landscape, whereas the isolation of early-successional and young patches has decreased.

Discussion

Changes in the Disturbance Regime and Their Ecological Implications

One hundred and fifty yr of postsettlement land use in the Oregon Coast Range has brought about a wholesale change in the spatial pattern and dynamics of the forest landscape. Historically, the Coast Range was a shifting landscape mosaic dominated by large patches of old-growth forest. Smaller fragments of old-growth forest were also widely distributed in blocks of younger forest.

In contrast, early-successional and young forests dominate the current landscape, with old-growth forest present only as small, fragmented patches. These changes have occurred because disturbance rates over the past 150 yr have been consistently higher than under the pre-European settlement disturbance regime. Increased burning in the late nineteenth and early twentieth centuries was at least partially the result of an increase in ignitions by settlers (Weisberg and Swanson 2003). Although fire suppression has greatly reduced the incidence of forest fires in the latter half of the twentieth century, disturbance from timber harvests in the modern landscape has occurred at a more rapid rate than did disturbance by fire in the pre-European settlement landscape. Clearcutting has typically removed the majority of trees from a site, in contrast to the variable numbers of remnant live trees left by the range

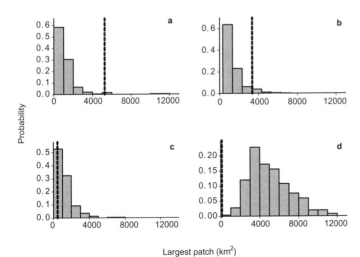

Largest patch (km²)

FIGURE 12.5. Simulated historical (pre-European settlement) ranges of variability for the largest patch of (a) early-successional, (b) young, (c) mature, and (d) old-growth forests in the Oregon Coast Range. Dashed lines represent the largest patch in the present-day landscape.

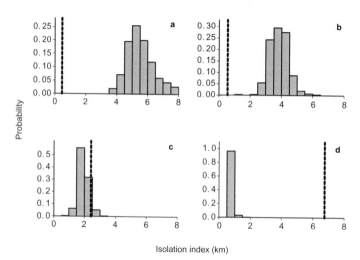

FIGURE 12.6. Simulated historical (pre-European settlement) ranges of variability for the isolation index of (a) early-successional, (b) young, (c) mature, and (d) old-growth forests in the Oregon Coast Range. Dashed lines represent the value of the isolation index in the present-day landscape.

of fire severities that characterized historical fire regimes. The small sizes and spatial dispersion of most harvested areas compared with the historical pattern of large fires have also altered the spatial pattern of the forest landscape.

These changes in landscape patterns have several ecological implications. Approximately one-third of all vertebrates and one-half of all amphibians inhabiting forests in western Oregon have been characterized as "closely associated" with old-growth habitats (Olson et al. 2001). Although many of these species are also found in younger forests, it is reasonable to hypothesize that the significant decline in their preferred habitat has affected their populations. Species that require large concentrations of mature or old-growth habitat, such as the northern spotted owl, may be particularly sensitive to the decreased size and dispersed pattern of patches of mature and old-growth forest (McComb et al. 2002). Although small patches historically represented only a minor portion of the total area of old-growth forest, they may still have facilitated the persistence of disturbance-sensitive, dispersal-limited species such as the lichen *Lobaria oregana* (Tuck.) Mull. Arg. This species can survive and reproduce in young forests, but is typically eliminated by stand-replacing disturbance and must recolonize regenerating stands through dispersal from undisturbed forest (Sillett et al. 2000). Loss of the numerous, scattered old-growth patches that were historically left by fires may greatly reduce the number of sources of propagules for reestablishing disturbance-sensitive, dispersal-limited species in young forests (Wimberly and Spies 2002).

Changes in the temporal and spatial patterns of stand-replacing disturbance may also affect broad-scale ecosystem processes, particularly the watershed-level dynamics of water, wood, and sediment. The large fires that occurred under the pre-European settlement disturbance regime often burned substantial areas within a watershed, reducing slope stability and initiating landslides and debris flows that delivered large pulses of wood and sediment to streams (Benda et al. 1998). In the short term, these disturbances likely destroyed habitat for Pacific salmon (*Oncorhynchus* spp.) and other native fish species. Over time, however, the wood and sediment inputs that occurred after fires provided raw materials for the development of diverse aquatic habitats. Long fire-return intervals provided time for fluvial processes to generate complex habitats that included pools, large pieces of in-stream wood, and a variety of different substrate types (Reeves et al. 1995). These fire-free periods also allowed forests on the surrounding slopes to reestablish themselves and develop into late-successional stands with large living and dead trees.

In contrast, the present-day disturbance regime is dominated by timber harvesting, which produces disturbances that are more frequent, smaller, and more widely dispersed than historical fires. Although individual timber harvests only affect small portions of a particular watershed, a larger number of disturbances occurs than takes place under the natural fire regime, and these disturbances are distributed more evenly in time and space than were historical fires. Consequently, wood and sediment inputs resulting from these disturbances have shifted from an episodic to a chronic process. Furthermore, the predominantly young forests produced by this disturbance regime have smaller trees and lower volumes of dead wood than the mature and old-growth forests that dominated the pre-European settlement landscape. These fundamental changes suggest that disturbance regimes driven by forest management activities aimed solely at maximizing timber production may not sustain the production of high-quality fish habitat (Reeves et al. 1995).

Prospects for Forest Management Based on Natural Disturbance Regimes

Current forest management practices in the Coast Range could be altered to better emulate the pre-European settlement disturbance regime. Timber harvests could be spatially clustered to generate large disturbed patches similar to those created by fire, while retaining contiguous blocks of uncut forest (Gustafson 1996; Cissel et al. 1999). However, changing only the spatial pattern of harvesting will not influence the distribution of forest age classes. Protecting old-growth and late-successional forests through long-term deferrals, reserves, and wilderness areas can limit further declines in the amount of older forests. In managed landscapes, longer rotation lengths could increase the proportion of older forests (Curtis 1997). Development of old-growth characteristics in young managed stands could be hastened through thinning, underplanting, or other silvicultural treatments (McComb et al. 1993; Tappeiner et al. 1997). It is also possible to retain some elements of old-growth structure in young stands by leaving residual trees, logs, and snags on the site after timber harvests in mature and old-growth stands, thereby emulating the effects of moderate-severity fires (Franklin et al. 1997).

Despite the many opportunities for modifying forest management practices, major ecological and social obstacles limit the potential for implementing management based on natural disturbance regimes in the Coast Range. The major ecological impediment is that our ability to manipulate landscape structures and composition in the near term is constrained by the past 150 yr of disturbance history. Fires and logging have already eliminated the majority of old-growth forests, and simply shifting the frequency and pattern of timber harvesting may do little to hasten the recovery of this forest type (Wallin et al. 1994; Baker 1995). Whereas old-growth forests can be rapidly converted into young forests through disturbance, the development of old-growth forests occurs over hundreds of years through the relatively slow processes of forest succession and stand dynamics. Thus, any attempt at landscape management based on natural disturbance regimes and historical landscape patterns must also consider the protection of existing old growth, as well as of mature forests that will develop into old growth over the next century.

The complex pattern of forest ownership is perhaps an even greater obstacle to implementing management based on natural disturbance regimes in the Coast Range. Forest management within each ownership type is influenced by a unique set of management goals and regulatory constraints, limiting the range of conditions that can be produced within a particular ownership class. For example, management on most private lands is driven by predominantly economic considerations, which prescribe clearcutting at fairly short rotations (40–60 yr) as the prevalent management practice. If present forest management regimes are extrapolated 100 yr into the future, private lands will continue to consist primarily of early-successional forest and young, closed-canopy forest (Spies et al. 2002).

In contrast, the U.S. Department of Agriculture Forest Service and the Bureau of Land Management operate under regulations imposed by the Northwest Forest Plan, which in the Coast Range allows only minimal timber harvests outside of late-successional and riparian reserves (Forest Ecosystem Management and Assessment Team 1993). If fire suppression policies continue to be effective, most federal lands will succeed to old-growth conditions. The Oregon State Department of Forestry intends to manage state forest lands in the northern Coast Range using a structure-based management approach, in which stands are harvested on 80- to 130-yr rotations, with active management to provide for large trees, snags, downed trees, and other old-growth structural components. Current plans call for eventually maintaining 20–30% of the land base in an older forest structure, which is intended to emulate the structural features and functional characteristics of old-growth forests (Oregon Department of Forestry 2001).

If Coast Range forests continue to be managed under existing policies, increasing the amounts of old-growth forest on public lands will move overall landscape patterns closer to the range of historical variability. However, the disparate disturbance regimes on federal and private lands may eventually leave few forests in the mature structure class. Harvest rotations on most private lands are too short to allow forests to reach the mature stage, and the continued exclusion of fire and timber harvesting on federal lands will allow most forests to pass through the mature stage and develop into old growth. Under current forest management policies, this divergence in the age structure of forests under public and private ownership is predicted to occur gradually over the next 100 yr (Spies et al. 2002). Some new mature forests will develop on state and private nonindustrial lands managed under long rotations, but the total area of mature forests will likely account for only a small portion of the Coast Range. If old growth is lost in the future as a result of fire, wind, or other natural disturbances, this gap in the forest age structure could limit the potential for rapid development of new old-growth habitats.

The spatial pattern of ownership also limits future possibilities for replicating large, historical fire events through timber harvesting. For example, the boundaries of the largest blocks of public land in the Coast Range correspond to those of large historical fires (see figure 12.1a, color plate 16b). The southern block of the Siuslaw National Forest includes a large area burned by the Siletz fire, and the northern block encompasses a significant portion of the Nestucca fire. The Tillamook State Forest includes the majority of land burned by the Tillamook fires, and the Elliot State Forest covers most of the area of the Coos Bay fire. Arbitrary human-imposed boundaries dominate in other areas, such as the checkerboard pattern of ownership in the southeastern portion of the Coast Range, where Bureau of Land Management lands and private lands are interspersed in 2.6-km² blocks. Also, many large blocks of private land actually have multiple landowners, each acting largely independently to manage their land.

Our present social, political, and regulatory systems do not provide a framework for management coordinated among multiple private landowners or between private and public ownerships. Therefore, the size of a particular ownership sets an upper boundary on the spatial scale of the forest management regime that can be applied therein. Because the largest ownership blocks in the Coast Range are artifacts of past fires, management strategies aimed at maintaining a range of forest structures in a single ownership block necessitate timber harvests that are considerably smaller than the largest historical burns. These management strategies may eventually produce novel landscape patterns that contain a habitat mosaic with finer resolution than was present under the pre-European settlement fire regime. The relative importance of habitat amount versus habitat pattern for species conservation is presently a subject of active scientific debate (Schmiegelow and Monkkonen 2002), and the implications of altering the spatial scale of the forest habitat mosaic are unclear.

Conclusions

The shift from a disturbance regime dominated by fire to one dominated by timber harvesting has had significant impacts on the pattern and dynamics of forest habitats in the Oregon Coast Range. Clearcuts are smaller and occur at shorter rotations than did historical fires. In addition, clearcutting removes a larger proportion of both live and dead trees from the site than do many wildfires. Whereas pre-European settlement fire regimes varied primarily along a climatic gradient, forest management policies vary among landowners. In turn, the pattern of land ownership reflects a legacy of past wildfires as well as the social, economic, and political history of the region. The present-day forest landscape has smaller areas of mature and old-growth forests than would be expected under the pre-European settlement disturbance regime. Whereas large (>100,000-ha) patches of old-growth forest were common in pre-European settlement landscapes, old growth is currently distributed in small (<650-ha) fragments that are isolated from other habitat types.

Any future forest management policy based on emulating natural disturbance in the Coast Range will necessarily represent a compromise between our scientific understanding of disturbance and its ecological effects, and the economic, social, and political realities of the region. Given the history of human land use and the constraints imposed by current ownership patterns, attempting to closely replicate the temporal and spatial patterns of natural disturbance may be an unrealistic goal, and could actually move landscape patterns further from the range of historical variability if the remaining old-growth forests are disturbed. Instead, a more tenable short-term goal might be to maintain key elements of historical landscape processes and structures at multiple scales. Examples of this approach include maintaining a few large patches of old-growth forest at the regional scale, numerous small patches of old growth at a landscape scale, and individual old-growth structures (e.g., large trees, snags, and downed trees) at a stand scale. Given the wide range of landowner goals and regulatory constraints, these broad-scale goals would have to be achieved through a mixture of strategies, including a combination of active management and reserve-based approaches.

Acknowledgments

We thank Janet Ohmann for providing data and feedback on earlier versions of the manuscript. Two anonymous reviewers also provided useful comments. David Boughton, Norm Johnson, Stephen Lancaster, and other researchers working with us on the Coastal Landscape Modeling and Assessment Study (CLAMS), provided valuable critiques throughout all stages of this project. Support for this work was provided by the Northwest Forest Plan Program of the U.S. Department of Agriculture Forest Service Pacific Northwest Research Station and by the University of Georgia.

CHAPTER 13

:: Using a Decision Support System to Estimate Departures of Present Forest Landscape Patterns from Historical Reference Conditions

An Example from the Inland Northwest Region of the United States

PAUL F. HESSBURG, KEITH M. REYNOLDS,
R. BRION SALTER, and MERRICK B. RICHMOND

Human settlement and management activities have altered the patterns and processes of forest landscapes across the inland northwest region of the United States (Hessburg et al. 2000c; Hessburg and Agee in press). As a consequence, many attributes of current disturbance regimes (e.g., the frequency, duration, severity, and extent of fires) differ markedly from those of historical regimes, and current wildlife species and habitat distributions are inconsistent with their historical distributions. Just as human-caused changes in ecological processes have led to alterations in landscape patterns, changes in patterns have produced alterations in ecosystem processes, and particularly in forest disturbance (Kimmins, chapter 2, this volume). Today's public-land managers face substantial societal and scientific pressure to restore landscape patterns of structure, composition, and habitats that will restore some semblance of natural processes and revitalize the productivity of terrestrial ecosystems. Motivations for restoration stem from genuine concerns over the functioning of ecological systems and aversion to the risks and uncertainties associated with current conditions, but our lack of knowledge of the ecosystem's former characteristics and variability limits our efforts. In this chapter, we present one approach to estimating the extent to which present forest landscape patterns have departed from the baseline conditions that existed before modern management began (around 1900). Our goal is to approximate the range and variation of these historical patterns and use that knowledge to evaluate present forest conditions and assess the ecological importance of departures.

Background

For a long time, ecologists asserted that ecosystem dynamics could be explained by theories of stable equilibria and "the balance of nature" hypothesis (e.g., Milne and Milne 1960; Lovelock 1987), but these explanations are no longer considered valid. Wu and Loucks (1995) proposed an alternative framework they termed the *hierarchical patch dynamics* paradigm, which comprises five elements. We briefly paraphrase these elements here, because our approach is based on their framework:

1. *Ecological systems can be viewed as nested, discontinuous hierarchies of patch mosaics* (Allen and Starr 1982; Ahl and Allen 1996). The clearly divisible scales associated with this hierarchical organization have been referred to as *loose vertical coupling* between levels. This loose coupling makes it possible to disassemble complex systems into their constituent levels for study without significant loss of information.

2. *The dynamics of ecosystems are an emergent property of the patch dynamics that occur at each level in the hierarchy.* However, most of the energy and material exchange occurs within individual levels of a hierarchy.

3. Across a broad range of spatial and temporal scales, *patterns enable and constrain ecological processes, and ecological processes create, maintain, modify, and destroy patterns.* Patterns and processes are tightly linked, and particular patterns and processes are linked to certain spatial and temporal scales.

4. At all spatial scales, *ecosystems exhibit transient patch dynamics and nonequilibrium behavior* because of the stochastic properties of the geological and climatic processes that support ecosystems.

5. *Lower level processes are incorporated into higher level structures and processes in the hierarchy of patch dynamics.* This incorporation integrates the effects of lower level (e.g., site) processes and higher level constraints imposed by the geological and climatic systems to generate quasi-equilibrium patch dynamics at all levels. These dynamics become manifest as a finite range of conditions that is somewhat predictable, as long as the underlying processes and higher-level constraints remain substantially unchanged.

Objectives

In our case study, we have developed an approach to estimating the quasi-equilibrium conditions associated with one level in a forest's hierarchy of patch dynamics. For simplicity, we have called these conditions the *reference conditions* and have called the typical variation in these conditions the *reference variation*. We chose the range on either side of the median that contained 80% of the historical values (hereafter, the *median range*) of metrics of spatial patterns as our estimate of reference variation because most historical observations typically clustered in this range.

Our study focused on forest landscapes and the spatial patterns of their structural classes (i.e., successional stages or stand development phases), cover types, and related conditions. We focused on these patterns because important changes in the dynamics of altered forest ecosystems are often reflected in the structures of the living and dead elements of the landscape (Spies 1998). The processes that underlie this focus are forest disturbances, including fires, forest diseases and insect outbreaks, unusual weather, and herbivory. Geological and climatic inputs constrain the forest patterns and related disturbances. We determined that environmental constraints occur at a subregional scale. (See the *Methods* section for the basis of these assertions.)

We describe a repeatable, quantitative method (outlined in table 13.1) for estimating the range and variation in historical forest vegetation patterns and in vulnerability to disturbance. Our objective is to estimate a reference variation that allows us to evaluate the direction, magnitude, and potential ecological importance of some of the changes we have observed in present-day forest landscape patterns. To automate our approach, we programmed estimates of reference variation for one ecological subregion into the Ecosystem Management Decision-Support (EMDS) system (Reynolds 1999a, 2001a). To illustrate this approach, we compared the current patterns for an example watershed from the same subregion with the estimates of reference variation, which allowed us to identify vegetation changes that lay beyond the range of the estimates. Changes that fell in the range of the estimates were assumed to lie within the natural variation of the interacting ecosystem's geological disturbance and climatic processes. Changes that lay beyond the range of the estimates were termed *departures* that could be explored in more detail to determine their potential ecological implications.

Methods

Stratifying Inland Northwest Watersheds

To identify sample landscapes constrained by similar environmental contexts, we used the ecological subregions of Hessburg et al. (2000a) to stratify subwatersheds (about 5000–10,000 ha) of the eastern Washington Cascades into geologic and climatic zones (figure 13.1). Subwatersheds (figure 13.2) were used as the basic sampling units because they provided a rationale for subdividing land areas that share similar climate, geology, topography, and hydrology, and enabled future use of our study data and results in integrated terrestrial and hydrologic evaluations of the landscape. Subwatersheds compose the sixth level in the established hierarchy of U.S. watersheds (Seaber et al. 1987). Lehmkuhl and Raphael (1993) showed that some spatial pattern attributes are influenced by the size of the area being analyzed when analysis areas are too small. We used subwatersheds or logical subwatershed pairs larger than 4000 ha to avoid this bias.

We selected the ecological subregion (ESR) 4, the Eastern Cascades, Warm/Wet/Low Solar Moist and Cold Forest subregion (hereafter, the Moist & Cold Forests subregion) as the geoclimatic zone in which we sampled and estimated reference conditions. Landscapes of this subregion are dominated by Moist (67% of the area) and Cold (21% of the area) forest potential vegetation types, with total annual precipitation of 1100–3000 mm (wet), generally warm growing season temperatures (mean annual daytime temperature,

TABLE 13.1. Approach for Estimating the Departure of Present Forest Landscape Patterns from Historical Reference Conditions[1]

Step	Action	References
1	Stratify the inland Northwest United States subwatersheds (5000–10,000 ha) into ecological subregions by using a published hierarchy	Hessburg et al. (2000a)
2	Map the historical vegetation of a large random sample of the subwatersheds of one subregion (the Moist and Cold Forests ESR4 subregion) from 1930s–1940s aerial photography	Hessburg et al. (1999b)
3	Statistically reconstruct the vegetation attributes of all patches of sampled historical subwatersheds that showed any evidence of timber harvesting	Moeur and Stage (1995)
4	Analyze spatial patterns for each reconstructed historical subwatershed by calculating a finite, descriptive set of class and landscape metrics in a spatial analysis program (FRAGSTATS)	McGarigal and Marks (1995); Hessburg et al. (1999b)
5	Observe the data distributions from the analysis of spatial patterns for the historical subwatersheds and define reference conditions, based on the typical range of the clustered data	Hessburg et al. (1999a,b)
6	Define reference variation as the range on either side of the median that contains 80% of the historical values of the class and landscape metrics for the sample of historical subwatersheds	Hessburg et al. (1999a,b)
7	Estimate reference variation for spatial patterns of ESR4 forest composition (cover types) and structure (stand development phases); model the accumulation of ground fuel (loading) and several attributes of fire behavior	Hessburg et al. (1999a,b); Huff et al. (1995); O'Hara et al. (1996); Ottmar et al. (in press)
8	Program the ESR4 reference conditions into a decision-support model (EMDS)	Reynolds (1999a,b, 2001a,b)
9	Map the current vegetation patterns of an example watershed (Wenatchee_13, from the Wenatchee River basin, also from ESR4)	Hessburg et al. (1999b)
10	Objectively compare a multiscale set of vegetation maps of the example watershed with corresponding reference variation estimates in the decision-support model	Hessburg et al. (1999a,b)

[1]The historical reference conditions are from around 1900.

5–9°C), and relatively low levels of solar radiation (frequently overcast skies, 200–250 W·m⁻²; Hessburg et al. 2000a). The subregion contains 93 subwatersheds. To map historical and current vegetation, we randomly selected 18 subwatersheds to represent about 20% (19.4%) of the total number of subwatersheds and about 20% (22.3%) of the subregion's area (figure 13.3).

One test of the validity and scaling of an ecological stratification is to evaluate how well the stratification reduces the variance of some of the ecological patterns and processes that it con-

strains. To evaluate the effectiveness of our subregion map at reducing variation in the area of fire regimes, we used GIS software to combine the subregion map with a predicted historical (about 1800) map of fire severity for the same areas (Hann et al. 1997). We then calculated the mean subwatershed area (percentage of total) and the variance of each severity class for the subwatersheds of six subregions of the eastern Cascades. The results showed that our subregions reduced the subwatershed variation in fire severity in all but a single instance (table 13.2). In that

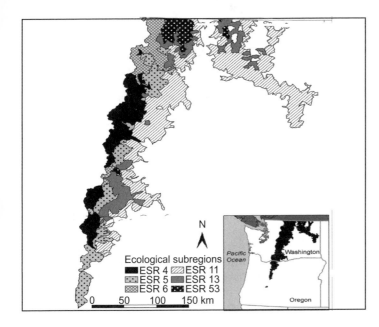

FIGURE 13.1. Map of the ecological subregions of the eastern Washington Cascades in the western United States. The ecological subregions (ESR) are: 4, Warm/Wet/Low Solar Moist and Cold Forests; 5, Warm/Moist/Moderate Solar Moist and Cold Forests; 6, Cold/Wet/Low and Moderate Solar Cold Forests; 11, Warm/Dry and Moist/Moderate Solar Dry and Moist Forests; 13, Warm and Cold/Moist/Moderate Solar Moist Forests; 53, Cold/Moist/Moderate Solar Cold Forests. Adapted from Hessburg et al. (2000c).

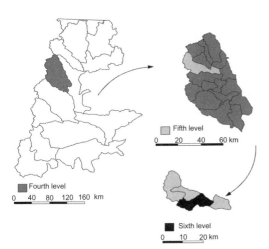

FIGURE 13.2. Hierarchical organization of subbasins (fourth level), watersheds (fifth level), and subwatersheds (sixth level) in the eastern Washington Cascades of the western United States (see also Seaber et al. 1987). The example shows the Wenatchee River subbasin at the fourth level, the Little Wenatchee River watershed at the fifth level, and our case study subwatershed (Wenatchee_13) at the sixth level (see also figure 13.4).

case (ESR11, the Dry & Moist Forests subregion), the variance of high severity fires exceeded the overall variance among subregions. This can be explained by the broad variation in topography and climatic gradients among subwatersheds; some had little or no high elevation terrain, whereas others had a considerable area of this terrain. For example, in subregion ESR11, dry and moist forests potential vegetation types either dominated the area of a subwatershed or were associated with high elevation cold forest types or low elevation dry shrublands and grasslands, both of which were historically areas with high severity fires.

Mapping Historical, Current, and Potential Vegetation

For each selected subwatershed, we mapped historical (1930s–1940s) and current (1990s) vegetation by interpreting aerial photographs. The resulting vegetation attributes let us derive forest cover types (sensu Eyre 1980), structural classes (sensu O'Hara et al. 1996; Oliver and Larson 1996), and potential vegetation for individual patches using the methods of Hessburg et al. (1999b, 2000b). The potential vegetation represented the most shade tolerant tree species that would provide the dominant cover in the absence of disturbance (e.g., Arno et al. 1985). Vegetation types were assigned to patches at least 4 ha in size by means of stereoscopic examination of color (current) or black and white (historical) aerial photographs. The scales of these photographs were 1:12,000 (current) and 1:20,000 (historical). Photointerpreters used available field inventory plot data to inform and correct errors in their visual interpretations. The attributes of the interpreted vegetation were the same as those reported

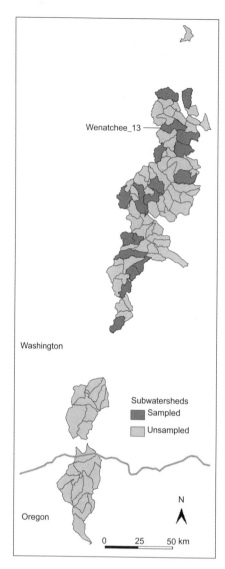

FIGURE 13.3. The subwatersheds sampled in the ESR4 (Moist and Cold Forests) ecological subregion. Note the location of Wenatchee_13, the subwatershed used in our case study.

by Hessburg et al. (1999b). Patches were delineated on clear overlays, and were georeferenced. Overlay maps were then scanned, edited, edge matched, and imported into GIS software to produce vector coverages with patch attributes.

Reconstructing the Attributes of Partially Harvested Historical Patches

Nine of the 18 historical subwatersheds (about 7.7% of the total area) showed evidence of timber harvesting, and nearly all this harvesting was light-to-moderate selection cutting. To re-

construct the preharvest vegetation attributes, we used Moeur and Stage's (1995) *most similar neighbor inference* procedure. Their algorithm is a multivariate procedure that identifies the patch that comes closest to matching a set of detailed design attributes and a set of broad-scale global attributes. This *stand-in* patch is chosen based on a measure of similarity that summarizes the multivariate relationships between the global and design attributes. Canonical correlation analysis is used to derive the similarity function. This analysis lets us define the historical values for each harvested patch as equal to the values for the design attributes of the corresponding stand-in patch.

The attributes of the reconstructed harvested patches comprised the total crown cover, overstory crown cover, canopy layers, and the overstory and understory size classes (seedlings/saplings, <2.7 cm dbh; poles, 12.7–22.6 cm dbh; small trees, 22.7–40.4 cm dbh; medium trees, 40.5–63.5 cm dbh; large trees, >63.5 cm dbh). The global attributes comprised the total annual precipitation (mm), mean annual daytime flux of shortwave solar radiation ($W \cdot m^{-2}$), and mean annual daytime temperature (°C) for 1989, which was considered a "normal" weather year (Thornton et al. 1997); the slope, aspect, and elevation; the potential vegetation group; and the landform feature (e.g., dissected glaciated slope, alpine glacial outwash, scoured glaciated slope, meltwater canyon or coulee). Modal values for each global attribute were assigned to individual patches in the maps in the GIS software. We obtained 2-km raster maps for the 1989 temperature, precipitation, and solar radiation from the University of Montana (Thornton et al. 1997) and a 1-km raster map of the potential vegetation groups developed by Hann et al. (1997) from the Interior Columbia Basin Ecosystem Management Project (www.icbemp.gov). Maps of slope, aspect, and elevation were derived from a 90-m-resolution digital elevation model by using standard methods. Landform features were photointerpreted from 1:12,000 scale color resource aerial photography by Wenatchee National Forest geology and soils personnel (Carl Davis, Wenatchee National Forest, Wenatchee, Washington, pers. comm.) and verified by sampling in the field.

Associating Fuel and Fire Behavior Attributes with Patches Based on Vegetation Characteristics

By using established classification rules (Ottmar et al. 1996; Schaaf 1996), we assigned vegetation patches to one of 192 fuel condition classes ac-

TABLE 13.2. Variance in the Percentage of Total Area $(S)^1$ for Historical Fire Severity Classes in Subwatersheds in and among Ecological Subregions of the Eastern Washington Cascades in the Western United States

Ecological Subregion	Fire Severity Class	Mean Subwatershed Area (%)	S^1	Quotient[2] $(S_{subregion}/S_{all\ subregions})$
ESR4: Warm/Wet/Low Solar Moist and Cold Forests (*n* = 93 subwatersheds)	Low	28.5	502	0.55
	Moderate	55.1	488	0.63
	High	16.4	240	0.40
ESR5: Warm/Moist/Moderate Solar Moist and Cold Forests (*n* = 80 subwatersheds)	Low	55.7	671	0.74
	Moderate	32.7	548	0.71
	High	11.6	172	0.28
ESR6: Cold/Wet/Low and Moderate Solar Cold Forests (*n* = 43 subwatersheds)	Low	18.2	302	0.33
	Moderate	66.8	339	0.44
	High	14.9	210	0.35
ESR11: Warm/Dry and Moist/Moderate Solar Dry and Moist Forests (*n* = 293 subwatersheds)	Low	58.8	756	0.83
	Moderate	7.4	139	0.18
	High	33.8	888	1.47
ESR13: Warm and Cold/Moist/Moderate Solar Moist Forests (*n* = 78 subwatersheds)	Low	68.1	590	0.65
	Moderate	15.0	202	0.26
	High	16.8	357	0.59
ESR53: Cold/Moist/Moderate Solar Cold Forests (*n* = 27 subwatersheds)	Low	29.4	784	0.87
	Moderate	51.9	666	0.86
	High	18.8	185	0.31

Source: Historical fire severity classes are from Hann et al. (1997).

^1S = the population variance associated with the mean area occupied by a fire severity class. $S_{subregion} = S$ for the subwatersheds of the indicated subregion. $S_{all\ subregions} = S$ for the subwatersheds of all six subregions.

[2]A quotient of 1 indicates that the variance associated with the mean area occupied by a fire severity class in subwatersheds of a subregion equals that of all subregions combined; a value <1 indicates that the variance in a subregion is less than the composite value; a value >1 indicates that the variance in a subregion is greater than the composite value.

cording to their cover type, structural class, and type (or absence) of prior logging. Fuel classes were used to compute the patch's fuel consumption, energy release component, particulate emissions (particulate matter [PM]2.5 and PM10 μ), crown fire potential, and fire behavior attributes for average prescribed burns and wildfires based on published procedures (Huff et al. 1995; Hessburg et al. 2000c). Equations for estimating fuel consumption for both burn scenarios were taken from the CONSUME 2.0 model (Ottmar et al. 1993, www.fs.fed.us/pnw/fera/products) and the First Order Fire Effects model (Reinhardt et al. 1996, www.frames.gov/tools/FOFEM). The attributes of fire behavior associated with each vegetation patch were the rate of spread, flame length, and Byram's fireline intensity (Rothermel 1983). We computed these attributes for an average wildfire scenario by using the equations of the National Fire Danger Rating System (Rothermel 1972; Deeming et al. 1977; Cohen and Deeming 1985).

Estimating the Reference Conditions

We used the FRAGSTATS software (McGarigal and Marks 1995) to characterize spatial patterns in 18 different maps of the historical subwatersheds of the Moist & Cold Forests subregion: (1) physiognomic conditions; (2) cover types; (3) structural classes; (4) cover types combined with structural classes; (5) potential vegetation types combined with cover types and structural classes; (6) canopy layers; (7) total crown cover; (8) patches of late-successional, old, and other forest; (9) patches with and without large remnant trees after wildfires; (10) fuel loading; (11–14) crown fire potential, fireline intensity, rate of spread, and flame length in an average wildfire; and (15–18) crown fire potential, fireline intensity, rate of spread, and flame length under prescribed burning.

We chose five class metrics to display the area and connectivity of the classes in each map: the percentage of the total area, patch density per 10,000 ha, mean patch size (ha), mean nearest-neighbor distance (m), and edge density (m·ha^{-1}). These metrics characterized the spatial patterns of individual classes in a landscape mosaic, such as the ponderosa pine (*Pinus ponderosa* Laws.) cover type in a landscape map of all cover types. Mean, median, full range, and reference variation statistics were computed for the 18 subwatersheds sampled for the subregion by using the S-PLUS software (Statistical Sciences 1993).

We characterized the features of each mapped landscape mosaic by using nine metrics. Landscape metrics characterized the spatial pattern relationships among the classes that composed the landscape mosaics (all cover types). Of the nine metrics for which we chose to display landscape patterns, six were already available in FRAGSTATS and three were added to the FRAGSTATS source code. We evaluated map patterns using the following parameters:

- *Patch richness:* patch richness (PR) and relative patch richness (RPR) (McGarigal and Marks 1995);
- *Diversity:* Shannon's diversity index (SHDI) (McGarigal and Marks 1995) and Hill's transformation of Shannon's index, N1 (Hill 1973), which is less sensitive to the occurrence of rare patch types than SHDI;
- *Dominance:* Hill's inverse of Simpson's lambda, N2 (Simpson 1949; Hill 1973), which combines measures of diversity and dominance and is the least sensitive of the three diversity measures to the occurrence of rare patch types;
- *Evenness:* a modified Simpson's evenness index (MSIEI) (McGarigal and Marks 1995), which measures relative to maximum evenness, the evenness of patch types, including rare ones, and Alatalo's evenness index (R21) (Alatalo 1981), which measures the evenness among the dominant patch types;
- *Interspersion and juxtaposition:* an index of interspersion and juxtaposition (IJI) (McGarigal and Marks 1995); and
- *Contagion:* an index of contagion (CONTAG) (McGarigal and Marks 1995), which measures dispersion and interspersion of patch types.

We also evaluated the influence of rare and dominant classes on our measures of diversity and evenness. We supplemented the FRAGSTATS source code with the equations for computing the N1, N2, and R21 metrics. The mean, median, full range, and reference variation were again computed for the abovementioned landscape

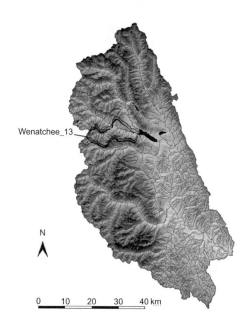

FIGURE 13.4. Map of the Wenatchee_13 subwatershed, a sixth-level drainage (hydrological unit designation: 170200111102) of the Wenatchee River subbasin of eastern Washington State in the western United States. (Refer to figure 13.2 for the geographical context for this figure.)

metrics for the subregion by using the S-PLUS software.

In the *Results and Discussion* section of this chapter, we use the estimates of reference variation to quantify spatial patterns in the departures of 18 different maps of an example subwatershed (Wenatchee_13) in its current condition. The 9553.7-ha Little Wenatchee River and Rainy Creek subwatershed, Wenatchee_13, lies in the Wenatchee River subbasin of the eastern Cascades of Washington (figure 13.4). The western edge of Wenatchee_13 abuts the crest of the Cascades, and the eastern edge abuts the dry and mesic forests of the lower Wenatchee drainage. We evaluated the current conditions in Wenatchee_13 against the estimates of reference variation by using a decision-support model. Our goal was to identify the subset of all vegetation changes that may have important ecological implications (departures). Finally, we conducted a transition analysis on the historical and current maps of cover type and structural class to discover the path of each change (tables 13.3 and 13.4). To conduct the transition analysis, we rasterized the maps of historical and current cover type (or structural class) to a pixel size of 30 m. These 30-m raster versions were combined in a

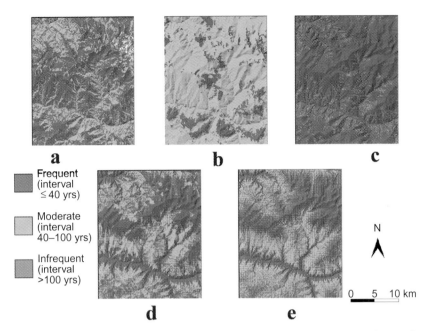

PLATE 1. Predicted fire interval maps generated by the five analytic combinations discussed in Keane, Parsons, and Rollins (chapter 5, this volume). (a) Classification-empirical; (b) simulation-stochastic; (c) statistical-stochastic; (d) statistical-physical; (e) statistical-empirical.

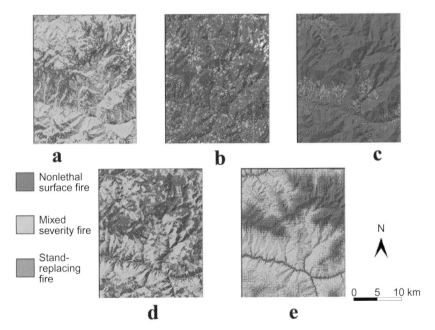

PLATE 2. Predicted fire severity maps generated by the five analytic combinations in Keane, Parsons, and Rollins (chapter 5, this volume). (a) Classification-empirical; (b) simulation-stochastic; (c) statistical-stochastic; (d) statistical-physical; (e) statistical-empirical.

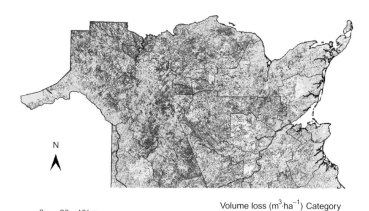

PLATE 3. Predicted volume loss caused by a severe spruce budworm outbreak using the Spruce Budworm DSS (MacLean et al. 2001). The area shown covers more than 3 million ha in northern New Brunswick, Canada. Higher levels of volume loss are equivalent to greater mortality levels caused by natural disturbance.

N

0 20 40km

Volume loss (m³·ha⁻¹) Category

	0–20	Low
	21–40	Moderate
	41–60	High
	>60	Extreme
	Nonsusceptible	
	Water	

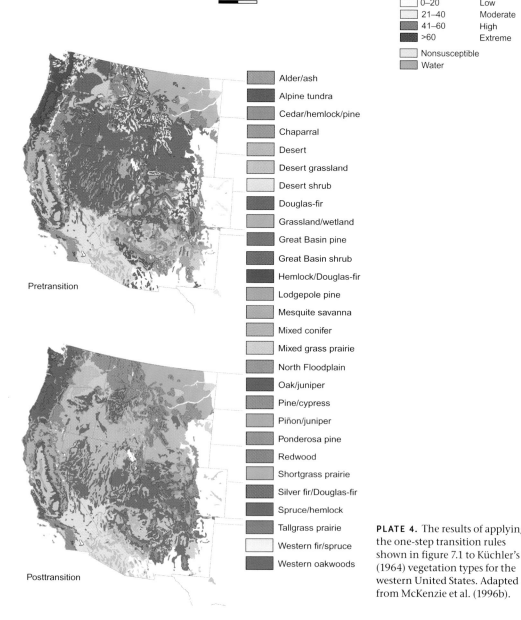

Pretransition

Posttransition

Alder/ash
Alpine tundra
Cedar/hemlock/pine
Chaparral
Desert
Desert grassland
Desert shrub
Douglas-fir
Grassland/wetland
Great Basin pine
Great Basin shrub
Hemlock/Douglas-fir
Lodgepole pine
Mesquite savanna
Mixed conifer
Mixed grass prairie
North Floodplain
Oak/juniper
Pine/cypress
Piñon/juniper
Ponderosa pine
Redwood
Shortgrass prairie
Silver fir/Douglas-fir
Spruce/hemlock
Tallgrass prairie
Western fir/spruce
Western oakwoods

PLATE 4. The results of applying the one-step transition rules shown in figure 7.1 to Küchler's (1964) vegetation types for the western United States. Adapted from McKenzie et al. (1996b).

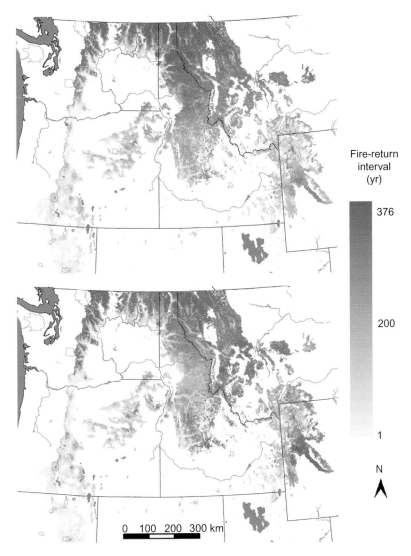

PLATE 5. Gridded maps of the predicted fire-return intervals from two multiple-regression models, using only high-quality fire history data (top) and using all fire history data (bottom). Reprinted, with permission, from McKenzie et al. (2000).

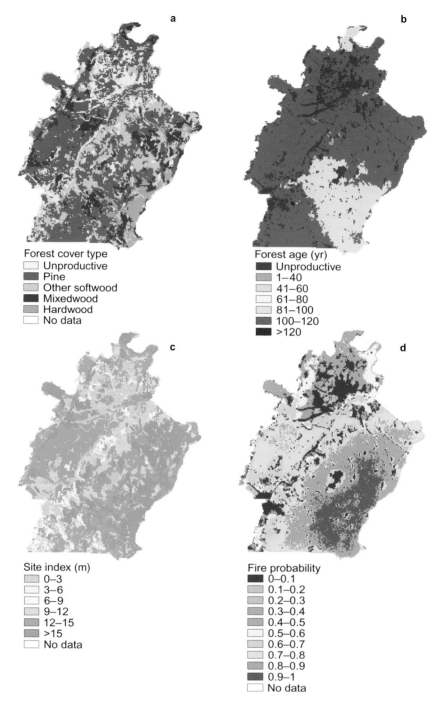

Forest cover type
- Unproductive
- Pine
- Other softwood
- Mixedwood
- Hardwood
- No data

Forest age (yr)
- Unproductive
- 1–40
- 41–60
- 61–80
- 81–100
- 100–120
- >120

Site index (m)
- 0–3
- 3–6
- 6–9
- 9–12
- 12–15
- >15
- No data

Fire probability
- 0–0.1
- 0.1–0.2
- 0.2–0.3
- 0.3–0.4
- 0.4–0.5
- 0.5–0.6
- 0.6–0.7
- 0.7–0.8
- 0.8–0.9
- 0.9–1
- No data

PLATE 6. The GIS data used as model inputs for the study area. (a) Forest cover type; (b) forest age; (c) site index (tree height in m). The data shown in (a–c) were used to produce a map of relative fire probability (d), constructed from fire regime simulations under natural conditions for the study area.

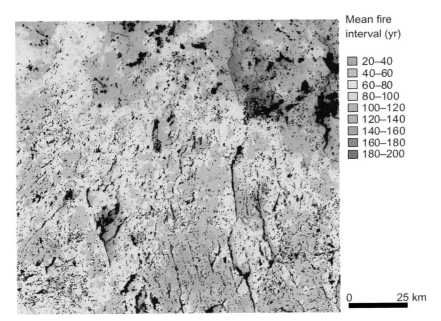

Mean fire
interval (yr)

- 20–40
- 40–60
- 60–80
- 80–100
- 100–120
- 120–140
- 140–160
- 160–180
- 180–200

0 _____ 25 km

PLATE 7. The spatial distribution of mean fire interval (the average interval between two consecutive fires for each 1-ha pixel) during the 200-yr simulation period ($n = 20$).

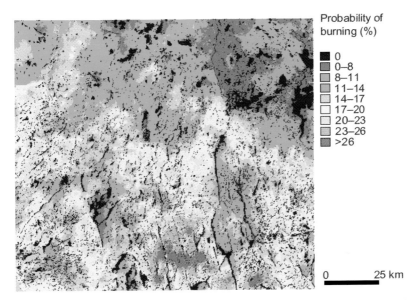

Probability of
burning (%)

- 0
- 0–8
- 8–11
- 11–14
- 14–17
- 17–20
- 20–23
- 23–26
- >26

0 _____ 25 km

PLATE 8. The spatial distribution of the probability of burning (the ratio of the number of times each 1-ha pixel burned to the number of opportunities that pixel was given to burn) during the 200-yr simulation period ($n = 20$).

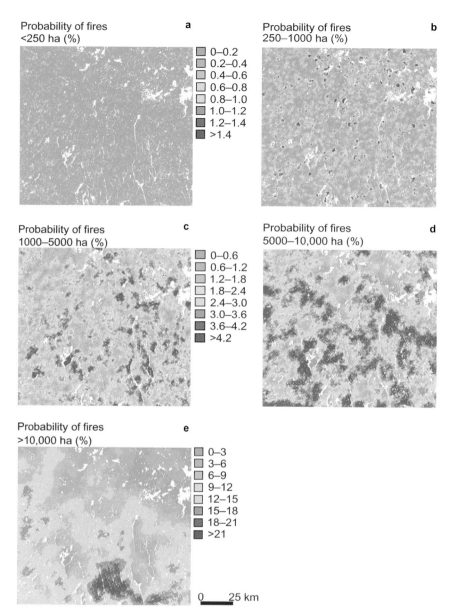

PLATE 9. The spatial distribution of the probability of spatial occurrence for fires of different sizes (the ratio of the number of times each 1-ha pixel was burned by a fire of a given size to the number of opportunities that pixel was given to burn) during the 200-yr simulation period ($n = 20$).

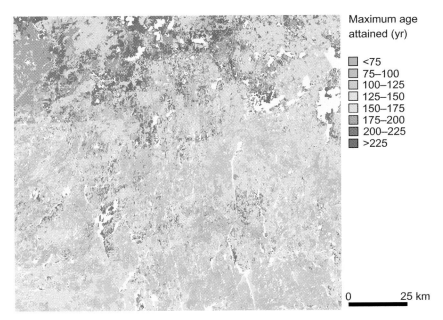

Maximum age
attained (yr)

- □ <75
- □ 75–100
- □ 100–125
- □ 125–150
- □ 150–175
- ▨ 175–200
- ▨ 200–225
- ▨ >225

0   25 km

PLATE 10. The spatial distribution of mean values for the maximum age attained by each 1-ha pixel during the 200-yr simulation period ($n = 20$).

0   10 km

PLATE 11. An example of the stochastic variability of a simulated characteristic of the fire regime. Different colors indicate the different values of the characteristic (e.g., mean maximum age in years), and the elevation ("surface roughness") indicates the stochastic variability (standard error; $n = 20$) of the predicted value.

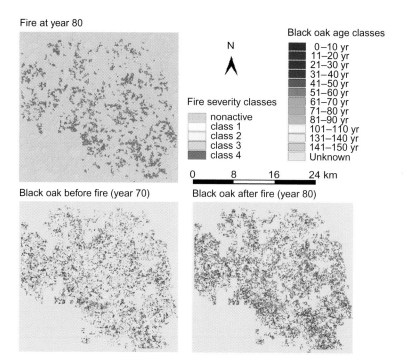

PLATE 12. A snapshot of the patterns of fire at year 80 of the simulation and the patterns for the black oak group before (year 70) and after (year 80) the fires. No fires of severity class 5 were observed at this point in the simulation. Stands in the "unknown" age class represent stands on private land, for which no age data were available. The map shows that fires interact with species in a spatially explicit manner.

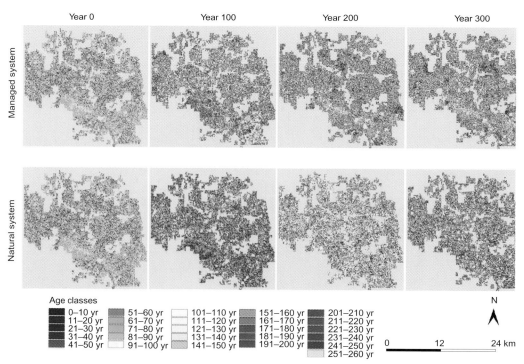

PLATE 13. The dynamics of age classes in the white oak group at the landscape level under a natural disturbance regime compared to a managed disturbance regime.

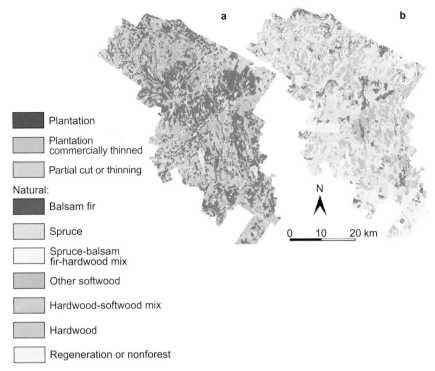

Plantation

Plantation commercially thinned

Partial cut or thinning

Natural:

Balsam fir

Spruce

Spruce-balsam fir-hardwood mix

Other softwood

Hardwood-softwood mix

Hardwood

Regeneration or nonforest

N

0   10   20 km

PLATE 14. Map of stand types in the Black Brook District in (a) 2001 and (b) 1945.

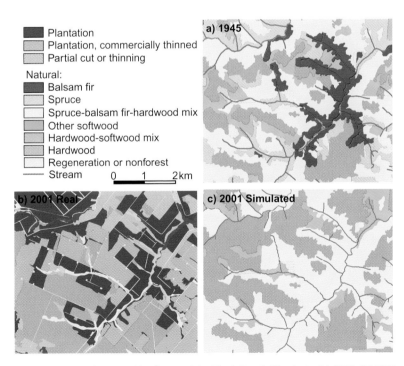

Plantation
Plantation, commercially thinned
Partial cut or thinning

Natural:

Balsam fir
Spruce
Spruce-balsam fir-hardwood mix
Other softwood
Hardwood-softwood mix
Hardwood
Regeneration or nonforest
Stream   0   1   2km

a) 1945

b) 2001 Real

c) 2001 Simulated

PLATE 15. Stand types for a 5 × 6-km² area of the Black Brook District in (a) 1945, (b) 2001, and (c) a simulated 2001 with no harvesting and with eastern spruce budworm outbreaks as the only disturbances since 1945.

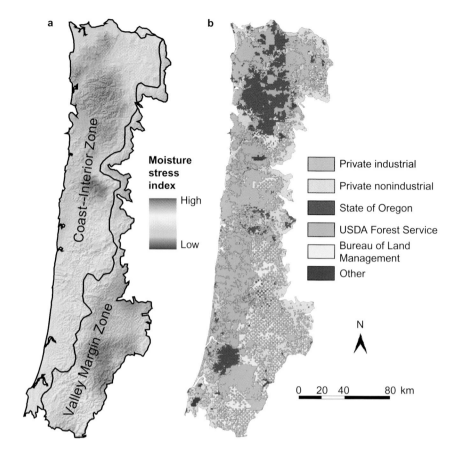

PLATE 16. Patterns of environmental variability and land ownership in the Oregon Coast Range. (a) Moisture stress index, computed as mean summer (May–September) temperature divided by total summer precipitation (Ohmann and Gregory 2002). Shading represents topographic relief. (b) Major land ownership classes.

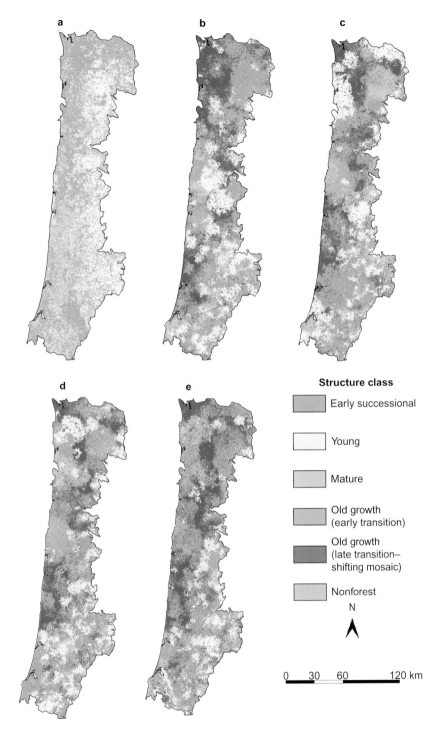

Structure class

- Early successional
- Young
- Mature
- Old growth (early transition)
- Old growth (late transition– shifting mosaic)
- Nonforest

N

0   30   60   120 km

PLATE 17. Distribution patterns for the five forest structure types in the Oregon Coast Range. (a) Present-day patterns. (b)–(e) Examples of simulated historical patterns from a single 1000-yr run of the LADS model. Because LADS is a stochastic model, the simulated landscapes are not intended to predict forest conditions at specific dates. Instead, they illustrate the range of potential landscape patterns under the historical disturbance regime.

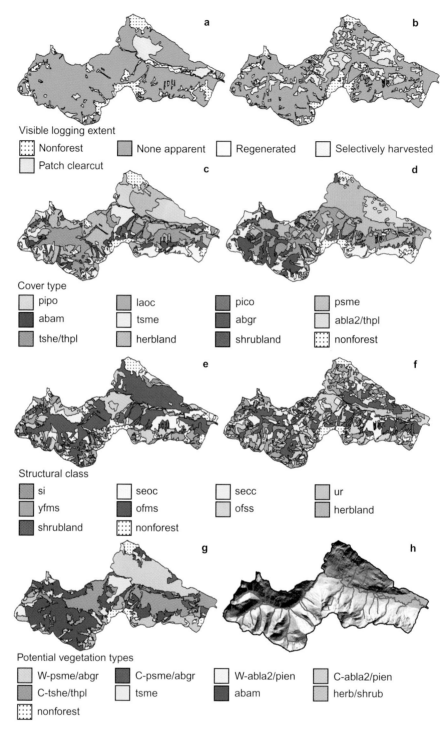

Visible logging extent

- ::: Nonforest
- None apparent
- Regenerated
- Selectively harvested
- Patch clearcut

Cover type

- pipo
- laoc
- pico
- psme
- abam
- tsme
- abgr
- abla2/thpl
- tshe/thpl
- herbland
- shrubland
- ::: nonforest

Structural class

- si
- seoc
- secc
- ur
- yfms
- ofms
- ofss
- herbland
- shrubland
- ::: nonforest

Potential vegetation types

- W-psme/abgr
- C-psme/abgr
- W-abla2/pien
- C-abla2/pien
- C-tshe/thpl
- tsme
- abam
- herb/shrub
- ::: nonforest

PLATE 18. Maps of Wenatchee_13, showing (a) historical (1949) and (b) current (1992) representations of visible logging extent, (c) historical (1900, after reconstruction) and (d) current (1992) cover types, (e) historical (1900, after reconstruction) and (f) current structural classes, (g) potential vegetation types, and (h) topography. Cover type classes and potential vegetation types are defined in table 13.3, and structure classes are defined in table 13.4. The letters W and C before a potential vegetation type indicate a warm–dry or cool–moist variant of the vegetation type, respectively.

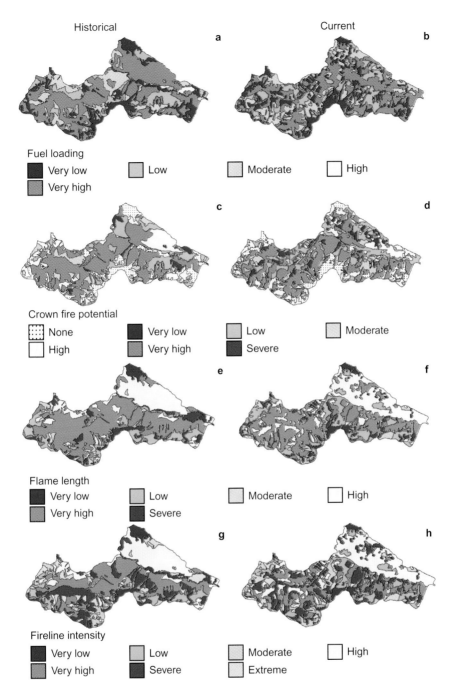

PLATE 19. Maps of Wenatchee_13, displaying historical (1900, after reconstruction) and current (1992) (a, b) fuel loading, (c, d) crown fire potential, (e, f) flame length, and (g, h) fireline intensity, respectively, under a wildfire scenario. Fuel loading (Mg·ha^{-1}) classes: very low, <22.5; low, 22.5–44.9; moderate, 45–56.1; high, 56.2–67.3; very high, >67.3. Crown fire potential classes are represented by an index. Flame-length (m) classes: very low, <0.6; low, 0.7–1.2; moderate, 1.3–1.8; high, 1.9–2.4; very high, 2.5–3.4; severe, >3.4. Fireline intensity (kW·m^{-1}) classes: very low, <173.0; low, 173.0–345.9; moderate, 346.0–1037.8; high, 1037.9–1729.6; very high, 1729.7–2594.4; severe, 2594.5–3459.2; extreme, >3459.2.

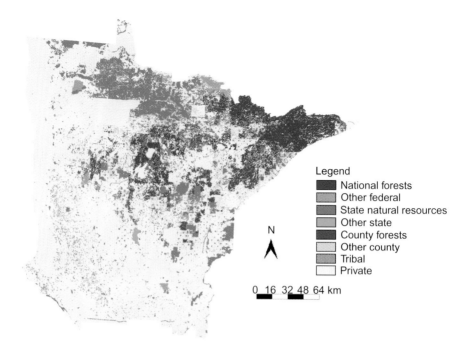

PLATE 20. Land ownership patterns in Minnesota (provided by the Great Lakes Forest Assessment program).

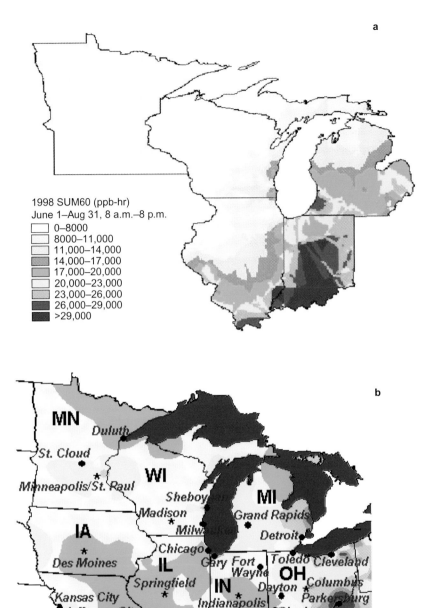

PLATE 21. Variation in trophospheric ozone in the Great Lakes region for different time scales. (a) Total exposure over an entire growing season (1998) shows greater concentration in population centers and industrialized areas to the south and lower exposure in the northern Great Lakes region of Minnesota, Wisconsin, and Michigan. (b) Ozone exposure map showing daily one-hour peak ozone concentrations for June 25, 2002. This map illustrates the short-term increases in tropospheric ozone concentration that can occur under particular weather conditions during the summer. Although these levels are not chronic and long-term, it is believed that they affect growth of sensitive tree species such as aspen and white pine. Color code: green 0–60 ppb; light yellow 61–79 ppb; dark yellow 80–99 ppb; brown 100–110 ppb; orange 111–124 ppb; red ≥125 ppb. Panel (a) provided by E. A. Jepson, Wisconsin Department of Natural Resources, Madison, Wisconsin. Panel (b) from the U.S. Environmental Protection Agency AIRNow web site.

PLATE 22. Fuel piles drying before burning or removal after a mechanized thinning operation carried out in 2001 to reduce fire hazard near developed areas in Grand Teton National Park.

TABLE 13.3. Transitions from Historical to Current Cover Type Conditions for Wenatchee_13

Historical Patch Type[1]	Current Patch Type											
	ABAM	ABGR	ABLA2	Herb	LAOC	Other	PIPO	PSME	Shrub	TSHE	TSME	Total[2]
ABAM	1.3		1.7	0.1		0.2		0.6	0.8	1.1	0.2	6.2
ABGR	0.1					0.1		0.3		0.5	0.1	0.9
ABLA2	1.3		8.3	0.9		0.7		0.3	0.2	0.6		12.4
Herb			0.6	1.0		0.4		0.7	0.1	0.1		2.8
LAOC	0.2							0.2				0.5
Other	0.4	0.1	1.5	0.2		4.0		1.9	0.2	0.6		8.9
PIPO						0.4	5.6	4.6				10.6
PSME	3.6	0.5	2.4	0.3	0.4	0.7	1.1	16.0	0.5	5.2	0.3	30.8
Shrub	0.2		0.5	0.2		0.2		0.4	2.5	0.4		4.4
TSHE	4.6	0.1	0.7	0.1		0.3	0.2	4.5	0.2	8.5		19.1
TSME			0.2					1.1		1.8	0.3	3.3
Total[2]	11.8	0.7	15.9	2.6	0.4	6.9	6.9	30.4	4.5	18.8	0.9	100.0

Notes: Wenatchee_13 is a subwatershed of ecological subregion ESR4 (Moist and Cold Forests). Values represent the percentage of the subwatershed area that transforms from one cover type in its historical condition to another cover type in the current condition, with values rounded to one decimal place.

[1]Cover types: ABAM, amabilis fir (*Abies amabilis* [Dougl.] Forbes); ABGR, grand fir (*Abies grandis* [Dougl.] Lindl.); ABLA2, alpine fir (*Abies lasiocarpa* [Hook.] Nutt.) and Engelmann spruce (*Picea engelmannii* Parry); Herb, all herbland cover types and structural classes combined; Other, all nonforest and nonrangeland and anthropogenic types combined; PIPO, ponderosa pine (*Pinus ponderosa* Laws.); PSME, Douglas-fir (*Pseudotsuga menziesii* [Mirb.] Franco); Shrub, all shrubland cover types and structural classes combined; TSHE, western hemlock (*Tsuga heterophylla* [Raf.] Sarg.); TSME, mountain hemlock (*Tsuga mertensiana* [Bong.] Carr.).

[2]Totals may not add to 100% because of rounding errors.

single coverage, so that each pixel had a historical and current cover type (or structural class). We computed the number of pixels for each unique type of historical-to-current transition, divided this number by the total number of pixels, and multiplied that result by 100 to derive a percentage of the subwatershed area.

Evaluating the Wenatchee_13 Landscape

We used the EMDS software (Reynolds 1999a, 2001b) to compare current spatial patterns in Wenatchee_13 with the reference conditions for the subregion. The current conditions were depicted by using 18 maps that represented a multiscale cross section of the conditions with respect to vegetation, fuel, and potential fire behavior. Each map was evaluated in EMDS against a corresponding set of estimates of reference variation.

Programming estimates of reference variation for the subregion

EMDS (version 3.0) is a decision-support system for integrated landscape evaluation and planning. The application provides support for landscape evaluation through logic and decision engines integrated with the ArcGIS 8.1 GIS software (Environmental Systems Research Institute, Redlands, California). In our application, the NetWeaver logic engine (Reynolds 1999b) evaluated data representing the current conditions of the landscape (which can be represented by data distributions, ranges of conditions, states, or mathematical functions) against a knowledge base we designed using the NetWeaver Developer System to derive interpretations of ecosystem conditions.

We had two main reasons for using EMDS in the present application. First, logic-based models accommodate large, multiscale analytical problems. In this study, for example, the class and landscape metrics that defined the reference conditions for the 18 evaluations were coded in NetWeaver using more than 2700 parameters. Second, although the knowledge bases evaluated by EMDS can be large and complex, the logic engine lets users trace the results using a browser interface that conveys the basis for each conclusion. Due to space limitations, we refer

TABLE 13.4. Transitions from Historical to Current Structural Class for Wenatchee_13

Historical Patch Type[1]	Current Patch Type										
	si	seoc	secc	ur	yfms	ofms	ofss	Herb	Shrub	Other	Total[2]
si	0.3	0.8		0.6	0.6		0.1	0.1	0.1	0.1	2.8
seoc	0.6	3.2	0.5	2.1	3.0	2.7	1.0	0.4	0.7	0.7	15.1
secc		0.1	0.2	0.3		0.7					1.4
ur	0.6	1.6	1.8	6.3	1.9	1.3	0.3		0.1	0.3	14.4
yfms	0.4	1.3		2.6	2.7	0.2	0.6	0.5	0.4	0.3	9.0
ofms	6.9	3.9	2.2	4.9	3.0	14.1	1.6	0.2	0.3	0.9	38.0
ofss	0.2	0.3	0.4	0.5	0.9	0.3	0.4		0.1	0.1	3.2
Herb	0.2	0.1		0.2	0.2	0.1	0.7	1.0	0.1	0.4	2.8
Shrub	0.1	0.3		0.4	0.1	0.4	0.2	0.2	2.5	0.2	4.4
Other	0.7	0.6		0.4	1.7	0.1	1.0	0.2	0.2	4.0	8.9
Total[2]	10.2	12.1	5.2	18.5	14.1	20.0	5.9	2.6	4.5	6.9	100.0

Notes: Wenatchee_13 is a subwatershed of ecological subregion ESR4 (Moist and Cold Forests). Values represent the percentage of the subwatershed area that converts from a structural class in the historical condition to another class in the current condition, rounded to one decimal place.

[1]Structural classes: Herb, all herbland cover types and structural classes combined; ofms, "old forest, multistory"; ofss, "old forest, single story"; Other, all nonforest and nonrangeland and anthropogenic types combined; secc, "stem exclusion, closed-canopy"; seoc, "stem exclusion, open-canopy"; Shrub, all shrubland cover types and structural classes combined; si, stand initiation; ur, understory reinitiation; yfms, "young forest, multistory."

[2]Totals may not add to 100% because of rounding errors.

readers to Reynolds (2001a,b) for detailed discussions of decision-support systems and the advantages of performing landscape evaluations in EMDS.

Results and Discussion

In our evaluation of patterns in the forest landscape, we examined the characteristics of the 18 landscape mosaics, as well as the spatial patterns of the component classes. Before we examined the products of the EMDS-based evaluations of Wenatchee_13, we first compared several different historical and current maps to assess obvious changes. We hoped to determine whether the most visible changes represented departures or were within the bounds of reference variation. We examined the following maps:

- Historical (1949, prior to reconstruction with most-similar-neighbor analysis) and current (1992) visible logging extent (color plates 18a and b, respectively);
- Historical (after reconstruction with most-similar-neighbor analysis) and current cover types (color plates 18c and d, respectively);
- Historical and current structural classes (color plates 18e and f, respectively);

- Potential vegetation types (color plate 18g);
- Topography (color plate 18h);
- Historical (after reconstruction with most-similar-neighbor analysis) and current (1992) fuel loading (color plates 19a and b, respectively); and
- Historical and current crown fire potential (color plates 19c and d, respectively), flame length (color plates 19e and f, respectively), and fireline intensity (color plates 19g and h, respectively) under an average wildfire scenario.

In the map of visible logging extent, selection cutting had affected about 7.7% of the area. Selection cutting had a relatively minor influence on structural conditions in the harvested areas, as fewer than half of the harvested "old forest, multistory" patches were sufficiently influenced to change their structural designation, and those that were affected were converted to "stem exclusion, open-canopy" or "old forest, single-story" structures. However, early selection cutting significantly modified cover types. Selection cutting apparently targeted large overstory ponderosa pine growing in the warm–dry and cool–moist

Douglas-fir (*Pseudotsuga menziesii* [Mirb.] Franco)/ grand fir (*Abies grandis* [Dougl.] Lindl.) potential vegetation types and converted much of the harvested area to a Douglas-fir cover type. An additional 23.7% of the current subwatershed area has been influenced by timber harvesting, and the most recent cutting has been patch clearcutting to promote regeneration in the same potential vegetation types.

The historical cover type and structural class maps (color plates 18c,e) show simple, contagious patterns of land cover and forest structure reflecting the relatively simple patterns of potential vegetation types (color plate 18g) and topography (color plate 18h). Both the simple patterns and the high historical area of "old forest, multistory" patches (38%, table 13.4) suggest that fires have been infrequent in Wenatchee_13. The current logging extent, cover type, and structural class maps (color plates 18b,d, and f, respectively) show a landscape fragmented into 49 regeneration units based on the type of harvesting (minimum, 2.25 ha; maximum, 302 ha; median, 11.8 ha). A significant area of the ponderosa pine cover type has been converted to Douglas-fir (4.6%), and 18% of the "old forest, multistory" area has been lost to the western hemlock (*Tsuga heterophylla* [Raf.] Sarg.)–western red cedar (*Thuja plicata* Donn), Douglas-fir, and ponderosa pine cover types (see table 13.3).

Historical maps of fuel loading (color plate 19a), crown fire potential (color plate 19c), flame length (color plate 19e), and fireline intensity (color plate 19g) depict a landscape that displayed large contiguous areas with a very high (>30 Mg·ha^{-1}) fuel loading and the potential for crown fires under an average wildfire scenario, high to extreme flame lengths (>6 m), and high to extreme fireline intensities (>1038 kW·m^{-1}). It is evident from looking at the historical fuel (color plate 19a), fire behavior (color plate 19c), and structural class (color plate 19e) maps that fires seldom burned in the Wenatchee_13 watershed, but when they did, they were probably significant: moderate severity fires with a large stand-replacement component (see table 13.2, ESR4). Current conditions suggest that past management activities in Wenatchee_13 have reduced the likelihood of stand-replacing fires, their attendant ecological effects, and the scales of those effects.

EMDS Evaluation of Wenatchee_13

For simplicity, the results of the complete EMDS analysis are summarized in graphical form (fig-

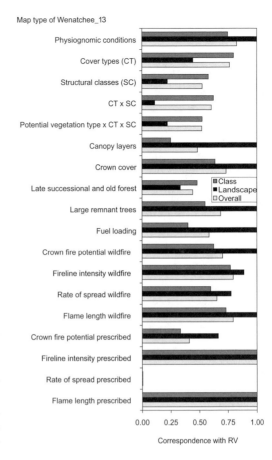

FIGURE 13.5. Summary graph of the results of using the EMDS system to analyze 18 maps representing the current conditions in the Wenatchee_13 subwatershed. Evaluations using five class metrics, nine landscape metrics, and the class + landscape metrics (overall comparison) compared with a map of current conditions with the corresponding reference variation (RV) estimates for the Moist and Cold Forests ecological subregion (ESR4).

ure 13.5). The x-axis shows the degree of correspondence between current Wenatchee_13 spatial patterns for a given map and the estimated reference variation. Intervals along the [0, 1] scale represent the following levels of correspondence between the two: 0 = no correspondence; [0.01, 0.25] = very weak; [0.26, 0.50] = weak; [0.51, 0.75] = moderate; [0.76, 0.99] = strong; and 1.00 = full correspondence. For example, in an evaluation of current physiognomic conditions, there is moderate correspondence when 75% of the values of each class metric for each physiognomic type fall within their respective ranges of reference variation, but full correspondence when 100% of the values for the landscape metrics fall within these ranges.

Evaluations of cover types

Evaluations of the current cover type conditions show strong overall correspondence; the correspondence is strong when the five class metrics are evaluated against estimates of reference variation for all cover types, and weak when the nine landscape metrics are evaluated (figure 13.5). Chief among the departures for the class metrics are the elevated patch densities for the Douglas-fir, grand fir, western hemlock–western red cedar, shrubland, and nonforest cover types. Weak correspondence between the current cover type mosaic and estimates of reference variation is a

result of elevated cover type richness, which is indicated by departures in patch richness and diversity (table 13.5).

Physiognomic conditions

Physiognomic conditions show strong overall correspondence; correspondence was full when landscape metrics are evaluated, and moderate when class metrics are evaluated (figure 13.5). When we traced the basis of the latter conclusion in EMDS, we learned that the patch densities of shrubland and nonforest and the edge density of forest (increased by the increased number of cutting units and their boundaries) are well above the limits of reference variation. Fragmentation of forest cover types is so widespread

TABLE 13.5. Comparison of Nine Metrics Representing the Current Combined Cover Type–Structural Class Landscape Conditions in the Subwatershed Wenatchee_13

Landscape Metrics[1]	Range Estimate		Wenatchee_13	
	Min	Max	Current Conditions	Historical Conditions
Richness and diversity				
RPR full range	20.48	61.45		
RPR 80% range (RV)	24.10	41.93	**55.42**	**46.99**
PR full range	17.00	51.00		
PR 80% range (RV)	20.00	34.80	**46.00**	**39.00**
SHDI full range	2.22	3.26		
SHDI 80% range (RV)	2.34	2.87	**3.32**	**2.96**
N1 full range	9.21	26.05		
N1 80% range(RV)	10.40	17.66	**27.66**	**19.30**
N2 full range	6.25	20.00		
N2 80% range (RV)	7.97	13.04	**25.00**	**14.29**
Evenness				
MSIEI full range	0.56	0.79		
MSIEI 80% range (RV)	0.62	0.76	**0.82**	0.71
R21 full range	0.59	0.81		
R21 80% range (RV)	0.63	0.77	**0.90**	0.73
Contagion and interspersion				
CONTAG full range	50.59	61.02		
CONTAG 80% range (RV)	50.91	59.35	**48.60**	53.25
IJI full range	64.37	77.12		
IJI 80% range (RV)	67.72	74.76	74.58	74.03

Notes: Corresponding reference variation (RV) and full range estimates developed for ESR4 B the Moist and Cold Forests subregion are also shown. Values in **bold** lie outside RV.

[1]CONTAG, contagion index; IJI, interspersion and juxtaposition index (see also McGarigal and Marks 1995); MSIEI, modified Simpson's evenness index; N1, Hill's N1 index = e^{SHDI}; N2, Hill's N2 index = $1/SIDI$; PR, patch richness; R21, Alatalo's evenness index = $(N2 - 1)/(N1 - 1)$; RPR, relative patch richness; SHDI, Shannon diversity index.

that it affected the edge density of the forest physiognomy.

Structural class conditions

Structural class conditions show only moderate overall correspondence between current conditions and estimates of reference variation, moderate correspondence when class metrics are evaluated, and very weak correspondence when the landscape metrics are evaluated. Key departures for the class metrics are the elevated patch density and reduced mean patch size. Weak correspondence between the current structural class mosaic and the estimates of reference variation is a result of elevated diversity, dominance, evenness, interspersion, and juxtaposition of structural classes, and of dramatically reduced contagion.

Combined cover type–structural class conditions

The combined map of cover type and structural class shows moderate overall correspondence between current conditions and estimates of reference variation; correspondence is moderate when class metrics are evaluated, and very weak when the landscape metrics are evaluated. Key departures for the class metrics are elevated patch densities and reduced mean patch sizes for most cover type–structural class combinations, but the mean nearest-neighbor distance is also reduced for most classes. In addition, there are departures in the percentage of total area in the ponderosa pine, western larch (*Larix occidentalis* Nutt.), grand fir, and amabilis fir (*Abies amabilis* [Dougl.] Forbes) stand-initiation structures, representing an expanded area of new forest, and in intermediate forest structures ("stem exclusion, open-canopy," "stem exclusion, closed canopy," and "young forest, multistory") associated with the ponderosa pine, grand fir, and western hemlock–western red cedar cover types.

Weak correspondence between the current combined cover type–structural class mosaic and the estimates of reference variation is a result of increased richness, diversity, dominance, evenness, interspersion, and juxtaposition of structural class patches and of reduced contagion. The combined potential vegetation type–cover type–structural class mosaic shows nearly identical results (figure 13.5).

Fragmented landscapes

The common finding among each of the 18 map evaluations is that the landscape had become highly fragmented, an observation that is consistently indicated by reduced contagion, elevated patch densities and mean nearest-neighbor distances, and reduced mean patch sizes. Nowhere is this better indicated than in the evaluations of fuel loading, crown fire potential, and fire behavior patterns. For example, fuel loading shows moderate overall correspondence between current conditions and estimates of reference variation; correspondence is full when landscape metrics are evaluated, and weak when class metrics are evaluated. When we traced the basis of the conclusions in EMDS, we learned that patch density and mean patch size in all fuel loading classes are well outside of the reference variation (color plate 19b).

Departures in Landscape Patterns for Wenatchee_13

Reduced area of old forest

Tables 13.3 and 13.4 display all cover type and structural class transitions from the historical to the current conditions. Early selection cutting apparently targeted large, easily accessed overstory ponderosa pine growing in warm–dry and cool–moist Douglas-fir–grand fir potential vegetation types, and large overstory Douglas-fir growing in cool–moist western hemlock–western red cedar and amabilis fir potential vegetation types. Most of the recent regeneration cutting has been in the same environments. Analysis of the transitions from historical to current structural classes (see table 13.4) shows a net 18% reduction in "old forest, multistory" structures and corresponding increases in stand initiation (6.9%), "stem exclusion, open-canopy" (3.9%), "stem exclusion, closed-canopy" (2.2%), understory reinitiation (4.9%), and "young forest, multistory" (3.0%) structures.

Reduced cover of large, early seral overstory trees

Similarly, analysis of the transitions from historical to current cover types (see table 13.3) shows a net 3.7% reduction in the ponderosa pine cover type and an increase (4.6%) in the Douglas-fir cover type. A combined cover type–structural class transition analysis showed a total transition of large, early seral overstory Douglas-fir cover equal to 11.7% of the subwatershed area to late seral cover comprising amabilis fir (3.6%), grand fir (0.5%), alpine fir (*Abies lasiocarpa* [Hook.] Nutt.)–Engelmann spruce (*Picea engelmannii* Parry) (2.4%), and western hemlock–western red cedar (5.2%) understory cover types. This transition can also be seen in table 13.3, and it reflects

TABLE 13.6. Comparison of Three Class Metrics Representing the Current Cover Type–Structural Class Combinations in the Subwatershed Wenatchee_13

	Percentage Area						Patch Density (10,000 ha)						Mean Patch Size (ha)					
			RV		Full Range				RV		Full Range				RV		Full Range	
SC[1]	C[2]	H	Min	Max	Min	Max	C	H	Min	Max	Min	Max	C	H	Min	Max	Min	Max
Ponderosa pine (PIPO)																		
si	**1.4**	0.2	0.0	1.1	0.0	3.0	**4.0**	1.0	0.0	2.0	0.0	12.0	34.4	19.4	0.0	38.1	0.0	74.9
seoc	0.5	0.6	0.0	2.1	0.0	4.8	3.0	**6.0**	0.0	5.0	0.0	24.0	15.3	9.3	0.0	38.6	0.0	129.2
secc	**0.7**	0.0	0.0	0.0	0.0	4.4	**4.0**	0.0	0.0	0.0	0.0	7.0	**16.1**	0.0	0.0	0.0	0.0	66.4
ur	0.4	0.0	0.0	2.1	0.0	6.2	2.0	0.0	0.0	2.0	0.0	11.0	20.7	0.0	0.0	45.7	0.0	327.7
yfms	0.6	0.2	0.0	1.9	0.0	2.4	**5.0**	2.0	0.0	3.0	0.0	10.0	11.1	10.8	0.0	87.8	0.0	139.9
ofms	**3.0**	**8.5**	0.0	2.3	0.0	8.5	2.0	**2.0**	0.0	1.0	0.0	15.0	**144.8**	**405.9**	0.0	34.3	0.0	405.9
ofss	0.3	1.1	0.0	1.3	0.0	3.0	1.0	2.0	0.0	3.0	0.0	7.0	26.6	**52.3**	0.0	33.3	0.0	56.8
Douglas-fir (PSME)																		
si	4.1	1.6	0.0	8.4	0.0	10.0	**44.0**	5.0	0.0	14.0	0.0	21.0	9.3	30.0	0.0	57.1	0.0	92.1
seoc	4.1	5.0	0.0	9.2	0.0	17.2	**23.0**	**18.0**	0.0	15.0	0.0	28.0	18.0	28.3	0.0	128.8	0.0	475.7
secc	3.6	0.1	0.0	14.9	0.0	20.5	**21.0**	1.0	0.0	13.0	0.0	18.0	17.1	7.8	0.0	195.7	0.0	425.3
ur	4.0	4.0	0.4	19.5	0.0	27.5	20.0	9.0	0.1	24.0	0.0	32.0	20.3	42.2	17.0	175.6	5.8	327.9
yfms	3.0	1.4	0.6	9.3	0.0	21.7	17.0	8.0	0.3	23.0	0.0	33.0	18.0	16.5	10.6	74.8	0.0	91.9
ofms	7.5	**16.9**	0.0	12.5	0.0	16.9	**17.0**	**17.0**	0.0	15.0	0.0	17.0	44.5	101.0	0.0	271.0	0.0	878.7
ofss	4.1	1.9	0.0	4.6	0.0	7.3	**4.0**	6.0	0.0	7.0	0.0	11.0	98.8	30.1	0.0	128.9	0.0	142.5
Western larch (LAOC)																		
si	**0.3**	0.0	0.0	0.2	0.0	1.3	**1.0**	0.0	0.0	1.0	0.0	2.0	**25.4**	0.0	0.0	17.6	0.0	74.3
ur	0.0	0.3	0.0	11.6	0.0	21.0	0.0	2.0	0.0	21.0	0.0	47.0	0.0	13.3	0.0	54.8	0.0	132.0
yfms	0.2	0.0	0.0	0.3	0.0	2.4	1.0	0.0	0.0	1.0	0.0	2.0	14.6	0.0	0.0	15.1	0.0	96.2
ofms	0.0	0.2	0.0	3.5	0.0	7.8	0.0	1.0	0.0	22.0	0.0	34.0	0.0	**20.5**	0.0	15.1	0.0	26.3
Grand fir (ABGR)																		
si	**0.1**	0.0	0.0	0.0	0.0	0.2	**1.0**	0.0	0.0	0.0	0.0	1.0	**4.6**	0.0	0.0	0.0	0.0	15.6
seoc	0.1	**0.8**	0.0	0.3	0.0	0.8	1.0	1.0	0.0	1.0	0.0	3.0	7.6	**73.2**	0.0	13.5	0.0	73.2
yfms	**0.3**	0.0	0.0	0.1	0.0	1.4	**2.0**	0.0	0.0	0.0	0.0	1.0	**15.4**	0.0	0.0	8.6	0.0	146.2

SC																		
ofms	**0.3**	0.0	0.0	0.0	0.0	0.0	**4.0**	0.0	0.0	0.0	0.0	0.0	**6.4**	0.0	0.0	0.0	0.0	0.0
Amabilis fir (ABAM)																		
si	**3.0**	0.3	0.0	2.6	0.0	6.2	**12.0**	2.0	0.0	8.0	0.0	10.0	26.4	12.8	0.0	43.5	0.0	62.1
seoc	1.1	1.7	0.0	4.5	0.0	6.1	8.0	13.0	0.0	14.0	0.0	23.0	13.6	13.6	0.0	84.6	0.0	173.9
secc	0.5	0.2	0.0	13.1	0.0	14.6	3.0	1.0	0.0	14.0	0.0	17.0	14.2	16.9	0.0	118.1	0.0	188.4
ur	3.4	2.2	0.0	10.0	0.0	13.9	12.0	9.0	0.0	18.0	0.0	22.0	29.4	23.8	0.0	91.4	0.0	173.8
yfms	1.7	1.0	0.0	6.1	0.0	12.5	10.0	7.0	0.0	14.0	0.0	30.0	15.9	13.0	0.0	73.4	0.0	100.4
ofms	1.9	0.8	0.0	2.4	0.0	10.6	2.0	4.0	0.0	4.0	0.0	7.0	**92.2**	19.4	0.0	57.2	0.0	281.4
ofss	0.2	0.0	0.0	2.5	0.0	21.8	2.0	0.0	0.0	4.0	0.0	6.0	10.8	0.0	0.0	90.0	0.0	461.5
Alpine fir–Engelmann spruce (ABLA2)																		
si	1.3	0.8	0.0	12.8	0.0	15.9	**15.0**	2.0	0.0	14.0	0.0	18.0	9.2	36.9	0.0	87.3	0.0	109.9
seoc	3.3	2.4	0.3	3.3	0.0	4.0	**16.0**	**16.0**	1.0	13.0	0.0	19.0	20.8	15.1	10.4	54.4	0.0	72.1
secc	0.2	0.2	0.0	2.5	0.0	3.1	**3.0**	4.0	0.0	4.0	0.0	5.0	6.8	3.6	0.0	75.1	0.0	248.9
ur	4.4	2.7	0.0	11.0	0.0	25.3	**20.0**	9.0	0.0	13.0	0.0	17.0	22.3	29.1	0.0	163.1	0.0	267.6
yfms	6.7	6.2	0.2	12.7	0.0	19.7	**30.0**	**26.0**	1.0	24.0	0.0	32.0	22.0	23.6	19.2	69.7	0.0	85.4
Western hemlock–western red cedar (TSHE/THPL)																		
seoc	**3.0**	**4.6**	0.0	1.1	0.0	4.6	**5.0**	**6.0**	0.0	4.0	0.0	6.0	**57.0**	**73.3**	0.0	47.9	0.0	73.3
secc	0.2	0.9	0.0	6.6	0.0	8.0	1.0	2.0	0.0	6.0	0.0	12.0	15.8	41.5	0.0	100.3	0.0	279.3
ur	5.9	1.8	0.0	8.7	0.0	31.4	**13.0**	**7.0**	0.0	6.0	0.0	13.0	46.6	24.3	0.0	288.8	0.0	660.3
yfms	**1.5**	0.3	0.0	1.2	0.0	3.4	**4.0**	2.0	0.0	3.0	0.0	11.0	36.2	13.9	0.0	40.5	0.0	94.5
ofms	**7.1**	**11.4**	0.0	1.8	0.0	11.4	**16.0**	**7.0**	0.0	5.0	0.0	7.0	44.9	**155.6**	0.0	74.1	0.0	155.6
ofss	1.2	0.2	0.0	4.7	0.0	7.2	**8.0**	1.0	0.0	3.0	0.0	8.0	14.7	18.3	0.0	143.5	0.0	259.7
Mountain hemlock (TSME)																		
secc	0.1	0.1	0.0	0.8	0.0	3.8	1.0	1.0	0.0	1.0	0.0	3.0	8.4	8.6	0.0	52.8	0.0	225.1
ur	0.4	**3.2**	0.0	2.3	0.0	3.2	2.0	2.0	0.0	3.0	0.0	9.0	16.6	**154.9**	0.0	83.7	0.0	238.7
yfms	0.2	0.0	0.0	2.5	0.0	7.3	1.0	0.0	0.0	5.0	0.0	13.0	17.5	0.0	0.0	31.3	0.0	74.4
ofms	0.3	0.0	0.0	0.4	0.0	2.1	1.0	0.0	0.0	0.0	0.0	2.0	**25.7**	0.0	0.0	12.9	0.0	111.9

Notes: Corresponding reference variation (RV) and full range estimates developed for the ESR4 Moist and Cold Forests subregion are also shown. Values in **bold** lie outside RV.

[1] SC (structural classes): ofms, "old forest, multistory"; ofss, "old forest, single story"; secc, "stem exclusion, closed-canopy"; seoc, "stem exclusion, open-canopy"; si, stand initiation; ur, understory reinitiation; yfms, "young forest, multistory." Species names are as defined in table 12.3.

[2] C, current; H, historical; RV, reference variation.

the removal of large Douglas-fir overstories from "old forest, multistory" patches.

In table 13.6, we compare current and historical values of three class metrics (percentage of total area, patch density, and mean patch size) for a partial list of cover type–structural class combinations with corresponding estimates of reference variation. For example, the estimated percentage area of the ponderosa pine stand-initiation class is 0.0–1.1% of the total area. In the current condition, this class occupies 1.4% of the area, and the current area is above the estimates for reference variation. This increased dominance of ponderosa pine stand-initiation structures resulted from regeneration harvests.

Also in table 13.6, we display historical values of the class metrics and the full range of historical values for each class and metric. Significant historical and current departures from reference variation are highlighted in bold. Class metrics for several current and historical cover type–structural class combinations lie outside the estimated reference variation. Structural classes of the ponderosa pine, Douglas-fir, and western hemlock–red cedar cover types exhibit the greatest departures, because these classes contain the greatest area of "old forest, multistory" structures, and old forests were targeted for early selection cutting, and more recently, for regeneration harvests (Hessburg et al. 1999b; Hessburg and Agee in press) (see tables 13.3, 13.4). In the historical condition, there was no area of "stem exclusion, closed canopy" or understory reinitiation patches in the ponderosa pine cover type, and the area in "young forest, multistory" patches was small (0.2%). A likely explanation is that historical surface fires and dry site conditions on sites with southern aspects maintained open-canopy rather than closed-canopy stem-exclusion structures (Agee 1993). The net effect was a simplified landscape mosaic on lower montane sites with southern aspects; ponderosa pine land cover was historically dominated by "old forest, multistory" and "old forest, single story" structures, and trace amounts of stand initiation, "stem exclusion, open canopy," and "young forest, multistory" structural classes (see tables 13.4, 13.6).

Working with unique or
borderline watersheds of a subregion

The estimates of reference variation for areas of ponderosa pine, Douglas-fir, and western hemlock–western red cedar "old forest, multistory" patch types ranged, respectively, from 0% to 2.3, 12.5, and 1.8%, but the historical areas of

these structures were 8.5, 16.9, and 11.4, respectively; these values lie well above the estimated range of reference variation. Wenatchee_13 is somewhat unusual among the historical subwatersheds we sampled because 38% of the watershed comprised "old forest, multistory" structures (see table 13.4), and this area was aggregated in a few large patches (table 13.6). This observation suggests that it may be appropriate to consider the full range of values for the class and landscape metrics when evaluating opportunities to restore the area, connectivity, and pattern of old forests and perhaps other attributes of some Moist and Cold Forest subregion landscapes. Unique fire ecology, landform features, or environments may make some landscapes of a subregion appear atypical. We discuss some reasons for this in our *Conclusions* section.

Relevance of Departures in Landscape Patterns

At relatively fine to broad spatial and temporal scales, the response of terrestrial species to landscapes and their patterns indicates whether these environments are more or less suitable to their particular needs. Changes in the patterns of vegetation at certain spatial and temporal scales may have a direct bearing on species migration, colonization, the availability of habitats and food, and the persistence of a species in a landscape (Wisdom et al. 2000). As patterns change significantly, different suites of species may be favored. Such changes also influence the spatial and temporal scales and parameters of disturbance regimes. As tables 13.5 and 13.6 show, variability in the spatial patterns of historical landscapes was commonplace. Variable landscape patterns at subwatershed, watershed, subregional, and regional scales probably provide alternating periods and patterns of plenty and need that may help native species to develop broad genetic and phenotypic diversity and necessary adaptations as long as habitats do not become overly fragmented or isolated in space or time (Swanson et al. 1994). Natural variability in vegetation patterns, climate, and geological systems is also linked to natural variation in disturbance regimes.

After evaluating the classes in the 18 maps with respect to reference variation, we characterized departures in spatial patterns for each landscape mosaic. For example, we compared the overall current cover type–structural class mosaic of Wenatchee_13 with corresponding estimates of reference variation (table 13.5). Current values of eight of the nine landscape metrics lie beyond these limits; historical values of five metrics also show evidence of departures. In the Wenatchee_13 historical condition, all richness and diversity metrics are above the estimates of reference variation for the subregion. Thus Wenatchee_13, at least in this temporal snapshot, displayed a richer and more diverse array of cover type–structural class patches than ordinarily occurred in similar subwatersheds of the subregion in the period for which we characterized reference variation. This was probably true of other watersheds and landscape attributes at other times. In the historical condition, Wenatchee_13 displayed nearly 47% of the total possible number of cover type–structural class combinations that were present in the entire subregion (relative patch richness$_{historical}$; table 13.5); in the current condition, more than 55% of those possibilities are displayed. The historical value of absolute patch richness exceeds the estimated reference variation by more than four cover type–structural class patch combinations, and seven cover type–structural class combinations developed in the landscape as a consequence of management activities.

In the historical Wenatchee_13, three indices registered as lying above the reference variation: Shannon's diversity index, which measures the proportional abundance of classes and the equitable distribution of area; N1, a transformation of Shannon's diversity index; and N2, the inverse of Simpson's lambda, a metric that represents dominance and diversity. Current values for the three metrics also lie above the limits of the full range of variation. Timber harvesting coupled with fire exclusion has created many new cover type–structural class combinations. For example, a comparison of historical and current values of N2 indicates that the number of dominant cover type–structural class combinations increased from about 14 to 25 (table 13.5).

A comparison of the historical and current values of the evenness metrics (modified Simpson's evenness index and Alatalo's R21 evenness index) shows significantly elevated evenness among the cover type–structural class combinations. The modified Simpson's index is sensitive to changes in the evenness of all classes, including rare ones; Alatalo's index is sensitive to changes in the evenness of the dominant classes. Harvesting activities increased the complexity and evenness of patterns in the cover type–structural class mosaic. Considering only the dominant 25 cover type–structural class combinations (N2$_{current}$), the current mosaic displays 90% of the maximum possible evenness for the number of classes in the subregion (R21$_{current}$). The historical mosaic displays 73% of the maximum possible evenness (table 13.5).

The current value of the contagion metric also lies outside the range of the estimates of reference variation, but unlike other landscape metrics, contagion decreased when compared with its historical value. Contagion was apparently reduced by timber harvesting (color plate 18b) and perhaps by fire exclusion, which fragmented the historical areas of "old forest, multistory" structures. Timber harvests had homogenized the simple, contagious patterns of forest structure in the historical landscape.

Conclusions

A scientific and social consensus is emerging that land managers must restore more natural conditions to forests. One approach focuses management and restoration efforts on emulating natural disturbance, but restoring the role of disturbance requires to some extent emulation of the range and variation in the vegetation conditions that support it. Before settlement of the region, fire played a dominant role in sculpting vegetation patterns and the associated processes. Before restoring the natural role of disturbance, managers must restore more natural variation in the spatial and temporal patterns of vegetation.

We estimated reference variation conservatively so as to define an approximate range of ecologically justifiable conditions (e.g., Landres et al. 1999; Parsons et al. 1999; Swetnam et al. 1999) and to identify ecologically important changes in pattern features (e.g., extents and patterns of old forest or early seral species). When preparing restoration prescriptions, reference conditions should be used as a general rather than a rigid guide, and restored landscapes should reflect broad variation in patterns rather than the modal conditions.

Our selection of a range statistic was arbitrary; other variance measures could be used. We used the median because the right-skewed distributions of reference variation required a measure of central tendency that defined a representative range of conditions. We excluded extremes by not using the full range of variation.

The sampling method used to define the reference conditions substitutes space for time. Broad sampling of spatial patterns of vegetation from similar environments with similar disturbance and climatic regimes should reveal a representative cross section of temporal variation in these patterns (Pickett 1989). In effect, variation observed over broad spaces and narrow times may be as effective as observing variation over broad times and narrow spaces; both let us infer variations in spatial pattern at a single location or across a single landscape over time. Particularly when we try to explain the influence of processes, sampling locations must have comparable biophysical and climatic conditions (Pickett 1989). We addressed this concern by stratifying our reconstructed historical subwatersheds into subregions with similar climate, geology, biology, and disturbance regimes. The remaining potential pitfalls include an inadequate time depth, locally incompatible disturbance and climate histories, convergent environmental histories, and nonhomogeneous environments.

Comparing current values of spatial pattern metrics with estimates of reference variation reveals ecologically important departures. Comparing historical values with these estimates reveals unique landscapes or landscape conditions that lie outside the typical reference conditions. Atypical cases should be frequent, because regionalizations define homogeneous ecoregions; in reality, these overlap somewhat, and each resembles neighboring ecoregions to some extent (e.g., see Hessburg et al. 2000a). It is difficult to map intergradations between the cores of ecoregions, where atypical landscapes and patterns often appear. To minimize this problem, we analyzed reference variation, but the full range of conditions could be used to evaluate apparently atypical conditions.

Regionally synchronous weather or disturbance (convergent environmental histories), which are often related, would simplify estimates of a region's reference variation. For this reason, estimates should ultimately include variations resulting from stochastic features and rare events. This can be done by temporally and spatially broadening samples where data are available and by process modeling (e.g., Keane et al. 2002b, chapter 5, this volume), in which simulated vegetation conditions contribute to computing reference variation. We simulated vegetation and disturbance conditions in a few ecoregions and found that estimates such as those presented in this chapter correspond reasonably well with the simulated results. However, the possibility for errors or uncertainties in the spatial data remains, and can lead to estimation and prediction errors.

Managers can use our approach to perform similar evaluations elsewhere. To do so, it is essential to associate estimates of reference variation with specific potential vegetation types, because distributions of cover type–structural class combinations vary significantly; fire, insect, and pathogen disturbances, which account for much of the natural variation in vegetation spatial patterns, strongly correlate with the environmental setting (Pickett and White 1985). Hessburg et al. (1999b) illustrated how reference variation can be computed for potential vegetation type–cover type–structural class combinations and how that information can identify biophysical environments and guide revisions to cover type–structural class patterns.

Empirical estimates of reference variation serve several useful functions. Managers can use them to:

- Evaluate current conditions and estimate potential consequences for native species and processes (Morgan et al. 1994; Landres et al. 1999);

- Assess scenarios that differ from reference conditions and evaluate the potential opportunities and risks for native species, processes, and ecosystem productivity;

- Develop and evaluate specific restoration goals (Allen et al. 2002);

- Develop strategies and conservation and restoration priorities at multiple geographic scales; and

- Monitor progress in ecosystem management at relevant geographic scales.

A decision-support system, such as EMDS, can automate landscape evaluations at several scales. At a regional scale, fully integrated knowledge bases can represent reference conditions for subregions. Landscape evaluations can reveal subregions that contain the land areas with important or extensive departures from natural ranges and thereby guide the strategic allocation of planning and restoration resources (e.g., Reynolds and Hessburg 2004). At a subregional scale, evaluating a few critical attributes of all watersheds can identify priorities for more complete evaluation. At a subwatershed or landscape scale, evaluations, such as the one in this chapter, can help map alternative restoration scenarios and contrast them with estimates of reference variation before choosing and implementing the most suitable approach.

Acknowledgments

The Sustainable Management Systems Program of the U.S. Department of Agriculture Forest Service's Pacific Northwest Research Station provided financial support for this study. The authors thank Dave W. Peterson, Jim Agee, Ed Depuit, and three anonymous reviewers for their helpful comments and reviews of earlier drafts.

CHAPTER 14

:: Changes in Tree Species Composition from Pre-European Settlement to Present
A Case Study of the Temagami Forest, Ontario

FRED PINTO and STEPHEN ROMANIUK

Knowledge of pre-European settlement forest conditions is important for providing an ecological basis for forest management plans in North America. Forest managers wishing to implement the forest management paradigm of emulating natural forest disturbance require a descriptive template of their forest that is derived from the natural disturbance regime (Kimmins, chapter 2, this volume). In Ontario, this template is used, for example, to develop mandatory objectives for biodiversity, evaluate restoration strategies, set targets for tree species thought to be changed in abundance relative to a chosen benchmark, and evaluate the effects of past forestry activities on the current abundance of tree species (see Mc-Nicol and Baker, chapter 21, this volume).

One approach to developing such a template involves describing the effects of historical disturbances on the forested landscape when the anthropogenic influence is assumed to be minimal or absent. Managers see pre-European settlement forests as a template against which they can compare changes in the forests of today and the future, because the low population density and concomitant light use of forest resources suggest that the pre-European settlement forests were primarily influenced by natural events (Driver 1970, cited by Williams 1989; Radeloff et al. 1999). This is the basis for arguments in favor of using the "historical range of variation" to inform management (Morgan et al. 1994).

There are many methods for discovering historical disturbance regimes (Suffling and Perera, chapter 4, this volume), including studies of the pre-European settlement period. Egan and Howell (2001) describe many of these methods. Land survey records offer one means of studying the

pre-European settlement conditions (Humphries and Bourgeron 2001), and specifically describe the composition of the forest cover that resulted from the prevailing natural disturbance regime. The pre-European settlement composition of many forests in the United States, and particularly those around the Great Lakes, has been reconstructed using land surveyor data (Dorney and Dorney 1989; Williams 1989; Abrams and Ruffner 1995; Schaetzl and Brown 1996; Barrett 1998; Ruffner and Abrams 1998; Radeloff et al. 1999; Zhang et al. 1999; Manies and Mladenhoff 2000; Black and Abrams 2001), but few studies have been undertaken in Ontario (Gentilecore and Donkin 1973; Clarke and Finnegan 1984; Jackson et al. 2000; Leadbitter et al. 2002).

Given this background, the goal of our study was to examine the utility of the pre-European settlement forest cover as a basis for emulating natural disturbance. Specifically, we used historical data from land surveyors to describe the pre-European settlement forest composition (on the assumption that anthropogenic disturbances were minimal during the periods covered by these surveys) and compared that data with the current forest cover that has resulted from anthropogenic disturbance to illustrate changes in tree species composition.

Methods

Study Area

We conducted our study in the Temagami Forest Management Unit (hereafter referred to as the *Temagami Forest*) in northeastern Ontario near the Quebec border (figure 14.1). The Temagami Forest lies in the Great Lakes–St. Lawrence forest

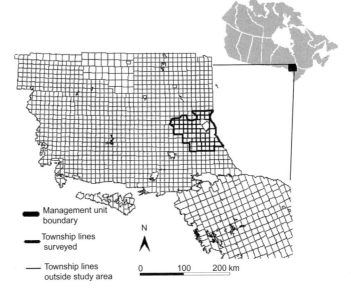

FIGURE 14.1. Outline of the township boundaries in the area that includes the Temagami Forest in northeastern Ontario, Canada.

Management unit boundary

Township lines surveyed

Township lines outside study area

N

0   100   200 km

region, primarily in section 9 but with some representation in section 4b (Rowe 1972). Section 9 is characterized primarily by eastern white pine (*Pinus strobus* L.), white birch (*Betula papyrifera* Marsh.), and white spruce (*Picea glauca* [Moench] Voss); by lesser components of red pine (*Pinus resinosa* Ait.), jack pine (*Pinus banksiana* Lamb.), trembling aspen (*Populus tremuloides* Michx.), and balsam fir (*Abies balsamea* [L.] Mill.) on slopes, and by tamarack (*Larix laricina* [Du Roi] K. Koch) and black spruce (*Picea mariana* [Mill.] B.S.P.) in lowlands. The surficial geology is characterized by bedrock outcroppings and shallow glacial tills, as well as by organic and podzolic soils. Section 4b has a boreal forest influence, with abundant black spruce, but sugar maple (*Acer saccharum* Marsh.), red maple (*Acer rubrum* L.), yellow birch (*Betula alleghaniensis* Britton), eastern hemlock (*Tsuga canadensis* [L.] Carr.), and eastern white pine are also prominent, frequently mixed with boreal conifers. The geology and soils are similar to those of section 9, but glaciofluvial deposits and localized lacustrine flats are more common. Both sections lie in the transitional portion of the boreal forest. A weather station at Bear Island on Lake Temagami receives 81.3 cm of mean annual precipitation, has a mean annual temperature of 3°C, and a mean growing season of 176 days (Rowe 1972).

Historical Land Survey Data

The pre-European settlement forest composition was reconstructed from Ontario land survey

notes for the Temagami Forest for the 1885–1958 period. The majority of the townships were surveyed between 1885 and 1912 by ten surveyors. Only six township boundaries were surveyed in 1958 (table 14.1).

Studies involving land surveys often select only one to three surveys within an area. The advantages of limiting the number of surveys include reduced variation in the recording style and increased ease of analyzing the more-consistent

TABLE 14.1. Summary of the Work Done by Crown Land Surveyors in the Temagami Forest, Ontario

Surveyor	Date(s) of Survey	Number of Township Boundaries Surveyed
A. Niven	1885, 1886	17
Unknown	1888	2
T. Scane	1903	1
R. S. Code	1903, 1912	5
W. J. Blair	1905	1
L. V. Rorke	1907, 1908	39
J. H. Smith	1908	1
C. H. Fullerton	1909	15
H. J. Beatty	1909	2
J. J. Newman	1910, 1911	21
E. W. Neelands	1958	6
Unknown	Unknown	1

survey data. However, using data from multiple surveyors can mitigate problems raised by the biases of any one individual (Schulte and Mladenoff 2001). Landscape-scale studies generally contain survey data from several surveyors collected over a long time, as the surveyors typically completed no more than about 20 km per day (Hutchinson 1988). This approach can obviously lead to discrepancies in the amount of post-European settlement disturbance and the species codes that were recorded. Most of the Ontario land survey data dates from after 1859, a time during which the split-line method was in use and the instructions to surveyors were fairly standardized (Clarke and Finnegan 1984). Adoption of the split-line method changed field notes from diaries into visual representations that can be easily interpreted. In contrast, more than 20 versions of the survey instructions were provided to surveyors working for the General Land Office in the United States between 1804 and 1902 (Galatowitsch 1990), greatly complicating the task of comparing data from different surveys.

There is no evidence in the Ontario survey notes to indicate that logging had occurred in the Temagami Forest before 1912. Survey notes for other forests have recorded logging activity when timber was removed, and the survey notes in these cases contained such comments as "pine cut" or "lumbered over." The variety of species designations used by surveyors, including subdominants and noncommercial species, suggests the absence of any bias that would have led to the exclusion of certain species. And although surveyors may have been biased toward mentioning pine, a highly valuable commercial species in the mid- to late nineteenth century, there is little evidence to support or refute this assumption. A comparable bias concerning commercial species may also be present in the province's forest resource inventory.

The Ontario survey notes contain a wealth of information collected in the field, including a description of the tree cover along the boundaries of each township, the location and extent of forest stands along these boundaries, and a list of the tree species and genera present within each stand. Stands were delineated based on the changes in composition (proportion accounted for by each taxon) or changes in the order in which tree species were listed. The survey notes provided no direct indication of the abundance of each taxon. However, based on the instructions issued to surveyors, we have assumed that the taxon codes were listed in order of decreasing abundance (Canada Department of Crown Lands 1862, 1867; Gentilecore and Donkin 1973). One aspect of the data supports this assumption: adjacent stands sometimes had identical species names, but their order differed, suggesting the use of a ranking system. To test this assumption, our study included an analysis based on the assumption of equal abundance for all species in each stand. The data in our study included only observations along the township boundaries and excluded data from lot and concession lines within the townships. Because full township surveys were not consistently completed across northeastern Ontario, the full set of survey data was not used. This approach also allowed us to include a larger, more consistent study area in our analysis.

Forest Resource Inventory Data

The Forest Resource Inventory data that describe the current species composition of the Temagami Forest are based on the interpretation of aerial photographs, supplemented by ground surveys. The last full survey for the Temagami Forest was carried out in 1989, but the data have been updated throughout the 1990s, with the most recent updates in 2000 and 2001 adding data on *free to grow* regeneration (trees taller than the competing vegetation and at a height where survival is likely—usually about 1 m in Ontario). The data for the Temagami Forest used in this chapter have been updated to include stand depletion forecasts to 2004.

The survey data report only the species composition along township boundaries, and unlike the data in the Ontario Forest Resource Inventory, do not cover the entire area within a township. To determine whether species abundance along the boundary of a township adequately predicted the abundance of tree species within the township, we transcribed the inventory data along the township boundary lines and compared the results with inventory data for the whole township. The strength of the similarity determines the validity of using the historical survey notes to estimate the species composition of entire townships and thus, the validity of comparing the historical data with the current data as a means of examining post-European settlement changes.

Data from most (85%) of the townships in the Temagami Forest were used to determine whether a sample along township boundaries adequately describes the composition of the associated forest. Data from the other 15% of the townships were incomplete and were thus excluded from our analysis. The data from the forest resource inventory were mapped to the level of individual townships using the Arcview 3.2 GIS software (ESRI 1996), and the area of forested polygons was used to calculate the percentage composition of the first-ranked species; this represents the dominant species and (in most cases) the working group species for the township. The resulting dataset represents the percentage composition for the whole township. We ran an intersect operation in Arcview to retrieve the stand descriptions from the forest resource inventory along township boundaries. This allowed us to establish a correspondence between the inventory data and the survey data. We then compared these intersected boundaries with the percentage composition for the whole township to determine how accurately the boundary descriptions described the overall forest of the township. We used a paired t-test to compare the species compositions of each of the 53 townships (townships that had Ontario land surveyor notes for at least three boundaries) after data transformations to improve normality. The results were pooled to obtain a single composition for each taxon.

Comparison of Species Abundance
between Historical and Current Data

Current (forest resource inventory) and historical (survey) data were compared along each township boundary using individual boundaries as the sampling unit (n = 128). Analyses were performed at two levels. The first comparison included only the first-ranked species in each stand and summarized the length of the township boundary occupied by each species as a percentage of the total length of the boundary. The second comparison calculated an importance value called the *ranked abundance* for the historical and current data. The ranked abundance represents the length of the boundary occupied by each species, but weighted based on the rank of the species (the order in which it was listed) in a stand; the first species was given a weighting of three times its actual length, versus two times for the second species and once for the third and subsequent species. Because this ranking system can overrepresent the first-listed species, we also calculated an importance value, called *equal importance,* by using the actual (unweighted) length for all species in each stand. The equal abundance value was only calculated for the historical data because the current data uses a ranked inventory system. We used the ranked- and equal-

importance values to calculate the percentage composition of each boundary being compared. Paired t-tests were run after data transformations. Statistical analyses were performed on the 128 boundaries, but the results presented in this chapter have been pooled for the whole Temagami Forest.

The frequency of taxa in the first three ranks was also calculated so as to detect any changes in species dominance over time. We analyzed the importance values for groups of associated species (intolerant hardwoods, tolerant hardwoods, and conifers) separately for all species groups, and individually for four groups (total pines, white pine, red pine, and unspecified pines) in the historical data. These additional analyses used all species based on their ranked abundance.

The surveyors often described only the genus of the trees observed along the township boundaries; this was the case for spruce, pine, maple, and birch. Only for the birch (*Betula*) genus were we able to analyze separately individual species, by using the logistic regression methodology outlined below. Different surveyors described this genus as birch, white birch, yellow birch, black birch, or *bouleau.* The latter two species were assumed to be yellow birch and white birch, respectively (table 14.2), based on the definitions of Jackson et al. (2000). Yellow birch appears darker in color when it is older, and for this reason, has commonly been referred to as *black birch* (Erichsen-Brown 1979).

Species assumptions were required for 5.6% of the stands in the historical data for the Temagami Forest, which had been recorded as *birch.* To separate the two birch species that could have been recorded under this category, we developed a model using logistic regression (SPSS 1999) on the current (to 2004) data on the presence or absence of both species in the Temagami Forest. We created a separate database for birch stands in the current data by using a rank column containing values of 0 (denoting white birch) and 1 (denoting yellow birch). If both species appeared in a stand, the stand with the higher ranking (i.e., closer to the beginning of the stand composition data string) was included and the other species was excluded. In addition, pure stands of a birch species were removed, because they did not contribute to the model. The frequency of the birch species in the current data determined the cutoff value used in the model. If the model showed greater than 70% overall accuracy in predicting the species

TABLE 14.2. Common and Scientific Names Associated with the Species Names and Codes Used in the Ontario Land Survey Notes for the Temagami Forest

Species in Survey Notes	Our Interpretation	Scientific Name
Ash	Black (AB), white (AW) and unspecified ash (A)	*Fraxinus nigra* Marsh, *Fraxinus americana* L., *Fraxinus* spp.
Balsam	Balsam fir (B)	*Abies balsamea* (L.) Mill.
Bouleau, w. birch, white birch	White birch (BW)	*Betula papyrifera* Marsh.
Black birch, yellow birch	Yellow birch (BY)	*Betula alleghaniensis* Britton
Cedar	Eastern white cedar (CE)	*Thuja occidentalis* L.
Tamarack	Tamarack (L)	*Larix laricina* (Du Roi) K. Koch
Unspecified maple	Maple species (M)	*Acer* spp.
Hard maple	Sugar maple (MH)	*Acer saccharum* Marsh.
Soft maple	Red maple or silver maple (MS)	*Acer rubrum* L. or *Acer saccharinum* L., respectively
Unspecified pine	White pine (PW) or red pine (PR) and unspecified pine (P)	*Pinus strobus* L. or *Pinus resinosa* Ait., respectively
Red pine, Norway pine	Red pine (PR)	*Pinus resinosa* Ait.
White pine, yellow pine	White pine (PW)	*Pinus strobus* L.
Banksian, pitch, or jack pine	Jack pine (PJ)	*Pinus banksiana* Lamb.
Poplar	Poplar species (PO)	*Populus* spp.
Unspecified spruce	Spruce species (SP)	*Picea* spp.
Black spruce	Black spruce (SB)	*Picea mariana* (Mill.) B.S.P.
Red spruce	Red spruce (SR)	*Picea rubens* Sarg.
White spruce	White spruce (SW)	*Picea glauca* (Moench) Voss
—[1]	Total maple (M, MR [red maple], MS, MH)	*Acer* spp.
—	Total pine (P, PR, PW)	*Pinus* spp. (excluding *P. banksiana*)
—	Total spruce (SB, SR, SW, SP)	*Picea* spp.

[1] The last three rows represent the groupings we used in our analyses.

based on current species associations, the species frequencies were changed by running the model again using the historic data and the result was transformed into a value between 0 and 1. Because the model accurately predicted 82% of the birch codes, we used it to separate the historical codings into yellow birch and white birch.

Other genera were lumped together with certain species for several of the analyses (table 14.2). When pine was mentioned in the historical data, we classified it as red pine and white pine; jack pine was always specified as jack pine, banksian pine, or pitch pine, and was thus not included in the "unspecified pine" or "total pine" groupings used in the analyses. All spruce and maple species were retained for genus-level comparisons. Additional working group species in the current and historical data that accounted for less than 1% of the forest composition and sample size were excluded from the analyses, although some of these species compositions are displayed in tables 14.3 and 14.4. These species included black ash (*Fraxinus nigra* Marsh.), willow (*Salix* spp.), alder (*Alnus* spp.), and cherry (*Prunus* spp.) in the historical data and black ash, beech (*Fagus* spp.), and red oak (*Quercus rubra* L.) in the current data.

In the analysis of categories for the trees, the historical and current working groups were regrouped into intolerant hardwoods (poplars and white birch), mid-tolerant-to-tolerant hardwoods (black ash, maples, and yellow birch), and conifers (balsam fir, cedar, tamarack, pines, and spruces) and compared based on the proportion of a township boundary's total length occupied by each category.

Forest fragmentation was analyzed by summarizing the historical and current datasets based on the land cover type (e.g., forest versus agricultural land). Some of these land cover types were not directly comparable, because of inherent differences in the data and in the classification schemes used in the two types of survey.

Results

The historical land survey data provided an acceptable measure of the overall composition of the Temagami Forest. That is, the forest cover recorded along the boundaries between townships predicted the overall forest composition with 95% confidence in all cases except for soft maple and black spruce, for which predictions differed significantly ($p = 0.05$) from overall compositions (table 14.3).

When comparing the historical and current data for the township lines, the differences were significant for all species groups except cedar, white pine, and total spruce (table 14.4). Curiously, the difference in white pine was not significant, even though the difference was significant for total pine. For this reason, we analyzed the white pine and pine groups separately to evaluate whether the individual white pine stands identified in the historical data represented stands with a high abundance of white pine, whereas stands recorded as pine had higher amounts of other species or species groups such as hardwoods. Because hardwoods grow readily in areas disturbed by fire, windthrow, logging, clearing of land, and so on, we reasoned that if hardwoods were present more frequently in pine stands, these areas would be more likely to have a higher abundance of hardwoods today. A paired t-test of equal-abundance importance values for hardwoods in the historical white pine and pine groups indicated a significant difference ($p = 0.012$; $n = 3$) (table 14.5). These results supported our hypothesis that hardwoods were more common in pine stands in general, but less common in white pine stands.

The total pine group decreased significantly from pre-European settlement times to the present, and maples increased significantly. The same trend is apparent for individual pine species. For example, the composition of red pine was historically higher than at present, even without considering the unspecified pine (P) grouping. Tamarack seems to have almost completely disappeared from the Temagami Forest (table 14.4). When species were grouped into functional categories, a significant ($p < 0.001$) decrease in

TABLE 14.3. Present Species Composition of the Temagami Forest for the Whole Township, Compared with the 2004 Inventory Based on Township Boundaries for Each Township, Pooled for Each Taxon for the Whole Forest

Working Group	Working Group Composition (%)	
	Based on FRI Township Area	Based on FRI Township Boundaries
AB	0.12[1]	0.01
B	2.89	2.28
BW	26.41	25.47
BY	0.81	1.22
CE	7.63	7.25
L	0.14	0.04
MH	2.28	2.87
MR	0.13	0.06
MS	1.15[2]	0.51
Total maple	3.55	3.43
PJ	10.12	9.94
PO	17.68	17.14
PR	2.09	2.39
PW	6.90	7.41
Total pine	8.98	9.80
SB	19.93[1]	22.04
SW	1.73	1.39
Total spruce	21.66	23.43

Notes: Present species composition is based on the 2004 Ontario Forest Resource Inventory (FRI) data. Composition is based on distance for the township boundaries and on area for the township as a whole. Because not all maple, pine (red or white), and spruce were identified to the species level in the land surveys, data have also been combined at the genus level (shaded rows). Abbreviations are explained in table 14.2.

[1] Species not analyzed.

[2] Significant at ($p = 0.01$) between 2004 FRI township area and FRI township boundaries ($n = 53$).

conifers and increase in intolerant and mid-tolerant-to-tolerant hardwoods was evident (data not shown).

Most species also changed significantly based on the ranked-abundance and equal-abundance importance values (table 14.6). Based on the ranked-abundance importance values, only cedar, poplar, and total spruce did not change significantly. Based on the equal-abundance importance values, only cedar and jack pine did not change significantly. The different ranking methods produced somewhat different results. For

TABLE 14.4. Comparison of the 1885–1958 Ontario Land Survey Data and the 2004 Forest Resource Inventory Township Boundary Data Based on Working Group

| Working Group | Working Group Composition (%) | | |
	Ontario Land Survey	Forest Resource Inventory	Difference
AB	0.15	0.01[1]	
B	5.94	2.16[2]	Decrease
BW	11.38	26.25[2]	Increase
BY	0.58	1.51[2]	Increase
CE	5.70	7.92	Not significant
L	1.64	0.03[2]	Decrease
MH	0.16	3.77[1]	
MR	—	0.05[1]	
MS	—	0.57[1]	
M	0.36	—[1]	
Total maple	0.53	4.40[2]	Increase
PJ	15.17	10.14[2]	Decrease
PO	11.19	14.51[2]	Increase
PR	5.18	2.76[2]	Decrease
PW	9.18	7.86	Not significant
P	3.96	—[1]	
Total pine	18.31	10.62[2]	Decrease
SB	21.29	21.44[1]	
SW	0.63	1.02[1]	
SR	0.28	—[1]	
SP	7.21	—[1]	
Total spruce	29.42	22.46	Not significant

Notes: n = 128. Abbreviations are explained in table 14.2.
[1]Species not analyzed because of insufficient sample size (AB) or a failure to distinguish all species (maple and spruce species).

[2]The difference between the 1885–1958 Ontario land survey data and the 2004 forest resource inventory township boundary data was significant ($p = 0.01$).

example, there was a significant difference in the effect of the different weightings for balsam fir, white birch, cedar, poplars and total maple, total pine, and total spruce (table 14.6). Regardless of the differences, the trends and significance coincided for all these working groups except total spruce and poplars.

The yellow birch, white birch, and total maple working groups increased significantly in dominance, whereas balsam fir, tamarack, total pine (i.e., stands with red pine and white pine), and total spruce decreased in dominance. For example, the total pine working group decreased from 18.31 to 10.62% as a dominant species, but increased from 7.49 to 15.5% as the third-ranked

TABLE 14.5. Proportion of Hardwoods in Pine versus White Pine Working Groups

Working Group	Hardwood (%)	Number of Stands
PW[1]	20.9	357
P	39.7	150

Notes: Percentages determined by using the equal abundance importance values calculated based on the Ontario land survey data. Hardwoods include black ash, white birch, maples, and poplars. Number of stands, the number of stands in which the named working group was the first-listed species. PW, white pine group, P, pine group.

[1]Significant at $p = 0.05$.

TABLE 14.6. Comparison of Working Groups Based on Calculated Importance Values between the Period Covered by the Ontario Land Survey Data (OLS, 1885–1958) and Current (to 2004) Forest Resource Inventory (FRI) Data

Working Group	Township Boundaries (OLS)		Township Boundaries (FRI)	Change between OLS and FRI
	Ranked Abundance	Equal Abundance	Ranked Abundance	
B[1]	12.11[2]	13.94[2]	7.23	Decreased
BW[1]	15.89[2]	16.81[2]	21.86	Increased
BY	0.49[2]	0.42[2]	2.59	Increased
CE[3]	6.34	6.94	7.70	Not significant
L	2.40[2]	2.17[2]	0.29	Decreased
MH[4]	0.17	1.09	2.63	
MR[5]	—	—	0.20	
MS[4]	0.02	0.09	3.05	
M[4]	0.94	1.32	—	
Total maple[1]	1.13[2]	2.50[2]	5.88	Increased
PJ	10.04[2]	8.55	7.57	Decreased
PO[1]	11.25	8.76[5]	11.38	Increased
PR[4]	5.94	7.02	3.15	
PW[4]	8.26	11.07	10.01	
P[4]	3.63	6.58	—	
Total pine[1]	17.83[2]	24.67[2]	13.16	Decreased
SB[4]	17.40	12.59	15.77	
SW[4]	0.67	0.65	6.15	
SR[4]	0.12	0.00	—	
SP[4]	4.00	1.11	—	
Total spruce[1]	22.20	14.34[2]	21.92	Increased

Notes: All values represent species composition (%). Because not all maple, pine (red or white), or spruce were identified to the species level in the OLS data, all entries for these working groups were grouped for analysis (shaded rows). Abbreviations are explained in table 14.2.

[1]Significant at $p = 0.01$ between ranked-abundance and equal-abundance importance values for OLS data.

[2]Significant at $p = 0.01$ based on data from OLS and FRI township boundaries.

[3]Significant at $p = 0.05$ based on ranked-abundance and equal-abundance importance values for OLS data.

[4]Species not analyzed.

[5]Significant at $p = 0.05$ based on OLS and FRI township boundaries.

species (table 14.7). White birch was ranked as the first species 11.38% of the time in the historical data, versus 26.25% in the current data; this represents the largest apparent change. These shifts in dominance are supported by the frequency data (table 14.7). Balsam fir and total pine both decreased in dominance in the first and second ranks, but were stable and increased (respectively) in the third rank. White birch also had the lowest frequency in the third rank of the current data, further illustrating its increase in dominance. We cannot compare the number of stands in each taxon between the historical and current datasets, because different resolutions were used to define the stands in each survey.

The land cover types in the historical and current data were very similar, particularly for forest and the category of "grass and meadows" (table 14.8). This suggests that forest fragmentation has not increased as a result of development,

TABLE 14.7. Summary of Forest Cover for the First Three Tree Species Mentioned in the Ontario Land Survey and Forest Resource Inventory Stand Descriptions

| Working Group | Ontario Land Survey Data | | | | | | Forest Resource Inventory | | | | | |
| | Rank 1 | | Rank 2 | | Rank 3 | | Rank 1 | | Rank 2 | | Rank 3 | |
	Percentage	No.	Percentage	No.	Percentage	No.	Percentage	No.	Percentage	No.	Percentage	No.
B	5.94	270	17.41	678	15.95	565	2.16	108	11.63	488	15.17	579
BW	11.38	494	21.41	903	22.23	816	26.25	1060	27.87	1209	16.57	716
BY	0.58	23	0.61	24	0.16	7	1.51	61	3.38	127	3.44	122
CE	5.70	317	4.96	222	6.52	240	7.92	378	7.97	343	9.20	375
L	1.64	93	3.73	171	3.57	106	0.03	2	0.49	24	0.92	41
Total maple	0.53	27	0.54	27	1.83	54	4.40	159	5.16	179	8.35	300
PJ	15.17	713	6.61	286	6.30	252	10.14	524	6.34	271	5.18	201
PO	11.19	551	10.39	389	11.90	391	14.51	641	9.87	439	8.89	372
Total pine	18.31	795	16.42	642	7.49	289	10.62	484	10.02	416	15.50	627
Total spruce	29.42	1427	17.65	767	23.51	864	22.46	1116	16.68	738	15.86	612
Total	99.85	4710	99.72	4109	99.45	3584	99.99	4533	99.40	4234	99.08	3945

Notes: For each species group, the forest cover data (percentage of total, based on weighted lengths) are shown for each of the first three species ranks (ranks 1, 2, and 3 in the table). No., number of times a working group occurs in each rank. Other abbreviations are explained in table 14.2.

TABLE 14.8. Proportion of Total Area in Each Land Cover Type Based on the Ontario Land Survey and Forest Resource Inventory Data

Land Cover Type	Percentage of Total Area	
	Ontario Land Survey	Forest Resource Inventory
Forest	72.4	71.2
Nonproductive forest (muskeg, alder)	8.9	5.2
Nonproductive forest (rock[1])	1.4	0.4
Developed agricultural land	n.a.	2.0
Grass and meadows	0.1	0.1
Unclassified land	2.6	3.5
Water	12.0	13.8
Burned areas	2.4	n.a.
Barren and scattered or not satisfactorily regenerated[2]	n.a.	3.7
No data	0.1	0.1

Note: n.a., not applicable (i.e., no land was classified in this category in the Ontario land survey or forest resource inventory data).

[1]Refers to forested or nonforested land containing exposed bedrock.

[2]Can include areas that have been harvested or burned and may be comparable to the area classified as burned in the Ontario land survey data.

deforestation, disturbance, or a change in land cover types. The "barren and scattered" and "not satisfactorily regenerated" categories in the current dataset include areas that have been harvested or burned, and may be comparable in part to the areas identified as burned, harvested, or windthrown in the historical data; there was no explicit mention of harvesting in the latter data, but several burns (some fires specified in 1908 and 1909) were mentioned. The 2.4% of the area that burned (table 14.8) represented fires that burned all trees, so that there was no regeneration growing in the burned area. An additional 4.3% of the forest land cover type had burned at least partially. There is a noticeable reduction in the area of nonproductive forest types (including alder swales, and treed and open muskeg), but the proportions of forest cover in the two datasets were very similar.

Discussion

Our assumption that tree species were ranked (recorded in order of abundance rather than randomly) in the Ontario land survey data did not appear to affect our interpretation of the changes in the abundance of taxa between the historical and current forest (see table 14.6). The only exception is for the total spruce working group, for which the ranked-abundance importance values

suggest no change over time, whereas the equal-abundance importance values suggest a significant increase.

The changes in abundance of the working groups and tree species abundance within working groups (based on the calculated importance values) between the historical and current datasets for the Temagami Forest suggest the occurrence of events that limited the regeneration and growth of conifers and favored the regeneration and growth of white birch, poplars, and maples. The most likely explanations for these changes are forestry operations, forest fires and their suppression, and insect infestations.

Forestry operations formerly emphasized the removal of economically valuable stems (in most cases, the best seed producers) of conifers and some hardwoods. This approach was common until recently, as indicated in the 1990 management plan (Maure and Dawson 1990) for the Temagami Forest. The removal of large seed-bearing trees by logging would decrease the regeneration potential of these tree species (Russell et al. 1993). The logging, particularly if intensive, may have destroyed established regeneration.

Fires after early logging may also have destroyed established regeneration. Cwynar (1977) states that the replacement of red pine and white pine throughout their range is not caused by fire or logging alone, but rather by a combination of the two that disrupts the balance between seed production and the frequency of fires. In the past, logging slash often supported fires that destroyed potential regeneration, further damaged or killed seed trees, and at times burned off the organic matter in the soil, leaving behind a barren wasteland (Lorimer 1977; Whitney 1987; Barrett 1998; Radeloff et al. 1999; Abrams 2001). Radeloff et al. (1999) also noted that salvage logging during budworm (*Choristoneura* spp.) outbreaks may aggravate this increase in fire potential. Whitney (1987) noted that in northern Michigan, harvesting of desired species (white pine initially, then eastern hemlock and better hardwoods) and fire suppression allowed shorter-lived successional species such as oak (*Quercus* spp.), trembling aspen, and jack pine to flourish and create a new, stable forest that was maintained for the pulp and paper industry in the 1950s. Further, Jackson et al. (2000) stated that the change in the boreal forest from conifer dominated stands to stands of intolerant hardwoods may be attributed to clearcut harvesting, something that was also evident in the Temagami Forest (figure 14.2). Cole et al. (1998) attributed the decline in the pine forests (~21%) of the Great Lake States to logging.

White birch and poplars are both well adapted to frequently producing large numbers of wind-dispersed seed, as well as sprouting and suckering vigorously, respectively, after harvesting or a fire (Burns and Honkala 1990). Past harvesting in maple stands targeted well-formed stems with little rot, and left poor-quality maple standing, which was able to regenerate areas to maple. Fire-suppression efforts over the past 80 yr would also have favored maples over associated species, such as yellow birch and eastern hemlock (Dahir and Lorimer 1996; Leadbitter et al. 2002). Fires often burn off the thick maple leaves that form a physical and chemical barrier to the regenera-

tion of other tree species but not to maple seeds, whose vigorous radicle penetrates easily into the soil and lets this genus regenerate sites beneath their parent trees despite a thick layer of undisturbed litter (Burns and Honkala 1990). Leadbitter et al. (2002) also noted a significant increase in the abundance of maple in Temagami and an increase of almost 17% in Algonquin Park from 1890 to 1990. This increase in maples and intolerant hardwoods is consistent with the findings of other studies based on surveyor records (Radeloff et al. 1999; Jackson et al. 2000) and on field surveys (Hearnden et al. 1992; Carleton and MacLennan 1994). Although the increased abundance of intolerant hardwoods could have occurred largely in response to fire, the increase in maples contradicts this hypothesis, because maples benefit from fire suppression. This suggests that logging may have played a more dominant role than fire in initiating these changes in the Temagami Forest.

Severe infestations by the eastern spruce budworm (*Choristoneura fumiferana* Clem.) in the 1980s or earlier would have reduced the area of balsam fir and the importance value of this species in stands. These reductions were indeed found in the Temagami Forest, suggesting that budworm defoliation may have played a role in the reduced dominance of balsam fir. Similarly, an outbreak of the larch sawfly (*Pristiphora erichsonii* Hartig), an accidentally introduced insect defoliator, may have played a dominant role in decimating tamarack populations in the Temagami area in the late 1800s and early 1900s (Leadbitter et al. 2002). Unlike in coniferous stands, outbreaks of defoliating insects such as the forest tent caterpillar (*Malacosoma disstria* Hbn.) in deciduous forests do not seem to have reduced the abundance of white birch, poplars, or maples.

The abundance of the yellow birch working group (see table 14.4) increased from 0.6 to 1.5%

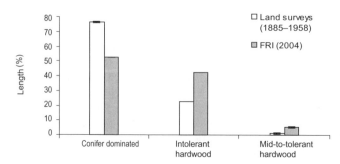

FIGURE 14.2. Forest cover of the three dominant groups of species in the Temagami Forest. The proportion of the total area is based on the total length of the township boundaries occupied by the species group in the 2004 forest resource inventory (FRI) and in the 1885–1958 Ontario land surveys; vertical bars represent the 95% confidence intervals.

between the historical and current periods. A similar study in the Temagami Forest using historical and current data (Leadbitter et al. 2002) produced contrasting results. The data show a decrease in yellow birch (from ~6 to ~1%) from 1890 to 1990; we are unable to explain the higher abundance Leadbitter et al. reported for the nineteenth century. Their dataset and assumptions concerning species not specified in the historical data were similar to ours, as was the amount of yellow birch in the 1990 forest resource inventory (0.9%), our data from the current inventory (2004), and Leadbitter's own data from 1990 (Farintosh 1999).

Between pre-European settlement times and the present, the proportion of stands dominated by white pine did not differ significantly in the Ontario land surveys and the forest resource inventory; however, stands composed of a mixture of red pine and white pine (our "total pine" working group) decreased significantly. Analyses of the importance values for hardwoods in the white pine and unspecified pine stands suggested that the latter group had a higher initial amount of hardwoods (see table 14.5).

There is no evidence in the Ontario land survey notes to indicate that logging had occurred in any of the areas surveyed before 1912. Survey notes for other forests do record logging activity, so this absence of logging records in the Temagami Forest data is not likely to be a simple omission. Only one boundary in our database was found to have been surveyed twice (once in 1916 and again in 1958). The change in tree composition evident between these two dates, with the majority of the pine having disappeared, suggests that the area may have been cut between the two surveys (figure 14.3).

Conclusion

The comparison of forest compositions in the Ontario land survey records with those in the current forest inventory showed changes in species composition that may differ from what might have resulted if natural disturbance was still the primary cause of change in the Temagami Forest. Forest managers in the Temagami Forest can use the pre-European settlement tree species abundances described in this chapter to help develop species diversity objectives and strategies for species restoration or reduction in their forest management plans, but must do so with caution. Some factors, such as climate and the effects of native and introduced insects and diseases, are likely to change over time. To use the template for pre-European settlement forest composition, managers must first understand the context under which the pre-European settlement forest established itself and the context under which their current forest will become established and grow.

Stand data from historical records provide a valuable source of information to help develop a template for the composition of the pre-European settlement forest. The historical records for the Temagami Forest provided a statistically adequate sample to describe the pre-European settlement abundance of all tree species except for the soft maple group, black spruce, and uncommon species, such as black ash.

Historical data from 1885 to 1958 for the Temagami Forest suggest that today's forest composition is significantly different from that in the past. Economically valuable species, such as white pine, red pine, and the spruces, as well as tree species with a shorter history of industrial

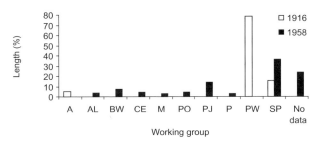

FIGURE 14.3. Proportions of the length of township boundaries occupied by each working group in the Temagami Forest along the same township boundary surveyed during the 1916 and 1958 Ontario land surveys. A, ash species; AL, alder; no data, no forest cover listed in the survey data for that stand for that year. See table 14.2 for definitions of the other working groups.

use, such as cedar and tamarack, have decreased in abundance. The changes in cedar may have arisen from differences between the ground-based Ontario land survey and the aerial photographs used to create the current forest resource inventory, as the latter approach may underestimate the presence of understory species and those with small crowns. The reduced abundance of tamarack was probably caused by the introduction of the larch sawfly, an insect defoliator new to North America, in the 1880s. We suggest that the changes in the abundance of economically valuable tree species have resulted from logging, fire suppression, and inadequate or nonexistent regeneration efforts. The reduction in the abundance of these valuable timber species and the concomitant increase in the abundance of intolerant hardwoods and maples are changes that might not have occurred if only natural disturbance had been responsible for the current landscape.

Acknowledgments

We thank the Ontario Ministry of Natural Resources' (OMNR) Southcentral Science and Information Section and Nipissing University, North Bay, Ontario, for project funding and support. We appreciate Dave Nesbitt's technical and database assistance and Brian Naylor's statistical support; both are with OMNR. We thank Alan Day of the Office of the Surveyor General, Crown Land Surveys, Peterborough, Ontario, for help with survey note collection and the Temagami area Sustainable Forest Licenses for their forest resource inventory data. Finally, we thank Marc Nellis, Marc Hebert, and their students at College Boreal in Sudbury, Ontario, as well as Jamie Geauvreau, OMNR, for data entry.

PART III

Applications: Perspectives, Practices, and Policy

A Conservation Perspective on Emulating Natural Disturbance in the Management of Boreal Forests in Ontario

DAVID L. EULER, CHRIS HENSCHEL, and TOM CLARK

From the perspective of conservation, the most important goal of forest management should be to conserve biodiversity. Conserving biodiversity does not mean that the manager tries to increase the diversity of plants and animals in the forest, nor does it mean that the status quo is always maintained. Rather, management planners should try to understand the diversity and natural changes that occur in an unmanaged forest, and use management techniques that keep the forest's diversity within the range of values (boundaries) that would occur in the absence of anthropogenic intervention. In this chapter, we consider the concept of emulating natural forest disturbance in forest management from the perspective of a conservation "ethic," in which maintaining the forest's biodiversity is a major goal. In using wording such as *should,* our goal is to emphasize that in addition to purely economic or practical considerations, managers should guide their management actions with ethical considerations that go beyond the merely pragmatic.

The conservationist's approach combines ethics and a practical approach to forest management. Science cannot offer everything that managers need to know, because not all questions are scientific, and ethics plays a role in decisions that affect society's values. The motives for a conservation perspective are twofold: first, it seems like the right thing to do (Hunter 1993) and second, as Aldo Leopold (1949, p. 214) noted, "a system of conservation based solely on economic self-interest is hopelessly lopsided. It tends to ignore, and thus eventually to eliminate, many elements in the land community that lack commercial value [but] that are (as far as we know) essential to its healthy functioning."

As Attiwill (1994, p. 248) pointed out, "an extensive literature supports the hypothesis that natural disturbance is fundamental to the development of [the] structure and function of forest ecosystems. It follows that our management of natural forest should be based on an ecological understanding of the processes of natural disturbance." Management based on emulating natural disturbance seems to be the most appropriate approach to forest management, especially as much remains unknown about forest ecology and the impacts of timber harvesting on forested ecosystems.

Forest management in Ontario provides examples to illustrate several points in this chapter, because the Crown Forest Sustainability Act (Statutes of Ontario 1995) requires Ontario's forest managers to emulate natural disturbance. In addition, as a result of the class environmental assessment for timber management on Crown lands in Ontario (Ontario Environmental Assessment Board 1994, Terms and Conditions 94b, p. 461), the Ontario Ministry of Natural Resources was required to "provide direction in relation to harvest layout, configuration and clearcut sizes."

The Environmental Assessment Board stated in their rationale for this statement (Ontario Environmental Assessment Board 1994, Terms and Conditions 94b, p. 461): "We conclude that clearcuts should be made in a range of sizes to emulate natural disturbances, and that—although extremely large clearcuts would likely be rare for practical reasons—limiting clearcuts strictly to small sizes would make it impossible to regenerate the Boreal Forest to its natural pattern of large even-age stands."

In this chapter, we elaborate on what emulating natural disturbances means in a conservation context that does not rely on clearcut size as a determining variable. We review some of the impacts of traditional forestry in the boreal forest region of Ontario, and consider how those impacts might be changed if our conservation-based definition of emulating natural disturbance is adopted as management policy. Our goal is to illustrate how forest management activities would differ from traditional forest management as a result of our suggested approach. We have not considered the economic impacts of our approach. Economic factors are important, of course, but they lie beyond the scope of this chapter and our expertise.

The interpretation of emulating natural disturbance that we advocate differs in a crucial way from the common interpretation in the literature, including that used in much of this book. That common interpretation involves emulating the characteristics of disturbances: their size, shape, periodicity, and spatial distribution. We contend that this interpretation misses the point of emulating natural disturbance and brings us no closer to the objective of maintaining the natural elements of the forest. Our alternative is to emulate the characteristics of the forest that result from a natural disturbance regime (Canadian Parks and Wilderness Society–Wildlands League 2002). This approach provides targets for the forest's condition (e.g., age, species diversity) that protect us from the elimination of certain elements that might occur in a system based solely on economic self-interest. It also protects us from unwittingly eliminating these elements through a well-intentioned but misguided focus on the characteristics of the disturbance themselves.

In a chapter of this length, we cannot address the operational limitations that will be encountered when managers implement our approach to emulating natural disturbance. Clearly, practical limitations constrain any management approach, and those raised by our approach will be important. However, we discuss the work of others who have instituted aspects of an approach to emulating natural disturbance similar to what we recommend, and address some of the practical limitations that will be encountered when a forester implements the approach we advocate. We hope that some managers will attempt to implement our approach and explore the practical limitations that arise at both the landscape and stand levels. Implementation of forest management always occurs at the stand level, but has impacts at both the stand and the landscape levels.

There is no experimental evidence based on rigorous scientific studies that demonstrates whether emulating natural disturbance results in less impact on biodiversity or wildlife than would be the case under current management practices in North America. However, neither are there rigorous studies that demonstrate that current management has a benign impact on biodiversity and wildlife. In fact, quite a few studies (cited later in this chapter) suggest that selected species and sometimes biodiversity as a whole have been affected negatively by modern forest management. Settling this question would require a comprehensive experiment, conducted at a broad scale and over several decades, to compare an approach based on emulating natural disturbance with a conventional approach. The ideas presented here could form the basis for conducting such an experiment.

Effects of Traditional Forestry in the Boreal Forest of Ontario

In this section, we discuss what is already known about the undesirable effects of traditional forestry on various aspects of biodiversity: species diversity, age-class distributions, landscape patterns, old-growth forests, and wildlife.

Changes in Tree Communities— Plant Species Diversity

The boreal forest has demonstrably been changed by modern forest management activities. Hearnden et al. (1992) undertook a study of how the boreal forest was regenerating following timber harvesting in Ontario. In 1991, they sampled areas harvested between 1970 and 1985 to detect changes in the species composition of the forest that was regenerating after harvesting. Although it was clear from their data that the forest was regrowing, the tree species composition differed from that in the original forest (table 15.1). For example, the proportion of hardwoods in the regenerating forest had increased dramatically.

Carleton (2000) examined the data from the report of Hearnden et al. (1992) and came to much the same conclusion based on data that he had collected independently. He concluded (p. 195): "The legacies of clear-cut logging have been a conversion from forests dominated by pioneer, fire-tolerant conifers to pioneer, fire-

TABLE 15.1. Pre- and Postharvest Percentages of Major Species Regenerating in Ontario, Canada

	Jack Pine	Spruce	Mixed Softwood (Pine and Spruce)	Mixedwood (Conifers and Hardwoods)	Hardwoods
Original forest before harvest (%)	10	18	29	36	6
New forest after harvest (%)	15	4	21	41	19

Source: Data from Hearnden et al. (1992).

Note: The data do not always add to 100 due to rounding errors.

tolerant broadleaved tree species. The legacy of clear-cutting on the understory vegetation is a greatly reduced beta diversity and the regional extirpation of rare species typical of needle-leaved evergreen forests."

Changes in the Age-Class Distribution of Trees

The age-class distribution of tree species in Ontario has also been changed by forest management. The two most economically important species in Ontario's boreal forest, jack pine (*Pinus banksiana* Lamb.) and black spruce (*Picea mariana* [Mill.] B.S.P.), have the age-class distributions shown in figure 15.1 (Ontario Ministry of Natural Resources 2002b). The area in age classes 21–40 and 41–60 for both these species is relatively low compared with the areas in other age classes. Moreover, these age classes correspond to the age of forests regenerated following the early modern years of industrial forestry in Ontario. Although small areas with such an age-class distribution could result after a fire or other disturbance, it is

unlikely that this distribution would occur over the entire boreal forest region of Ontario under a natural disturbance regime. The observed age-class distribution would not occur across the entire province under natural conditions because it would be impossible to produce areas of older age classes that are larger than the area of younger stands that gave rise to the older stands.

The age-class distributions in figure 15.1a,b do not mimic the distribution in unharvested boreal forest in Ontario or Quebec (Van Wagner 1978; Harvey et al. 2002). Even theoretical distributions of age classes under a variety of different burning conditions and at different landscape scales, as illustrated in Boychuk et al. (1997), do not simulate the current age-class distribution of trees in Ontario's boreal forest. The observed distribution has resulted from a combination of concentrating the timber harvest on jack pine and black spruce, poor regeneration of these species after harvests that occurred 20–60 yr ago, and a fire suppression policy that has reduced

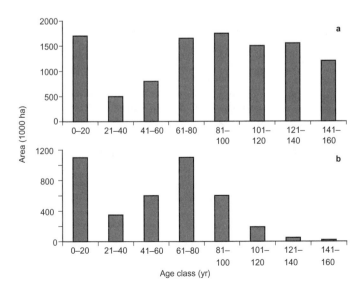

FIGURE 15.1. A summary of the age-class distribution of (a) black spruce and (b) jack pine in Ontario, Canada. Adapted from the Ontario Ministry of Natural Resources (2002b).

the natural abundance of young stands dominated by these fire regenerated species.

Changes in Landscape Patterns

The landscape patterns created by timber harvesting also differ from those created by fire. The forest that regenerates after harvesting is generally more spatially complex than postfire forests, has more than 20 times as much edge, is less dispersed across the landscape, and is more likely to lie adjacent to younger forests (Perera and Baldwin 2000). Schroeder and Perera (2002) examined the broad-scale vegetation patterns that arose after clearcuts and fires in Ontario and concluded that postfire landscapes had larger and less numerous patches than postclearcut landscapes, and were also less heterogeneous. In addition, McRae et al. (2001) reviewed the empirical evidence that illustrates the differences between the results of wildfire and forest harvesting. They found many differences, including that harvesting does not maintain the natural distribution of stand ages associated with wildfire in many regions and that the forest after a wildfire is a complex mosaic of stand types and ages that differs strongly from the postharvesting mosaic.

Although fires were once the major form of disturbance in the boreal forest, timber harvesting now accounts for a more significant proportion of the total area disturbed. The total area clearcut within the managed forest of Ontario increased from 0.5 million ha between 1951 and 1960 to more than 2 million ha between 1981 and 1990. In contrast, the area burned within this region has been constant at approximately 0.5 million ha per decade from 1951 to 1990 (Schroeder and Perera 2002). Despite considerable efforts to control forest fires, the area burned by fires does not seem less now than it was a half century ago.

Changes in the Abundance of Older Age Classes

Based on research conducted in Quebec's boreal forest, similar forests in Ontario would be expected to have 30–40% of the land area older than the average disturbance interval. When disturbance intervals are relative short, as they are in northwestern Ontario, these older forests may not be what the public considers classic old-growth forests. However, forests older than the interval between disturbances are significant components of the landscape and perform important ecological functions, whatever the actual age of the stands.

The abundance of older forests in the landscape varies with changes in the interval between fires, but around one third of a typical forested area will have stands older than the disturbance interval (Y. Bergeron et al. 1999). Askins (2000) cited research from Maine that suggests that such old-growth comprised about 27% of the land area of northeastern Maine prior to settlement.

Figure 15.2 shows data for red pine (*P. resinosa* Ait.) and eastern white pine (*P. strobus* L.), jack pine, upland mixed conifers, and lowland mixed conifers, based on data from the most recent State of the Forest report for Ontario (Ontario Ministry of Natural Resources 2002b, pp. 3–21). This figure illustrates the current percentage of the "planning area" in Ontario that contains older forests. The forest management regime of the past half century has left much smaller areas of old forest in some species than would have been left in a forest disturbed primarily by fire. For example, only 3–4% of jack pine is present in the form of old forests, and upland mixed conifers, red pine, and white pine contain much smaller amounts of old forest than would occur in a disturbance-based management scenario. However, in contrast, the proportion of lowland mixed conifers classified as old forests is very high. This could have occurred because of fire protection, but has more likely been caused by the very long intervals between fires in this forest type. What is clear is that the sum of all forest management activities has produced, on average, a forest very different from what would exist if fires had been the dominant form of disturbance (see also Wimberly et al., chapter 12, this volume).

Impacts of Changes on Wildlife

The overall picture that emerges of the current, managed boreal forest is increased diversity at the regional scale (*gamma* diversity) and reduced diversity at more local scales (*beta* diversity), with the total area disturbed by timber harvesting exceeding the area disturbed by fire. Less land area is occupied by older forests than would be expected under a natural disturbance regime. Under these circumstances, moose (*Alces alces*) and white-tailed deer (*Odocoileus virginianus*) populations, which require stands in early successional stages, flourish if hunting by humans is controlled (Voigt et al. 2000). In a study of moose habitat in northwestern Ontario, Rempel et al. (1997) found that the progressive clear-

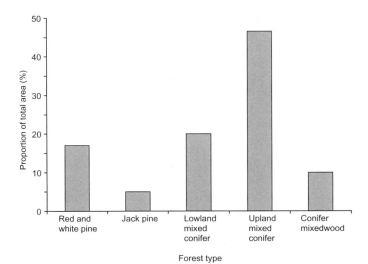

FIGURE 15.2. Percentage of old growth area by forest type in Ontario, Canada. Adapted from Ontario Ministry of Natural Resources (2002b).

cutting practiced there produced excellent moose habitat, and moose populations were increasing. In areas suitable for white-tailed deer, similar habitat is produced by clearcutting. Schmiegelow and Monkkonen (2002) compared forest fragmentation by logging in boreal Finland and boreal Canada and presented data from Finland that show that populations of common bird species often associated with early-successional habitats increased between 1940 and 1990. In a comparison of logged and burned black spruce stands in the boreal forest of Quebec, bird species that thrive in early-successional habitats were benefited by logging (Imbeau et al. 1999). This observation is not surprising, because logging generally returns a stand to earlier successional stages, and the observation has also been supported for broader areas of the boreal forest by Askins (2000). Predators associated with ungulates and birds would also be expected to thrive under these conditions because of the increased availability of food.

Populations of woodland caribou (*Rangifer tarandus*), however, have declined, and their range now lies farther north than when Ontario was settled (Thompson 2000). This change has been correlated with increased timber harvesting, although a causal effect has not yet been clearly established. Woodland caribou were hunted extensively, and serve as prey for the abundant wolves (*Canis lupus*) that were present during this time period (Voigt et al. 2000). The decline in woodland caribou in the southern boreal forest region could be a result of extensive timber harvesting in one of their important habitat types,

jack pine. However, research on this topic is also not definitive, and although it is clear that this caribou's range has been reduced, the cause of the reduction may also be related to hunting earlier in the century accompanied by predation by wolves.

For stands of eastern white pine, Thompson (2000) reviewed the literature and his own studies and concluded that two *Myotis* species of bats, four species of carabid beetles, and four species of birds were found more frequently in old-growth white pine forests than elsewhere. He concluded that "white pine–dominated mixed woods in general and old-growth pine forest in particular maintain discrete animal communities." With the loss of white pine and red pine forests, particularly in the older age classes, that has occurred since the advent of European settlement in North America, these wildlife communities may well be considerably reduced compared with their historical levels in both abundance and diversity.

Imbeau et al. (2001) compared the impact of intensive forest management on birds in the boreal forests of Quebec and Fennoscandia. Several species of birds in Fennoscandia, mostly cavity nesters or other nonmigratory birds, showed population declines as a result of intensive forest management in that region. Askins (2000) reviewed the literature and concluded that a number of bird species are threatened by large-scale logging. Birds identified by both studies included the boreal chickadee (*Parus hudsonicus*), yellow-bellied flycatcher (*Empidonax flaviventris*), gray jay (*Perisoreus canadensis*), red-breasted nuthatch

(*Sitta canadensis*), brown creeper (*Certhia americana*), golden-crowned kinglet (*Regulus satrapa*), Cape May warbler (*Dendroica tigrina*), blackburnian warbler (*D. fusca*), and bay-breasted warbler (*D. castanea*). Some woodpeckers are also vulnerable; for example, Imbeau and Desrochers (2002, p. 30) stated that "among boreal forest bird species, the three-toed woodpecker (*Picoides tridactylus*) is closely associated with old-growth forest (>120 years) and possibly the most negatively affected by long-term changes induced by commercial forestry in eastern Canada."

Drapeau et al. (2000) studied forest conditions before industrial logging, in the presence of industrial logging, and in the presence of natural disturbance, and found significant differences in the bird species composition in the different areas. They found that "the mean number of mature-forest bird species was significantly lower in the industrial and pre-industrial landscapes than in the natural landscape." They concluded that "overall, our results indicated that the large-scale conversion of the southern portion of the boreal forest from a mixed to a deciduous cover may be one of the most important threats to the integrity of bird communities in these forest mosaics" (p. 423). In Saskatchewan, Hobson and Bayne (2000) came to a similar conclusion.

There is ample evidence to suggest that the changes in species composition, the age-class distribution of the forest, and the decline in abundance of older forests are affecting wildlife populations. These characteristics of the forest are therefore strong candidates for guiding the emulation of natural disturbance. In addition to the birds mentioned above, mammals, such as the American marten (*Martes americana*) (Strickland and Douglas 1987; Thompson 1994) and woodland caribou (Voigt et al. 2000) require large core areas of habitat that exist in a natural forest governed by a natural disturbance regime, and these large cores are threatened by traditional forest management. Therefore, emulating the patterns of natural disturbance should include a primary focus on maintaining adequate supplies of these habitat types over time.

Emulating Natural Disturbance from a Conservation Perspective

Unfortunately, the concept of biodiversity provides little guidance on how managers should approach forest management so as to conserve biodiversity and achieve ecological sustainability. The usefulness of the concept thus depends entirely on how it is interpreted and applied. Emulation of natural disturbance must include more than the characteristics of disturbance; it must also include the characteristics of the forests that have developed under a natural disturbance regime. Viewed in this broader sense, emulating natural disturbance means, for example, managing to preserve the old-growth forest that develops naturally in the absence of stand-replacing fires, as well as to create the young forest that resulted from these fires. It is also important that emulating natural disturbance focus on characteristics that are practical to emulate and represent important ecosystem functions. Disturbances by fire, wind, and insects in the boreal forest have a large number of measurable characteristics that could be emulated. Forest managers must focus on some subset of these characteristics that can serve as indicators of the conservation of biodiversity and ecological sustainability. Obviously, it is important to choose the correct characteristics. From a conservation perspective, characteristics of a natural disturbance and of the forest that results from that disturbance regime are both important for the conservation of biodiversity (see Thompson and Harestad, chapter 3, this volume).

From the many characteristics that could be measured, we believe that the following are most important, are practical to emulate, and perform important ecosystem functions:

- Changes in diversity, and particularly in plant species diversity;
- Age-class distributions of major tree species;
- Landscape patterns, including undisturbed patches;
- Extent and distribution of old-growth forest within the landscape; and
- Trends in wildlife populations.

This is not a definitive list: we may have missed other important parameters. For brevity, we concentrate mostly on the landscape level in this chapter, but characteristics at other scales—most notably the within-stand structure—are vitally important to wildlife and other ecosystem functions and should also be considered within emulation-based management. Our list represents a starting point for managers who are evaluating forest management plans, and illustrates what would be involved in implementing the concept of emulating natural disturbance in forest management. These items also are reflected in the Canadian Council of Forest Ministers' list of criteria and indicators of sustainability (Canadian Forest Service 2000).

An approach to forest management based on emulating natural disturbance must consider how these characteristics have changed and the importance of the changes, and then try to emulate natural levels of the important characteristics through management. If managers could successfully manage a forest by emulating the natural processes that give rise to these characteristics, this would benefit biodiversity. It is, of course, important to verify these benefits through monitoring and to identify additional characteristics that must be emulated or maintained.

Forested ecosystems are always in a state of flux, and disturbance has played an important and probably necessary role in the structure and function of all forests (Attiwill 1994). That these changes in ecosystems are natural means that a single, clear target for each characteristic is not possible for every area. For example, the proportion of old-growth forest in an unharvested landscape will not always be the same in every area, age-class distributions are never static, and diversity ebbs and flows in response to many kinds of events, including fire, windstorms, insect defoliation, diseases and other parasitic organisms, and human influences. It is a mistake to believe that anyone can manage a forest to preserve a given characteristic at a specified level for long periods of time, or can maintain rare species if evolutionary forces are pushing them toward extinction. The most important idea is not that a single state should be maintained against all forces of change, but rather that managers should create a set of changing conditions that keep the limits of change within reasonable bounds, defined as much as possible by the natural range of variation in these conditions. The goal is not to maintain static states, but rather to ameliorate the impacts of human intervention.

How Would the Conservation Ethic Change Forest Management?

The change from traditional forestry practices to a new approach based on disturbance ecology has only just begun, and we have no illusions that the change will be easy or simple. Much remains to be learned—our scientific understanding of forest ecology is not complete. The motive for change is the need for us to do our best to work toward sustainability and, in the absence of complete knowledge, use the patterns and processes of ecological events to guide our decisions. Success is not guaranteed, but is more likely than

if we simply continue to follow traditional forestry practices.

A conservationist would strive to maintain important characteristics of the forest and the forested landscape within the context of the disturbance regimes that have occurred in the past. Several approaches have been developed to meet this objective. Yves Bergeron and colleagues (e.g., Harvey et al. 2002) have illustrated how logging operations can be conducted to maintain an age structure that is consistent with the age structure of unharvested forests. Burton et al. (1999) have shown how forests can be harvested for economic benefits while still conserving old-growth forests. Askins (2000) and Imbeau et al. (2001) have provided guidance on wildlife and forest management, considered bird species deemed vulnerable to timber harvesting, and suggested ways of dealing with that issue.

Other approaches to emulating natural disturbance have focused on emulating landscape patterns. Andison (1999) and Andison and Marshall (1999) illustrated how a computer model can be used to simulate the impact of forest harvesting and compare the results to a landscape that evolved in the absence of timber harvesting. Baker (1994) provided a method of restoring a landscape altered by fire suppression to a landscape that more closely resembles the landscape produced by a natural disturbance regime. Boychuk et al. (1997) and Schroeder and Perera (2002) provided useful background information on vegetation patterns and the impact of fire and logging at the landscape level. Boychuk and Perera (1997) illustrated the use of a computer model that provides insight for managers planning to include old-growth stands. Seymour et al. (2002) recently developed an index that has the potential to measure the sum total of forest management activities and compare them with an index of natural disturbance. This index potentially provides a quantitative expression of how closely a management plan approaches the natural disturbance regime.

What Results Might Come from Emulating Natural Disturbance in Forest Management?

Few studies have been conducted specifically to illustrate how implementing an approach based on emulating natural disturbance would change the economics of timber harvesting, the response of wildlife to the new forest, or the responses of other ecosystem functions, such as carbon storage. Studies should be conducted on those subjects to improve our understanding of the consequences of disturbance and thereby improve the process of implementing changes in forestry practices.

Despite the lack of specific studies in some areas, it remains possible to predict the likely outcomes for some wildlife populations, based on existing knowledge of their habitat requirements. For example, if we protect younger stands so they will mature into old-growth stands, early-successional species would probably decline from current levels and old-growth specialists would likely increase. Schmiegelow and Monkkonen (2002) estimated that about one third of the birds in Canada's boreal forest depend on the presence of forests older than the disturbance interval. Moose and deer populations would likely decline as younger stands are replaced by older ones, whereas populations of woodland caribou might be expected to increase as older stands of jack pine are allowed to develop. In a study in Alberta, Hobson and Schieck (1999) compared postharvest bird species composition to that after wildfires. Species with higher densities after the fire included the Connecticut warbler (*Oporornis agilis*), brown creeper, and American robin (*Turdus migratorius*), whereas populations of Lincoln's sparrow (*Melospiza georgiana*), the alder flycatcher (*Empidonax ainorum*), the Tennessee warbler (*Vermivora peregrina*), and the mourning warbler (*O. philadelpha*) increased in the postharvest area.

The Forest Management Guide for Emulating Patterns of Natural Disturbance

The Ontario Ministry of Natural Resources recently produced a set of forest management guidelines for emulating patterns of natural disturbance (Ontario Ministry of Natural Resources 2002a; McNicol and Baker, chapter 21, this volume). Because these are one of the first formalized sets of guidelines for implementing the emulation of natural disturbance, there will be many lessons to be learned from the experience of working with these guidelines. After examining their adequacy in the context of our discussion of a conservation ethic, we believe that the guidelines have two major shortcomings. First, there is a disproportionate focus on landscape patterns rather than a more balanced focus on maintaining a range of important forest characteristics. Second, within its focus on patterns, the guide concentrates not on emulating an overall landscape pattern of disturbed and undisturbed forests, but rather on the size of clearcuts. These guidelines let managers justify clearcuts as large as 10,000 ha, as long as 80% of the clearcuts in the area being managed are smaller than 260 ha. Provisions are also made for leaving residual trees and cavity trees, and for preserving islands and peninsulas within cut areas. However, by focusing mostly on the disturbances, the guideline de-emphasizes the older, undisturbed forests and associated species compositions that would arise under a natural disturbance regime. The most likely result of implementing these guidelines is a forest with a much greater quantity of edge ecosystems, smaller patches, and less old-growth forest than would be expected under a natural disturbance regime.

Although the guideline does provide general directions that require the retention of a natural range of age classes and species diversity, these directions are optional and too vague compared with the very clear directions provided for desired and permissible clearcut sizes. The developers of the guidelines were right to seize upon emulating natural disturbance as a promising concept, but their focus is insufficiently comprehensive. In addition, as Hunter (1993) has pointed out, extremely large clearcuts can be justified as a management tool, but it is very difficult to persuade the public to accept them. Furthermore, E. A. Johnson et al. (2001) have demonstrated that large fires will continue to occur in the boreal forest, despite intensive fire protection, and adding clearcuts the size of the largest burns may produce more large disturbed areas than would be natural or desirable under an approach based on emulating natural disturbance.

Summary

In the past, the philosophy of sustainability has focused on preserving the ability to harvest forests in perpetuity. Although this is unquestionably an important economic goal, it neglects the values of other stakeholders in our forests, and thus raises the ethical issue of whether an approach can be considered truly sustainable if it fails to sustain other important values, such as maintaining wildlife, biodiversity, and the existence of wilderness for its own sake. In this chapter, we have proposed a conservation ethic that includes these neglected values within the overall approach of sustainable forest management.

Ecological studies and current forest inventory results in Ontario make it clear that forest management in the twentieth century has caused major changes in biological diversity, age classes of major forest trees, forest patterns at the landscape level, and populations of wildlife that inhabit the boreal forest. Old-growth forests have been significantly reduced in size and abundance, and the wildlife species associated with them have also declined, whereas populations of early-successional species have increased. Relatively few scientific studies of ecosystem functions, such as carbon storage, nutrient conservation, and energy flow, have been conducted in the context of the managed forest of the twenty-first century. From both ethical and practical perspectives, it seems prudent to manage the boreal forest based on the emulation of natural disturbance, as we have interpreted it in this chapter. The goal is to conserve ecosystems similar to those that exist naturally to provide a safeguard against future changes and promote the sustainability of all forest values.

CHAPTER 16

Consequences of Emulating Natural Forest Disturbance
A Canadian Forest Industry Perspective

DARYLL HEBERT

The increasing emphasis on sustainable forest management appears to be in direct conflict with the decades of sustained yield management practiced across Canada. The former emphasizes sustaining all values provided by the forest, whether or not they have economic value, whereas the latter emphasizes mainly the production of economic value in the form of timber. Of the many possible approaches for implementing sustainable forest management, one that is currently receiving much attention is that of emulating natural forest disturbance. As resource managers attempt to incorporate natural disturbance as the basis for maintaining ecological function, they appear to be struggling with the transition to a new paradigm in which recognition of these natural disturbance attributes forms one key decision factor in developing management strategies. Canada's *National Forest Strategy 1998–2003* (Canadian Council of Forest Ministers 1998, p. xii) suggests that our goal is "to maintain and enhance the long-term health of our forest ecosystems, for the benefit of all living things both nationally and globally, while providing environmental, economic, social and cultural opportunities for the benefit of present and future generations." The authors support this statement by suggesting that "our forests will be managed on an integrated basis, supporting a *full* range of uses and values."

One tool to achieve these goals for sustainable forest management is based on the emulation of natural disturbance, ecological processes, forest composition, and forest structure at several scales (Kimmins, chapter 2, this volume). Fire appears to be the major disturbance to be emulated, but insects, diseases, floods, blowdown, and gap dy-

namics also play interacting roles. Fire can be characterized in terms of its frequency (the fire-return interval), size, shape, and severity, as these characteristics determine many, if not most, ecological processes (e.g., succession). In addition, fire determines the species composition (e.g., vegetation, vertebrates, invertebrates), landscape patterns, and structural attributes of stands, and affects most forestry practices, such as rotation age, block size, and variable retention within the block (Stuart-Smith and Hebert 1995). These attributes are reflected in the rotation age, the size and shape of cut blocks, and the severity of the disturbance that results from harvesting.

The baseline levels of the ecological attributes created by recurring fires (e.g., the number of snags per hectare, the frequency of disturbance) can be used to establish harvesting levels and the resulting stand structure during harvest planning and operations, thereby achieving many of the goals of sustainable forest management. Currently, government policy and forestry operations in several Canadian provinces embrace the goal of managing forests in terms of their patch-size distribution, shape, and degree of retention of residual trees (e.g., McNicol and Baker, chapter 21, this volume).

Based on these concepts and goals, implementing sustainable forest management that emulates natural disturbance will have significant consequences for all aspects of sustained yield management, from education to application and cost. Many aspects of these management approaches have been examined in detail in other chapters of this book. This chapter examines the transition from forest management based primarily on sustained yield to forest management based on emulating natural disturbance from the perspective of the forest industry, one of the more visible forest stakeholders. In particular, I address some of the challenges and potential economic consequences of this approach from the industry perspective, using examples from western Canada.

Traditional Sustained Yield Management

The current status of Canada's forested lands is outlined in *The State of Canada's Forests 1997–1998* (Canadian Forest Service 1998). Canada is largely a forested country, with half its land mass covered by a variety of ecosystems with different tree species and in different seral stages. Of this land, approximately half is classified as productive–commercial, and most of it is still publicly owned. Thus, programs based on sustained yield actually manage only around 50% of the land base. Most of that land is administered by provincial forest ministries, with most timber harvesting undertaken by private industry. The legislation, regulation, and policy governing this relationship has largely been based on the principles of sustained yield and economic sustainability, but has evolved somewhat independently in each province. "For much of the first half of the 20th century, forest policy was driven by two imperatives: revenue generation and economic development" (Canadian Forest Service 1998, p. 53). Tenures in Crown forest evolved based on negotiations between governments and the forest industry to ensure that supply would meet demand, while generating revenue for both the Crown and the industry. During much of this period, the harvesting rate was unregulated and reforestation, largely a Crown responsibility, was inadequate.

During the past several decades, sustained yield has become a more efficient and effective tool for improving fiber production and economic sustainability. In British Columbia, for example, the Chief Forester sets the annual allowable cut in accordance with a set of criteria that reflects the rate of timber production that can be sustained based on the forest's composition and growth rate, the re-establishment rate after harvesting, silvicultural treatments, timber utilization standards, and allowances for losses (to decay, waste, and breakage).

Most of these criteria support enhanced fiber production, and the revenues generated by these activities support an economic network throughout society that provides health and education services. This efficient development of sustained yield currently supports some 337 communities across Canada, employs 837,000 Canadians, pays Can$11.1 billion annually in wages, ships products valued at Can$71.4 billion (1995), and contributes exports of Can$32.1 billion to the net balance of trade (1996), almost as much as energy, fishing, mining, and agriculture combined (Canadian Council of Forest Ministers 1998).

British Columbia's Chief Forester also considers the short- and long-term implications of alternative harvesting rates and the timber requirements of established and proposed processing facilities, as well as the economic and social objectives of the government (Binkley et al. 1994). However, from an industry perspective, increases in delivered wood costs ($\cdot m^{-3}$) can result in de facto reductions in annual allowable cut, because some wood becomes too expensive to obtain, even if it is technically "available" under the permitted harvesting rates.

The harvesting rate is based on the productivity of the forested land and the ability to optimize growth and yield through management. Provinces are continually developing enhanced forest management techniques to maintain or increase annual allowable cut. In Alberta, the Enhanced Forest Management Task Force (1995–1996) included a joint government and Alberta Forest Products Association committee, whose responsibility was to recommend and facilitate the implementation of enhanced forest management.

Throughout the years, sustained yield management has included several goals: maintaining or increasing the harvesting rate; improving utilization standards, site classification, growth and yield, planting stock, and stocking standards; reducing regeneration delays; mixed species planting; even-aged management; shortening rotations; and reducing site disturbance; among others.

Sustained yield management has evolved to include the goal of maintaining the forests, not just the economy. The initially restricted economic view of the value of forests led to enhanced practices for fiber extraction and regeneration. As such, the transition to approaches based on emulating natural disturbance should recognize the intensive nature of sustained yield management and its economic contribution to society, and develop a transition process that carefully measures the trade-offs inherent in the new form of management. Reinhardt (1999) suggests that managers must go beyond the question of whether it pays to be green; the answer is, of course, that it depends. With environmental questions, the right policy depends on the circumstances confronting the company and the strategy it has chosen. Although ecological management and management based on emulating natural disturbance may solve some environmental problems, these approaches do not automatically create economic opportunities for the industry. In fact, in most cases, the cost of these approaches is significantly higher, and offers no proven market premium to compensate for the additional costs. Thus, managers should begin to look at forest management problems as business issues that require a full assessment of the benefits and trade-offs, as suggested by Binkley et al. (1994).

Transition to Forest Management Based on Emulating Natural Disturbance

Almost all approaches to sustained yield have attempted to increase growth and yield and thus, to increase annual allowable cut. To assess the expectations for and consequences of management based on emulating natural disturbance, I use sustained yield as the traditional baseline against which to assess the impact of changing to the new paradigm. However, approaches based on emulating natural disturbance can and should be practiced over the entire land base (whether harvestable or not), whereas sustained yield has only been practiced in the harvestable land base. Depending on the ecosystem, the nonharvested land base may compose as little as 10–15% of the total land base or as much as 60% of the total.

Within British Columbia's sustained yield system, and undoubtedly in other provinces as well, sustainable forest management is ensconced in the system's methodology. In British Columbia, this falls under criterion 6 of the methodology ("other pertinent information") and in the statement of the economic and social objectives of the government (British Columbia Ministry of Forests 1995b). Sustainable forest management in general, and approaches based on emulating natural disturbance in particular, are being forced to fit within the existing sustained yield model, without a clear, independent view of a sustainable forest management framework. Sustained yield management is ultimately assessed by means of annual allowable cut, along with various procedures and practices used at a variety of scales, whereas sustainable forest management can be based on the maintenance of ecological function in relation to an appropriate annual allowable cut. Using British Columbia as an example, Binkley et al. (1994) suggested that the economic impacts of reductions in annual allowable cut have significant negative consequences, and that the social costs in terms of unemployment and community stability or survival are even higher. They concluded by stating that decisions concerning harvesting must comprehensively reflect the costs and benefits in terms of both timber and nontimber values.

Implicit in the change from sustained yield to an approach based on emulating natural disturbance is the need for government policy to reflect the needs of the people, based on a careful calculation of both the benefits and the costs of any new regulations. Dismantling a paradigm that plays such a pivotal role in our society must be done carefully. Although the *National Forest Strategy 1998–2003* indicates that we must manage for a *full* range of uses and values, the actual levels of these values, their presence or absence in specific circumstances, and how much is enough must be clearly assessed by evaluating the benefits and trade-offs of each approach. In doing so, trade-offs between the values provided by the old and new approaches must be done within and across scales, and the economic and ecological functions must be quantified somehow.

To date, the move toward the newer paradigm has included a seemingly random dismantling of traditional practices, with an incompletely justified ecological rationale, whose positive and negative consequences have not been quantified. The transition from old to new espouses adaptive management systems, trials of these systems, and monitoring of the results, but little such evaluation is being done for individual ecological or economic functions or (more importantly) for the tradeoffs and impacts of management choices.

Assessing the expectations for and consequences of management based on emulating natural disturbance depends on the state of development of the traditional system that is to be changed, its economic contribution and status, and its social acceptance.

Impetus behind Management Based on Emulating Natural Disturbance

The move from sustained yield to management based on emulating natural disturbance has taken a circuitous path over the past several decades. At some time during the 1960s and 1970s, part of society realized that sustained yield and intensive forest management were not satisfying its nontimber needs (e.g., wildlife habitat, recreation) and might in fact be detrimental to such needs as the preservation of various species and ecological components of the forest. To make matters more difficult, wildlife biologists were also promoting sustained yield for the management of ungulate populations based on various optimization techniques. Attempts to optimize both fiber production and ungulate populations changed some forest practices, but also increased conflicts with other values and increased the confusion over "ecological" management. During this period, the requirements for most nonungulate species were not included in the management regimes and were not necessarily consistent with those for ungulate species.

The initiation of sustainable management based on emulating natural disturbance began with this increased pressure for ungulate habitat; for example, guidelines targeted black-tailed deer (*Odocoileus hemionus sitkensis*) on Vancouver Island, elk (*Cervus elaphus*) throughout the western provinces, moose (*Alces alces*) in central and northern Canada, and caribou (*Rangifer tarandus caribou*) in Ontario, British Columbia, and Alberta. The increased recognition of the importance of biodiversity (including ecosystem, habitat, and species values) led to increased pressure to develop new and innovative management systems.

Role of Adaptive Management

Because it is unlikely that a management system based on emulating natural disturbance will provide the correct outcomes from the outset, the ability to monitor and test the system within an adaptive management strategy is essential. In general, any forest management system should produce a testable response or outcome (figure 16.1).

Walters (1986) suggested that active adaptive management involves deliberately altering management plans to observe the effects of these alterations and thereby gain knowledge about the system. This approach requires the ability to implement management plans that are not necessarily expected to be the right ones—in some instances, they may be suspected of being the wrong ones—so as to gain information about how the forest resource responds to different strategies, and adapt subsequent plans to account for these results. Although the forest industry espouses continuous learning and science-based approaches to forest management, neither the industry, the provincial governments, nor academia have done much work to investigate active

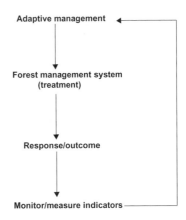

FIGURE 16.1. An adaptive management system that permits testable outcomes.

adaptive management. This form of management can replace prescriptive regulation with innovation that uses alternative management treatments and relies on questions that are defined based on the results of these treatments (Fisher 2002). Baker (2000) suggested that companies that use their knowledge to develop innovative management systems could provide an appropriate model for the agencies responsible for ecosystem-based resource management. However, the additional costs of active adaptive management have only been crudely estimated at this time, and cost sharing arrangements are only in the early stages of development.

A conceptual illustration of such a management system (figure 16.1) includes multiscale forest components (e.g., seral stage, forest composition), landscape patterns, and habitat elements (e.g., stand structural attributes) that are developed based on specific management strategies, assessed by modeling the trade-offs and various indicators of success, and tested at multiple scales by means of adapting the management approach.

Because the sustained yield system, including its planning methodology and operational practices, has become well established, modifying this system by incorporating management approaches based on emulating natural disturbance can add significant costs to the overall task of managing the resource. Monitoring fine filter (i.e., species level) indicators and developing management programs for the structural components of stands are both new activities for the industry, and will thus be expensive initially, and will require additional education. Development of these programs requires additional forest inventory data, the assessment and development of baseline levels against which a program's success will be measured, and the development of a range of permitted values for indicators of success, and may require the development of new models and modeling techniques. In addition, it is unclear whether such indicator species as the pileated woodpecker (*Dryocopus pileatus*) and the red-shafted flicker (*Colaptes cafer*) are actually representative of the more than 40 species of birds and mammals that nest in cavities in residual trees and snags; similarly, it remains unclear how well other indicator species represent the particular habitat element they are intended to represent. Currently, these questions require additional research that is expensive and different from the type of research conducted by the forest industry in the past. It is also unclear what roles the government and industry should play concerning the development of new forest inventory data and the performance of research on and trials of adaptive management.

Challenges during the Transition

Over the past decade, forestry practices began changing to include such new approaches as two-pass harvesting, reduced cut-block size, the creation of rectangular openings, green-up restrictions, adjacency requirements, and buffer strips. Block size has continuously decreased from sizes greater than 100 ha to smaller cuts that now average as little as 17 ha in many areas. As the requirements for wildlife species were included in silvicultural prescriptions, conflicts among competing habitat requirements (e.g., seral stage, pattern, structure) increased, especially with respect to ungulate management. Increased concerns over forest fragmentation began to contradict regulations intended to reduce cut-block sizes, and the concept of connectivity between ecosystems became confused with the need to create wildlife corridors and establish connections between old-growth stands. Old-growth forest was presumed to be in short supply or on the verge of extinction, further confusing the issue of providing the early seral stages required by ungulates.

Planning has become a subjective mix of changes in practices and various interpretations of untested concepts. The entire process during this period of transition has become adversarial, as well-meaning biologists have attempted to integrate ecological components into (or sometimes remove them from) the current sustained yield system, rather than develop an ecologically

based sustainable forest management system. The approach has been ad hoc at best, and poorly planned at worst, and has unnecessarily increased harvesting costs in many cases while also increasing uncertainty and insecurity for both the ecological and economic aspects of management. The result has been a loss of trust between the various forest stakeholders and decreased credibility of the approach because of large shifts in the management of critical habitats. The "command and control" type of operation imposed by provincial regulations produced little opportunity to try alternative prescriptions. The ongoing process of advocating and subsequently revising new approaches is somewhat ironic, given the resistance to active adaptive management.

In general, most regulatory systems have provided rules, pathways for following these rules, and penalties for failure to do so. British Columbia's *Forest Practices Code* (British Columbia Ministry of Forests 1995b) and *Biodiversity Guidebook* (British Columbia Ministry of Forests 1995a) have followed this approach. Unfortunately, the ability to test the proposed treatments by modifying the prescriptions and observing a variety of responses is missing from the approach. Current policies advocate systems based on emulating natural disturbance, but current monitoring systems are inadequate for assessing coarse filter (ecosystem level), intermediate filter (habitat level elements), and fine filter (species level) indicators. In addition, the merits of randomized versus stratified monitoring systems are still being debated. The Alberta randomized monitoring system, which uses a 20-km grid, may cost up to Can$20,000 per plot, and it remains unclear how government and industry will cover these costs.

Consequences of Forest Management Based on Emulating Natural Disturbance

Sustainable forest management based on emulating natural disturbance can probably be equated to motherhood and apple pie: who would not want such obvious benefits as productive and sustainable ecosystems? Unfortunately, the confusion and contradictions of the transition to this newer approach were compounded throughout the 1990s by increased references to managing natural disturbance based on science. Fitzsimmons (1996, p. 221) suggested that "of all the recent ecosystem-based policy actions taken by the [U.S.] federal government none offers a better insight into the difficulties of shifting the ecosystem concept from being a scientific paradigm to becoming a basis for public policy than the proposed rule to manage the National Forest System according to ecosystem management principles." He further described the confusion that still faces the industry today: "the ecosystem-based aspects of the proposed rule reflect all the uncertainty, confusion and imprecision of the ecosystem concept." The regulations establish by fiat that "the capacities of natural systems" will serve as a benchmark for determining which human uses of the National Forest System are to be permitted (U.S. Department of Agriculture Forest Service 1995). They do so without defining either natural systems or their capacities and without demonstrating any appreciation of the enormous ambiguities that surround the definition of what is natural.

Fitzsimmons (1996) also suggested that the proposed U.S. regulations recognize that ecosystems exist at multiple spatial scales and are infinite in number, but offer no guidance on how individual ecosystems are to be sustained or how spatial scales are to be assessed. In short, he remarks that "the proposed rule [emulating natural disturbance] calls for the Forest Service to oversee the National Forest System in order to sustain undefined conditions on undefined landscape units that exist in limitless numbers in undefined locations and that are dynamic and constantly changing over time and space in unclear ways. Moreover, there are no standards by which to measure the undefined landscape conditions the rule is intended to guarantee" (p. 221).

Ecosystem management and natural disturbance systems are not established by legislation in Canada, except in Ontario (Statutes of Ontario 1995), but they are demonstrated in British Columbia's *Forest Practices Code* (British Columbia Ministry of Forests 1995b) and *Biodiversity Guidebook* (British Columbia Ministry of Forests 1995a) and in Alberta's *Forest Conservation Strategy* (Government of Alberta 1997). Although Canada is proceeding somewhat more slowly than is the U.S. National Forest System, we face the same confusion and lack of multiscale system planning, which leads to old regulations being interpreted in new contexts, continuing uncertainty, and a significant increase in harvesting costs. To a large degree, management based on emulating natural disturbance is being instituted on a trial and error basis. Implemented in such a piecemeal fashion, we have no indication of what is still to come, no process for evaluating the trade-offs, and no idea how the future system will relate to the current system.

Binkley et al. (1994) suggested that a reduction in the forest industry's activity produces four categories of impacts: (1) direct, indirect, and induced impacts; (2) impacts on investment; (3) impacts on the transportation and wholesale sectors; and (4) impacts on industry activities that support other industries. He suggested that the estimated impact of a 10% reduction in annual allowable cut in 1989 in British Columbia would be an employment reduction of 2.2% (31,570 people) and a decline of 2.5% in provincial GDP (Can$1 billion). Compared with the proposed baseline of sustained yield, reductions in annual allowable cut would occur at all scales, and most of these reductions have not been assessed across scales for duplication or for their cumulative costs, even though they are likely to be additive. Alternative scenarios are seldom, if ever, assessed quantitatively. Most importantly, there is no proven market premium to serve as a reward for managing the forest resource under a system based on emulating natural disturbance. Thus, in a free market economy, this form of management may not be a good or service that society wants or can afford (i.e., society may not be willing to accept the cost).

Habitat Supply

Adjusting to a system based on emulating natural disturbance will affect habitat supply as well. The largest single effect on annual allowable cut can occur through adjustment of the age class distribution (seral stages) from the distribution that exists under sustained yield management (even-aged stands managed under shortened rotations) to a system with a broader range of age classes, in which old-growth stands must be retained at levels approaching 12–20% of the total managed area (British Columbia Ministry of Forests 1995b). In addition, recruitment of old-growth stands must come from adjacent mature stands, approximating similar retention values for both age classes. Current proposals are to protect around 12% of the land base to maintain conservation values, although the relationships between intensive, extensive, and protected area management have not yet been assessed (Hunter 1990). The proportions, spatial distribution, and ecological and economic costs of intensive and extensive management through the emulation of natural disturbance and of the management of protected areas have not been qualitatively or quantitatively assessed.

At a landscape level, cut-block size may increase under management based on emulating natural disturbance, but in many jurisdictions, restrictions on the adjacency of cut blocks still play a role in decreasing annual allowable cut. In many (if not most) jurisdictions, regulations that dictate forest retention apply to the harvestable land base, but ignore the ecological contribution of the nonharvestable land, which often composes up to 50% of the total land base. Assessment of the partially and fully constrained land base has only recently been initiated, but the results have not yet been integrated with the harvestable land base to reduce duplication. Ecological and economic trade-offs across scales (i.e., across site series, clusters, and variants) are still required to assess the risk of this approach.

At a stand level, the incorporation of most or all structural elements of habitats within forestry operations and planning would play a significant role in reducing annual allowable cut. Although efforts spanning decades have been devoted to increasing and improving fiber utilization in the sustained yield approach, the requirements to retain coarse woody debris under management based on emulating natural disturbance can range from 60 $m^3 \cdot ha^{-1}$ in pine (*Pinus* spp.) and Douglas-fir stands (*Pseudotsuga menziesii* [Mirb.] Franco) to more than 1400 $m^3 \cdot ha^{-1}$ in coastal Douglas-fir stands (Caza 1993). For example, postharvest objectives for the retention of large woody material can be as high as 24–37 pieces per ha with a minimum length of 9 m at diameters of 40–56 cm in Oregon's Blue River ranger district (Caza 1993). At larger diameters (107–152 cm), 20 pieces per ha at 2.4-m length (plus 2–7 smaller pieces per ha) are required.

Snags have been shown to be a necessary component for up to 40 species of cavity-nesting birds. Stands in British Columbia's interior Douglas-fir and montane spruce biogeoclimatic zones may contain from 50 to more than 200 snags per ha (S. Viszlai, Riverside Forest Products, British Columbia, pers. comm.). In most Canadian provinces, worker safety requirements often significantly reduce the ability to retain natural densities of snags, because of the hazards they pose to workers. Thus, in most situations, snags can only be produced and maintained by means of variable-retention harvesting. This approach may require planners to set aside 5–10% of the managed area to provide 6–10 snags per ha (Bunnell et al. 1999), depending on the ecosystem in which the approach is applied.

Similarly, riparian habitat has been shown to be valuable for a wide range of vertebrates (up to 50% of the species in an area) and invertebrates.

However, riparian habitat is generally managed by means of linear buffer strips, which often ignore key ecological attributes (e.g., terrain, moisture regimes) and disturbance dynamics. For example, in Riverside Forest Products' Tree Farm License 49 in British Columbia, the fully constrained riparian land base defined under the *Forest Practices Codes* accounts for about 1% of the licensed area, and the partially constrained riparian land base may account for an additional 3–4% of the area. In the case of Alberta Pacific Forest Industries, the constrained riparian land base, including some intermittent streams, is estimated to account for up to 7% (S. Wasel, Alberta Pacific Forest Industries, Alberta, pers. comm.) of the company's forest management agreement area. Since it has been shown that trees at the edge of boreal streams are the same age as adjacent upland trees (Harper and Macdonald 2001), it is unlikely that linear buffer strips are a component of natural disturbance systems.

Economics

In addition to affecting habitat supply, approaches based on emulating natural disturbance will affect the industry's costs and the available wood supply. The Ontario Ministry of Natural Resources (1997) found that 2–10% of burned areas were covered by residual islands of trees after wildfires. Alberta Pacific Forest Industries retains an average of 6% of the managed area in residuals (as per the variable-retention guidelines) in its postharvest cut blocks, with the corresponding volumes ranging from 2 to 22 $m^3 \cdot ha^{-1}$. The estimated cost of this retention ranges from as little as Can\$20·m^{-3} to as much as Can\$100·m^3, depending on the province in which the operations occur and whether logging costs and royalties (stumpage) are included.

British Columbia imposes significant additional constraints in the form of spatially fixed old-growth management areas and the province's requirements for early-successional stands (e.g., reduced stocking, delayed stocking, reduced brushing), broadleaved stands, and adjacency, all of which further reduce annual allowable cut. Public pressures have also caused the introduction of subjective visual requirements, in which partial cuts are used to minimize the visual impact of harvesting, but at the expense of further reducing annual allowable cut.

Binkley et al. (1994) predicted that many regions of British Columbia would experience annual allowable cut reductions of 10–30% based on the timber supply reviews that had been completed at the time. He questioned whether the various constraints on harvesting were indeed beneficial in terms of their ultimate environmental outputs. These levels of harvesting reduction could reduce employment by up to 6% (92,000 people), reduce provincial GDP by up to 6.5% (Can$4.9 billion), and proportionately increase the government's budget deficit, with direct effects on the budgets for health, education, and general government services, including employment.

A comparison of the sustained yield baseline with values developed for management based on emulating natural disturbance suggests a significant impact on annual allowable cut and on the resulting economics of forestry operations. The biggest problem involves the lack of a complete economic analysis that would evaluate the ecological benefits and economic costs for nontimber uses associated with these reductions in annual allowable cut. As a result, proceeding with new approaches on a trial and error basis fails to consider the possible duplication of efforts on different scales and the associated ecological benefits. In addition, there is no understanding of the consequences of alternative harvesting levels and the potential trade-offs, nor has any risk analysis been performed (see also McKenney et al., chapter 17, this volume).

Work by the Forest Engineering Research Institute of Canada (Mark Ryans, FERIC, Pointe-Claire, Quebec, pers. comm.) has shown that implementing alternatives to traditional clearcutting can significantly increase harvesting costs (table 16.1). In addition to the increased production costs, issues related to worker safety, forest health, and windthrow all produce secondary cost increases.

TABLE 16.1. Cost Increases Compared with Clearcutting for Various Alternative Types of Harvesting

Harvesting Approach	Cost Increase (%)
Large patch cuts	5
Harvesting with protection of small stems	10
Shelterwood	15
Selection cuts	30
Sensitive sites	50

Source: Unpublished data from Mark Ryans, FERIC, Pointe-Claire, Quebec.

At this point, uncertainty is high, regulations are constantly changing, a system that addresses issues at multiple scales is nonexistent, costs are increasing significantly, the trade-offs have not been evaluated, and measurement of the outcome has been inadequate.

Expectations for Forest Management Based on Emulating Natural Disturbance

The idea of sustaining ecosystems based on the emulation of natural disturbance appears to provide a muddled basis for policy. The transition is also complicated by the need to preserve stability in the economics of forestry operations, the stability of communities and the standard of living of Canadians, and highly variable and dynamic ecological processes and ecosystems. Thus, it is not surprising that sustainable forest management based on emulating natural disturbance is an elusive target and an expensive one to achieve.

In general, we must learn to incorporate ecological parameters in a cost effective manner by measuring the desired ecological benefits and analyzing the trade-off between ecological and economic goals. This analysis must include ecosystem diversity (ecosystem representation) at a coarse filter level, habitat elements at an intermediate filter level, and a system for monitoring indicator species at a fine filter level. In addition, forest inventories must be expanded to include a new set of indicators for planning (coarse and medium filters) and monitoring. Data management systems must be developed to include all forest-related values.

As no one is opposed to sustainability, what should the forest industry expect in the future? It should expect more initiatives based on results, with less onerous regulations and more responsibility and accountability for results over the entire land base. Management policies, regulations, tenure agreements, and stumpage rates may all have to be adjusted to achieve this goal. Results-based policies and procedures could be more effective in achieving responsible participation by the forest industry. The current "command and control" process should be adjusted proactively with industry participation. Most importantly, the industry should be a full partner in developing the science and performing the planning necessary for the transition from sustained yield to management based on emulating natural disturbance. We should all expect a much better systems-oriented approach, so that duplication, uncertainty, and costs are reduced and ecological benefits can be measured and optimized. The use of active adaptive management should allow an assessment of the trade-offs as well as opportunities for flexible implementation tailored to the needs of both the industry and the environment. In short, it should provide the most benefits for the least cost. The pathway to this goal might include the following steps:

- Recognition of the concept of natural disturbance;
- Quantification of this concept (e.g., the range of variability);
- Testing of the concept at multiple scales;
- Initial application of the concept;
- Adaptive management to fine tune the results;
- Analysis of the economic impacts;
- Analysis of the trade-offs between economic and ecological goals;

- Analysis of the economic, ecological, and social risks; and
- Application of newly developed and optimized procedures.

As Reinhardt (1999) suggested, managers should look at environmental problems as business issues. He indicated that environmental management is often perceived as a zero sum game, and for every winner in such a contest, there is a loser. However, this belief makes the assumption that the "contest" is inevitably adversarial, and that compromises cannot be reached that would optimize the benefits for all stakeholders. A better recipe for success would involve a win-win process, in which our ability to flexibly assess these opportunities and trade-offs by better quantifying the outcomes will determine both the success of implementing management by emulating natural disturbance and the required management practices.

CHAPTER 17

An Economic Perspective on Emulation Forestry and a Case Study on Woodland Caribou–Wood Production Trade-Offs in Northern Ontario

DANIEL MCKENNEY, AL MUSSELL, and GLENN FOX

The pursuit and practice of forestry based on the emulation of natural disturbance is of growing interest to resource managers. This approach to forestry attempts to manage commercial wood harvests in a manner that mimics natural disturbance processes (see Perera and Buse, chapter 1, this volume). However, considerable controversy remains over the ecological effectiveness of such approaches (e.g., McRae et al. 2001; Thompson and Harestad, chapter 3, this volume).

Notwithstanding these debates, there are also economic implications that should be considered. This chapter begins with a review of some basic principles of forest economics that are relevant to the economics of this strategy. We then provide an example of how to apply the techniques of economic analysis to this type of forest management. The example takes the form of a case study of the problem of managing forests for both woodland caribou (*Rangifer tarandus*) and wood supply in northwestern Ontario. Wildlife managers believe that large disturbances are required to maintain populations of woodland caribou in the region (Racey et al. 1992). The proposed timber management guidelines (Racey et al. 2002) for this uncommon, nomadic species could radically change timber harvest planning in the approximately 6,000,000-ha region by affecting the spatial pattern and timing of timber harvests. The guidelines recommend the use of large clearcuts to mimic the effects of large fires, the most common form of disturbance in this region prior to recent efforts at fire suppression. This practice, along with management constraints that would not allow harvesting to occur in adjacent blocks for 20 yr or more, will affect the nature and viability of timber harvest-

ing operations in the region. Thus, the case study is a particularly germane example of the economic implications of implementing management based on emulating natural disturbance.

Our presentation of some results of the trade-offs between habitat management for woodland caribou and wood supply in northwestern Ontario exemplifies many of the issues embodied in emulation forestry, but is by no means intended as an exhaustive representation. We present a spatial representation of this problem and estimate some of the policy's opportunity costs in terms of the potential wood production values that would be forgone. (An *opportunity cost* represents value that is lost as a consequence of making a particular choice, usually to obtain another value.) We also discuss the general implications of our results. These include the importance of properly assessing the spatial variation in the value of standing timber and of quantifying the costs of alternative silvicultural options and yield responses that arise specifically because of the emulation objective.

Some Forest Economics Concepts

Wood Production and Stand Level Models

The models discussed in this section form the basis of the economics of silviculture as well as of emulation forestry. Much of modern forestry economics has focused on the determination of the economically optimal harvest age. The standard model, usually attributed to Martin Faustmann (1849), the nineteenth-century German forest economist, chooses the harvest age so as to maximize the net present value of either a representative hectare of forest or a stand (Samuelson

1976; Johansson and Lofgren 1985; Bowes and Krutilla 1989). The *net present value* represents the value, in today's dollars, of a future amount net of (less) all costs. If only the value of wood as a source of fiber and the establishment costs are considered, the optimization problem for a representative hectare of forest land can be written

$$\text{Max } \beta = \frac{PV_{Te}^{-iT} - C}{(1 - e^{-iT})}, \qquad (17.1)$$

subject to $V_T \leq V(T)$

where β is the net present value of the forest land ($\cdot$ha^{-1}), P is the standing timber or stumpage value ($\cdot$m^{-3}) net of harvesting costs and transportation costs to the mill, V_T is the timber volume at harvest age T (m$^3 \cdot$ha^{-1}), C is the stand-establishment cost ($\cdot$ha^{-1}), T is the rotation or harvest age (yr), i is the real discount rate (%, expressed as a decimal), and $V(T)$ is the timber production function.

The optimal harvest age T maximizes the net present value of wood production over an infinite number of rotation periods. An optimal solution occurs when the rate of growth in the value of the asset, which includes the wood and the land, equals the real discount rate. The *real discount rate* is the rate of return from other investments in the economy. The optimal rotation period depends on the stumpage value, the real discount rate, and the costs of establishing and managing the stand. Economic analysis of the timing of harvesting indicates that if the stand is not growing in value at a rate at least as great as the real discount rate, then it should be harvested. This type of model can be modified to reflect the stochastic nature of both biological and economic parameters. Nevertheless, the basic construction of the model clearly identifies the key economic dimensions of the wood production problem—growth and yield estimates through time, establishment costs, the value of standing timber, and the application of a discount rate to reflect the "time cost" of investments.

Connection to Emulation Forestry

Emulation forestry could have impacts on each of these elements, although the specific effects in a given context will depend on exactly how emulation forestry is implemented. For example, projected timber volumes may decline if there are fewer postestablishment interventions or if the operations involve single-tree harvesting that results in increased within-stand competition among the remaining trees. These types of changes could occur whether the silvicultural prescription were motivated by the attempt to emulate natural disturbance or by other objectives. Per-hectare establishment and maintenance costs may not vary significantly under an emulation regime, but again, much depends on how the emulation is implemented. Forest management treatments with minor postharvest interventions could involve smaller investments in future stands than may otherwise occur.

For a given stand, the standing timber or stumpage values are not likely to change because of the emulation practices per se, as timber prices are determined administratively or in world markets and are unlikely to be affected by changes in silvicultural practices in a single stand or even in a single region. An exception could be the enhanced marketing opportunities made possible by an "ecologically friendly" forestry label that increased access to export markets as a result of the application of emulation forestry. However, as is demonstrated in the case study below, standing timber values vary spatially because of transportation costs. All else being equal, stands farther from mills are less valuable than stands close to mills because of the increased transportation (haul) costs (Hyde 1980). If emulation policies influence the spatial pattern of harvesting over a large region, the effects on transportation costs can have important economic implications.

Economics of Nonwood Values

Forest economists have long recognized that forests produce other outputs and services besides wood fiber, and that these nonfiber products have economic value (e.g., Randall and Peterson 1984). The economic analysis of forest management has evolved to address these issues —at least at a theoretical level. Hartman (1976) proposed an extension of the Faustmann model in which other outputs of standing forest, such as habitat or recreation, have value. With the Hartman model, a standing forest is viewed as producing a flow of so-called *amenity benefits* from establishment to harvesting, in addition to the outputs (fiber production) captured under the Faustmann model. A general version of the Hartman model is:

$$\text{Max } \beta = \frac{PV_{Te}^{-iT} - C + \int_{t=0}^{T} a(t)e^{-it}\, dt}{(1 - e^{-iT})}, \qquad (17.2)$$

subject to $V_T \leq V(T)$

where $a(t)$ is the amenity (nontimber) value at age t ($\$\cdot ha^{-1}$) and all other terms are as defined in equation 17.1.

With the Hartman model, a rotation period is chosen that maximizes the net present value of both the wood and the amenity flows. Hartman showed that there could be circumstances in which it is optimal to never harvest a stand and capture its wood fiber value. Much depends on which nonwood values are included in the formulation of the problem and the nature of the function that describes amenity values over time. The result also depends on the initial composition and age structure of the stand being analyzed.

Application of the Hartman model requires estimates of the values of all of the relevant inputs (e.g., establishment and management costs) and outputs (e.g., standing timber, water flows, recreation, biodiversity). However, there are substantial difficulties involved in obtaining objective measures of the values of some inputs and outputs. There is a spirited debate in the economics literature about the reliability and even the validity of nonmarket values obtained through surveys, interview-based techniques, and other means. Thus, the maximization of net present value set out in equation 17.2 is often not only difficult to apply, but also controversial. Nevertheless, it is often helpful to use an economic cost–benefit analysis as a framework to help set out and understand the interplay of economic, ecological, and other relationships. One practical approach is to examine the effect of implementing a particular management regime on net present values—the opportunity cost of the proposed policy. The effect is usually negative, and at a policy level, the question becomes whether the loss in net present value is justified by the benefits of the proposed regime. This approach can be used to identify the least-cost strategy for achieving the benefits sought and to highlight some of the trade-offs involved. Sensitivity analysis for the economic values, the nonpriced values, and the ecological relationships is critical to a thorough understanding of the trade-offs and key technical relationships that are driving the results.

Extrapolation to Forest-Wide Analyses

An additional complication is, of course, that actual forests consist of a multitude of stands of varying ages and types. It is common to analyze forest-wide planning problems by solving equations containing several variables in an effort to optimize the solution in the face of specific constraints; this approach is called *linear programming* (Buongiorno and Gilless 1987; see also Hof 1993).[1] For example, the province of Ontario uses the Strategic Forest Management model, which is based on linear programming. These types of models can be constructed to include alternative silvicultural options and costs. The usual convention is to maximize the net present value, subject to such constraints as maintaining minimum harvest levels. Including and then excluding the constraints can provide insights as to the costs. Usually such models only calculate the net present value over a single rotation, rather than the infinite series of rotations set out in the Faustmann and Hartman models. Although not ideal from a theoretical perspective, this type of calculation generally does not change the results of the model in northern climates because of the long rotations and the resulting minimal present values of future harvests. We have described the Faustmann and Hartman models primarily so readers will be aware of the theoretical ideal.

Although developed for forest-wide, multi-stand applications, the tracking of individual stands in linear programming models is often problematic, because topology is not recorded. (*Topology* represents the spatial configuration of the landscape—the relationship of individual stands to each other.) Hence, modeling problems that require a spatial dimension, such as the example discussed in the next section, can be difficult. The remainder of this chapter illustrates the use of opportunity-cost analysis for a problem that has an important spatial dimension. We used a simulated annealing model of forest planning (Lockwood and Moore 1993): *simulated annealing* is a stochastic optimization technique that is better suited to the case study than simple linear programming, because it can more easily represent stand topology. In our case

---

[1]Bowes and Krutilla (1989) extended the model of a "representative hectare" to a hypothetical multiple stand dynamic model in which nonwood values were interdependent among stands. Interdependencies are more compelling for the nonwood benefits from forests (e.g., habitat for a species may comprise many different forest types and stands). Interdependencies significantly complicate empirical work; hence, much of applied forest economics still uses the notion of a representative hectare and inherently linear models simply for practical reasons.

study, we use a model that calculates the net present value of the stands and then aggregates these values across all stands while simultaneously trying to achieve an explicit target for the aggregate harvesting level and while not harvesting adjacent areas for at least 20 yr.

Woodland Caribou Management in Northwestern Ontario

Setting

Woodland caribou are a species of large, nomadic deer native to northern Canada and Alaska. In northwestern Ontario, woodland caribou are relatively rare. Only 2000–5000 animals are thought to exist over an area of approximately 7,000,000 ha (figure 17.1) (G. D. Racey, Ontario Ministry of Natural Resources, Thunder Bay, Ontario, pers. comm.; Cumming 1998). The region includes 17 forest management unit administrative areas. Over the past century, woodland caribou's range in Ontario has been receding northward (Darby et al. 1989), roughly corresponding with the northernmost boundary of current commercial logging operations (Cumming 1998; see also Cumming and Beange 1993), although the exact reasons for this change are not well established. In recognition of the decline in woodland caribou numbers, the Ontario Ministry of Natural Resources, the primary land manager in Ontario, has initiated a strategy with a number of goals (Racey et al. 1992; Ontario Ministry of Natural Resources 1994). These include the maintenance of current population levels and habitat, the provision of future harvesting opportunities, and the provision of public viewing opportunities.

The Ontario Ministry of Natural Resources recommends that timber management operations create a long-term mosaic of habitats with large areas of specific tree species, age classes, and stocking rates. The mosaic should consist of harvest blocks at least 100 km², scattered throughout large areas of the caribou's range, thereby providing sufficient winter habitat for woodland caribou over time. The larger cutovers are intended to simulate the effects of large fires, which were more frequent prior to modern fire suppression activities and thus, were historically the primary force behind habitat renewal for the woodland caribou in this region. In the current management guidelines, blocks of timber adjacent to a clearcut cannot be harvested for 20 yr or more. One concern of the forest industry is that large tracts of merchantable timber may have to be bypassed as a result of this constraint, forcing them to harvest in more remote and perhaps less desirable areas (Mussell 1995). This may increase road construction and maintenance costs, and there is some fear that bypassed stands may be lost to fire or blowdown. However, individual clearcuts under the proposed guidelines would be much larger than under current practices.

There are some concerns that the habitat mosaic for woodland caribou may be detrimental to other wildlife species and to other forest uses. For example, the larger cutover areas would reduce the amount of forest edge relative to a comparable harvested area composed of smaller cutovers, making these areas less attractive as moose (*Alces*

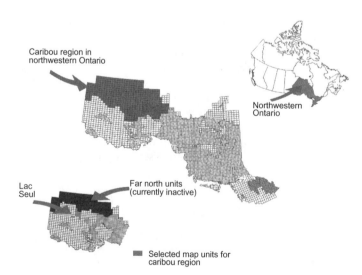

Caribou region in northwestern Ontario

Northwestern Ontario

Lac Seul

Far north units (currently inactive)

Selected map units for caribou region

FIGURE 17.1. A map of northwestern Ontario and of the woodland caribou region used in this chapter as a case study area.

TABLE 17.1. Anticipated Effects of the Woodland Caribou Guidelines in Ontario

Provision of Woodland Caribou Guidelines	Anticipated Effects
Bypass of currently used woodland caribou habitat by timber harvest operations; harvesting directed to other areas	Increased loss of timber to natural factors such as fire and insects; increased size of the primary road network; increased fiber transportation costs
Increase in size of individual cutovers	Negative public perceptions; potential decrease in unit harvesting costs
Restriction of access	Reduced human impact in habitat areas; increased quality of wilderness canoeing and camping
Decrease in the amount of forest edge; regeneration to coniferous species	Decrease in the moose population and in moose hunting opportunities

alces) habitat. The woodland caribou guidelines thus conflict with another provincial policy intended to increase moose numbers, and could have adverse consequences on remote communities that benefit from activities related to moose hunting. Larger cutovers may also increase soil erosion, which is detrimental to freshwater fisheries and aquatic life. Woodland caribou management may also adversely affect the wilderness tourism and fly-in fishing industries. Another complication is that the large areas affected by the guidelines can cross current administrative boundaries for harvesting activities. Thus, the proposed policy clearly presents a diverse, complex, and contentious set of issues and potential trade-offs. Table 17.1 summarizes the anticipated effects of the woodland caribou guidelines. These types of trade-offs are likely to be representative of the range of complications that could arise in the pursuit of many forms of emulation forestry.

There have been relatively few case studies of the implications of trade-offs between wildlife habitat and wood supply. Montgomery et al. (1994) investigated the debate over the northern spotted owl (*Strix occidentalis caurina*) in the Pacific Northwest of the United States, and Hyde (1989) examined the opportunity cost of conserving colonies of the red cockaded woodpecker (*Picoides borealis*) in North Carolina. McKenney and Lindenmayer (1994) examined the cost of a nest box program for an endangered marsupial in southeastern Australia compared with the cost of logging bans. In the study most closely related to the case that we present in this chapter, Mussell et al. (1996) used rotation ages in a nonspatial Hartman model to estimate the opportunity cost of implementing the woodland caribou guide-

lines in northwestern Ontario. Their results indicated a cost (in terms of the decreased net present value of wood production) of approximately Can$500 million. Our study is distinguished from the Mussell et al. (1996) study by the spatial nature of the problem formulation.

A Spatial Model of the Trade-Offs between Woodland Caribou Habitat and Wood Supply

Simulated annealing is an optimization technique applicable to large combinatorial problems (those characterized by nonlinear and discontinuous constraints). It uses an iterative-improvement algorithm (Kirkpatrick et al. 1983) that searches through the solution space randomly, looking for continual improvements. The solution space can be thought of as the range of technically feasible combinations of harvest patterns, rotation ages, and so on. The random element is large enough that the algorithm is unlikely to become stuck at a local optimum. The random element starts off quite large and then gradually decreases to zero. Penalty cost functions are used to simulate the effects of constraints in linear programming models, but allow for freer movement through the potential solution space. Thus, certain constraints can be made more important than others. For a detailed description of simulated annealing, see Otten and van Ginneken (1989).

Simulated annealing has seen relatively limited use as a forestry planning tool (see McKenney et al. [2000] for an application in the design of nature reserves). Lockwood and Moore (1993) describe the development and application of a tool based on simulated annealing (MIDAS) in the context of forest management. Their particular problem was the determination of a sequence of harvests that would achieve a target timber volume while complying with specified exclusion periods and restrictions on maximum clearcut size. Their analysis focused solely on harvest levels in a forest composed of approximately 6148 stands and covering 240,000 ha, rather than examining the economic implications.

Lockwood and Moore's problem formulation has some important similarities to the woodland caribou management problem: trying to achieve an overall harvest target while incorporating adjacency constraints. In our case study, all the management units in the region were grouped together and a target harvest level per decade was calculated. The models were run over a 90-yr planning horizon divided into 10-yr periods as in a typical linear programming model. The use of 10-yr periods was justified, because this is the period over which the 100-km² blocks will be harvested. Standing timber values for each spatial unit were assigned a different dollar value, based on the unit's distance from the mill. Our approach estimated the cost of the proposed spatial adjacency constraints by running the model with and without the adjacency constraints and comparing the resulting net present values. The difference between the two values indicates the minimum amount that people would have to be willing to pay to justify the policy on economic efficiency grounds. The cost of different harvest block sizes is not examined. All else being equal, larger harvest blocks should provide economies of scale in harvesting costs, but at the time of our study, no data were available to quantify this.

Data and Methods

Ontario's Forest Resource Inventory contains summaries of the state of the forest, including such parameters as the stocking level, the percentage of the total number of trees accounted for by each component species, the site class, the stand's age class, and the area of the stand (Trudell 1996). Information on nonforested land types and areas reserved (excluded) from harvesting (e.g., protection forest) were also included in our analysis. To simplify construction and interpretation of the model, the data were aggregated into five major species associations, based on the dominant species present in the stand: white birch (*Betula papyrifera* Marsh.), jack pine (*Pinus banksiana* Lamb.), poplar (*Populus* spp.), spruce (*Picea* spp., or black spruce [*Picea mariana* (Mill.) B.S.P.], plus white spruce [*Picea glauca* (Moench) Voss]. Those stands that do not consist of at least

60% of one of these five major species were labeled as *mixed* stands.

The study area is heavily dominated by spruce at least 90 yr old. However, there is also a large amount of 50–90-yr-old jack pine and mixed forest. According to local experts, the best woodland caribou habitat in the region is spruce forest at least 80 yr old. The latitude and longitude of each individual timber stand were not available for the entire region in a GIS-compatible format. However, stand positions were digitally summarized at the base map level for the region. These spatial units are essentially 100 km², which conveniently matches the size of the proposed harvest blocks and therefore made our analysis possible. There were approximately 600 of these spatial units in the analysis region.

Growth and yield projections were obtained from other forest models used in the region to study woodland caribou at the management-unit level (Arlidge 1995). Growth and yield estimates describe the changes in volume for different forest types at different ages or through time. Very little is known about the spatial variation in growth rates across northwestern Ontario; hence, the same growth and yield estimates were used for each map unit. Note that volumes decline over time as these forests become overmature. This is a contributing factor to the results presented later in this chapter.

An important component of the study was that the values of standing timber in each of the spatial units would be adjusted based on the unit's distance from the nearest mills. Although these values will vary spatially because of the effect of distance on haul costs, data on the precise nature of the change in costs are often not available. All else being equal, the farther a stand is from a mill, the less valuable it will be to that mill because of the increased haul costs (Hyde 1980; Nautiyal et al. 1995). *Residual valuation* is an approach that quantifies the difference between the final product value and the cost of producing the product. In the case of lumber, the value of standing timber would equal the market value of the lumber less the cost of harvesting, transportation to the mill, processing, and an allowance for profit. (In Ontario, there are also forest renewal charges to consider.) This number represents the maximum amount the firm would be willing to pay for the right to harvest the standing timber.

Using the principle of residual value, standing timber values were calculated for each map unit based on:

$$\text{MWTP} = \text{Starting Value} - \text{Haul Cost}, \quad (17.3)$$

where MWTP stands for mill willingness to pay ($\$\cdot m^{-3}$), and Starting Value represents the maximum amount a mill would be willing to pay for standing timber if it were situated right next to the mill. Three values for MWTP were used in our analysis: CAN$15, $30, and $40 per m^3. The CAN15\cdot m^{-3}$ cost follows the values used by van Kooten et al. (1995) and Mussell et al. (1996). The other two values represent the higher values for standing timber that some readers may consider more appropriate. Because these values are not determined competitively in Canada, there is some difficulty and controversy in determining the value of standing timber in economic analyses of forest management problems. That being the case, a range of values should be used.

Haul costs are the costs per m^3 of transporting wood fiber from the harvest site (i.e., from each map unit) to the mill. The farther wood is from a mill, the more it will cost to haul the wood. Haul costs were calculated based on Nautiyal et al. (1995):

$$\text{Haul Cost} = \text{Distance to Mill (km)} \cdot 0.0772\ (\$\cdot m^{-3}\cdot km^{-1}) \quad (17.4)$$

A digital map of the road network and a planimetric outline of the map units and current mill locations were used to calculate haul distances and costs in the Arcinfo GIS software (ESRI 1994), using an algorithm created to calculate the perpendicular distances from the centroids of each map unit to the nearest major road, then along that road to the major mill associated with that forest management unit. Table 17.2 summarizes the estimated average values of the standing timber for each forest management unit in the study after adjusting for transportation costs. Notably, with a MWTP of Can15\cdot m^{-3}$, none of the management units have positive average timber values. At Can30\cdot m^{-3}$, all but two management units have positive values, and at Can40\cdot m^{-3}$, only one management unit has a negative value. Stumpage fees (the actual prices paid for the standing timber) in the region ranged from Can$6 to Can$10 per m^3 at the time of the analysis. Table 17.2 also provides some insights

TABLE 17.2. Average Values of Standing Timber for Each Forest Management Unit in the Study Area for Three Values of MWTP

Forest Management Unit Number	Value of Standing Timber (Can$\cdot m^{-3}$)		
	MWTP = 15	MWTP = 30	MWTP = 40
20	−12.12	2.88	12.88
30	−7.28	7.72	17.72
120	−5.77	9.23	19.23
171	−10.53	4.47	14.47
172	−18.93	−3.93	6.07
173	−11.83	3.16	13.16
174	−1.79	13.21	23.21
240	−10.26	4.73	14.73
241	−11.59	3.41	13.41
242	−6.75	8.25	18.25
431	−50.17	−35.17	−25.17
432	−4.17	10.83	20.83
447	−8.55	6.45	16.45
470	−13.46	1.54	11.54
684	−9.67	5.43	15.43
840	−2.65	12.34	22.34
869	−3.83	11.17	21.17

Note: MWTP, mill willingness to pay (Can$\cdot m^{-3}$).

about mill locations in relation to the current primary road network: the more negative the number, the farther the mill lies from its current supply of wood. This analysis assumes that the primary road network and existing mill allocations remain in place over the planning horizon.

A harvest target of 4,431,286 m³·yr⁻¹ (44.3 million m³ per 10-yr planning period) approximates the current harvest levels in the region. This value was determined by examining mill receipts in the region for the past 5 yr and through discussions with regional experts and planners; thus, it is loosely related to the annual allowable cut. Additional analyses using target harvests of 30 million m³ and 50 million m³ per planning period were also simulated. The results for the scenario with 30 million m³ are not reported, because this target was easily achieved both with and without the adjacency constraints.

Following the convention used in other forestry cost–benefit analyses, 4% and 6% real discount rates were used for our calculations of present value (e.g., Row et al. 1981; McKenney et al. 1997). Several scenarios were analyzed, based on different values for standing timber, discount rates, and harvest targets. The scenarios were intended to represent a range of reasonable interpretations about the standing value of timber and discount rates. Each scenario was simulated with and without the spatial-adjacency constraints.

Results and Discussion

Out of the myriad of results produced by our analysis, we have chosen to focus on two issues: the imputed cost of the spatial-adjacency constraints, and the degree to which the harvest targets can be maintained over time.

Table 17.3 presents some summary results for present values and harvest volumes for each planning period in the constrained and unconstrained analyses for the scenario with a Can$30·m⁻³ MWTP and a 4% real discount rate. *Unconstrained* refers to turning off the spatial-adjacency constraints. The harvest targets were applied in all models. In both the Can$30·m⁻³ and Can$40·m⁻³ (not shown) models, the total present values of both constrained and unconstrained models were positive; however, the value of the timber harvests was negative in some periods. This is not unexpected, given the average values of the standing timber associated with each management unit. At Can$30·m⁻³ as the starting value, the harvest value of the standing timber in the region is Can$817.2 million in

the unconstrained model and Can$558.6 million in the constrained model, using a 4% real discount rate and current harvest targets. Increasing the harvest target to 50 million m³ per decade increased the present value slightly in the unconstrained analysis to Can$824.0 million and decreased the present value in the constrained analysis to Can$546.1 million (a difference of Can$277.9 million).

When standing timber values were Can$15·m⁻³, long-run harvest values were negative even without the adjacency constraints. Values were positive only in the first two periods of the unconstrained analysis. The implication is that the value of standing timber would have to increase in real terms to justify harvesting activities. Whether the correct long-run value is Can$15, $30, or $40 per m³ depends on the productivity of the industry in the region and on the actual stumpage fees charged, because such charges clearly affect residual values and MWTP for the standing timber.

The aggregate harvest target for the region (44.3 million m³ per decade) does not appear to be achievable beyond six planning periods (60 yr), even in the absence of the woodland caribou guidelines. This is an important result for the forest industry in the region, irrespective of the debate over woodland caribou and the effectiveness of the emulation objective. With the woodland caribou constraints, the aggregate harvest target does not appear attainable after the third planning period (30 yr). Given that stands are aggregated to the 100-km² map units, some stands will be scheduled for harvest before the age when they reach their maximum mean annual increment (i.e., the target age for maximizing the physical wood yields from the stands). The currently perceived minimum possible rotation length (60 yr) may have to become the norm to help meet harvest targets in the longer run.

Table 17.4 summarizes the estimated opportunity costs of the policy in terms of the value of timber forgone. The spatial-adjacency constraints in the woodland caribou guidelines reduce net present values by Can$129.6 million to Can$295.1 million, depending on what starting value for standing timber and what discount rate are used. The cost of these guidelines can be construed as large or small, depending on one's perspective. If the beneficiaries of the guidelines are deemed to be the ca. 10 million residents of Ontario, then these present values (lump sums) can be recast in terms of the cost per person. On this basis, the costs equal an annual

TABLE 17.3. Present Values and Harvest Levels in the Scenario with MWTP of Can$30·m^{-3} and a 4% Real Discount Rate

Planning Period	Present Value (million Can$)		Harvest Volume (million m^3)	
	Unconstrained	Constrained	Unconstrained	Constrained
1	520.9	420.3	44.4	44.3
2	280.4	150.9	44.3	44.3
3	−71.3	−63.6	44.3	42.6
4	87.0	31.6	44.3	35.5
5	−40.9	−4.9	44.3	35.5
6	36.7	12.1	44.0	32.7
7	0.1	9.8	36.4	31.1
8	−0.6	−0.7	21.9	12.6
9	4.9	3.1	18.6	12.5
Total	817.2	558.6	342.5	291.1

Notes: Unconstrained means that the spatial-adjacency constraints were not included in the analysis. MWTP, mill willingness to pay.

TABLE 17.4. Estimated Opportunity Costs Resulting from the Woodland Caribou Habitat Management Guidelines

Scenario (MWTP, real discount rate)	Opportunity Cost (million Can$)	Equivalent Annual Cost (million Can$)	Annual Cost per Ontario Resident (Can$)
Can$15·m^{-3}, 4%	186.3	7.5	0.74
Can$15·m^{-3}, 6%	155.1	9.3	0.92
Can$30·m^{-3}, 4%	258.9	10.4	1.03
Can$30·m^{-3}, 6%	129.6	7.8	0.77
Can$40·m^{-3}, 4%	295.1	11.8	1.17

Note: MWTP, mill willingness to pay.

value of Can$0.74 to Can$1.17 per person. Mussell et al. (1996) estimated an annual cost of Can$2.53 per person and a reduction in net present value of more than Can$500 million (using a discount rate of 5%). However, their estimates did not explicitly represent adjacency constraints. The results presented here suggest that the cost of the guidelines could be 30 to 75% lower than what Mussell et al. suggested.

Whether Ontario residents are willing to pay these amounts and whether the benefits of maintaining this population of woodland caribou at least equal this amount is a separate economic question. Whether spatial-adjacency constraints, as implied by this particular strategy for emulation silviculture, are the most cost-effective means of achieving the policy goal is also an important question.[2]

[2]The MIDAS model does not "grow" the forest after a harvest. This type of model is therefore not appropriate for examining certain issues, such as optimal silvicultural expenditures or rotation periods. Note, however, that planned rotation periods are at least 60 yr but more likely to be 80–120 yr. Hence, second-growth forests would have a very small to minimal impact on the net present value calculations or the ability of the model to reach harvest targets in the last two to three periods. If less-intensive silvicultural prescriptions are used because of the emulation objectives, longer (not shorter) rotations will be the norm. Given the slow growth rates of trees in northern Ontario, we think that the results provide an indication of the sustainability of the currently desired harvest levels for the next several decades and the opportunity costs of the recommended approach to creating woodland caribou habitat.

Summary and Concluding Comments

In this chapter, we reviewed basic forest economics principles relevant to the emulation of natural forest disturbance and presented a case study that we believe typifies many of the economic and ecological challenges associated with this strategy. For example, habitat management for such species as the woodland caribou involves management over a much larger land area than is traditionally considered for both economic analysis and forest management planning. The case study also highlighted the difficulties of obtaining numerical estimates of the factors that are required for either economic or ecological modeling. This case study was, by necessity, constructed to assess opportunity costs rather than the net benefits of the proposed policy. In the context of woodland caribou, it is difficult to estimate the benefits of preservation. These caribou are not a game species in Ontario and no data exist to provide insights as to how much people would be willing to pay to preserve the population of this species. Benefits could include hunting opportunities, a subsistence food source for native peoples in northern Ontario, and other values related to the existence of the species and its value to the ecosystem's functioning. Deriving such values for use in environmentally adjusted cost–benefit analyses is clearly difficult and also controversial. Our approach was to estimate an opportunity cost for habitat management in terms of the potential timber forgone due to implementing a policy of woodland caribou habitat management. Such an approach would be useful for any forestry policy problem, not just those related to emulating natural disturbance.

Our spatial model estimated opportunity costs 30–75% below those produced by a stand-level Hartman model (Can\$129.6 million to Can\$295.1 million, or Can\$0.74–1.17 annually per Ontario resident). This indicates the need for caution in applying stand-level models if spatial patterning is an important issue—as would likely be the case for many problems related to emulating natural disturbance. Care must be taken to assess spatial variation in the value of standing timber, the costs of alternative silvicultural options, and yield responses.

At a more fundamental level, controversies and uncertainties about the economic values of model inputs or outputs interact with the biological uncertainties. If wildlife biologists cannot agree about the impact of, say, the implementation of the woodland caribou guidelines in northwestern Ontario on populations of woodland caribou or moose, what relationships should be incorporated in the economic analysis? One approach would be to abandon modeling altogether and make policy decisions through nonquantitative means. We suggest a middle ground between "analysis paralysis" and completely qualitative policy development. Economic models of proposed forestry practices, whether representative-hectare Hartman models or spatial models, can be indispensable in synthesizing and making explicit both ecological and economic information. Possible outcomes and trade-offs can then be examined in a more objective and repeatable manner.

Acknowledgments

We thank Gerry Racey, Mick Common, and two anonymous referees for their comments on earlier versions of this manuscript. Any remaining errors are the responsibility of the authors.

CHAPTER 18

Developing Forest Management Strategies Based on Fire Regimes in Northwestern Quebec

SYLVIE GAUTHIER, THUY NGUYEN, YVES BERGERON,
ALAIN LEDUC, PIERRE DRAPEAU, and PIERRE GRONDIN

In the boreal forest, fire is the main natural disturbance that initiates succession and creates a mosaic of forest stands of different ages and compositions, in conjunction with the physical configuration of landscapes (Johnson 1992; Gauthier et al. 1996). Until recently, it was generally assumed that the North American boreal forest was characterized by short fire cycles (the time needed to burn a total area equivalent to the size of the study area), resulting in forest mosaics composed of even-aged stands (Heinselman 1981; Johnson 1992). This generalization has often been used to justify forest management based on clearcutting and short rotations (the age at which the forest is harvested). However, long fire intervals that allow for changes in canopy dominance and the development of uneven-aged forests have also been reported, particularly in the eastern boreal forests of North America (Cogbill 1985; Foster 1985; Frelich and Reich 1995; Bergeron 2000; De Grandpré et al. 2000; Gauthier et al. 2000; Lesieur et al. 2002; Lefort et al. 2003). Bergeron et al. (2001) have shown that in four regions of eastern Ontario and western Quebec, more than 50% of the area is occupied by stands that burned more than 100 yr ago. Therefore, using a short industrial forest rotation can lead to a dramatic decrease in stand diversity (composition and structure) at the landscape level (Gauthier et al. 1996), with potentially significant consequences for biological diversity (Hunter 1999; see also Thompson and Harestad, chapter 3, this volume).

Forest ecosystem management based on an understanding of natural disturbance regimes has been suggested as a means of maintaining biological diversity and productivity in forest ecosystems (Attiwill 1994; Bergeron and Harvey 1997; Angelstâm 1998; Bergeron et al. 1999a, 2002). Management strategies aimed at maintaining stand and landscape compositions and structures similar to those that characterize natural ecosystems should favor the maintenance of biological diversity and essential ecological functions (Franklin 1993; Gauthier et al. 1996; Hunter 1999; MacNally et al. 2002). Hence, forest characteristics such as the age class distribution, the stand composition and structure, and the spatial arrangement of stands in natural landscapes should be key indicators for the implementation of sustainable forest management. Our knowledge of these indicators must also include knowledge of forest succession, a key process that structures ecological diversity and determines the availability of timber, habitat for wildlife, and other resources (see Kimmins, chapter 2, this volume).

Recently, Y. Bergeron et al. (1999) proposed an approach in which diverse treatments can be applied to maintain the forest's structural and compositional diversity without lengthening the timber rotation. The main objective of this chapter is to show how knowledge of natural forest dynamics can be simplified to create a management strategy in boreal Canada. Zasada et al. (chapter 19, this volume) provide a similar discussion for the Great Lakes forests of the United States. We illustrate the development and initial implementation of this strategy using a pilot study in the black spruce (*Picea mariana* [Mill.] B.S.P.)–feathermoss (*Pleurozium schreberi* [Brid.] Mitt.) ecoregion of northwestern Quebec.

Framework and Approach

The development of the management strategy was based on four different research themes. The first theme documented the characteristics of the region's fire regime, and particularly, the variability in fire frequency, size, and severity (see also Suffling and Perera, chapter 4, this volume). The variability imposed by permanent site features that influence the combination of thermal, hydric, nutritional, and disturbance regimes is responsible for the variety of forest habitats in a region and thus determines the coarse filter on which to base the maintenance of biodiversity (Attiwill 1994; Bergeron et al. 2002). The second theme documented the natural forest dynamics that occur on the region's main surficial deposits. To assess whether key differences in patterns of biodiversity exist between managed and naturally disturbed forests, and how critical these differences are to the maintenance of biodiversity, we required baseline data on the responses of organisms to the natural dynamics of forest mosaics. In consequence, the third theme had the objective of defining these responses with respect to various components of forest stands, including structural elements (e.g., forest composition and structure, coarse woody debris). These first three research themes allowed us to define a number of objectives (e.g., landscape, stand ecological type, stand level) that could serve as targets at different phases of management planning. The fourth theme examined the integration of these objectives within management planning.

Study Area

The data used in the development of the management strategy were collected in a subsection of the Matagami (6a) ecoregion (Saucier et al. 1998). This ecoregion is part of the black spruce–feathermoss bioclimatic domain, a subregion of Quebec's boreal forest. The area belongs to the Northern Clay Belt forest region (Rowe 1972), which is characterized by forest stands generally dominated by black spruce and jack pine (*Pinus banksiana* Lamb.). There are also mixed stands of trembling aspen (*Populus tremuloides* Michx.), balsam poplar (*Populus balsamifera* L.), white birch (*Betula papyrifera* Marsh.), white spruce (*Picea glauca* [Moench] Voss), and balsam fir (*Abies balsamea* [L.] Mill.) on coarser tills and alluvial deposits along rivers and lakes.

The regional climate is continental, with a mean annual temperature ranging between –2.5 and 0°C; the January and July temperatures average –17.5 and 17.5°C, respectively. Annual precipitation averages about 1000 mm. The ecoregion lies within the James Bay watershed, and plains form the dominant landscape, with occasional hills; the average altitude is 284 m above sea level. The area's surficial deposits are predominantly clays and silts associated with the glacial Lake Ojibway. Subsequent readvance of a glacial lobe created a clay-textured Cochrane till in the northwestern part of the region.

The management strategy described in this chapter is currently being implemented in part of the Lac Grasset (119) regional landscape unit at the western end of the Matagami ecoregion. The Lac Grasset landscape is characterized by relatively flat topography (an average slope of 1%) and predominantly hydric and subhydric soil moisture regimes (Robitaille and Saucier 1998). As a result, more than 60% of its soils are classified as organic (an organic horizon >40 cm), and the better-drained sites are associated with the Harricana moraine that crosses the unit from south to north and has scattered buttes. The management area covers about 4750 km^2 (from 49°37′30 to 50°22′30 N and 78°30′00 to 79°30′00 W) and includes part of a forest management unit licensed by Tembec Industries and the Nexfor division of Norbord Industries.

According to the forest inventory conducted by Quebec's Ministère des Ressources naturelles, only about 50% of the area is covered by commercial forest, and large areas are covered by bogs. This commercial forest is primarily black spruce stands (87%), with occasional jack pine (8%) and trembling aspen (5%) stands on the better-drained sites. Three major fires have occurred in the past 40 yr (1962, 1976, and 1997), and harvesting operations have taken place since 1984. The regenerating forests represent about 30% of the land area occupied by the commercial forest, and about half of the regenerating forest originates from harvesting activities. Although industrial forestry only began around 20 yr ago in this region, the territory has been partially accessible for about 30 yr due to the construction of primary roads to two mines in the area. However, many parts of the territory are only accessible in winter, because the extensive bogs and major rivers can only be crossed when they are frozen. Fire suppression was minimal before 1970, as the territory was mostly inaccessible and the usage of "water bombers" only started at that time. Moreover, the area was only included within the province's zone of intensive fire protection once harvesting activities began.

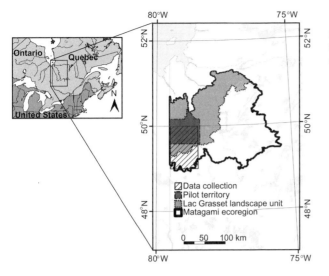

FIGURE 18.1. Locations of the areas used for data collection and for the pilot management study in Quebec.

Knowledge of Natural Dynamics

Variability in Natural Disturbance

Data on the region's fire history and forest dynamics were mostly gathered in the southwestern subsection of the Matagami ecoregion (from 49°00′ to 50°00′ N and 78°30′ to 79°30′ W; figure 18.1). Using archival data, air photos, and ground-truthing techniques, the fire history for the past 300 yr has been reconstructed; see Bergeron et al. (2001) and Gauthier et al. (2002) for a complete methodology. This reconstruction shows that large areas are still covered by stands that have not burned for a considerable period of time; 57% of the data collection area is composed of forest that burned more than 100 yr ago, and 20% is older than 200 yr (figure 18.2). This suggests that under a natural fire regime, a significant proportion of the stands were undisturbed by fire for long periods.

However, fire cycles have changed in the past 300 yr (Bergeron et al. 2001). From a cycle of about 100 yr before 1850, fire cycles increased to 130 yr by around 1920, and have increased since then to an estimated 400 yr (table 18.1). As this change started before European settlement, it appears to be related to climate change, which became less conducive to large fires after the end of the Little Ice Age (Bergeron et al. 2001; Lefort et al. 2003). The more recent increase in fire cycle (from 1920) may have also resulted from human activities, such as road construction and fire suppression.

As table 18.1 shows, the mean stand age also increased from 1850 to the present, but more slowly than the increase in fire cycle. The age structure of the forests has a certain inertia with respect to changes in the fire cycle. Moreover, the mean stand age integrates the effects of the changing fire cycle over the past 300 yr. For these reasons, we suggest that the average stand age in the study area (expressed in terms of time since the last fire) can be used as a baseline for planning harvesting activities in order to estimate the desired proportion of each age class to be maintained by means of different silvicultural

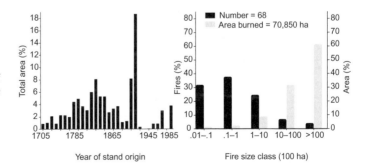

FIGURE 18.2. The postfire origin of stands in the study area (percentage of total area, left panel) and fire size distribution (1945–1998, right panel) in the southwestern section of the Matagami ecoregion.

TABLE 18.1. Fire Cycle and Mean Stand Age Estimated in the Southwestern Section of the Matagami Ecoregion

	Era		
	Before 1850	Before 1920	Today
Fire cycle (yr)[1]	103 (80–131)	133 (106–167)	398 (302–524)
Mean age (yr)[1]	87 (66–114)	102 (85–123)	183 (154–218)

Note: See figure 18.1 for the location of this region.

[1]Range in parentheses gives the 95% confidence interval.

treatments. This is discussed in more detail later in this chapter in terms of the cohort model of Y. Bergeron et al. (1999, 2002). For the study area, the mean age was estimated at 151 yr.

We used the data from the same area to study the fire size distribution (figure 18.2). Whereas the majority of fires were smaller than 1000 ha, these fires were only responsible for a small fraction of the total area burned (Bergeron et al. 2002). Consequently, large fires (>1000 ha) were responsible for most of the area burned, and therefore created much of the age structure and configuration of the landscape in terms of stand composition (Johnson et al. 1998). Bergeron et al. (2002) suggested that the characteristic size of these fires varied from 950 to 20,000 ha, and that regeneration areas should therefore fall within this range of sizes.

It must be recognized that severity varies within any single fire, especially if the fire covers very large areas and burns for longer than a day; as a result, the fire will leave patches of living trees behind (Van Wagner 1983; Turner and Romme 1994; Kafka et al. 2001; Bergeron et al. 2002; Ryan 2002). Although the areas that totally escape burning (preserved "islands") appear to be relatively constant from year to year, at around 5% of the total area (Eberhart and Woodard 1987; Bergeron et al. 2002), the zones of low fire severity depend greatly on seasonal weather. In fact, low-severity zones, which include areas that totally escaped the fire and those that only sustained intermittent crown fires, may occupy up to 50% of a burned area, depending on the type of forest, especially when the prevailing weather conditions are relatively mild (Bergeron et al. 2002). The presence of these lightly burned zones suggests that the pattern of mortality generated by fire is very distinct from that which results from conventional forest harvesting. Trees that survive a fire not only appear

to play a determining role in regenerating burns (Greene and Johnson 2000), but also represent refuge (shelter) habitats for wildlife in the regenerating forest and increase the spatial heterogeneity of the forest mosaic that results from the fire.

Forest Dynamics

Many studies on vegetation dynamics have been conducted in the Matagami ecoregion (Y. Bergeron et al. 1999; Gauthier et al. 2000; Harper et al. 2002, 2003). Although the data on the vegetation were obtained from chronosequence studies, the congruence between these results and the results from permanent sample plots (Lesieur et al. 2002) in another region of western Quebec's boreal forest supported our interpretation of successional trends. The results of these studies are summarized in the next section and table 18.2. After a fire, stands are recolonized mainly by three species (black spruce, trembling aspen, and jack pine), and these species show some preference for specific types of surficial geology. For instance, jack pine and trembling aspen are more common on rock, sand, or clay till sites than on organic sites, whereas black spruce is present on all sites.

In the first 100 yr, the majority of stands on organic soils are dominated by black spruce (table 18.2) (Gauthier et al. 2000; Harper et al. 2002), although some stands are dominated by jack pine or trembling aspen. As the time since the last fire increases, the presence of both pioneer species tends to decrease considerably and the importance of black spruce increases. In black spruce stands, the vertical structure becomes irregular over time; for example, stand height and density both tend to decrease with increasing time since the last fire, even as the abundance of canopy gaps is increasing (Harper et al. 2002). On other site types, canopy dominance changes

TABLE 18.2. Mean Importance Values for Eight Tree Species for the Main Surficial Geology Types

Surficial Geology	Time Since Fire (yr)	Number of Stands	Black Spruce	Jack Pine	White Birch	Balsam Fir	White Spruce	Trembling Aspen	Balsam Poplar	Larch[2]
						Mean Importance Value[1]				
Rock	50	5	29.7 (18.9)	46.8 (29.1)	9.2 (13.6)	2.8 (6.2)	0.4 (0.9)	9.0 (20.1)	0.0 (0.0)	2.2 (5.0)
	150	2	64.0 (0.3)	34.2 (2.8)	1.8 (2.5)	0.0 (0.0)	0.0 (0.0)	0.0 (0.0)	0.0 (0.0)	0.0 (0.0)
Clay till	50	12	23.9 (25.1)	52.3 (35.0)	6.3 (13.6)	6.3 (13.8)	2.2 (6.5)	9.1 (24.3)	0.0 (0.0)	0.0 (0.0)
	150	4	41.2 (36.6)	18.1 (36.2)	7.1 (8.2)	1.4 (2.9)	0.0 (0.0)	22.5 (29.8)	7.9 (15.7)	1.8 (3.7)
	250	1	1.8 (—)	95.3 (—)	0.0 (—)	0.0 (—)	0.0 (—)	2.9 (—)	0.0 (—)	0.0 (—)
Sand	50	9	26.0 (20.6)	58.7 (23.8)	9.3 (17.8)	0.9 (1.6)	1.5 (3.1)	3.6 (8.0)	0.0 (0.0)	0.0 (0.0)
	150	3	45.5 (27.5)	20.8 (19.5)	6.0 (10.3)	22.8 (39.5)	0.0 (0.0)	5.0 (8.6)	0.0 (0.0)	0.0 (0.0)
Organic	50	35	61.8 (31.6)	22.8 (27.4)	1.7 (7.2)	4.3 (11.7)	3.4 (12.6)	4.2 (11.5)	0.0 (0.0)	1.8 (6.3)
	150	20	85.7 (24.8)	5.4 (18.0)	0.9 (2.8)	6.6 (12.8)	1.0 (4.4)	0.0 (0.0)	0.0 (0.0)	0.5 (1.4)
	250	17	92.4 (17.3)	0.0 (0.0)	0.8 (3.4)	6.1 (14.1)	0.0 (0.0)	0.0 (0.0)	0.0 (0.0)	0.6 (1.9)

[1]Importance value = (Relative frequency + Relative basal area)/2. Standard deviations are in parentheses.
[2]*Larix laricina* (Du Roi) K. Koch.

from deciduous stands or stands dominated by jack pine to stands dominated by black spruce when the interval between fires is long enough. For instance, on rocky and sandy site types, jack pine tends to dominate the young stands, but black spruce becomes dominant in stands older than 100 yr. Moreover, if the time between fires is sufficiently long, the pioneer species may disappear from the stand. Because the fire cycle in this region typically exceeds 100 yr, changes in both species composition and stand structure represent a significant component of the vegetation dynamics in the area (Carleton and Maycock 1978; Gauthier et al. 2000; Harper et al. 2003).

Harper et al. (2003) also conducted a stand-level study of structural changes along a chronosequence, and found that the stand-level structural characteristics of old-growth forests differed among site types. On organic sites, the abundance of snags and logs was highest in the oldest stages, as would be expected for old-growth forests (Kneeshaw and Burton 1998), whereas snags and logs were least abundant in the oldest stages

on clay and sand site types; the peak value for these two structural attributes was observed at around 150–200 yr. As stands converge to pure black spruce, both the canopy cover and the canopy height tend to decrease (Harper et al. 2002). The old-growth phase is also characterized by an increasing thickness of organic matter and increased richness and cover of sphagnum (*Sphagnum* spp.) compared with younger sites (Boudreault et al. 2002; Harper et al. 2003).

Responses of Organisms to Natural Dynamics

The nonvascular and vascular understory species composition, as well as that of epiphytic lichens, have been studied along a stand chronosequence defined in the forest dynamics section of this chapter (Boudreault et al. 2002; Harper et al. 2003). For these three groups, the composition of the species assemblages changed along the chronosequence, mainly in response to changes in the degree of canopy closure, time since the last fire, or depth of organic matter in the soil. The nonvascular terricolous assemblages responded mainly to changes in the stand composition, time since the last fire, and depth of organic matter. The floor of the mature forest, with its high basal area of trees and relatively low depth of organic matter, was characterized by a carpet of mosses typical of closed boreal forests, such as *Pleurozium schreberi* (Brid.) Mitt., *Ptilium crista-castrensis* (Hedw.) De Not., *Polytrichum commune* Hedw., and *Dicranum polysetum* Sw. (Boudreault et al. 2002; Harper et al. 2002). Such a species composition persists through the early stages of old-growth development (>100 yr). When tree basal area begins to decline, at about 150 yr after a fire, the species richness and percent cover of *Sphagnum* spp. increase together with the increasing thickness of the organic matter horizons.

Differences in the species composition of understory vascular plants were also strongly related to the basal area of live trees and the time since the last fire (Harper et al. 2003). Young and mature forests, which are closed-canopy stands, exhibit greater cover and richness of herbs and low shrubs, as well as greater fern richness, than in older forests. With the increasing dominance of black spruce, and decreasing tree basal area in the absence of fire, the understory became dominated by ericaceous species. Finally, epiphytic communities responded mainly to variations in the stand composition, time since the last fire, and tree age. The total number of species of epiphytic lichen was greater in forests in the early stages of old-growth development and where trembling aspen or jack pine were still present in the canopy (Boudreault et al. 2000, 2002). The abundance of epiphytic lichens was closely and positively related to mean tree age.

Drapeau et al. (2003) have studied bird communities in the same area along a 300-yr chronosequence. They showed that bird communities are also responding to changes in the degree of canopy closure: more species associated with closed-canopy forests were observed in forests that are in the early stages of old-growth development (100–120 yr), whereas in the oldest forests (>200 yr), bird assemblages were dominated by species associated with or tolerant of more open canopy conditions. Moreover, primary and secondary cavity-nesting birds associated with standing dead wood (e.g., woodpeckers, nuthatches, creepers) were also more abundant in landscapes that had not burned for 100–120 yr than in our oldest forests (>200 yr). This result is not surprising, given the low densities of snags observed in the oldest forests of our study area (Drapeau et al. 2001; Harper et al. 2003). Old-

growth forests in their early stages retain a closed canopy with large trees while snags start to accumulate. This combination of canopy closure with attributes that generate structural heterogeneity (snags and downed woody debris) meets the habitat requirements of many forest dwelling birds. Such conditions are lost as the forest becomes older than 200 yr. Bird communities in old-growth forests thus change considerably as the forest ages (from 100 to 200 yr), and the composition of these species assemblages closely reflects the changes that take place in the structure of the vegetation.

Finally, forest fires in this region extend over very large areas and produce large areas covered by stands with a similar age and structure. Such a coarse grained landscape mosaic also affects the distribution patterns of organisms. Several studies have shown that mobile organisms, such as birds, are affected by the nature of the landscape that surrounds their habitats in the boreal forest (Edenius and Elmberg 1996; Schiemegelow et al. 1997; Drapeau et al. 2000). In the black spruce ecoregion, the spatial adjacency of similar structural conditions over large areas (10–100 km^2) is an important factor affecting the composition of bird communities (Drapeau et al. 2003).

Overall, our studies show that regardless of the taxonomic group selected, many species reach their peak of abundance in the early stages of old-growth development (between 100 and 200 yr after a stand replacing fire), when highly diversified structural conditions occur. This phase corresponds to the start of decadence in the post-fire cohort, which begins to die and be replaced in the canopy by trees recruited from the understory (see the next section for details). These conditions do not persist in forests older than 200 yr, so species associated with more open habitats occupy these forests. For this reason, the portions of the chronosequence that cover old-growth development prior to 200 yr after a fire are key habitats from the perspective of biodiversity for the black spruce forest ecosystem. Because these forests cover more than 20% of the land base in the study region, they may play a strategic role in the population dynamics of many forest dwelling species by providing source habitats.

Conceptual Model of Natural Disturbance and Forest Dynamics

As suggested by Gauthier et al. (1996) and discussed by Y. Bergeron et al. (1999, 2002), maintaining the region's observed biological diver-

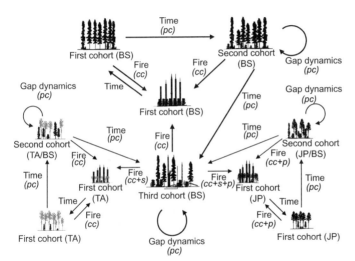

FIGURE 18.3. Conceptual model of natural forest dynamics proposed for Quebec's Matagami ecoregion and the corresponding management strategy. BS, black spruce; JP, jack pine; TA, trembling aspen; cc, clearcutting; p, planting; pc, partial cutting; s, selective cutting.

sity requires the development of management strategies that will maintain a forest mosaic similar to that observed under natural conditions. To accomplish this goal, the proportions of the area occupied by different stand types (based on composition and structure) must be kept similar to those observed under the forest's natural dynamics, the spatial patterns of harvesting must resemble natural spatial patterns, and a variety of disturbance severities must be maintained within harvesting areas. The information presented in the previous section can serve as a basis for designing such a strategy.

As a first step, we have illustrated a simplified version of forest succession in the pilot study area (figure 18.3). Three main pathways have been defined, depending on the relative importance of the main species that form the postfire cohort: jack pine, trembling aspen, and black spruce (the most common species in our study area). The postfire cohort forms the first cohort, and if a fire occurs while the stand is in that stage, cyclical succession would occur, as is observed in many boreal regions (Johnson 1992). Given that the fire cycle is relatively long in the study region, the interval between fires at many sites is longer than the normal longevity of individual trees in the postfire cohort, and these sites succeed to the old-growth forest stage (Kneeshaw and Gauthier 2003). With a long fire interval, the species composition (or stand structure) changes from that of a forest regenerated by a shade intolerant postfire cohort (trembling aspen or jack pine) to stands recolonized by black spruce. These changes reflect the replacement of individuals that became established immedi-

ately after a stand-replacing fire and that initially formed the stand's canopy (the first cohort) by individuals that previously occupied the understory (the second cohort). Moreover, in the continuing absence of fire, gap dynamics perpetuates the replacement of individuals from these first two cohorts by individuals from later cohorts (collectively called the third cohort, because the individual cohorts gradually become harder to distinguish). Note that the rate of these transitions may vary among site types and types of transition. For instance, Bergeron (2000) and Cumming et al. (2000) have shown that trembling aspen is capable of recolonizing gaps and maintaining itself in stands for more than 200 yr.

These three developmental stages (cohorts) cover a gradient in the time since the last fire (*temporal* cohorts) that is also associated with gradients in a stand's horizontal, vertical, and compositional structure. Thus, the three stages can also be described using different stand structural attributes (*structural* cohorts). More specifically, forest stands associated with the first cohort are usually closed canopied and have a relatively simple vertical structure that lacks distinct canopy layers. When shade intolerant species are present (which depends on the composition of the initial regeneration), these trees are found in the dominant canopy layer of stands in the first cohort. In stands of the second cohort, the canopy is often semi-open and usually has several distinct canopy layers. These stands are composed of both shade intolerant and shade tolerant tree species, when the stand's composition is mixed. Finally, stands in the third cohort are primarily composed of shade tolerant species, have an open canopy, and possess a continuous vertical structure (i.e., numerous indistinct canopy layers in which trees follow an inverse-J diameter distribution).

Our results have also shown that the species assemblages of forest flora and fauna respond to changes in stand structure, mostly in terms of changes in the abundance of individual species, with few species restricted to a particular stand age. Consequently, we have chosen to use structural cohorts in our operational model to simplify the implementation of a management strategy that more closely resembles the effects of natural disturbance, because a stand's structural attributes are easier to measure in the field than is stand age; in addition, species respond more to changes in structure than to changes in stand age per se, and silvicultural treatments directly modify stand structure. For these reasons, we have simplified the patterns of natural disturbance and forest dynamics in the Matagami ecoregion by using the cohorts illustrated in figure 18.3. In this model, the first cohort has been divided into two distinct developmental stages (regenerating and mature), which have significant differences in stand structure. Figure 18.3 also uses different successional series to illustrate the differences in composition observed among site types. All three successional series converge to the primarily shade tolerant third cohort of black spruce, but this convergence may take different lengths of time, depending on the successional series or site type. In all, our conceptual

model of forest dynamics includes 10 main stand types.

This conceptual management model assumes that we can manage for the 10 main stand types by means of several types of silvicultural intervention (figure 18.3). The proposed mixed (even-aged and uneven-aged) management strategy uses clearcutting to initiate stand regeneration and partial harvesting techniques (partial cutting or selective cutting) to maintain or establish the structural and compositional characteristics of later successional stages (Y. Bergeron et al. 1999). Thus, it aims to preserve the integrity of the forest ecosystem through the maintenance of its different ecological elements and processes.

Implementation of a Forest Management Model Based on Natural Disturbance

Landscape-Level Objectives

Implementation of the model based on natural disturbance that we developed for the Matagami ecoregion is currently being tested in a pilot study area in the Lac Grasset regional landscape unit (see figure 18.1). In this area, the forest management strategy presented in figure 18.3 has been integrated into different levels of forest management planning (Tittler et al. 2001) to assess the feasibility and possible repercussions of implementing the management model. At the strategic level, regional objectives for cohorts in the study area were established as follows: 62% of the area would be covered by stands in the first

TABLE 18.3. Annual Allowable Cut and Proportions of Treatments and Area of the Land Base in the Pilot Study Occupied by Three Cohorts for Different Management Strategies

	Management Strategy		
	CPRS	CPPTM14	CPPTM16
AAC ($m^3 \cdot ha^{-1}$)	130,000	132,000	132,000
		Percentage of area	
Clearcut	100	76	74
Partial cut	0	24	26
Commercial area suitable for partial treatment	0	21	17
First cohort	75	69	65
Second cohort	19	22	26
Third cohort	6	9	9

Notes: Study results are shown for the end of the simulation horizon (year 150). AAC, annual allowable cut; CPRS, clearcutting with the protection of regeneration and soils; CPPTM, harvesting with the protection of small merchantable stems (14 and 16 cm breast height diameter limits, respectively). The timber supply analysis used the SYLVA II model (Ministère des Ressources naturelles du Québec 1997).

cohort, 21% by stands in the second cohort, and 17% by stands in the third cohort. As suggested by Bergeron et al. (1999a), these proportions were established using the predicted negative-exponential curve for age distribution (Van Wagner 1978), with an average stand age of 151 yr (determined for the pilot study area by using the fire history map described previously) and with transition ages of 150 and 275 yr between the first and second cohorts and the second and third cohorts, respectively. The proportions of the total area in each cohort represent overall objectives for the three successional series. Note also that specific objectives for each forest composition can be derived using the same procedure.

The ability of different management strategies to achieve the regional cohort objectives was compared by using a timber supply analysis (Nguyen 2000). At the start of the analysis, the three cohorts occupied 58, 28, and 14% of the total area (first, second, and third cohorts, respectively). The first strategy is an even-aged management system, based on clearcutting with the protection of regeneration and soils (the equivalent of Quebec's CPRS, coupe avec protection des sols et de la régénération). The two others are mixed management systems, based on clearcutting and partial cuts with a 14-cm diameter limit (the equivalent of Quebec's CPPTM, coupe avec protection de petites tiges marchandes, CPPTM14) or with a 16-cm diameter limit (CPPTM16). The timber supply analyses indicate that at the end of the 150-yr simulation period, the mixed management systems come closer than the even-aged management system to meeting the cohort objectives (table 18.3). The analyses also suggest that a mixed management system would not significantly affect annual allowable cuts, which remain at the same level for all scenarios (table 18.3). Note, however, that in the mixed management strategies, only 15–20% of the managed forests is considered suitable for partial harvesting.

Spatial Patterns of Harvesting

The region's fire regime, which is dominated by large fires, is responsible for the large areas covered by stands with similar ages and structures. Old-growth forests are therefore extensive and represent unique and significant structural features of the landscape. This contrasts with the patterns in western Canada, where old forests are mostly remnant patches that have not burned in recent fires (Johnson et al. 1998). At the planning level, we devised a strategy for determining the

spatial and temporal distribution of harvesting activities inspired by those natural patterns. The strategy is built on two main elements: management *zones* that represent the individual management units to be managed over time and management *complexes* that determine the spatial distribution of harvesting activities within a management zone. Management zones represent the individual disturbances that have historically shaped the forest landscape, whereas management complexes correspond to the patterns of disturbance severity historically observed within the area affected by each disturbance.

Figure 18.4 illustrates a portion of the initial strategy proposed for the area of the pilot study based on a preliminary analysis of the potential agglomeration of harvesting activities (management complexes) in different management zones. It proposes three types of management zone: even-aged management, based primarily on clearcutting; uneven-aged management, based primarily on partial cutting; and light management, based on limited harvesting in areas dominated by nonproductive forests. These zones range from 5000 to 40,000 ha.

Table 18.4 describes the management complexes for three of these zones and figure 18.4 illustrates the spatial distribution of the harvesting systems, which is quite distinct for the three zones. Each zone will be managed over a period of 5–10 yr. Following the strategy illustrated in figure 18.4, 63% of the pilot area consists of even-aged management zones over the course of a rotation, versus 9% of the area in uneven-aged management zones and 28% in light management zones. Compared with the regional cohort objectives, this strategy yields a deficit in the second cohort and a surplus in the third cohort. This discrepancy reflects the difficulty of properly identifying stands eligible for partial harvesting exclusively from current forest inventory maps (Nguyen 2001, 2002). Planning of a mixed management strategy will benefit from the ongoing development of complementary analytical tools.

Silvicultural Field Guide

The final planning level at which the mixed management strategy must be integrated is the operational level. This requires the development of field recognition guides for the 10 main stand types represented in the conceptual model in figure 18.3. It also entails the development of silvicultural field guides that will be used by forest managers to establish the appropriate silvicultural prescription for a given stand. Finally, it

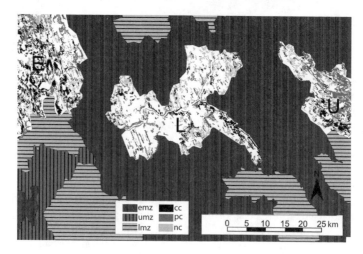

FIGURE 18.4. Initial proposed strategy for the agglomeration of harvesting activities (management complexes) in three management zones. Types of management zones: emz, even-aged management zone; lmz = light management zone; umz, uneven-aged management zone. Harvesting blocks in a given management zone: cc, clearcut harvesting; nc, no harvesting–unexploitable; pc, partial-cut harvesting. One zone in each category has been selected to show the array of treatments possible. These units would be affected entirely by treatments in 5–10 yr. The white areas are noncommercial forest areas that would remain untouched.

requires the definition of criteria that can be used to evaluate the success of the different silvicultural interventions.

Monitoring

The forest management model in figure 18.3 assumes that partial harvesting will maintain or establish forests that possess the ecological elements associated with older forests. To begin validating this assumption, several partial-cutting trials have been performed or are currently under way in the Matagami ecoregion. The dendrological, floristic, and faunal data collected prior to and following these interventions will serve to document and test the assumptions. Although the overall study is not yet complete, preliminary results from an analysis of the dendrological data collected in the Maskuchi partial-cutting trials (Morasse 2000) suggest that partial harvesting (in this case, a 17-cm diameter-limit cut) can successfully maintain the irregular (uneven-aged) stand structure associated with the second and third cohorts. Similar results have been reported by MacDonell and Groot (1996) in diameter-limit cuts performed in black spruce stands of the nearby Ontario Claybelt. Monitoring the responses of organisms in these trials and comparing their distributions with those in natural stands in the second and third cohorts are required before we can assess the effectiveness of such a forest management strategy for maintaining biodiversity.

TABLE 18.4. Proportions of the Area in Each Management Zone Occupied by the Different Types of Harvesting Blocks

	Proportion of Area in Management Zone (%)		
Type of Harvesting	Even-Aged Management	Uneven-Aged Management	Light Management
Clearcut (cc)	50.7	16.5	27.9
Partial cut (pc)	20.5	56.3	22.0
No harvesting (nc)	28.8	27.2	50.1

Notes: The management zones are illustrated in figure 18.4.

Conclusions and Future Improvements

The implementation, begun in 1999, is already providing promising results in terms of forest sustainability. For instance, preliminary simulations of annual allowable cut suggested that the strategy can be implemented with very little effect on timber supply while helping to maintain the forest's structural and compositional diversity. Moreover, preliminary results are also beginning to show that treatments can be developed to maintain the structural characteristics of mature, overmature, and old-growth stands, at least in the short term. It appears that the scientific knowledge of the natural ecosystem's dynamics on which we based the development of the management strategy was sufficient to initiate the implementation of such a strategy. Moreover, the ongoing input provided by new results from the first three research themes facilitates and improves the development of tools and guidelines that stem from the fourth research theme. In this sense, the monitoring of managed forests forms an integral part of the strategy and can easily be incorporated in the different research themes.

On an operational level, there is still a need to continue developing silvicultural treatments and planning tools. In fact, considerable work will have to be devoted to the development of practices that have ecological effects resembling those of fires. The work done on ecosite classification by the classification group of Quebec's Ministère des Ressources naturelles will help to refine the silvicultural prescriptions, as has been done for the boreal mixedwood forest (J. F. Bergeron et al. 1999; Harvey et al. 2002). Moreover, information on stand structure, which is difficult to derive from current forest resource inventory data, will become easier to derive by using tools that are currently in development (Boucher et al. 2003). We do not yet have wood supply models that can simulate partial or selective cutting. However, our management knowledge and tools are rapidly improving.

It is at the social and institutional levels that successful implementation of the strategy remains uncertain. The use of management strategies based on natural disturbance and of mixed (even-aged and uneven-aged) management systems in the boreal forest is not yet fully acknowledged in Quebec's forestry policies and regulations. The social acceptability of these strategies and systems (at the aboriginal, local community, regional, and provincial levels) also remains to be assessed. Increased public participation in the strategic planning process and an adaptive management approach would facilitate the implementation of the proposed strategy in the province's forest management units.

However, there is an urgent need to develop and implement strategies for the conservation of biodiversity. Current forestry practices, which target mostly mature and overmature forest, will considerably modify the proportions of these types of forests in the land base, particularly in eastern Canada, where they represent large, continuous areas. This will in turn affect the biodiversity of these forest regions. We trust that in many regions of Canada, disturbance-based strategies will offer a coarse filter solution that could mitigate these predicted changes in biodiversity and that can be developed and implemented quickly. It is urgent to do so, even though our knowledge is incomplete, because it will be easier to implement such strategies as the one we have proposed while we still have virgin forest than it will be to restore vanished ecosystems, as is presently the case in Europe (Kuuluvainen et al. 2002).

Acknowledgments

This chapter is a contribution of the Sustainable Forest Management Network (Natural Science and Research Council)–UQAT/UQAM Industrial Chair in sustainable forest management. We acknowledge many of our students and colleagues for their contributions at many different stages during the development of this strategy, particularly Catherine Boudreault, France Conciatori, Louis De Grandpré, Réjean Deschênes, Louis Dumas, Gilles Gauthier, Martin Gingras, Karen Harper, Louis Imbeau, Jean-Pierre Jetté, Victor Kafka, Gérard Laforest, Martin Landry, Gaetan Laprise, Patrick Lefort, Michel Lessard, Pierre Ménard, Emmanuel Milot, Jacques Morissette, Antoine Nappi, Pierre Paquin, Jean-Pierre Savard, and Daniel Spalding. This work was made possible by the support of NSERC, Quebec's Fonds forestiers, the Programme Volet 1 of Quebec's Ministère des Ressources naturelles, the Canadian Wildlife Service, the Lake Abitibi Model Forest, Norbord Industries, and Tembec Industries.

CHAPTER 19

Emulating Natural Forest Disturbance

Applications for Silvicultural Systems in the Northern Great Lakes Region of the United States

JOHN C. ZASADA, BRIAN J. PALIK, THOMAS R. CROW, and DANIEL W. GILMORE

The forests of the northern Great Lakes area of Minnesota, Wisconsin, and Michigan (figure 19.1) are a valuable resource for residents of the region and visitors who benefit from the products and services provided by these forests. The forests were heavily exploited in the late 1800s and early 1900s. As a result of this disturbance and subsequent forest use, there is little similarity between pre-European settlement forests and the present forest. In many ways, the forest ecosystems of the region are still recovering from historical severe exploitation.

The many sources of natural forest disturbance prior to European settlement of the Great Lakes region interacted to varying degrees. Fire and wind are usually considered the main causes of disturbance, but insects, diseases, frost, herbivores, and ice and snow damage also affected forest structure and composition, which, in turn, affected fuel conditions and fire behavior (Heinselman 1973; Mladenoff and Pastor 1993; Frelich and Reich 1995; Frelich 2002).

In addition to natural disturbances, the normal daily activities of Native Americans influenced the composition and structure of the regional forest. Their use of fire affected individual plants, stands, and landscapes. Native Americans used many plant materials, and although the effects of their gathering activities were less dramatic than those of fire, those activities continued for centuries. This affected plant species and communities to varying degrees, depending on the size and distribution of the Native American population (Densmore 1974; Meeker et al. 1993). Numerous accounts of the logging, burning, and other events that so drastically changed

these forests provide a detailed description of pre-European settlement forest conditions and the chronology of events that changed them (e.g., Eyre and Zehngraff 1948; Whitney 1987; Pyne 1988; Johnson 1995; Andersen et al. 1996).

In this chapter, we describe the silvicultural systems that have evolved to the present day, and the forest composition and structure that have resulted from these systems; describe constraints that appear to be particularly important for future forest management in the region and that will influence the ability of land managers and owners to emulate natural disturbance through silvicultural systems; and describe how silvicultural systems and prescriptions could be amended so as to emulate natural disturbance. We also consider how progress toward the goal of incorporating natural disturbance into the development of silvicultural systems in the northern Great Lakes forests can be assessed. Gauthier et al. (chapter 18, this volume) provide a similar assessment for boreal Canada. The regional landscape ecosystems covered in this chapter are Albert's (1995) sections VII (Northern Lacustrine-Influenced Lower Michigan), VIII (Northern Lacustrine-Influenced Upper Michigan), IX (Northern Continental Michigan, Wisconsin, and Minnesota), and X (Northern Minnesota).

In nature, the frequency, severity, and size of disturbance events act together to affect structure and drive the spatial and temporal development of these forests, and this continuum of disturbance processes overlap at scales ranging from genes to the region itself. Because we cannot pay proper attention to all scales in this chapter, we focus mainly on neighborhoods (groves of

FIGURE 19.1. Location of the northern Great Lakes region within the United States.

trees 10–20 m across) and stands (1–100 ha), with some consideration of individual trees and landscapes (collections of stands totaling >1000 ha), as defined by Frelich (2002). We believe that all scales are important and must be considered at every step of the planning process to determine how best to incorporate natural disturbance within silvicultural systems.

Conceptual Framework

Fundamental ecological information provides the basis for developing adaptive silvicultural programs that incorporate treatments emulating natural disturbance. The essential information needed to move in this direction includes the autecology of trees and associated plants, a multifactor ecological classification system, an understanding of the management objectives of landowners, an understanding of the importance of scale in forest management, and an ethic that appreciates the need for a long-term view of forest management. The available information is incomplete in each of these areas, but is most complete for tree autecology and ecological classification systems. Considering forests at scales beyond the tree and stand is critical for achieving wildlife and other goals, but coordination of

efforts among public and privately owned lands is required to achieve management goals at the landscape scale.

These goals are changing as recreation, wildlife habitat, water quality, biodiversity, and ecological services have become increasingly important objectives for the 15–20 million people living within a 10-hour drive of the region's forests. This brings considerable pressure on landowners and managers to provide multiple values from these forests. Meeting the objectives of most stakeholders in the region will require the development of more sophisticated silvicultural systems than are presently used.

A useful framework for addressing disparities between the outcomes of silviculture and natural disturbance could take the form of a conceptual model that arrays stands along a linear gradient of management intensity (figure 19.2). This model and the next section draw on existing work (e.g., Zasada 1990; Seymour and Hunter 1992; Franklin et al. 1997; Palik and Engstrom 1999; Crow et al. 2002; Palik et al. 2002; Seymour et al. 2002). At one extreme of management intensity lies unmanaged, pre-European settlement, or other benchmark conditions characterized by diverse composition and structural complexity.

FIGURE 19.2. A linear conceptual model for evaluating the difference in compositional and structural complexity between managed forests and benchmark conditions.

These may be the desired conditions for the management of ecological reserves and basic ecological services (e.g., water quality and quantity). The benchmark reflects conditions in which natural disturbances are the primary drivers of forest development. At the other extreme lies production management, such as a short-rotation, single-cohort plantation of an exotic or genetically modified species, for which maximum fiber production is the objective. Between these extremes lies a range of methods we call multiple objective management. This range includes extended-rotation, single-cohort plantations and multiple-cohort, mixed species stands. It is in this part of the range that natural disturbances will most likely be emulated to varying degrees.

In this model, it is important to consider the differences between managed stands and ecological benchmarks. These differences must be measured before managers can attempt to reduce them—if that is the management goal. It is equally important to understand the distance from the intensively managed end of the gradient. In other words, one must understand both the distance yet to go to approach the benchmark condition and the extent to which management has improved the complexity of stand structures and compositions relative to the benchmark condition.

A more complicated but more realistic model would include the dimensions of time since disturbance and variability in the degree of complexity (figure 19.3). In this model, the gray area depicted in the figure represents the system's natural variability (Landres et al. 1999), as defined by combinations of the time since distur-

bance and the degree of structural and compositional complexity. This area represents the array of benchmark conditions for the system, and includes not only old-growth systems with various degrees of complexity but also young postdisturbance systems. Even immediately after a stand-replacing natural disturbance, young stands typically have high structural and compositional complexity compared with managed systems of the same age (Franklin et al. 1997).

The dashed rectangles in the figure represent examples of structural and compositional variability for management scenarios that differ in the degree to which they achieve multiple objectives such as balancing wood production and the sustainability of native species diversity. Domain *a* in figure 19.3 might represent a plantation of an exotic species managed intensively for fiber production. The variability in complexity is both small and outside the range of natural variability. Domain *b* might include systems managed for large-diameter sawlogs and structural complexity by developing two-cohort stands. The range of variability is still relatively small, but lies in the range of natural variability. Domain *c* might represent systems managed for maximum similarity to the benchmark condition, with only limited removal of trees. An implicit assumption of this conceptual model is that stand domains move closer to the benchmark with increasing efforts to emulate natural disturbance.

Managers considering a gradient analysis such as the one shown in figure 19.3 must determine how closely the benchmark conditions must be matched for the management method to be considered successful. The answer is driven by

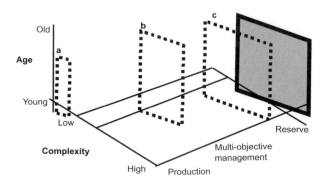

FIGURE 19.3. A multidimensional conceptual model for judging the differences in compositional and structural complexity between managed forests and benchmark conditions. The gray area represents the natural variation in complexity of the benchmark condition as a function of the time since a stand-initiating disturbance. Dashed domains *a, b, c* represent variability in complexity as a function of the time since a stand-initiating disturbance for various management scenarios.

the management objectives (Palik and Engstrom 1999; Palik et al. 2002). In ecological reserves, the goal will probably be to approach the benchmark condition as closely as possible. Conversely, when managing for maximum fiber yield, approximating the benchmark condition is probably not a consideration. In multiple-objective management, approximating the benchmark condition is important, but may often be done in ways that are constrained by the need to meet other objectives. Thus, the goal will often be to model the benchmark condition as closely as real-world constraints (e.g., managing for wood production) permit. In this case, the objective is to shift stands toward the right on the management axis (figures 19.2, 19.3) through innovative silvicultural practices that emulate natural disturbance to varying degrees while maintaining timber production as a primary objective.

Incorporating a Natural Disturbance Paradigm into Practice

The conceptual framework outlined in the previous section uses ecosystem structure, composition, and function as response variables to be compared with the benchmark conditions created by natural disturbance. In reality, ecosystem functions are at best difficult to measure and may thus be impractical for managers to assess. Fortunately, a forest's structure and composition often relate strongly to ecosystem function, and are much easier to measure; thus, structure and composition can act as surrogates for function (Franklin et al. 1997; Palik and Engstrom 1999). For example, the species richness and reproductive success of certain types of birds and small mammals depend on the presence of large, persistent snags. Rather than measuring faunal responses directly, a manager can focus on retaining or creating snags in managed stands.

Thus, in practice, a manager will focus on the structural and compositional outcomes of disturbance. The practical goal is then to create outcomes that reflect the structural and compositional characteristics of natural disturbance. In this section, we focus on three silvicultural activities (retention, partial canopy disturbance, and disturbance of the understory and forest floor) that can create structures and compositions that better reflect the outcomes of natural disturbance.

Silvicultural Retention

In practice, silvicultural retention focuses on retaining patterns and those quantities of struc-

tures and plants in place during a regeneration harvest that would reflect the outcomes of natural disturbance for a particular ecosystem. Ecological theory suggests that these retained components, often referred to as biological legacies (Franklin et al. 1997), connect the pre- and post-disturbance forest temporally, providing continuity of functions that otherwise might be lost during the early stages of forest development. These legacies include living trees, snags, downed dead wood, and residual understory and ground vegetation. Studies of natural disturbance have demonstrated that substantial amounts of biological legacies survive even seemingly catastrophic disturbance (Foster et al. 1998). For instance, some residual trees survive catastrophic crown fires in jack pine (*Pinus banksiana* Lamb.) ecosystems, resulting in multiple-cohort stands after the fire (Abrams 1984). Thus, silvicultural retention attempts to preserve forest components in a matrix of harvested forest (figure 19.4a): the operational focus is as much on what is left behind as on what is removed.

Retention is perhaps of greatest importance in systems that naturally sustain large catastrophic events and have traditionally been managed to produce single-cohort structures. In these systems, including oak (*Quercus* spp.), aspen (*Populus tremuloides* Michx. and *P. grandidentata* Michx.), jack pine, red pine (*P. resinosa* Ait.), and spruce (*Picea mariana* [Mill.] B.S.P. and *P. glauca* [Moench] Voss) in the Great Lakes region, the prevailing traditional silvicultural systems (clearcutting, shelterwood, and seed-tree) remove all or most of the overstory during regeneration harvesting. No deliberate attempts are made to retain understory and ground vegetation. Moreover, traffic by harvesting machines may fell snags and disturb large downed wood. As a result, the structure and composition of the postdisturbance forest often differ measurably from the results of natural stand-replacing disturbances. In contrast, retention silviculture deliberately preserves large living trees, individually or in groups, as well as snags, downed logs, and patches of understory or ground vegetation.

Although retention silviculture is often viewed as a contemporary idea, the concept of retention in this region goes back to the Morris Act of 1902—one of the first serious attempts to legislate forest management activities in the United States. This act mandated retention of 5% of the total volume (increased to 10% in 1908) of red pine and eastern white pine (*Pinus strobus* L.) to serve as a seed source in the harvested areas of

FIGURE 19.4. A diagram of (a) silvicultural retention and (b) partial disturbance.

[L.] Carr.)–hardwood ecosystems, and white pine dominated ecosystems were all heavily weighted toward partial smaller to mid-sized canopy disturbances (Frelich and Lorimer 1991; Seymour et al. 2002). This is not to say that stand-replacing disturbances do not occur in these systems, but they are rarer (return interval of >2000 yr) than in other forest types of the region (Canham and Loucks 1984). In contrast to retention, partial disturbance focuses on what is removed within a matrix of mostly uncut forest (figure 19.4a,b).

In forests typified by partial disturbance, wind, insects, lightning, and fire open single-tree and small-group gaps in the canopy. The size of the gaps determines the shade tolerance of the regenerating plants, with larger openings favoring less tolerant species. Gap sizes frequently range from 0.02 to 0.20 ha, and the disturbance has return intervals of 50–200 yr (Runkle 1982; Frelich and Lorimer 1991; Seymour et al. 2002). On average, about 1% of a stand's area is opened by new gaps each year (Runkle 1982). Operationally, a manager might attempt to create a range of gap sizes that total no more than 1% of the total stand area disturbed per year. However, because the actual cutting cycle might use a 10-yr re-entry interval, 10% of the stand's area would be harvested at each entry.

Disturbance of the Understory and Forest Floor

Prescribed fire is the most frequently used silvicultural model of a natural disturbance for treating the forest floor and understory, but disturbance also occurs to varying degrees in areas affected by windstorms. Forest floor conditions are affected by both stand-replacing fires and low intensity surface fires. The resulting forest floor and understory mosaic is an important factor in determining the relative importance of sexual and asexual reproduction, as well as the plant interactions associated with these forms of regeneration. Low intensity fires also thin dense areas of regeneration and kill some mature trees.

In the contemporary Great Lakes landscape, prescribed fire is not commonly used, and an opportunity exists for increasing its use as an analog for natural surface fires. In many cases, prescribed fire will initially be used as a restoration tool to reduce dense populations of competing shrubs and late-successional trees, such as hazel (*Corylus cornuta* Marsh. and *C. americana* Marsh.) and balsam fir (*Abies balsamea* [L.] Mill.) in pine forests (Buckman 1962a, 1964; Dickmann 1993) and red maple (*Acer rubrum* L.) in oak forests. Eventually, fire could be used to facilitate

the newly created Chippewa National Forest. Although the act is generally believed to have been unsuccessful in increasing regeneration, particularly of red pine, one can still find some of these old trees in the region, an interesting legacy of the pre-European settlement pine forests (Eyre and Zehngraff 1948). Although our focus on retention deals more with increasing structural complexity, legacy trees will also be an important seed source—the original intent of retention a century ago.

Partial Canopy Disturbance

Considering the spatial and temporal patterns of overstory removal is particularly relevant in systems naturally characterized by partial disturbance. Partial disturbance leaves substantial amounts of a forest intact, resulting in two-cohort and multiple-cohort stands (figure 19.4b) (Seymour et al. 2002). In the upper Great Lakes region, natural disturbance regimes for northern hardwood ecosystems, bottomland hardwood ecosystems, eastern hemlock (*Tsuga canadensis* natural regeneration of native species by preparing seedbeds (Van Wagner 1971; Benzie 1977) and maintaining structural and compositional similarity to that found in benchmark conditions. Prescribed natural fire is applied in some wilderness areas in an effort to reintroduce stand-replacing fires to these areas (Romme et al., chapter 20, this volume).

The reality for the Great Lakes region is that prescribed fire will never be used at a frequency or scale reflective of conditions in the pre-European settlement forest, except perhaps in wilderness areas. Thus, we must consider how other types of understory and forest floor manipulations might replace fire. For instance, selective herbicide application can accomplish similar objectives by reducing competition from aggressive plants and by allowing less aggressive species to become established. Mechanical disturbance of the forest floor can reduce competition and expose mineral soil, creating a range of microsites and providing seedbeds for species with smaller seeds that are unable to become established in thick litter and duff layers (Orlander et al. 1990). Although these practices are not complete ecological replacements for surface fires, they can still shift management outcomes closer to benchmark conditions. This approach also highlights the need to consider the goal of silvicultural emulation of natural disturbance: the goal is not the activity itself (i.e., the use of fire), but rather the outcomes the activity produces (Christensen et al. 1996). In our example of prescribed fire, it may not always be necessary, and certainly is not always practical, to replicate the natural disturbance. We must perform the research necessary to understand how to alter a natural disturbance regime while still achieving the desired objective of moving management outcomes closer to benchmark conditions (Palik et al. 2002).

Interactions among Interventions

Although the preceding discussion of retention, partial disturbance, and disturbance of the understory and forest floor treated each intervention independently, interactions should obviously be expected. The types of natural disturbance that each attempts to emulate were mixes of these conditions, and their relative importance differed among sites. Kimmins (chapter 2, this volume) discusses the idea of decoupling various ecosystem processes—for example, returning overstory conditions to an earlier successional stage, as in retention or partial disturbance, but without the changes in forest floor conditions

that often accompany these overstory changes in response to a natural disturbance. As we describe in more detail below, new harvesting and site preparation technologies permit a large variety of combinations of overstory and forest-floor conditions—some of which closely resemble the effects of natural disturbance.

Silvicultural Practices in Relation to Emulating Natural Disturbance

Natural and human disturbances are ongoing aspects of all forest ecosystems. Managers have numerous options for developing silvicultural systems that create different compositional, structural, spatial, and temporal patterns at the desired scales. In the northern Great Lakes region, silvicultural systems have been evolving over the past 75 yr. That evolution continues today, as new management goals are identified, new information becomes available, and new technology is developed. In this section, we consider silvicultural systems for aspen, red pine, and northern hardwood forest types in terms of the management changes currently occurring and how the emulation of natural disturbance might be incorporated into new silvicultural systems.

We have chosen red pine and aspen because historically, they have been managed as single species and single-cohort stands, two conditions generally believed to differ greatly from those in pre-European settlement forests that are often considered desirable benchmark conditions. Both forest types include other species and could be managed to include additional species and cohorts. These forest types are examples of ecosystems in which substantial changes in species composition and forest structure will likely occur under silvicultural systems that attempt to incorporate selected aspects of natural disturbance. They also provide good examples of the interaction between new harvesting technologies and the manager's ability to modify silvicultural systems and incorporate aspects of natural disturbance. In contrast, northern hardwoods provide an example of a forest type that has sometimes been managed by using a silvicultural system that emulates natural disturbance more closely. Incorporating natural disturbance into northern hardwood silviculture will thus produce subtler changes than will be seen with aspen and red pine.

Our intent is to illustrate how managers can creatively use their understanding of their particular forest systems and the principles discussed above to pursue multiple management goals. At this point, our lack of experience with such silvicultural systems in the northern Great Lakes region of the United States argues for operational trials of a variety of treatments and treatment combinations, based on our understanding of basic ecological principles.

In this discussion, it is not possible to provide the in-depth consideration of the site-specific silvicultural prescriptions that will be necessary to achieve various stand and landscape management objectives. We mention some specific examples for each forest type to provide an idea of the necessary considerations in developing site-specific systems that incorporate natural disturbance.

Aspen

The choice of a suitable benchmark for current aspen forests is somewhat more complex than for red pine and northern hardwood forests. Although aspen does attain ages of up to 100 yr on some sites, 60–80 yr is more common. Forest succession will result in the replacement of aspen by a more shade-tolerant species, but the time frame depends on the site conditions, natural disturbance, logging history, seed availability, and understory species.

Silvicultural systems for aspen have already begun to incorporate retention (intentional) and partial disturbance (sometimes inadvertent, but arising from retention) as forest managers move away from total removal and the single-cohort management that maximized suckering and dominance by aspen (Perala 1977; Puettmann and Ek 1999). Forest floor disturbance, other than what occurs during harvesting, has received less attention but will be critical if additional species are to be introduced. The distribution and composition of retained species has not been particularly well planned to date. The many possible patterns range from completely systematic single-tree distributions to random distribution of multiple-tree groups, as schematically indicated in figure 19.4. Each pattern has an associated degree of partial disturbance, forest floor disturbance, and percentage of the site in which machinery operates. Each pattern also has ecological, economic, and social implications and provides different options for the landowner.

Adding compositional diversity to aspen systems requires consideration of a combination of factors and balancing trade-offs between them. Stone (1997) provides a decision tree for aspen sites and discusses the overstory composition necessary to work with the existing stand composition so as to reduce the presence of aspen and approach a benchmark that comprises more tolerant species. Although Stone's decision tree considers mainly aspen–hardwood stands, the same type of decision tree can be developed for aspen–conifer mixtures. Increasing the presence of any species will require reducing the vigor of aspen suckering. The density of aspen suckers can be reduced by retaining mature aspen (Ruark 1990; Stone et al. 2001), but even so, the density will likely be too great for intolerant and mid-tolerant species to regenerate naturally or for planted seedlings to survive. Aspen density and vigor will also be affected by soil compaction, which is in part determined by the amount of machine activity on the site, and prescriptions that reduce the potential for compaction will also affect stand structure and composition (Stone 2001).

Another change in the aspen silvicultural system is increased precommercial (8–10-yr-old) and commercial (25–35-yr-old) tree thinning. The initial reasons for increased thinning were to increase the quantity and quality of wood and shorten the rotation, salvage future mortality and increase total stand yield during the rotation, and provide higher quality wood from younger trees. However, thinning is also currently being considered as a means to reduce aspen dominance and improve conditions for regeneration and growth of other species, thereby increasing species diversity and the variety of potential products. From the perspective of emulating natural disturbance, this technique has potential for managing succession.

Precommercial thinning in aspen has mostly involved mechanized row thinning. Initially, a skidder or other logging equipment was used to push down the aspen, and stems were not severed. More recently, a hydro-ax brushcutter or similar equipment was used to sever stems and mulch the material in the thinned strips. The resulting stand structure consists of a 1.5–2-m-wide thinned strip alternating with 1.5–2.5-m-wide unthinned strips. In northern Minnesota, approximately 10,000 ha of 8–10-yr-old aspen has been precommercially thinned (Chris Peterson, UPM-Blandin, Grand Rapids, Minnesota, pers. comm.).

Commercial thinning in aspen has been considered for many decades (Zehngraff 1947), but economic conditions have not been favorable, and existing harvesting equipment was not suitable (Perala 1990). The introduction of processing at the stump (cut-to-length) harvesting technology has resulted in renewed interest in commercial thinning. The use of cut-to-length equipment allows careful harvesting with less damage to the residual stand and understory. Furthermore, the effects of harvesting on soils (e.g., rutting and compaction) can be minimized by placing nonmerchantable material, such as the stems and crowns removed during processing, on the travel corridors, thereby reducing the impact of harvesting machinery on the soil.

Although the aspen silvicultural system has been driven mainly by the need to supply fiber to the forest industry, there has also been significant effort to harvest aspen so as to maintain early-successional forest for wildlife habitat—in particular, ruffed grouse (*Bombus umbellus*) and white-tailed deer (*Odocoileus virginianus*). For grouse, the rapid regrowth of aspen following clearcutting is essential and is still the main stand-level goal; there is also a very important landscape component. The habitats needed for successful grouse management are a mix of aspen age classes that provide everything from dense young stands for protecting young birds from predation to older stands of mature male clones that provide catkins for food in late winter and spring (Gullion 1990). These conditions can only be met by landscape-scale planning.

The development of a regional aspen silvicultural system has taken a "one size fits all" approach. This simplification does not do justice to the complexity of the aspen forest—there are significant differences in, for example, site and ecosystem productivity, species composition, susceptibility to insects and diseases, tree longevity, and susceptibility of soils and aspen root systems to logging impacts. These and other considerations will require site-specific silvicultural systems to achieve the desired variety of benchmark stand conditions. Alban et al. (1991) and Stone (2001) provide good descriptions of variation in the aspen forest type.

Red Pine

The silvicultural objective for red pine in the northern Great Lakes region has traditionally been single-cohort, single-species stands—not unlike that for aspen silviculture. The prevailing silvicultural system is based on planting followed by one or more thinnings to create the desired stand conditions and products. No silvicultural system is based on natural regeneration (Eyre and Zehngraff 1948; Buckman 1962b; Benzie 1977; Johnson 1995). The trend has generally been to simplify the composition and structure

of red pine forests, and this continues to be the case. Historically, the red pine forest type consisted of mixtures of red pine, white pine, and jack pine, along with various hardwoods, in single and multiple cohorts (Palik and Zasada 2002). The relative importance of each species varied across the region and within specific localities. Thus, there is historical precedent for developing silvicultural systems with single- and multiple-species stands and cohorts.

Periodic low intensity surface fires and stand-replacing fires with longer return intervals have both played an important role in the regeneration and development of red pine forests. Lightning was the primary uncontrolled cause of fire, whereas Native Americans used fire for specific management purposes. The relative importance of each form of fire likely varied across the region. The interest in prescribed underburning as a means of returning fire to this forest type has wavered over the past 50–60 yr. There was significant interest in both underburning and broadcast burning of logging residues from 1950 to 1970, but interest in that approach declined (Buckman 1962a, 1964; Buckman and Blankenship 1965; Dickmann 1993). The main objective of burning was to manage forests for wood production—specific objectives were to reduce the presence of hazel in the understory and prepare sites for coniferous regeneration. During the past decade, there has been a renewed interest in the use of understory burns. The objectives for burning now focus less on timber management and more on the reintroduction of fire into red pine forests to attain other benefits that accrue from burning (Dickmann 1993). Earlier interest in the use of fire aimed mainly at stand-level management, but more recent interest aims at the landscape level as well. Understory burning in red pine is currently being carried out in the Chippewa National Forest (by the U.S. Department of Agriculture Forest Service), Voyageurs National Park (by the U.S. Park Service), and Itasca State Park (by the Minnesota Department of Natural Resources). The amount of area burned will not likely approach that typical of burns prior to European settlement.

Because the current silvicultural system usually includes one or more intermediate thinnings, it is possible to begin incorporating some aspects of natural disturbance during stand establishment or during one of the entries for thinning. Two possible scenarios are described here to illustrate how retention, partial disturbance, and understory disturbance might be incorporated.

Scenario 1

Beginning with a well-stocked, mature red pine plantation, 60–80 yr old, an initial regeneration harvest might reduce basal area to a stand-level average of 10–15 m²·ha⁻¹. Within the stand, trees would be retained in spatial patterns that range from aggregated to dispersed, thereby providing variable retention (Franklin et al. 1997; Palik and Zasada 2002). The objective of this initial harvest is to create a two-cohort structure by providing environments favorable for new establishment of red pine and other desirable species, and to retain structural complexity in the stand. Prescribed fire or chemical and mechanical site preparation could be used to reduce competition and prepare seedbeds for pine regeneration. If seed sources for other pines are unavailable, these species could be planted. The silvicultural outcome, although perhaps not as compositionally and structurally complex as in some old-growth, pre-European settlement pine forests, still likely falls within the range of natural variability for the system.

Scenario 2

These forests could also be harvested more completely if the primary objective is timber production. Operationally, the new stand would be considered a single cohort, but with a concerted effort to retain structural features from the pre-disturbance stand. Retained elements include a few large individual or aggregated pines, as well as snags, downed logs, and patches of undisturbed understory in association with residual overstory trees. Conceptually, this silvicultural outcome is shifted to the left on the management axis compared with the previous example (see figure 19.3), but still falls within the range of natural variability for the system's complexity.

Northern Hardwoods

The northern hardwood forest type is less uniform in terms of composition and management history across the region than are the aspen and red pine types. Here we consider those sites on which sugar maple (*Acer saccharum* Marsh.) tends to dominate. Common species associated with sugar maple are yellow birch (*Betula alleghaniensis* Britton), basswood (*Tilia americana* L.), white birch (*Betula papyrifera* Marsh.), aspen, white ash (*Fraxinus americana* L.), black ash (*F. nigra* Marsh.), ironwood (*Ostrya virginiana* [Mill.] K. Koch), red maple, balsam fir, and eastern

white pine (Tubbs 1977; Eyre 1980; Burns and Honkala 1990; Niese and Strong 1992).

Silvicultural systems have been developed both for stands with multiple species and multiple cohorts and for stands with multiple species and single cohorts. Available guidelines provide a number of options for managing structure and composition (Arbogast 1957; Tubbs 1977; Erdmann 1986, 1987; Niese and Strong 1992; Strong et al. 1995; Crow et al. 2002). These guidelines have been applied successfully in northern Wisconsin and Michigan, but have received less attention in Minnesota. Niese and Strong (1992) and Strong et al. (1995) have shown how different silvicultural systems can be used to move primarily single-cohort, second-growth stands toward a multiple-cohort benchmark by using a planned series of harvests over a 30–40-yr period. This approach would move stands significantly changed by harvesting in the early 1900s toward a benchmark condition containing multiple cohorts.

The current silvicultural system for multiple-cohort management of northern hardwoods comes closer to emulating some aspects of natural disturbance than do the systems for other forest types. Guidelines for converting second-growth, single-cohort, multiple-species stands recommend thinning the stand (based on crown-cover targets) and the creation of small gaps. Gap creation is desirable, because the tendency for sugar maple to dominate these stands makes it difficult to regenerate mid-tolerant species, such as white ash, basswood, yellow birch, and red oak (*Quercus rubra* L.). Gaps emulate the creation of single- and multiple-tree gaps by the mortality of individual trees and by wind damage (Erdmann 1986, 1987).

Concerns about wildlife habitat and the development of stands with old-growth characteristics have led to several changes in the standard silvicultural system. To produce larger trees, the maximum diameter limit was increased to allow trees to attain a larger size and thus be utilized by more animal species, particularly those needing larger-diameter trees. To simulate old-growth conditions and provide wildlife habitat, trees with cavities and the potential for cavity production that would have been removed under the former guidelines were retained; some of these trees will not be removed and will ultimately die and provide the large woody materials required by some wildlife. The creation of multiple-tree gaps in older multiple-cohort northern hardwood stands is also a part of this new silvicultural system (Crow et al. 2002; Terry Strong, U.S. Department of Agriculture Forest Service, North Central Research Station, Rhinelander, Wisconsin, pers. comm.). These practices come very close to creating stands that resemble the multiple species and cohorts found in the benchmark conditions.

Constraints on Modeling Silvicultural Practices on Natural Disturbance

Regardless of forest type, there is increasing interest in the management of forests for recreation, nontimber forest products, wildlife, esthetics, and other values and services. We have not addressed these goals in this chapter. However, we believe that silvicultural systems must be developed to address these values alone or in the context of multiple values.

The rate and degree of incorporation of a natural disturbance paradigm into silvicultural systems is determined by social, economic, and ecological factors acting alone or in combination. Constraints fall into two broad categories that we have labeled as global and regional factors or as local factors. Global and regional factors include considerations that affect a manager's ability to return forest ecosystems to a pre-European settlement or other benchmark conditions and to predict long-term patterns of development for the existing forest, whether disturbed or not. Although a return to pre-European settlement conditions is not generally the stated goal, it is often proposed as one possible benchmark toward which forest succession should aim. Global and regional factors also include long-term changes, often driven by past events and current human activities and economics that are usually outside the control of local groups. In contrast, local factors include conditions that tend to be more locally controlled than is true of global factors. That is, solutions to these considerations often arise from the range of options (e.g., socioeconomic, technological, operational) that are controlled locally. There is certainly overlap and interaction between the two categories, and their relative importance varies over time and among geographic regions. Here we summarize aspects of several of these factors.

Condition of the Current Forest

The region's forests differ greatly from pre-European settlement conditions, and most of this change resulted from settlement by immigrants a century or more in the past and from the philosophy of land use they brought with them. The pre-European settlement condition, with its attendant disturbance regimes, is considered by some to be the state to which forests should be restored (see Suffling and Perera, chapter 4, this volume). Certainly, understanding this earlier forest is important for a full appreciation of the former ecosystem's structure, composition, and productivity. There is little doubt that the diversity of tree species in some forest types has been intentionally simplified by current silvicultural systems and that increased species diversity in these stands would be beneficial. However, the factors summarized below make it unlikely that much of the region can ever be returned to its pre-European settlement condition. Instead, the forest that is present today, or that is possible based on site conditions, natural disturbance, and species biology, will provide the benchmark conditions for emulating natural disturbance.

Land Ownership

Table 19.1 illustrates the mix of ownership across the region, and color plate 20 shows the pattern of ownership in Minnesota. The fragmentation that has occurred dilutes the occurrence of any disturbance that requires an intact forest for its generation and spread (Frelich 2002). Thus, the current level of fragmentation will, in many cases, restrict disturbances to local neighborhoods and stands, with little chance for the disturbance to expand to the landscape scale. The trend on nonindustrial private lands is toward further subdivision and thus, toward further reductions in the likelihood of such disturbances as fire expanding to larger scales, as they once did. Other disturbances, such as windstorms and insect outbreaks, will occur without regard for ownership patterns.

Associated with the land ownership pattern is the system of roads that developed in the region. Although some areas have no roads, these make up a small fraction of the total land area (figure 19.5). In addition to providing easy access to most of the region, the road system has been a major factor in minimizing the area affected by human-caused and natural fires.

Global Climate and Atmospheric Change

Increases in carbon dioxide (CO_2), tropospheric ozone, acid rain, mercury deposition, and other variables are well-known consequences of human development in this region, as well as globally. The contrast between CO_2 and tropospheric

TABLE 19.1. Land Ownership in the Great Lakes Region of the United States

State	Federal	State	County	Native American	Forest Industry	Individual	Corporate	Reference
Minnesota	16	23	21	4	6	26	4	Miles et al. (1995)
Wisconsin	14	5	19	4	10	4	44	Schmidt (1996)
Michigan	16	22	1	<1	10	39	12	Leatherberry and Spencer (1996)

Notes: Minnesota: Aspen–Birch and Northern Pine units included, total area 4.8 million ha; Wisconsin: Northeastern and Northwestern units included, total area 3.9 million ha; Michigan: Northern Lower Peninsula and Eastern and Western Upper Peninsula included, total area 6.3 million ha.

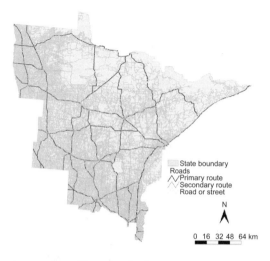

State boundary
Roads
Primary route
Secondary route
Road or street

N

0 16 32 48 64 km

FIGURE 19.5. The network of primary and secondary roads in Minnesota. These roads, along with the many forest roads and trails, make much of the forest in Minnesota accessible. A similar situation exists in Wisconsin and Michigan.

search has shown the complexity of their effects on forest development and disturbance regimes (Isebrands et al. 2001; Longauer et al. 2001).

Global change is generally accepted to be occurring, but the nature and magnitude of the potential changes remain unknown. Some predictions suggest that wind, fire, and other historically important disturbances may differ in frequency and severity in the future. More subtle changes are likely to affect individual species. Isebrands et al. (2001) provide an example of the effects of CO_2, ozone, and their interaction on the development of trembling aspen ramets and clones within aspen stands. Their results illustrate the complexity of predicting the effects of multiple factors that are changing simultaneously. Longauer et al. (2001) have shown that air pollution may affect the persistence of sensitive alleles with as yet unknown adaptive value to future forest development and succession in Norway spruce (*Picea abies* [L.] Karst.), European silver fir (*Abies alba* Mill.), and European beech (*Fagus grandifolia* Ehrh.).

Introduced Pests and Diseases

The impact of introduced insects and diseases on forest composition and structure are well documented (Campbell and Schlarbaum 2002) for some high-profile species, such as white pine blister rust (*Cronartium ribicola* Fish.), gypsy moth (*Lymantria dispar* [L.]), Dutch elm disease (*Ophiostoma ulmi*), and larch sawfly (*Pristophora erichsonii*). The effects of less prominent insects are not as well documented, but all introduced organisms will affect long-term forest ecosystem development in some way.

White pine management provides one example of how an introduced disease, blister rust, has

ozone illustrates the differences between some of the changes caused by human development. Whereas CO_2 is relatively uniformly distributed in the atmosphere, gradients of tropospheric ozone exist. Color plate 21 provides an idea of the variability in ozone concentration. The highest ozone concentrations occur in industrial areas and large metropolitan areas. Depending on weather conditions, ozone concentrations can vary significantly from day to day, especially in regions with relatively low ambient concentrations that are far removed from centers of production with higher concentrations. The effects of these changes are poorly understood, but re-

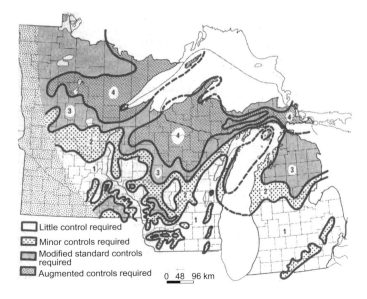

FIGURE 19.6. White pine blister rust hazard zones in Minnesota. Variability in the risk of blister rust infection is a function of the effects of landform and weather on the dispersal of fungal spores and on tree biology. This is but one example of the interaction between plant biology and the environment that exists across the region and must be considered in silvicultural systems that incorporate aspects of emulating natural disturbance. From Anderson (1973).

Little control required
Minor controls required
Modified standard controls required
Augmented controls required

0 48 96 km

affected forest composition and structure and the forest manager's view of the species as a component of the forest. Eastern white pine was once the premier species in much of this region and a primary target of early forest harvesting. Blister rust was introduced to this region in the early twentieth century, causing damage ranging from partial dieback to mortality. The probability of infection by blister rust varies significantly across the region (figure 19.6) (Anderson 1973; Stine and Baughman 1992). Land managers thus responded differently to the management problem posed by blister rust. The Menominee Reservation in eastern Wisconsin has always included eastern white pine as a main component of its forest management. This was due to a combination of management history, commitment of its landowners, and managing in an area of lower probability of infection. Management of eastern white pine on Menominee lands has thus been viewed as a significant success in the northern Great Lakes region. However, the general attitude toward management of eastern white pine in Minnesota was different. When serious efforts to reforest cutovers began in the 1930s, eastern white pine was more or less ignored compared with the effort devoted to regeneration and management of red pine. It has only been in recent years that the species regained acceptance in Minnesota and planting programs were developed to restore the species on appropriate sites (Stine and Baughman 1992; White Pines Strategies Working Group 1996; Rajala 1998).

The current attitude is that eastern white pine is too valuable to ignore, and despite the problems posed by blister rust, its presence needs to be increased.

Local Factors

When planning to incorporate techniques and practices that emulate natural disturbance into silvicultural systems, managers must consider local biological, ecological, and socioeconomic concerns. Across the Great Lakes region, there are significant differences in ecosystems, forest industry requirements for raw material, and public perceptions. Obvious differences exist between the states in this region, but even within states, there can be substantial differences among regions. As a result, generalizations are difficult, but the following sections summarize some important considerations.

Emulating Variability and Complexity

Modeling silviculture based on natural disturbance is challenging due to the variability and complexity of natural disturbance regimes and the outcomes of disturbance in terms of forest structure and composition. Variability makes the use of a single model for a silvicultural prescription problematic. Often, several types of natural disturbance affect a particular ecosystem, resulting in different structural outcomes or different pathways that lead to similar structures. A manager may find it difficult to duplicate this variability through silviculture. Even when the disturbance model is clear, the outcomes are impossible to duplicate fully with existing technology. Thus, a willingness to accept some differences is the key to developing silvicultural approaches that balance trade-offs among different objectives. It may be impossible or even unnecessary to duplicate all components of the natural disturbance regime (Palik et al. 2002). Natural disturbance regimes reflect the interactions of multiple components, including disturbance type, intensity, severity, frequency, seasonality, and spatial pattern. Some components have a greater impact on compositional and structural outcomes than others, and some are easier to emulate, suggesting that a manager need not always adhere to a strict natural disturbance-based model.

Socioeconomic Considerations

Modeling silviculture on natural disturbance is even more challenging when the socioeconomic constraints of forest management are considered. Silviculture, unlike natural disturbance, relies on human willingness to achieve a desired future condition. Currently, few foresters possess the knowledge required to incorporate a natural disturbance paradigm into their forest management. The costs of silviculture, particularly when part of commercial forest management, must be kept low. Natural disturbance-based silviculture, unlike natural disturbance itself, faces operational constraints, such as minimizing the impacts of equipment on soil conditions (Stone 2001) and moving equipment between residual trees, snags, and dead and downed wood, that increase management costs. Moreover, the widely varying goals of landowners, governments, and the public make it difficult to develop a solution that satisfies all stakeholders.

These challenges point to the need to balance the various, often competing, goals of silviculture. The balance may shift toward one goal or another at different times or locations, depending on the manager's objectives and ability to act independently. In all cases, the ultimate objective is to facilitate the implementation of silviculture based on emulating natural disturbance without ignoring the economic goals and social constraints inherent to modern forest management and the interests of stakeholders on different sides of the issue (Palik et al. 2002).

Conclusions

The current composition and structure of the forests of the northern Great Lakes bear little resemblance to those conditions in the forests prior to European settlement of the region. Developing silvicultural systems that emulate natural disturbance must address a variety of challenges across the range of sites and current forest conditions. The model that we propose as a step in this direction arrays stand conditions along a gradient of management intensity and desired future conditions. Progress toward achieving a forest that resembles the desired benchmark by incorporating the emulation of natural disturbance will likely be slow in most cases. Assessing progress toward a desired benchmark must include measures of both how close we are to the desired benchmark and how far we have progressed from the initial conditions (e.g., how far we have progressed from the intensively managed forest).

We believe that three practices provide a means for us to begin incorporating natural disturbance in silviculture:

- The retention of trees in numbers and patterns that are representative of the results of natural disturbance for a given ecosystem;
- Partial disturbance that considers the spatial and temporal patterns characterizing natural disturbance; and
- Disturbance of the understory and forest floor that creates patterns of microsite conditions suitable for regeneration.

Silvicultural systems have been evolving over the past 100 yr or so. Although there are preferred silvicultural systems today, these are now changing, driven in part by a decision to retain trees in areas that would normally be clearcut. Retention, partial disturbance, and forest floor disturbance are further steps in the evolution of silvicultural systems that will meet the needs of landowners and society for the various goods and services that forests can provide.

Acknowledgments

We thank David Cleland, Edward Jepson, and Tina Scupien for their help with the figures. The book editors and two anonymous reviewers provided helpful comments on an earlier draft.

CHAPTER 20

Emulating Natural Forest Disturbance in the Wildland–Urban Interface of the Greater Yellowstone Ecosystem of the United States

WILLIAM H. ROMME, MONICA G. TURNER,
DANIEL B. TINKER, and DENNIS H. KNIGHT

The Greater Yellowstone ecosystem encompasses Yellowstone and Grand Teton National Parks, plus adjacent national forests, other public lands, and extensive tracts of private lands in the American states of Wyoming, Montana, and Idaho (figure 20.1). The area is dominated by high, mountainous terrain, but also includes several lower elevation river valleys and portions of the plains surrounding the mountains (Hansen et al. 2002). Although the ecosystem contains one of the largest tracts of wild country in the continental United States, it also has one of the country's fastest growing human populations and economies (Riebesame et al. 1997; Rasker and Hansen 2000; Hansen et al. 2002). Population growth has been driven by migration from nearly all other regions of the United States, as highly mobile people are attracted to the region's outstanding natural amenities, including the scenery, wildlife, and outdoor recreational opportunities (Hansen et al. 2002). The major areas of economic growth are associated with services, real estate, and sources of nonlabor income rather than the traditional economic activities of mining, agriculture, and timber harvesting—a pattern typical of emerging "New West" economies throughout the Rocky Mountain region (Wilkinson 1993; Power 1998). Associated with this rapid population growth and a shifting economic base is a striking change in land use patterns, as formerly agricultural and grazing land is being converted to low density exurban or rural residential development (Theobald 2000; Hansen et al. 2002).

Research is beginning to demonstrate that these new patterns of land development pose serious threats to the biodiversity and the charismatic wildlife of the ecosystem, in part because the most productive and biologically rich areas are concentrated on or near private lands at lower elevations (Hansen et al. 2000, 2002). Even the large, protected national parks are threatened by the changing land use, because many species of migratory wildlife, such as elk (*Cervus elaphus*), rely on lands outside the parks for critical winter habitat, and populations of other species in the parks are actually maintained largely by dispersal of individuals from more productive habitats outside the parks; for example, this is true of yellow warblers (*Dendroica petechia*) (Hansen et al. 2002).

Another issue that has barely begun to be addressed involves the effects of changing land use patterns on such fundamental ecosystem processes as the natural disturbance regimes of the Greater Yellowstone ecosystem. As we explain below, fire is essential to the diversity and ecological functions of the natural ecosystems in this area. Management policies for the large national parks and wilderness areas emphasize that natural processes should be allowed to occur with minimal human interference. However, implementation of this policy grows ever more challenging as vulnerable homes and other structures become more numerous on private lands adjacent to the nature reserves. In particular, natural fires that would otherwise be allowed to burn in parks or wilderness areas must now be suppressed to prevent the fires from spreading to private lands. Although people are beginning to question the relative responsibilities of the government and the landowners for protecting private property against wildland fires, the western United States has a long tradition of government

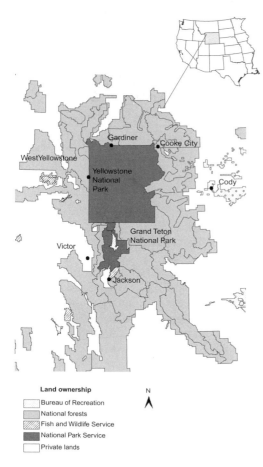

Land ownership

☐ Bureau of Recreation
☐ National forests
▨ Fish and Wildlife Service
■ National Park Service
☐ Private lands

N
Λ

FIGURE 20.1. Location of the Greater Yellowstone ecosystem, United States. Private lands occupy approximately 1.2 million ha, and another 5.5 million ha are included in national parks, national forests, and other public lands, for a total area of approximately 6.7 million ha. Modified from Rasker and Hansen (2000).

ponderosa pine (*Pinus ponderosa* Laws.) forests of the southwestern United States—mechanized thinning of dense forest canopies followed by low intensity burning is an effective strategy for reducing wildfire hazard and simultaneously restoring natural forest structures and processes (e.g., Hardy and Arno 1996; Covington et al. 1997; Friederici 2003). This strategy is being implemented widely throughout the western United States (e.g., Lynch et al. 2000; Arno 2002; Farnsworth and Summerfelt 2002), and is a key component of the National Fire Plan (U.S. Department of Agriculture Forest Service and U.S. Department of the Interior 2000) that lays out the federal government's overall strategy for wildland fire management over the next decade (e.g., Rains and Hubbard 2002). Unfortunately, natural disturbance regimes in the Greater Yellowstone ecosystem are dominated not by frequent, low intensity fires, but rather by infrequent, high intensity fires, as described later in this chapter. Thus, the mechanized thinning and low intensity prescribed burning that work so well in other kinds of forests fail to emulate the natural disturbance regimes of the very different forests of the region. Consequently, these approaches are unlikely to succeed in this region for a variety of operational and ecological reasons. In this chapter, we first describe the region's natural fire regimes and demonstrate why it would be both impossible and inappropriate to eliminate fire from this ecosystem. We then identify the major problems associated with attempting to apply a southwestern ponderosa pine model to the region's forests. Finally, we suggest an alternative fire management strategy that would more closely emulate the natural disturbance regimes of the region while reducing fire hazards to private property.

Natural Vegetation Patterns and Disturbance Regimes

The Greater Yellowstone ecosystem ranges in elevation from 1500 to 4000 m above sea level (asl), encompasses a variety of geological substrates from crystalline Precambrian rocks to Quaternary lake sediments, and includes rugged, glaciated mountains, as well as broad plains. Consequently, the area is very ecologically diverse (Romme and Knight 1982; Despain 1990; Glick et al. 1991). Broad river valleys and basins at lower elevations (<1800 m asl) are generally dominated by semi-arid sagebrush grasslands (characterized by *Artemisia tridentata* Nutt., *Agropyron spicatum* [Pursh] Scribn., and *Festuca idahoen-*

agencies—notably the U.S. Department of Agriculture Forest Service—assuming primary responsibility for preventing damage to human life or property caused by wildland fires (Pyne 1982). That basic policy appears likely to continue (U.S. Department of Agriculture Forest Service and U.S. Department of the Interior 2000). Thus, an urgent land management question in the region and other similar regions in the American West is how to maintain such natural disturbance processes as fire while reducing wildfire damage to the dispersed homes and other vulnerable structures being developed so rapidly on private lands in these fire prone natural ecosystems.

In ecosystems formerly characterized by frequent low intensity fires—particularly in the *sis* Elmer.) on upland sites, with cottonwood–willow woodlands (*Populus* spp., *Salix* spp.) along streams. Foothills and higher valleys (~1800–2300 m asl) support forests of Douglas-fir (*Pseudotsuga menziesii* [Mirb.] Franco), trembling aspen (*Populus tremuloides* Michx.), limber pine (*Pinus flexilis* James), and lodgepole pine (*Pinus contorta* var. *latifolia* Engelm.). The extensive volcanic plateaus of the region are covered by lodgepole pine forests, whereas higher mountain slopes (2300–3000 m asl) are dominated by forests of lodgepole pine, Engelmann spruce (*Picea engelmannii* Parry), alpine fir (*Abies lasiocarpa* [Hook.] Nutt.), and whitebark pine (*Pinus albicaulis* Engelm.).

Fire is the major natural disturbance agent in most of the vegetation types of the region (Romme and Knight 1982; Despain 1990). Fires of mixed severity, with patches of high tree mortality interspersed with patches of low mortality, occurred at intervals of 20–50 yr prior to the twentieth century and maintained a mosaic of Douglas-fir forests that included dense, closed-canopy stands and open savannalike stands (Houston 1973; Arno and Gruell 1986; Littell 2002). Some of these stands, especially at the lower forest–grassland ecotone, have increased in density over the past 100 yr—a scenario similar to that in southwestern ponderosa pine forests (e.g., Covington et al. 1997). However, this pattern of a natural fire regime dominated by frequent, low intensity fires, subsequently disrupted by twentieth-century fire exclusion efforts, is a conspicuous exception to the region's more general pattern. Most of the forested area in the Greater Yellowstone ecosystem, including lodgepole pine, spruce–fir, and whitebark pine forests, was influenced primarily by infrequent, but large, stand-replacing fires before the twentieth century, and this natural fire regime has not been altered substantially by recent fire exclusion efforts (Romme 1982; Romme and Despain 1989). Ignitions occur every summer, but most are extinguished naturally by wet weather. Only a handful of ignitions, caused by either lightning or humans, occur during weather dry enough to permit fire spread, but these few fires account for most of the area burned during any given time period (Renkin and Despain 1992). Differences in fuels related to stand age and site productivity influence fire activity during relatively moderate fire weather, but have little effect on fire behavior or its effects during extreme weather conditions—as occurred in 1988 (Renkin and Despain 1992; Turner and Romme 1994).

Based on the fire history during the past 250 yr, Romme and Despain (1989) concluded that twentieth-century fire suppression has had little effect on the infrequent large fires that dominate the fire regime of the lodgepole pine forests in the Greater Yellowstone ecosystem.

Fire Effects and Ecological Responses to Fire

Although fires in the Greater Yellowstone ecosystem tend to be severe and stand replacing, the natural ecosystems are very resilient to fires of this kind. Much of our understanding of fire effects in the area comes from studies initiated after the 1988 fires, which burned for more than 4 months, often under extreme fire weather conditions, and ultimately affected nearly 300,000 ha (Christensen et al. 1989). The fires burned at variable intensities and created a complex mosaic of patches through crown fires, severe surface fires, and low intensity surface fires, although forested areas were affected mainly by stand-replacing fires (Turner et al. 1994a). Research has focused mostly on lodgepole pine forests and grasslands, which are among the most extensive vegetation types in the Greater Yellowstone ecosystem. Postfire responses in other forest and nonforest types are probably similar to the responses described later in this chapter for lodgepole pine and grassland ecosystems.

Burned grasslands resprouted from surviving belowground plant parts, quickly restored the prefire cover in most places, and exhibited increased aboveground productivity for at least 2–3 yr after the fire (Van Dyke et al. 1991; Turner et al. 1994b; Wallace et al. 1995). Forest floor herbs (e.g., *Epilobium angustifolium* L., *Lupinus argenteus* Pursh.) also survived as rhizomes and roots, even where intense fire consumed nearly all aboveground herbaceous vegetation and litter (Anderson and Romme 1991). These survivors sprouted after the fire, and contributed most of the living biomass during the first 2 yr after the fire (Turner et al. 1997b). A flush of flowering and seedling establishment occurred in many forest floor species 3–5 yr after the fire and combined with the continued sprouting to rapidly increase total plant cover. Because so many individual prefire plants survived the fire, species composition was remarkably similar in burned and unburned forests—aside from several native annual species (e.g., *Collinsia parviflora* Lindl., *Gayophytum diffusum* T.&G.) that are rare in mature forests but flourished for a few years in the early postfire forests (Turner et al. 1997b). Most areas were overwhelmingly dominated by native species. Some nonnative species (e.g., *Cirsium arvense* L. Scop., *Lactuca serriola* L.) became established in the burned forests, but their distribution was patchy, apparently because of the general scarcity of nonnative seed sources in Yellowstone Park at the time of the fire (Turner et al. 1997b).

Lodgepole pine's seeds were released from cones in the burned canopy, and became established in the greatest numbers during the first 2 years after the fire. The resulting seedling density ranged from more than 10,000 seedlings ha^{-1}, where a high proportion of the burned canopy trees had serotinous cones and the fire burned at moderate intensity, to fewer than 100 seedlings ha^{-1}, where no trees were serotinous and the fire was very severe (Tinker et al. 1994; Romme and Turner, in press). Despite this variability, seedling density was sufficient to restock most of the burned forest at densities equaling or exceeding those under the prefire conditions. Burned trembling aspen forests resprouted after the fire (Romme et al. 1995a), and aspen seedlings were unexpectedly discovered in burned coniferous forests—many in places outside the prefire range of the species (Romme et al. 1997).

Associated with the rapid recovery of the original plant cover and species composition was a rapid recovery of ecosystem functions. Within 10 yr after the fire, some dense stands of lodgepole pine saplings (>60,000 stems · ha^{-1}) exhibited aboveground net primary productivity and leaf area index equal to that in mature lodgepole pine forests (Reed et al. 1999). Although lower density postfire stands have lower productivity and leaf area, the rapid accumulation of new biomass occurring throughout nearly all of the burned area also resulted in the conservation of nutrients. We detected very low nitrate concentrations (0.04–0.12 mg · L^{-1}) in the streams draining several severely burned watersheds 8 yr after the 1988 fires, as well as 1 yr after a smaller but nevertheless stand-replacing fire that occurred in 1996 (Romme and Turner, in press). Less is known about the specific soil changes and biogeochemical cycling processes after fire, but we have initiated research to investigate these aspects of ecosystem function.

Over the long term, many of the new forests that emerged after the 1988 Yellowstone fires may remain very dense for many decades or centuries before natural mortality begins to occur extensively in the canopy, in the absence of fire. It is important to recognize that dense lodgepole

pine stands are a normal part of postfire succession in this ecosystem (Romme and Despain 1989; Despain 1990)—and are decidedly not an artifact of twentieth-century fire suppression, as are the dense stands of southwestern ponderosa pine and some other forest types (e.g., Covington et al. 1997).

Alternative Fire-Management Strategies

Given the natural role of fire in the forests of the Greater Yellowstone ecosystem, how can or should we attempt to manage this powerful ecological process? In this section, we briefly discuss two possible strategies that we believe are inappropriate for both operational and ecological reasons, and then suggest the broad outlines of a preferred management approach that would emulate this ecosystem's natural disturbance processes (Swetnam et al. 1999; Romme et al. 2000).

Inappropriate Strategy 1: Complete Fire Exclusion

Federal land management agencies have attempted to exclude fire from most western ecosystems throughout much of the twentieth century, and achieved some success early on (Pyne 1982). However, complete fire exclusion over the long term now appears unattainable from an operational standpoint, and is ecologically undesirable as well.

Operational problems

Recent severe fire seasons have demonstrated unequivocally that fire cannot be excluded for long in the Greater Yellowstone ecosystem and many other ecosystems. The vegetation and climate of the Greater Yellowstone ecosystem are such that fire is inevitable, even with modern fire suppression technology. We can control fires during relatively moderate fire weather, but have little influence on fire behavior or spread during the severe fire weather that recurs periodically, as happened in 1988 and 2000 (Despain and Romme 1991).

Ecological problems

Even if we were capable of suppressing all fires, we probably should not do so. As explained earlier in this chapter, fire does not threaten the natural ecological integrity of the Greater Yellowstone ecosystem. On the contrary, it appears to be essential for long-term persistence of at least some important species, communities, and ecosystem functions. Several species are known to be more or less fire dependant; for example, Bicknell's crane bill (*Geranium bicknellii* Britt.) appears to persist primarily as dormant seeds and to germinate only in response to a fire induced stimulus (Abrams and Dickmann 1984), and the three-toed and black-backed woodpeckers (*Picoides tridactylus* and *P. arcticus*) breed primarily in recently burned forests (Hutto 1995). Large, infrequent fires also shape the landscape's dynamic mosaic of patches, which influences wildlife habitat, hydrology and other important ecosystem properties (Romme and Knight 1982; Christensen et al. 1989; Knight and Wallace 1989). Periodic large, severe fires may have additional subtle effects that have only begun to be recognized. For example, sexual reproduction in trembling aspen and many forest floor herbs appears to occur primarily in the aftermath of severe fires (Romme et al. 1995b, 1997). Although these species persist between fires via asexual sprouting, the fire stimulated infusion of new genotypes into these populations of long-lived clonal plants may be critical for persistence of the species in the face of climatic and environmental variability on the scale of centuries (Eriksson 1992; Tuskan et al. 1996). Thus, we argue that a primary goal of fire management in the Greater Yellowstone ecosystem should be to let fire play its natural ecological role to the greatest degree possible, recognizing that fire is a key process in maintaining the resilience and ecological integrity of this ecosystem.

Inappropriate Strategy 2: The Southwestern Model of Thinning and Low Intensity Burning

Mechanized thinning of dense stands followed by low intensity prescribed burns in fall or spring has been shown to accomplish the goals of ecological restoration and mitigation of the fire hazard in southwestern ponderosa pine forests and other forest types characterized by pre-1900 fire regimes of frequent, low intensity fires. This approach is called the *southwestern model* (Covington et al. 1997). However, the approach is not likely to succeed in the forests of the Greater Yellowstone ecosystem, where infrequent, high intensity fires have long been the norm, for both operational and ecological reasons.

Operational problems

Prescribed burning in lodgepole pine forests, the most extensive forest type in the Greater Yellowstone ecosystem, is a challenging enterprise. Because of typically late snowmelt and frequent summer rain showers, suitably dry conditions for ignition occur only infrequently (Brown 1991; Renkin and Despain 1992). Once the fuels do dry,

however, only a small amount of additional drying is needed to push them into a state where crown fires become likely. Unlike the fuel bed in ponderosa pine forests, that in lodgepole pine forests is not conducive to extensive spread of surface fires. On the contrary, fires in lodgepole pine forests tend to move via crown-to-crown spread and wind-driven spotting, rather than spreading diffusely through surface fuels (Brown 1991). Thus, burning extensive areas of lodgepole pine forest with low intensity controlled burns would be very difficult, and perhaps impossible.

Ecological problems

Mechanized thinning of dense lodgepole pine forests would be problematic for at least two reasons. First, lodgepole pine is usually not very windfirm, and most of the residual stand left after a heavy thinning could easily blow down over a few years after this treatment. Second, many herbs and low shrubs on the forest floor are adapted to shady understory conditions and may be unable to tolerate the increased light, wind, and temperature variation associated with an open canopy structure; for example, this is true of the fairyslipper orchid (*Calypso bulbosa* L.), whitevein pyrola (*Pyrola picta* Sm.), and twinflower (*Linnaea borealis* L.). Although the native species are well adapted to infrequent, severe fires, they may not fare well under a novel disturbance regime composed of frequent, low intensity fires. Juvenile and adult individuals of lodgepole pine, Engelmann spruce, and alpine fir are not very fire resistant, and are easily killed by even a low intensity fire. Most forest floor herbs and shrubs would readily resprout after a single fire, but they may be incapable of repeated sprouting after repeated fires. The resprouting response probably requires substantial energy reserves, and we do not know how long it takes for roots and rhizomes to replenish these reserves after a fire. The density and vigor of the postfire sprouts of gambel oak (*Quercus gambelii* Nutt.), a common perennial shrub in the southwestern United States (but not present in the Greater Yellowstone ecosystem), were substantially reduced following repeated, short interval burning (Harrington 1985). Thus, frequent, low intensity burning could potentially cause initial stimulation but may ultimately lead to local extirpation of at least some of the forest floor species of the Greater Yellowstone ecosystem. Loss of these native herbs and shrubs, coupled with increased light and nutrients in burned forests, would create a condition in which the burned forests were vulnerable to invasion by nonnative plant species—which thrive in environments with high availability of light and nutrients and frequent disturbance (e.g., Mooney and Drake 1986). Much of this discussion is somewhat speculative, as research to test these hypotheses has not been conducted. However, this analysis does indicate that hasty implementation of mechanized thinning and low intensity prescribed burning over large areas could lead to very undesirable results, in part because such a disturbance regime would be so different from that which the organisms have experienced in their evolutionary history.

Preferred Strategy:
Emulating the Natural Fire Regime

We suggest that the only fire management strategy likely to be effective over the long term will be one that accepts and incorporates infrequent, large, severe fires throughout much of the landscape. Such fires are a natural part of this ecosystem, they probably are unavoidable, and they have the ecological benefits described earlier in this chapter. Naturally ignited fires in the extensive wilderness portions of the Greater Yellowstone ecosystem can probably be allowed to burn with minimal interference under most conditions. However, an overall strategy must also deal with the risk posed by such fires to homes, watersheds, and other human property and values. Full development of this kind of fire management strategy will require much additional work on the part of managers, researchers, and the public, but here we sketch out three of the strategy's major elements.

Create defensible spaces around homes and other vulnerable elements

The quickest and most effective method for reducing losses of homes and property to wildfire is to reduce fuels in the immediate vicinity of vulnerable structures. Although local fuel reduction does not guarantee full protection, thinning dense forests within 30–60 m of a home can aid firefighters attempting to protect the home and at least reduces the probability of serious damage or loss when fire occurs (Cohen 2000; Arno 2002). To be effective, thinning treatments must be accompanied by the use of fire resistant building materials, the removal of pine needles from roof gutters, and other similar actions. Although we argue against broad-scale thinning of forests in the Greater Yellowstone ecosystem, we believe that the use of thinning to create a defensible

space around vulnerable structures is appropriate, and is necessary before any prescribed burning can be conducted safely (Brown 1991). Specific information and guidelines for protecting homes and other structures from wildland fire are readily available (e.g., Firewise, undated; Dennis, undated). Moreover, homeowners can take responsibility for accomplishing the work themselves or by contracting with private companies, and the overall costs that result from this approach are relatively low for individuals and society. Color plate 22 illustrates a recent fuel reduction project in Grand Teton National Park of the kind that we advocate here. This treatment is unlikely to stop the advance of an intense fire across the landscape, but it will probably increase the survivability of buildings in the immediate vicinity, which is the explicit and appropriate objective of the treatment.

Clear or burn larger areas near vulnerable structures

A large, intense forest fire burning into a housing development during extreme fire weather may cause great damage, even if homeowners have taken appropriate steps to reduce fuels around their homes. Therefore, it may be appropriate in some locations to treat the surrounding vegetation (especially dense forests) so as to reduce the intensity of any future fire that burns into the area (Brown 1991). Where homes are surrounded by dense Douglas-fir forests, which had a more open structure before 1900, mechanized thinning and low intensity prescribed burning (something like the southwestern model) may be feasible and appropriate. However, lodgepole pine forests probably should not be treated in this way over extensive areas. Rather, effective treatment of lodgepole pine forests should probably involve either clearcutting or prescribed burning.

Clearcutting is relatively easy and effective in reducing fuels and retarding fire spread, and clearcut patches can be designed to mimic the shape and size of natural fire created openings. However, the ecological effects of clearcutting are very different from the effects of natural fire with respect to coarse woody debris, biogeochemistry, soils, and wildlife habitat. The focus of this chapter is on fire rather than timber harvesting, but the idea of making harvesting more closely emulate the ecological effects of natural stand-replacing fire in the Rocky Mountains is explored in Romme et al. (2000).

Prescribed burning in proximity to houses and other structures will require great skill and understanding, given the nature of fire in this eco-system, and specific treatments will need to be modified according to local conditions. Brown (1991) provides guidelines for this kind of pre-scribed burning in the context of protecting vis-itor centers and other developments in Yellow-stone National Park, but the principles apply more broadly. Young lodgepole pine stands de-veloping after a clearcut or a stand-replacing burn are less likely than older stands to ignite or allow fires to spread under all but the most ex-treme fire weather (Renkin and Despain 1992; Turner et al. 1994a). Such stands would provide at least some protection from uncontrollable wildfires for several decades (Brown 1991). We stress that such treatments as the prescribed burn-ing we propose would be most appropriate in close proximity to developed areas. In the exten-sive wilderness portions of the Greater Yellow-stone ecosystem, naturally ignited fires can and should be allowed to burn with minimal inter-ference under most conditions. The exception would be fires that show a high likelihood of encroaching into developed areas; for example, fires that are burning during extreme fire weather or in the absence of natural fire barriers.

Encourage responsible development patterns

Most land development in the western United States is minimally regulated by government or other entities, and fire hazard is rarely a major consideration in decisions about where or how to build homes and other structures in wildland areas. Consequently, many homes are located in places that are extremely vulnerable, such as near the top of a steep, brushy slope at the head of a shallow ravine and facing the prevailing wind direction. Such structures probably cannot be protected from wildfire, even by using the meas-ures outlined above. The best strategy would be to not build homes in such vulnerable locations in the first place—or if the homeowner insists on building at this site, then it should be clearly understood that the community has minimal obligation to attempt to save the house from wild-fire. A variety of means are available to encour-age responsible development patterns in wildfire prone ecosystems. These include formal zoning regulations, tax incentives by governments, and reduced insurance rates for building at safer lo-cations and creating a defensible space around homes. Clustering of homes could also be en-couraged in an effort to increase the cost effec-tiveness of treating surrounding vegetation and protecting homes when a wildfire does occur.

Clustered homes, with protected tracts of open space between clusters, would also reduce the fragmentation of natural vegetation (Romme 1997) and create greater opportunities for fires to burn in more natural patterns over greater areas, thereby enhancing fire's beneficial effects on bio-diversity and ecosystem processes.

Conclusions

Fire is a key natural process in the Greater Yel-lowstone ecosystem, and is essential to the eco-system's long-term biodiversity and ecological integrity. Living with fire in this system is espe-cially challenging for the human members of the community, however, because the most ecologically important fires are infrequent, large, and severe. As the human population continues to grow in the Greater Yellowstone ecosystem and similar places throughout the American West, wildfire hazards to life and property will also grow, and fire management will become increas-ingly urgent and difficult. In this chapter, we evaluated and rejected two possible strategies for fire management in the Greater Yellowstone Eco-system: (1) complete fire exclusion and (2) a south-western model, based on mechanized thinning and low intensity prescribed burning of dense forests. Instead, we recommend a strategy that emulates the region's natural disturbance regime by accommodating stand-replacing fires, while acting locally to protect lives and property.

The technical challenges associated with our recommended approach are daunting: fires in this ecosystem are inherently difficult to con-trol, and the art and science of conducting pre-scribed, stand-replacing fires in lodgepole pine forests are in their infancy. The sociopolitical challenges are also substantial. Collaboration among many agencies, both governmental and nongovernmental, will be required. Many local residents have a poor understanding of the fire regimes in this ecosystem, as well as unrealistic expectations about the government's ability to protect them from fire and a general resistance to formal regulation of land use. We believe that the measures we call for can be accomplished by means of incentives from governments and per-haps from the insurance industry, rather than by regulation. Moreover, we envision that the people of the American West can gradually move toward a view that living in such a grand land-scape as the Greater Yellowstone ecosystem is not merely a right, but also carries with it a re-sponsibility to protect ecological integrity for the benefit of residents and nonresidents alike.

As Aldo Leopold (1934) wrote, "every landowner is the custodian of two interests . . . the public interest and his own." We believe that maintaining at least the major elements of a natural fire regime in the Greater Yellowstone ecosystem is in the public's long-term interest, and will ultimately prove to be the most effective way of protecting human life and property while maintaining biodiversity and ecological integrity.

Acknowledgments

We thank the many colleagues, agencies, and student research assistants who have made our Yellowstone research possible over the past 20 years. In particular, we acknowledge Jay Anderson, Don Despain, Bob Gardner, Jerry Tuskan, and Linda Wallace for research collaboration, stimulating discussions, and insights into the fire ecology of the Greater Yellowstone ecosystem. Our research has been supported by the National Science Foundation, U.S. Department of Agriculture, U.S. National Park Service, the Universities of Wyoming and Wisconsin, and the National Geographic Society. The University of Wyoming–National Park Service Research Center and Yellowstone and Grand Teton National Parks have provided invaluable logistic and administrative assistance. Bill Baker and an anonymous reviewer gave us helpful comments on an earlier version of the manuscript.

CHAPTER 21

Emulating Natural Forest Disturbance
From Policy to Practical Guidance in Ontario

JOHN G. MCNICOL and JAMES A. BAKER

The concept of basing resource management on knowledge about natural ecosystem processes has existed since at least the 1960s (Landres et al. 1999). Natural processes that influence trees and stand-level forest composition and growth had traditionally been incorporated into silvicultural treatments to regenerate and grow new forests. By the 1990s, this consideration for natural processes was being extended in North America to include broad-scale landscape dynamics caused by fire and insects (Holling and Meffe 1996). A consequence of this new concept was the recognition that landscape patterns and processes driven by fire and insects would have to become an integral component of how forests are managed (Hunter 1990). In effect, resource management agencies and forest managers must develop approaches that moved from the conceptual to the practical.

In Ontario, as in many other jurisdictions, public values associated with forests appear to have changed considerably in the late 1980s and the 1990s. Two major public consultation processes resulted in changes in Ontario's approach to forest management. The first of these processes was the class environmental assessment hearings for timber management (Ontario Environmental Assessment Board 1994). This process examined all aspects of forest management, and the final report made numerous recommendations for improving the management of forests on publicly owned lands. These included a recommendation that clear guidelines be provided to regulate the size and distribution of clearcuts. A second and somewhat parallel process was the work of the Ontario Forest Policy Panel (1993), which conducted broad consulta-

tions with experts and the general public to determine their views on improving forest management. This process led to the implementation of the Crown Forest Sustainability Act (Statutes of Ontario 1995) to govern how forests on Crown land are managed. One of the clauses in this act stipulated that forest practices be used that emulate natural disturbances and landscape patterns.

All forest management activities on Crown land in Ontario are regulated through a set of guidelines. These guidelines prescribe, for example, the acceptable silvicultural options, and the protection of cultural heritage values and habitat for various wildlife species. Forest planners must consult these guidelines and document how they have implemented them in their forest management plans.

In this chapter, we present the development of Ontario's *Forest Management Guide for Natural Disturbance Pattern Emulation* (Ontario Ministry of Natural Resources 2002a) with reference to some of the challenges, so that others contemplating this path can learn from our experience. We examine the legal and legislative background to the guide's development, the process of formulating its direction, and future challenges. The development of a guideline for emulating natural disturbance flowed from the two parallel (but linked) paths of public participation that the Ontario Ministry of Natural Resources undertook during the late 1980s and throughout the 1990s (figure 21.1). One of these paths emerged from the recommendations generated by the class environmental assessment, whereas the other emerged from the Crown Forest Sustainability Act.

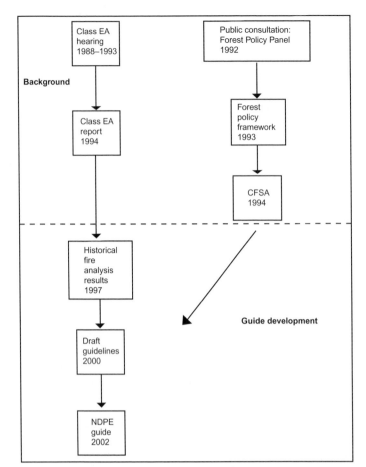

FIGURE 21.1. Guidelines used in Ontario's natural disturbance pattern emulation (NDPE) guide (Ontario Ministry of Natural Resources 2002a) flowed from two parallel but linked public consultation processes: the class environmental assessment (EA), and the Crown Forest Sustainability Act (CFSA).

Background: 1980–1994

Featured-Species Guidelines

Much of the forested land in Ontario is Crown owned and is therefore public land. As a result of concerns raised by the public over the impacts of logging, several management guidelines for forest habitat were developed in Ontario during the late 1980s with the goal of creating and maintaining suitable habitat conditions for a number of species that seemed to warrant particular attention. Much of this concern was directed at endangered, threatened, or vulnerable species, but it also included such species as white-tailed deer (*Odocoileus virginianus*) and moose (*Alces alces*). This approach to habitat management was referred to as *featured-species management.* Increasing concern that this approach might not be enough to maintain biodiversity pushed Ontario toward embracing the coarse filter and fine filter approaches to habitat management.

Coarse and Fine Filter Approaches

The coarse filter approach to habitat management, which was popularized by Hunter (1990), directs forest managers to maintain landscape-level patterns that provide habitat for a broad array of species. The fine filter approach requires specific actions, often in the form of guidelines to maintain specific ecosystems, such as communities of rare plants or breeding areas for rare species. Often, the fine filter approach is site specific and requires some form of protection either by setting aside areas in parks where forest harvesting cannot occur, or by restricting disturbance in areas next to sensitive sites in managed forests, such as areas near the nests of the bald eagle (*Haliaetus leucocephalus*).

From Featured Species to Landscapes

The guidelines for Ontario's featured-species approach to habitat management had characteristics of both the coarse filter and fine filter approaches, but placed more emphasis on the fine filter approach, because most guidelines dealt with habitat features for particular species. For example, the guidelines for moose habitat (Ontario Ministry of Natural Resources 1988) provided recommendations for landscape-level patterns that addressed the specific habitat conditions required by moose. However, resource managers were concerned that landscape features or ecosystem types might not be explicitly protected within a strict interpretation of the featured-species approach.

Resource managers became increasingly interested in managing forested landscapes to create and maintain a diversity of patch sizes that would satisfy the spatial and other requirements of a diverse range of wildlife species. Of particular concern was the need to maintain large patches of continuous forest cover to accommodate the requirements of larger species (Harris 1984). Hunter (1990) synthesized the results of a number of studies conducted during the 1980s and proposed that forest managers create and maintain a variety of patch sizes by using such silvicultural methods as selection cuts and clearcutting. Moreover, he proposed that the patch sizes created by natural disturbance be used as a model for establishing the range of patch sizes to be created and maintained by clearcutting.

Observations of Ontario Landscapes

Although the data were not published until the late 1990s and early 2000s, observations of broad-scale fires and clearcut landscapes began to reveal some interesting and disturbing trends. Rempel et al. (1997) reported that landscapes composed of small clearcuts and that followed the moose habitat guidelines (Ontario Ministry of Natural Resources 1988) were creating considerably different landscape patterns than those created by fires and contiguous clearcuts. Furthermore, the mosaics of small cuts were not having the desired effect of increasing local moose populations; in fact, these landscapes had lower moose densities than the landscapes that resulted from fire or contiguous clearcutting.

At the scale of Ontario's boreal forest, an examination of the patterns of disturbance revealed obvious changes to landscape patterns as a result of forest harvesting over the previous 45 yr. This analysis demonstrated that the landscape patterns created by harvesting in the boreal forest since the 1950s were quite different from the patterns created by the stand-replacing fires (Perera and Baldwin 2000) in terms of the amount of edge created; in this context, *edge* is defined as the border between two ecologically dissimilar patches. These differences had accelerated since the early 1980s, whereas no increase in edge density has been observed for landscapes affected only by wildfire (Perera and Baldwin 2000). This increase in the amount of edge coincided with the implementation of the moose habitat guidelines in 1988, which recommended clearcut sizes in a range of 80 to 130 ha. An unfortunate side effect of increasing the number of small clearcuts was a corresponding increase in the amount of edge, since a single large clearcut has less edge than a comparable area composed of several smaller clearcuts.

Ontario's Class Environmental Assessment

At the same time as the featured-species guidelines were being developed and implemented, timber management in Ontario fell under the scrutiny of a class environmental assessment—a decisionmaking process used to promote good environmental planning in Ontario by assessing the potential effects of certain activities on the environment. In Ontario, this process is defined and finds its authority in the Environmental Assessment Act (Statutes of Ontario 1975). The purpose of this act is to provide for the protection, conservation, and wise management of Ontario's environment. To achieve this goal, the act ensures that environmental problems or opportunities are considered and their effects are anticipated before the activities are allowed to occur. Individual assessments are performed for one or a few activities with limited scope, whereas class assessments are for activities, such as forestry, which are carried out routinely and have predictable and mitigable environmental effects.

In 1988, public hearings began on the class environmental assessment submission on forest management from the Ontario Ministry of Natural Resources. The public hearings before an Environmental Assessment Board were designed to determine whether "timber management planning for building forest access roads, cutting timber, regenerating the new forest and tending and protecting it over 385,000 square kilometres of public lands in northern Ontario pass the test of approval under the Environmental Assessment Act" (Ontario Environmental Assessment Board 1994). The board heard evidence from the Ontario Ministry of Natural Resources, the forest industry, experts testifying on behalf of nongovernmental organizations, representatives of other governmental agencies on behalf of the public, and ordinary citizens from across the province. The board's approval for forest management on Crown lands in Ontario was granted by the Minister of the Environment in May 1994 (Ontario Environmental Assessment Board 1994).

During the hearings, the size of clearcuts and whether a silvicultural system based on clearcutting was an environmentally safe forest management tool were the focus of much testimony from experts on all aspects of the issue. The Ontario Ministry of Natural Resources argued that natural disturbances such as fire occurred in a range of size classes and thus, that forest harvesting should not be restricted to one size class. The Board decided that clearcuts could exceed 260 ha only where an identified ecological rationale supported this exception. The Ontario Ministry of Natural Resources was directed to create guidelines that governed the shape, size, and distribution of clearcuts, based on how the forest was naturally disturbed.

Ontario's Crown Forest Sustainability Act

As the class environmental assessment hearings were proceeding, Ontario's forest policy panel was gathering public input about forest management. The Crown Forest Sustainability Act (Statutes of Ontario 1995) provides the Ontario Ministry of Natural Resources with the authority to manage publicly owned forests. Much of the content of the act is based on public input received by the Ontario Forest Policy Panel that was established in 1991. This seven-member panel solicited public input about how forests should be managed through the release of a discussion paper (Ontario Forest Policy Panel 1992). More than 3000 people responded. On the basis of this public input, the panel produced a policy framework that contained five principles (Ontario Forest Policy Panel 1993). One of the five principles was that "forest practices, including clearcutting and other harvest methods, will emulate, within the bounds of silvicultural requirements, natural disturbances and landscape patterns" (p. v).

The Crown Forest Sustainability Act replaced the old Crown Timber Act, which had been enacted in 1849 (Statutes of Upper Canada 1849) and which continued to be amended until 1990 (Statutes of Ontario 1990). Introduction of the Crown Forest Sustainability Act heralded much broader goals for sustainable forest management,

as indicated in the two guiding principles of the act (p. 3):

1. Large, healthy, diverse, and productive Crown forests and their associated ecological processes and biological diversity should be conserved.
2. The long-term health and vigor of Crown forest should be provided for by using forest practices that, within the limits of silvicultural requirements, emulate natural disturbances and landscape patterns while minimizing adverse effects on plant life, animal life, water, soil, air, and social and economic values, including recreational values and heritage values.

Note that the emulation of natural disturbance and landscape patterns must take into account "the limits of silvicultural requirements" while "minimizing adverse effects" on ecological, social, and economic values. These qualifiers would prove important in subsequent discussions leading to the development of the province's guide to emulating natural disturbance.

Development of the Natural Disturbance Emulation Guide: 1994–2002

Rationale for Emulating Natural Disturbance

There was considerable public input on and a high degree of acceptance of the concept of emulating natural disturbance during the parallel processes of the class environmental assessment and development of the Crown Forest Sustainability Act (figure 21.1). A strong legal rationale existed for Ontario to develop a guide to help forest managers implement practices that emulate natural disturbance and landscape patterns. The first decision that policy implementers had to address was which natural disturbances they should emulate. This decision was based on data availability and knowledge of the chronology of disturbance events.

Analysis of Data on Natural Disturbance

Ontario had collected 30 yr of fire data (from 1920 to 1950), a period when anthropogenic influences on the natural fire regime were much less than in the subsequent five decades. This mapped data set, developed by Donnelly and Harrington (1978), documented all recorded fires in Ontario that covered 200 ha or more during that 30-yr period. Although there are now similar broad-scale historical provincial records of other natural disturbances, such as forests killed by insect pests (Fleming et al. 2000), these records were not available at the time the analysis for the guidelines was completed. Furthermore, wildfire events often follow and subsume such forest disturbances. Because fires and clearcutting were both believed to remove most standing trees, it seemed reasonable to create a guideline that would emulate the patterns of disturbance caused by fire.

Donnelly and Harrington's data were digitized and then analyzed on the basis of an ecological classification system (site regions) developed in Ontario by Hills (1966). The analyzed data were presented in the form of bar graphs for each site region that depicted the area and frequency of fires in various size classes from 1920 to 1950. These templates (figure 21.2) and an analysis of

42 individual fires in terms of such stand-level parameters as the amount, type, and location of residual patches were released in a 1997 document (Ontario Ministry of Natural Resources 1997).

These data obviously had limitations and biases. The 30-yr period represented only one snapshot of Ontario's fire disturbance landscape; the landscape picture may have been quite different in different periods. It was suggested that, instead of empirical data, models should be used to generate many possible landscape scenarios. In support of this argument, it was noted that the historical data included anthropogenic influences in the form of fire ignition and suppression. However, no models were available at the time to deal with such large areas, and experts

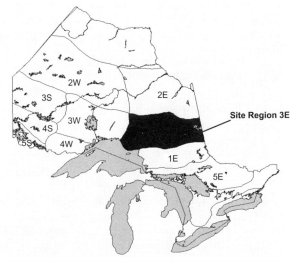

Size class (ha)	Total number fires	Total area burned (ha)	Annual frequency	Annual area burned (ha)
10–130	46	1996	3	111
131–260	16	3085	1	171
261–520	52	19,257	2	642
521–1040	50	36,608	2	1220
1041–2500	41	69,303	1	2310
2501–5000	24	79,486	1	2650
5001–10,000	10	72,942	0.3	2431

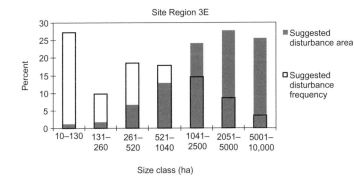

FIGURE 21.2. A sample template used in the first draft of Ontario's emulation guide for the area (size class) and frequency (size class) of disturbances in the province's Site Region 3E. The template was primarily based on wild fires that occurred between 1920 and 1950.

believed that the anthropogenic influences were minimal.

First Practical Guidelines for Emulating Fire Disturbance

The intent of the analysis (Ontario Ministry of Natural Resources 1997) was to provide background information to forest management planners, whose responsibility it was to "show movement towards emulation of natural disturbance frequency by size class" (Ontario Ministry of Natural Resources 1996a, p. A63) by the manner in which they allocated harvest blocks in their proposed forest management plans. This planning requirement had come into effect as a result of the Forest Management Planning Manual (Ontario Ministry of Natural Resources 1996a), revised based on the regulations of the Crown Forest Sustainability Act. The act incorporated the new planning requirements arising from the class environmental assessment (Ontario Environmental Assessment Board 1994). Planning teams endeavoring to meet the act's requirements needed some idea of what a "natural disturbance frequency by size class" might have been within their forest management units. The frequency and area (size class) of disturbance and their patterns, based on historical fire data (figure 21.2), could be used by forest managers as the basis for comparison with current disturbance patterns (fire plus clearcuts). The suggested frequency of disturbance and area of disturbance (figure 21.2) were used as the potential targets for a given site region. Forest managers were expected to account for current fire disturbance rather than simply assuming that clearcuts would replace fires. For example, site regions that had recently sustained large fires might have few or no large clearcuts, whereas other site regions that had sustained few large fires might be permitted to have several large clearcuts.

A Legal Challenge to the Policy of Emulating Fire Disturbance

Forest management planning teams, lacking detailed knowledge of historical fire patterns to provide guidance on the frequency and size classes of natural disturbance, quickly adopted the "natural" templates generated by the analysis of historical results (Ontario Ministry of Natural Resources 1997). However, because this document was not a guideline, several environmental organizations objected to the approach of patterning the allocations of clearcuts by using the site region templates. They particularly objected to the tendency of planning teams to allocate a significant percentage of their annual allowable cut into a few large clearcuts, even though this seemed to follow the spirit of emulating the size and frequency of natural disturbance. Their position was that until there was an approved guideline that governed the shape, size, and distribution of clearcuts, the class environmental assessment guidelines were intended to limit most clearcuts (often misconstrued as "all" clearcuts) to a maximum of 260 ha.

These objections culminated in a legal challenge to one forest management plan. The case, brought before the Ontario Ministry of the Environment, in part accused the Ontario Ministry of Natural Resources of ignoring the Environmental Assessment Board's decision to limit the size of clearcuts. Although the legal challenge was not successful, the Ontario Ministry of the Environment ordered the Ontario Ministry of Natural Resources to produce clearer guidelines.

The Forest Management Guide for Emulating Natural Disturbance Patterns

Following this legal challenge, the Ontario Ministry of Natural Resources recognized that the development of clearer guidelines required addressing the concerns of all stakeholders from the beginning of the development process. A multidisciplinary writing team was appointed, with one representative each from the environmental and academic communities, one representative from the forest industry, and three Ontario Ministry of Natural Resources staff. It was clear from the outset that the team disagreed in their interpretations of the class environmental assessment results and on the direction provided by the Crown Forest Sustainability Act. Some believed that the guide should focus strictly on governing the shape, size, and distribution of clearcuts; others interpreted the act's guiding principle of emulating natural disturbance and landscape patterns to mean that all aspects of the forest that would be affected by natural disturbance (hence, forest management activities that affected the forest's composition and age-class distribution) should be dealt with in the guide.

Ultimately, a compromise was reached that included a decision on the shape, size, and distribution of clearcuts but did not incorporate all aspects of forest composition and age-class distribution, as most of these aspects were being addressed through other initiatives (e.g., management of old-growth forest was addressed by implementing an existing policy; access controls were dealt with in the current requirements for forest management planning; forest composition was addressed in current forest management plans at the local level; and proposed subregional planning at scales above the forest management unit). Two drafts of the guide were produced and made available for 60-day public review periods.

First Draft

The first draft of the guide provided guidelines for emulating the structural attributes of fire disturbance based upon the templates produced by the analysis of historical fire data (Ontario Ministry of Natural Resources 1997). These guidelines governed the distance between disturbances, their size, and the distance between clearcuts within the larger disturbed area, with larger clearcuts requiring more separation (table 21.1). The intention of the guidelines was to ensure that the cumulative number and spacing of clearcuts more closely emulated the legacy of a large fire, after which some areas have been burned clear, some partially burned, and others have not been burned at all. Clearcuts represented the burned areas, unlogged areas between clearcuts represented the areas not burned, and retention of residual patches of trees and individual trees represented the partially burned areas. Patches of

trees were to be retained within and along the edges of cuts to emulate the "island and peninsula" patterns observed after historical fires. The size of the residual patches of trees was based on the percentage of the area clearcut. Internal patches could cover from 2 to 7% of the area of the clearcut. Peninsular patches could cover from 8 to 27% of the area of the clearcut, depending on the forest type and whether the cuts were in the boreal or Great Lakes–St. Lawrence forests. Individual living or dead trees had to be retained at a density of 25 per hectare in all clearcuts.

When the first draft was released for public commentary, it generated mostly negative reviews. Some reviewers commended the progress made on retention of residual patches and individual trees, but others criticized the use of templates for determining clearcut frequency by size class. In their view, the templates violated the class environmental assessment decision, because the guide recommended a wide range of clearcut sizes that went far beyond the recommended 260-ha maximum. There was also much criticism of clearcutting, which many believed would destroy wildlife habitat and soils, alter drainage patterns, and cause erosion. Permitting a few large clearcuts exacerbated these concerns. Some nongovernmental organizations criticized the draft guide because it used clearcutting as a

TABLE 21.1. Standards for Disturbances and Individual Clearcuts for the Boreal and Great Lakes–St. Lawrence Forests Submitted in the First Draft of Ontario's Natural Disturbance Pattern Emulation Guide

		Forest Disturbance[1]	Individual Clearcuts
Maximum size (ha)	Great Lakes–St. Lawrence	3000	260
	Boreal	10,000	1000
Time frame for creation (yr)		≤20	≤3
Distance between		200–5000 m, depending on size of disturbance	Great Lakes–St. Lawrence: 200–300 m
			Boreal: 200–500 m
Time between adjacent harvest areas (yr)		20 (from most recent cut)	20, or until 3-m-tall regeneration achieved
Acceptable break between (i.e., intervening vegetation)		10–10,000 ha: at least 200–300 m of young forest, 3 m in height	Great Lakes–St. Lawrence: 300 m, young forest 3 m in height
		1001–10,000 ha: at least 70% break in immature forest (≥6 m in height)	Boreal: 500 m, of which at least 300 m is young forest 3 m in height

Source: Unpublished data from the Ontario Ministry of Natural Resources.

[1]Contains one or more individual clearcuts.

means of emulating patterns of fire disturbance. The draft guide clearly recognized that fire is a primarily chemical process, whereas cutting is a mechanical process. Therefore, critics argued, if the underlying processes differed, how could patterns created by clearcuts emulate those maintained by natural disturbance? However, abandoning the clearcut silvicultural system was not an option, because its use in Ontario had previously been examined and approved.

Some nongovernmental groups advocated that all new logging roads should be closed to public traffic, because wildfires do not produce roads, along with the associated traffic and human activities. However, imposing a universal prohibition on public access to new logging roads would require a major policy decision that would involve an additional consultation process outside the scope of the guidebook exercise.

As a result of these criticisms, a different approach to governing the frequency and size classes of clearcuts had to be developed. The templates for frequency and size class were heavily criticized during the 60-day review period and had always been an issue for the guideline writing

team. Some on the team perceived an environmental impact and others feared the potential reaction of foreign markets to the fact that the guide would allow a range of clearcut sizes up to 10 000 ha.

Second Draft

The second draft of the guide proposed that 80% of clearcuts in the boreal forest region and 90% of clearcuts in the Great Lakes–St. Lawrence forest region should be smaller than 260 ha. This proposal was designed to quantify the Environmental Assessment Board's general guidelines (table 21.2). There was much criticism of these percentages by some members of the writing team, who wanted to have 80% of the total area of clearcuts accounted for by cuts smaller than 260 ha. This proposal ignored the intent of the class environmental assessment direction and evidence that in fire-disturbed landscapes, most of the area is burned by a few large fires (Hunter 1990).

Other members of the writing team believed that there was too much emphasis on small clearcuts, particularly in the boreal forest region. They

TABLE 21.2. Guidelines for Clearcuts in Ontario, as Proposed in the Final Draft of Ontario's Natural Disturbance Pattern Emulation Guide

Clearcut Parameter	Parameter Requirement or Description
Size and frequency	• Boreal Forest: 80% must be <260 ha
	• Great Lakes–St. Lawrence Forest: 90% must be <260 ha
Timeframe for creation	• 5 yr
Separation distance (edge to edge) between clearcuts with different sizes[1]	• If 3-m-tall regeneration is not reached in an adjacent clearcut and the clearcut is <20 yr old, 10–260-ha clearcuts should be separated by an average of 200 m (and a minimum of 100 m).
	• For every 100-ha increase in clearcut size, an additional 50 m of separation is required from clearcuts of similar size
	• Dissimilar-sized clearcuts should be separated by the distance indicated for the smaller clearcut.
Time before harvesting of adjacent areas can occur	• 20 yr or 3-m-tall regeneration in the earlier cut, whichever occurs first
Characteristics of the forest separating two clearcuts or shelterwood cuts	• Within the forest separating the two cuts, at least 200 m of the gap will include forest ≥3 m in height and with a 0.3 stocking level
	• For larger distances between clearcuts (≥600 m), at least 70% of the intervening land must be forest, 6 m in height with a 0.3 stocking level

Source: Ontario Ministry of Natural Resources (2002a).

[1]Separation distances represent the averages for similar-sized clearcuts. Spacing of the proposed cuts will be considered and assessed during the spatial allocation of the harvest area during planning of the allowable cut over a 5-yr period.

suggested that there should not be an arbitrary limit on clearcut size, because size could now be decided in each forest management plan after considering the management options for other forest values, the fire history, and inputs from the general public and local citizens' committees. Members of the citizens' committees are appointed by the district manager for each Ontario Ministry of Natural Resources administrative district, and are represented on each forest management planning team.

The percentage rule took a slightly different approach to determining the spacing between clearcuts that was necessitated by the move away from the old templates. Another significant change was to harvest the trees in a clearcut area in one pass that spanned no more than 5 yr (table 21.2), rather than harvesting over an extended 20-yr period, as had been proposed in the first draft. This change was intended to emulate closely the temporal component of a fire, as the disturbance caused by even a large fire occurs over a matter of days or (at most) weeks. Guidance provided for the type, amount, and location of residual patches of trees and the density of individual residual trees and snags remained the same as in the first draft, because there had been little controversy over these parameters.

The public review of the guide's second draft also produced much criticism. Many ignored those changes made to accommodate criticisms of the first draft: the dropping of the templates and closer adherence to the Environmental Assessment Board's decision on clearcut size. Instead, they focused on the absence of a limit on clearcut size (which the templates had provided, to some degree). Forest industry organizations, which had said little publicly at the time of the release of the first draft, now publicly indicated their concerns over potentially reduced wood supplies resulting from the guidelines contained in the second draft. However, after considering the comments received on the second draft, the final draft of the guide (Ontario Ministry of Natural Resources 2002a) closely reflects the second draft (table 21.2).

The final version of the guide achieved a compromise by trying to address the concerns raised by the critics of clearcutting. Nobody expected that the final guide would satisfy the highly polarized, disparate views of the various critics, particularly given the uncertainty over how best to emulate natural disturbance. The extent to which the compromise will meet the goal of maintaining biodiversity in managed landscapes remains to be seen.

Future Challenges

Modifying forest management techniques to emulate natural patterns of disturbance is a complex and controversial topic. One outcome of the process for developing practical guidelines in Ontario is the realization that, despite considerable public involvement and acceptance of the concept of emulating natural disturbance, there is considerable disagreement about how the concept should be implemented. What, then, are some of the challenges whose solutions might reduce the controversy and lead to wider acceptance of practical guidelines?

Developing Hierarchical Guidance

Future improvements to the process of developing practical guidelines for emulating natural disturbance should include a consideration of the spatial and temporal variation in the frequency of disturbance and in the area disturbed. Ecological systems are hierarchically organized (Allen and Starr 1982), with nested, interdependent processes from fine scales (individual trees) to broad scales (forested landscapes); emulation of natural disturbance should be similarly organized. A "one size fits all" approach to setting standards for clearcut size and frequency does not account for the natural spatial and temporal variation in the boreal forest. Both the frequency of fires and the average area burned vary considerably across the boreal forest (Bergeron et al. 1998; Johnson et al. 1998). The disturbance templates used in the first draft of the guide failed to deal with this variation, but a hierarchical guide would account for this variation and set guidelines for disturbance accordingly. For example, it should prescribe the nature of disturbance at broad spatial scales and over long time periods. The guidelines would vary among ecological regions in accordance with the natural fire cycles and range of disturbance sizes in each region. Finer-scale guidelines for landscape patterns should be nested within the broader-scale guidelines both spatially and temporally. Fine-scale decisions on silvicultural systems and harvesting location, which are based on local topography and soil conditions, should be left to the forest manager. Within these landscapes, further guidance should be provided to protect rare ecosystems and habitats for particular species. Both coarse filter and fine filter management would more likely be achieved by this kind of hierarchical guide.

Using Null Models

Before stakeholders can produce acceptable guidelines for dealing with spatiotemporal variation, they must develop a better understanding of the dynamics of fire-driven ecosystems. Carefully constructed and evaluated null models can accomplish this by describing baseline conditions of ecological and disturbance processes in these systems (Perera and Baldwin 2000). These null conditions are those that would exist in the absence of human interference. However, these approximations can be compared with the spatiotemporal patterns caused by human disturbance. Such null models are now being developed for Ontario (e.g., Perera et al., chapter 9, this volume) and will provide a defensible approach to developing guidelines for emulating natural disturbance regimes.

Evaluating Risk

There is little doubt that the landscape patterns produced by a proliferation of small clearcuts had increased the amount of edge in the boreal forest during the 1980s and early 1990s (Perera and Baldwin 2000, Schroeder and Perera 2002). The original intention of the Ontario Ministry of Natural Resources guide (2002a) was to take a more conservative approach to landscape management than a management policy that favored particular wildlife species and economic efficiencies for accessing and harvesting wood. Using natural patterns to guide landscape management should reduce the risk of long-term impacts for edge-avoiding species, such as woodland caribou (*Rangifer tarandus*) (Racey et al. 2002). Because the guide favors small clearcuts, the amount of edge created in the managed landscape is likely to increase in the future. What appears to be at issue for the public and for environmental groups is the relative risk to biodiversity posed by small clearcuts and increased amounts of edge versus the risk posed by a range of clearcut sizes that permits occasional large clearcuts. In particular, these groups consider the large clearcuts to pose unacceptable risks. This issue of relative risk will continue to be controversial whenever clearcutting becomes associated with emulating natural disturbance.

Improvements to the guidelines for emulating natural disturbance must deal effectively with perceived risks to biodiversity and other values. Good environmental policy requires reliable predictions of how to mitigate human impacts. Knowledge of the relative degree of mitigation that can be achieved by emulating natural disturbance by using a range of clearcut sizes is incomplete and at this point in time appears insufficient to sway deeply held assumptions about the environmental risks posed by large clearcuts.

Another challenge is to test hypotheses about relative risk. It is difficult, if not impossible, to conduct the large-scale experiments required to demonstrate the effects of management based on emulation. The best option for meeting this challenge appears to be a combination of null models and empirical evidence from retrospective studies. Such studies can be conducted for historical clearcuts and areas disturbed by natural fire (Baker 2000). Even so, information on the relative risks of emulating natural patterns and the impacts on biodiversity and other values may not change deeply held opinions. One need look no further than the heated debates over genetically modified foods or the use of lawn pesticides, which are never resolved, despite strong scientific evidence of small relative risks to the environment and people.

Considering Policy and Ecological Outcomes

In this case study, policy development preceded a full scientific understanding of the landscape dynamics responsible for the patterns and processes of fire. The public's desire that forest management should emulate natural patterns and processes was logical, and was accepted by forest managers. However, the public's perceptions and the forest managers' interpretations of this goal have been at odds. It is not unusual that public policy reflects a desire to mitigate a problem that science or technology is not yet ready to solve. Inclusion of the methodology of science in public forums remains a challenge. For example, computer simulations of long-term changes to landscapes, based on the proposed guidelines for clearcut size, might have revealed consequences that could not have been foreseen by a simplistic tabulation of written guidelines.

Other alternatives could also be simulated. For example, Bergeron and Harvey (1997) have suggested a different method for emulating natural disturbance in western Quebec's boreal forest, which is ecologically similar to northeastern Ontario's boreal forest. They suggest that it is highly unlikely that we can emulate broad-scale natural disturbance regimes—and indeed, it is impractical to attempt to do so. Instead, their approach emulates the temporal variation in disturbance from both broad-scale fires and fine-scale gap-phase dynamics by managing for a variety of forest types and ages in smaller areas, and then replicating these areas across an ecological region (see Gauthier et al., chapter 18, this volume). This is only one example of the possible alternatives. The challenge will be to consider alternative policy options and the predicted ecological outcomes of each alternative. This is the rationale for adaptive management, in which regulations permit the development of prescriptions based on the best available knowledge, with the implicit understanding that some of these prescriptions may initially fail and have to be revised (Walters 1986).

Lessons Learned from Ontario's Experience

We conclude with a brief discussion of some lessons we have learned that might help other agencies that are contemplating or are in the process of developing a policy on emulating natural disturbance.

- Ensure Interactive and Adaptive Public Consultation
  Although the extensive public consultation conducted through two parallel exercises did a commendable job of articulating public views, it was not until the guide was being developed that the public received feedback on the government's interpretation of their opinions of the emulation policy. The interpretation of complex ecological theory and management practice through public consultation processes clearly requires extensive and ongoing interaction to prevent such surprises in the future. Thus, stakeholders should carry the analysis of alternatives far enough to provide sufficient grounds for broad agreement about the potential outcomes. This process may still not overcome deeply held views, but it at least provides a means of exposing the evidence to support or refute various viewpoints (Lee 1993).

- Develop a Monitoring Program to Evaluate Policy Effectiveness
  To ensure that the goals of the process have been met, the government must commit to evaluating the impacts of a policy after it has been implemented. Accordingly, the Ontario Ministry of Natural Resources has committed itself to monitor the impacts of implementing the emulation guide (Ontario Ministry of Natural Resources 2002a). Making such a long-term commitment is beneficial to the long-term interests of the government agency and the public. At some point, the impacts of the policy can be evaluated based on evidence rather than on speculation.

- Address Possible Inconsistencies in Policy Consequences
  One of the major criticisms of emulating natural disturbance through clearcutting was that fire and harvesting represent completely different processes and therefore forest management based on clearcutting could not emulate all aspects of fire. The public seemingly interpreted this observation to mean that it made little sense to emulate large fire disturbances by means of large clearcuts. Although the government attempted to explain that it is the patterns that are being emulated, not the processes, the apparent logical inconsistency suggested a potential environmental risk that outweighed the logic of the explanation. Thus, agencies must be prepared to explain why using clearcuts to create and maintain a diversity of patch sizes is not inconsistent with the differences between fire and clearcutting.

- Communicate Proactively about New Policies
  Changing forest management practices as substantially as would be the case when emulating natural disturbance is as much a communication and education exercise as it is a scientific and technical one. Agencies contemplating new policy in this area should use appropriate communication tools to inform stakeholders—particularly the public—about ecological processes and the best current scientific knowledge of these processes. However, regardless of the evidence to support or refute alternatives, government efforts to change public opinion are viewed with great suspicion. One of the most consistent criticisms during the public review process was that creating larger clearcuts meant accelerating harvesting and the ensuing destruction of the forest. This confusion could not have been entirely eliminated, but a communications package that anticipated some of the potential criticisms and misunderstandings would have helped the public to better understand and accept the new guidelines.

- Strengthen Stakeholder Partnerships
  It is doubtful that forest management can ever fully emulate natural disturbance patterns or processes. However, with

improved knowledge, forest managers can better understand the impacts and mitigate the undesirable effects of forest harvesting. Such knowledge will not be gained cheaply or immediately, but over the longer term, ongoing consultation with the public and various nongovernmental agencies to create a true partnership will promote a better understanding of how to mitigate the risks and greatly increase the likelihood of reaching a compromise that is acceptable to all stakeholders.

PART IV

Conclusion

CHAPTER 22

Emulating Natural Forest Landscape Disturbances

A Synthesis

AJITH H. PERERA, LISA J. BUSE, MICHAEL G. WEBER, and THOMAS R. CROW

In this chapter, we present a synthesis of the state of knowledge about emulating natural disturbance in northern North America and provide a view of the future of this forest management approach. We summarize the information presented in the other chapters of this book to describe what natural disturbance emulation is, why one might do it, methods of understanding natural disturbance, and practical approaches to implementing this concept. In addition, we present the status of emulating natural disturbance from the perspective of forest practitioners and stakeholders, and describe the general challenges they face in doing so. Finally, we highlight future needs that we view as critical to the success of implementing the emulation of natural disturbance in forest management.

Summary of the Book

Considerable interest in emulating natural disturbance by forest management exists across northern North America. From the Cascades in Oregon and Washington in the western United States to the east coast of Canada, examples can be found for most northern forest types, including the montane and coastal regions of the west coast of the United States (Wimberly, Spies, and Nonaka, chapter 12; Hessburg et al., chapter 13) and Canada (Kimmins, chapter 2), the foothills of the Rockies (Li, chapter 8), the intermontane region (Romme et al., chapter 20), as well as the central hardwood (He et al., chapter 10), boreal (Perera et al., chapter 9; Gauthier et al., chapter 18; McNicol and Baker, chapter 21), Great Lakes–St. Lawrence (Pinto and Romaniuk, chapter 14; Zasada et al., chapter 19), and Acadian (MacLean, chapter 6; Porter, Hemens, and MacLean, chap-

ter 11) forest regions. Relevant applications are apparent in policy development, land use planning, and the exploration of implementation approaches, and we have witnessed the evolution of a multitude of methods for characterizing disturbance regimes. In addition, there have been early attempts at economic analysis and evidence of interest from different stakeholders. In this section, we provide a synopsis of the ideas and knowledge relating to the emulation of natural disturbances that have been presented in the book.

What Is the Emulation of Natural Disturbance?

As offered in the conceptual definition (Perera and Buse, chapter 1), emulating natural disturbance is not a single practice, but rather a very broad concept that encompasses a continuum of management practices. It is an approach that can be put into practice in many ways, ranging from modifying silvicultural systems (e.g., Gauthier et al., chapter 18; Zasada et al., chapter 19) where forests continue to be managed for resource extraction, to not suppressing disturbances; for example, by letting fires burn uncontrolled (Romme et al., chapter 20). The emulation of successional trajectories (Kimmins, chapter 2) provides a good practical basis for implementation. The emulation criteria and the degree to which they emulate natural disturbance, as well as the spatial and temporal scales of the intended emulation (Perera and Buse, chapter 1), can characterize this range of practices.

The question of how to define *natural* seems to have no objective answer (Suffling and Perera, chapter 4). The assumption that pre-European settlement conditions are a reasonable indication of the results of natural disturbance is widespread (e.g., Hessburg et al., chapter 13; Pinto and Romaniuk, chapter 14), but this assertion is debated (Suffling and Perera, chapter 4). Adding to the debate is the question of whether excluding human influences from the definition of natural disturbance truly represents the natural condition. The *variety* of natural disturbances addressed in this paradigm range from frequent disturbances, such as fires, pests, and pathogens, to infrequent disturbances, such as hurricanes, all of which occur at a variety of spatial and temporal scales and interact with one another (Kimmins, chapter 2; Suffling and Perera, chapter 4). The disturbances that are currently being emulated in North America are primarily fire (e.g., Keane, Parsons and Rollins, chapter 5) and insect pests (MacLean, chapter 6; Porter, Hemens, and MacLean, chapter 11), although others factors, such as landslides (Kimmins, chapter 2), wind, and diseases are also being explored (Zasada et al., chapter 19).

Why Emulate Natural Disturbance?

The main *ecological premise* for emulating natural disturbance is the hypothesis that disturbance-driven ecosystems are intrinsically resilient to those disturbances, and if forest management activities can be made to emulate these natural disturbances, they would ensure the long-term resilience of the ecosystem and its sustainability (Perera and Buse, chapter 1). Moreover, the ability of species assemblies to adapt to natural disturbance strongly supports the principle that embracing the emulation of natural disturbance in forest management will help to conserve biodiversity (Thompson and Harestad, chapter 3), and conservationists (Euler, Henschel, and Clark, chapter 15) advocate this as a conservative approach to forest management. Societal pressures may also move us toward the emulation of natural disturbance, which is seen as "softer forestry" (Kimmins, chapter 2). In addition to the ecological rationale, there may be social and economic motivations. Certainly, the breadth of *expectations* is considerable, ranging from the ecological (Thompson and Harestad, chapter 3) to the forest industry's expectations (Hebert, chapter 16) and those of conservationists. However, it is also evident that the potential economic losses and gains associated with emulating natural disturbance remain unknown (McKenney, Mussell, and Fox, chapter 17) and must be compared with those attending other alternatives, so as to meet expectations of economic sustainability (Hebert, chapter 16).

How Might Natural Disturbance Be Emulated?

An array of methods is being used to understand natural disturbance regimes, as discussed by Suffling and Perera (chapter 4), as well as specific disturbance types, such as fire (Keane, Parsons, and Rollins, chapter 5; McKenzie et al., chapter 7) and insect pests (MacLean, chapter 6). Three general groups of methods are presented in the book:

- The use of evidence from the distant past (e.g., Hessburg et al., chapter 13; Pinto and Romaniuk, chapter 14; Gauthier et al., chapter 18);
- The use of evidence from the recent past (e.g., Porter, Hemens, and MacLean, chapter 11; Zasada et al., chapter 19); and
- The use of simulation models (e.g., Li, chapter 8; Perera et al., chapter 9; He et al., chapter 10).

Two distinct *conceptual approaches* to portraying natural disturbances are apparent. One uses evidence of disturbances that *did* occur, either based on direct historical evidence (e.g., Hessburg et al., chapter 13; Pinto and Romaniuk, chapter 14; Gauthier et al., chapter 18) or based on simulated historical scenarios (Wimberly, Spies, and Nonaka, chapter 12). The implicit premise of this approach is temporal extrapolation that assumes what occurred in the past is a good window into the future; that is, what did happen is what will happen again. The other approach uses evidence synthesized from scientific information, either from generalized empirical evidence of disturbances (e.g., Porter, Hemens, and MacLean, chapter 11) or a combination of empirical evidence and mechanistic knowledge of disturbance dynamics (e.g., Li, chapter 8; Perera et al., chapter 9; He et al., chapter 10), to develop simulation models. This premise assumes that the synthesized information is a good indication of what could happen and describes the potential for natural disturbances.

In addition, emulation can be practiced at a variety of spatial scales, including the level of stands (Gauthier et al., chapter 18; Zasada et al., chapter 19), watersheds (Hessburg et al., chapter 13), subregions (Wimberly, Spies, and Nonaka, chapter 12), and regions (Romme et al., chapter 20). In addition, many criteria can be used as the basis for emulating disturbance. Some use, or advocate the use of, the forest structure and composition that result from natural disturbance (e.g., Pinto and Romaniuk, chapter 14; Zasada et al., chapter 19); others use the age-class distribution or temporal patterns (e.g., Gauthier et al., chapter 18), spatial patterns (Porter, Hemens, and MacLean, chapter 11), or combinations thereof (McNicol and Baker, chapter 21). Still others propose a suite of criteria based on the disturbance regime's broad-scale spatial patterns (Hessburg et al., chapter 13), broad-scale age patterns (Wimberly, Spies, and Nonaka, chapter 12), comprehensive age and spatial and temporal characteristics (Li, chapter 8; He et al., chapter 10), and spatial and temporal probabilities (Perera et al., chapter 9). Also apparent is the desire for a formal comparison of these approaches with other management approaches in terms of their economic, social, and ecological costs and benefits (e.g., Hebert, chapter 16; McNicol and Baker, chapter 21).

The Present State of Emulating Natural Disturbance

In this section, we describe the present state of the art in emulating natural disturbance in terms of its practice and applications. Our views are based in part on the literature, as well as on our own observations and a series of discussions with a diverse group of forestry professionals and stakeholders; these include policymakers from government agencies, practicing foresters from both the private and the public sectors, environmental nongovernmental organizations, and a representative of the Canadian Aboriginal community.

Natural Disturbance Emulation Is Practiced Now

The emulation of natural disturbance is currently being practiced in various forms throughout North America. In some cases, such as in shade tolerant northern hardwoods, such natural disturbance processes as tree gap dynamics have been used as the implicit null model in designing management practices for a long time. Moreover, these practices may not necessarily have evolved from policy directives or guidelines. As one forester told us, "the key to good forest management is to learn from nature." Practicing foresters are generally well attuned to local disturbance dynamics, and are capable of formulating management practices that emulate those disturbances. It appears that these practices have

been primarily restricted to finer scales, predicated on knowledge of autecology and stand-level disturbance. Interest in emulating broader-scale disturbance has emerged only recently. One such example is the ongoing debate about the value of fire (or its absence) in restoring the "health" of fire-driven forest ecosystems. Some practitioners view emulating natural disturbance by allowing natural fires to progress unimpeded as a way of restoring forest systems to their natural state after decades of active fire suppression. Others see this practice as a means of reducing the risk of catastrophic fires. Another example of the current application of disturbance emulation is the growing focus on finding historical benchmarks for forest composition and landscape patterns, which can be used as templates to be followed in restoring the "natural" processes and patterns in forest landscapes. In addition, some regions are already moving toward or have developed policies relating to the emulation of natural disturbance.

We group the various *templates* that forest practitioners use as guides for implementing natural disturbance emulation by forest management into five categories of approaches:

- The simplest and most commonly used template is the landscape pattern created by the last major disturbance. The spatial pattern of the ensuing forest stands, with distinct age classes, is used as the pattern to be emulated by harvesting along these stand boundaries. In other words, managers attempt to recreate the last known natural disturbance. The use of this form of template implicitly assumes that natural disturbances are spatially and temporally constant, and will always create identical patterns.

- The second approach uses a known historical landscape pattern that has resulted from a series of previous natural disturbances. One such benchmark is the pre-European settlement period (also referred to as the "time of first contact"). The templates created in this approach are based on one or several of the many approaches for the reconstruction of historical patterns through inferential analysis of historical surveys and other records. This approach assumes that the inferred landscape patterns are the net result of a series of prior natural disturbances modified only by pre-European settlement populations, and that what did happen is a good indication of the intrinsic potential of natural disturbance regimes (i.e., of what will happen).

- The third approach, which is a variant of the second, involves broadening the temporal span of the historical period that is examined. Although this approach follows the same logic as the previous one, it has the added advantage of estimating the variability associated with disturbance regimes, which is not possible using a single point estimate of a benchmark.

- The fourth approach derives historical templates through simulation modeling of historical scenarios. The philosophy behind this approach resembles those of the first three approaches, but construction of the templates involves inductive reasoning and has the general advantages and disadvantages associated with simulation modeling.

- The fifth, and relatively recent, approach to template development involves spatially based simulation modeling of the potential natural disturbance regimes (i.e., modeling of scenarios). This method differs from the other four in that it focuses on what *could* happen without referring to historical conditions. Although this approach has the disadvantages of being somewhat abstract, simulating these scenarios has the advantages of capturing both the empirical and the mechanistic variability that are innate to natural disturbance regimes and of being spatially explicit.

Emulating Natural Disturbance Is Complex and Challenging

In our discussions, we discovered near unanimous agreement that emulating natural disturbance is a worthy goal in forest management. Only a few practitioners opposed the concept, believing it to be a purely social construct and a leap of faith that has no sound ecological basis. However, even among those who accepted the paradigm, disagreements emerged once actual practices were addressed. This divergence in views may be due to the inherent *complexity* of the concept.

First and foremost, the associated terminology is confusing. Among the terms used to describe the emulation of natural disturbance are *emulation silviculture, ecological forestry, natural disturbance forestry, variable-retention forestry, historical* range of variation or *historical range of variability, natural range of variation* or *bounds of natural variation,* and *multicohort forestry.* Although each of these terms has a specific meaning and may represent different facets and degrees of the same concept, the details are not always articulated explicitly, causing confusion among practitioners and stakeholders.

Second, the complexity of the emulation task increases when the management scale extends beyond the level of the stand. As the variety of natural disturbances operating at those broader scales increases, their interactions become more complex. In addition, the economic goals and social land-use values broaden and multiply, leading to an increasingly difficult task of setting and optimizing management goals. Possibilities for conflicts in land ownership also increase at broader scales and further hinder the implementation of this paradigm.

Finally, stakeholder expectations begin to diverge and conflict at the implementation stage. Forestry practitioners generally believe that emulating natural forest disturbance will lead to sustainability of forest products while conserving biodiversity. Environmental special-interest groups believe that the primary goal is to conserve forest biodiversity, and that this should not be compromised or constrained by economic goals. Social expectations add cultural values, including the restoration of historical forest conditions, to this mix.

Many Implementation Challenges Exist

The perceived challenges that face the emulation of natural disturbance in resource management are many. Some of the conceptual and application challenges facing those intending to implement the approach are described here.

Conceptual challenges

As discussed above, misconceptions about the meaning of emulating natural disturbance are common. Ideas range from making forest harvesting patterns more "natural" by using uneven cut boundaries and islands of residual trees scattered within the cuts to other approaches, such as cutting along the boundaries of recent past disturbances (e.g., burn boundaries in the boreal forest), creating gaps (to emulate tree fall and windthrow in tolerant hardwood forests), using more prescribed fire in forest management, and (in extreme cases) letting natural fires burn. Without a clear picture of what emulating natural disturbance is, it is difficult for practitioners and stakeholders to agree on how to incorporate it into management planning or practice.

In addition to misconceptions about the concept, this approach is seen as not supporting all three aspects of sustainability equally. Some argue that the approach primarily supports the ecological aspect but compromises the social and economic aspects. Ironically, although some view emulating natural disturbance as a panacea that will solve such current resource management issues as conserving biodiversity, maintaining wildlife habitat, and decreasing forest fragmentation, others fear that the emphasis on landscape-level objectives will result in finer-scale objectives, such as the needs of rare species, being overlooked. There is also discomfort with the inherent uncertainty that surrounds natural disturbances, which are seen as being unpredictable in space, time, severity, and intensity. As a result, some forest stakeholders are concerned that this is too risky an experiment to be conducted on such a broad scale. While they may agree that "natural" is good, and that "emulating nature" is the right thing to do, stakeholders want proof that the outcome will not only be what they expect, but also an improvement over the results of current management efforts. Others argue that this approach is too conservative, based on the precautionary principle, and is thus more restrictive than management interventions not based on natural disturbance regimes. It is clear from these opposing views that the concept is still a poorly understood approach with which practitioners and stakeholders are not yet familiar.

Another major challenge for those opting to emulate natural disturbance is overcoming the conviction that disturbances are bad, because they are unpredictable and affect public safety. Fire especially is seen as destructive, and something to be avoided at all costs. North Americans are increasingly isolated from the extremes of the natural environment, and have developed a sense that everything can be moderated or controlled. However, natural disturbances can be catastrophic in disturbance-driven ecosystems. Oft-cited examples of this in the northern United States include the 1871 Peshtigo (Wisconsin) fire that consumed more than 500,000 ha of forest and a 1910 fire in the northwestern United States that consumed 1.2 million ha of forest. Little wonder, then, that concern exists about emulating such disturbances as fire.

At the same time, the importance of fire in maintaining ecosystems such as the native prairies and savannas of the midwestern United States, and fire adapted ecosystems in the boreal forest, is receiving more attention. In addition, it is increasingly evident that if disturbance-driven ecosystems are protected from one type of disturbance, such as fire, they may become more susceptible to another equally damaging disturbance, such as insects or pathogens. Thus, evading disturbance may not be an option in the long run.

Concerns about emulating disturbance are compounded by the impression that harvesting is detrimental and emulating natural disturbance simply means creating larger harvested areas. Forest harvesting is often associated with destruction, and its effects are viewed as long lasting and irreversible. Objections arise concerning the immediate impacts on such forest uses as trapping, hunting, and recreation, even though these effects are relatively short term. Allaying these concerns is made more difficult because the time frame for recovery from any type of disturbance is long in human terms—often longer than an individual's lifetime. To successfully emulate natural disturbance, the challenge lies in balancing short term social needs with longer term ecological requirements.

Application challenges

Even those who are comfortable with the concept of emulating natural disturbance question how it will be implemented. The challenges of implementing this approach are likely to be temporary, as is the case for any new concept, and many may be overcome by, for example, agreeing on the expected outcomes and developing land use policies that are compatible with this new paradigm.

To date, there have been no general guidelines on whether or how to implement the emulation of natural disturbance in practice. As discussed above, expectations among practitioners and stakeholders not only differ but may conflict. The forest industry wants a sustained timber supply from the forest, trappers want animals and undisturbed forests for their trap lines, cottage owners want vistas and mature forests for recreation, and anglers and hunters want access roads and wildlife habitat. Given these varied expectations, setting clear management goals that satisfy all stakeholders will be difficult. This problem is compounded because these goals are normally set at the level of the management unit, whereas disturbance emulation should be implemented at broader scales that require co-ordination across management units and even across regions. Only by setting very clear objectives at the outset, based on some consensus or compromise, would this approach even begin to meet the diverse expectations.

Another implementation challenge is the possibility for conflicts involving land use and ownership. If only part of the landscape can be managed using this approach, does this necessarily negate the validity of the approach over the whole landscape? Mixed and fragmented ownerships complicate implementation where such broad-scale disturbances as fire are being emulated. For example, multiple ownership and conflicting land use objectives in adjacent areas can arise among neighbors with differing objectives. Questions arise over property loss and public safety if one policy consequence is an approach in which natural fires are allowed to burn. Can fires be allowed to burn where private land is interspersed with public land? What is the role of parks and protected areas in this management scenario? How are private landowners involved in setting landscape-level management goals and objectives?

As with any management paradigm, clear policies are required to govern the implementation of this approach. For example, if the approach becomes widely accepted and applied, managers might attempt to emulate large disturbances even where these disturbances are not inherent or natural. Guidance is required to ensure that emulation will be applied only when and where it is appropriate. Policies associated with this paradigm must also be flexible. Given that this is an umbrella approach that could be applied in many different forest types with any number of expected outcomes, an inflexible prescriptive approach is unlikely to succeed.

From an operational perspective, the discomfort associated with the transition to emulation of natural disturbance is understandable. The complexity of the techniques and magnitude of the scales of application are overwhelming for practitioners. Most often, several different natural disturbances affect a given forest ecosystem. For example, disturbances in the coniferous forests of the coastal Pacific Northwest include broad-scale stand-replacing fires, windstorms that create small gaps in the canopy, and low- to moderate-intensity fires that kill patches of trees. The rates of change and degree of alteration in forest structure vary among these types of disturbance. The questions for practitioners attempting to model silvicultural practices on natural disturbance then multiply: What is the appropriate model? Which disturbances should be emulated, where, at what scale, and with what frequency? Once they have answered these questions, practitioners must still determine how to go about implementing the approach in the landscape. These and other implementation-related challenges require a better knowledge base to inform their decisions, as well as more management and decision-support tools to increase their comfort levels with making decisions and predicting the potential outcomes.

Even after the practical and policy related uncertainties have been resolved, the challenge remains of gaining public support for the landscape-level practices that are likely to be associated with many forms of emulating natural disturbance, especially related to disturbance size and frequency. Of all the challenges, this one is perhaps the most difficult to resolve.

Future Directions

As with any emerging management paradigm, emulating natural disturbance has its share of challenges, as discussed above. The approach's promise lies in its basis on a conservative and logical premise, the testable null model it offers to address the uncertainties surrounding long-term forest sustainability, and its intuitive appeal to most forest stakeholders. For these reasons, the approach merits further examination. In this section, we limit our discussion to issues that we deem essential at this early stage in the evolution of emulating natural disturbance. In particular, we believe that immediate effort is necessary to advance our conceptual understanding and bring clarity to this approach to forest management and its application.

The Concept Must Be Clarified

Once again, we emphasize that emulating natural disturbance is an umbrella concept that encompasses a variety of specific forest management practices: it is imperative that this diversity be articulated to all stakeholders in a clear manner. Fundamentally, there are two broad motivations for emulating natural disturbance. The less common one is based on the premise that periodic natural disturbances are essential to the functioning and sustainability of some ecosystems, and that we have successfully suppressed these disturbances for long enough to degrade these ecosystems. The ensuing practice is to allow natural disturbances to progress unimpeded as a means of restoring and revitalizing such forest ecosystems. The more common motivation is based on the premise that forest harvesting is a socioeconomic reality, and therefore it is important to minimize its ecological impacts and ensure the sustainability of the managed forest ecosystems. This may be accomplished through forest management practices that strive to emulate the structural and functional aspects of natural disturbance.

Both these broad approaches can be captured by the conceptual definition we offered in the first chapter of this book: *Emulating natural disturbance* is an approach in which forest managers develop and apply specific management strategies and practices, at appropriate spatial and temporal scales, with the goal of producing forest ecosystems as structurally and functionally similar as possible to the ecosystems that would result from natural disturbances and that incorporates the spatial, temporal, and random variability intrinsic to natural systems. Further definitions of specific practices for emulating natural disturbance may be developed under this conceptual umbrella. Moreover, since a myriad of structural and functional criteria for describing natural disturbances can guide forestry practices when we opt to emulate disturbance through management, any given set of emulation practices can be positioned on a sliding scale with many dimensions. Figure 22.1a illustrates a simplified example of three *emulation dimensions,* in which any given practice, present or future, can be positioned along three axes: spatial (emulation) scale (from fine to broad), emulation criteria (from few to many), and degree of emulation (from low to high). Many more dimensions could be added to define this description space, including the complexity of the disturbances and their intensity. Although using this form of ordination may not change or refine the application of a particular practice, it helps to clarify the conceptual premise, assess the complexity of the practice, and articulate the management expectations. As shown in figure 22.1b, any approach to emulating natural disturbance can be positioned within this framework. We illustrate this by using three hypothetical practices:

- Unsuppressed natural disturbances encompassing a range of scales, many criteria, and a high degree of emulation (the outer region enclosed by dashes in figure 22.1b);

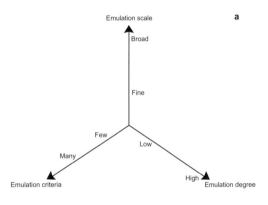

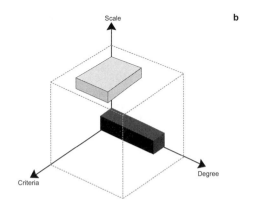

FIGURE 22.1. (a) A simplified ordination scheme for practices designed to emulate natural disturbance using dimensions of scale, number of criteria, and degree. (b) Three examples of scenarios positioned within this ordination: the region enclosed by dashes represents letting broad- and fine-scale disturbances occur; the dark rectangular region represents emulating a fine-scale disturbance using few criteria to a high degree; and the light region represents emulating a broad-scale disturbance using multiple criteria to a moderate degree.

- Emulation of few criteria, at fine scales, to a very high degree of emulation (the dark rectangular region), and
- Emulation of multiple criteria at broader scales and to a moderate degree (the light region).

Goals and Expectations Must Be Articulated

Given that the goals of emulating natural disturbance can range from the restoration of degraded ecosystems to producing a sustained supply of a desired forest resource, one way to understand this multitude of goals is to examine their broad roots in the major land use philosophies. The economic expectations of sustained forest resource use when emulating natural disturbance are similar to Gifford Pinchot's resource-based land use ethic, the ecological expectations of the sustainability of forest biodiversity are similar to Aldo Leopold's conservation-based land use ethic, and the social expectations of emulating nature are similar to John Muir's preservation-based land use ethic. Although none of these philosophies is universally inappropriate, our expectations for practices that emulate natural disturbance must be set judiciously for each specific circumstance, and articulated explicitly to all stakeholders. The various ecological, economic, social, and even political expectations must be addressed and clearly expressed a priori. In essence, the goal (what?) and the reasons (why?) should be made explicit to all stakeholders. We perceive this clear communication as an essential prerequisite to the successful application of emulation practices. Not only will it assist in managing the expectations of stakeholders and avoid disinformation, but it will also help to dispel any beliefs that this approach is a remedy for all forest management problems.

Natural Disturbance Emulation
Approach Must Be Deliberated

Because emulating natural disturbance is not a panacea, careful forethought is necessary to its adoption, whether in policy or in practice. We suggest that practitioners consider the following decision steps in sequence (see figure 22.2) as a heuristic guide in implementing the emulation of natural disturbance.

Is natural disturbance emulation appropriate?

For the natural disturbance emulation approach to be appropriate, evidence must exist that the forest ecosystem in question is primarily disturbance-driven, compared with other causal factors. Moreover, forest conservation and ecological sustainability of the whole forest ecosystem must be a management goal, and because the emulation approach is accompanied by a high degree of uncertainty about the long-term ecological impacts of management practices on biodiversity, a precautionary and conservative approach is necessary. A critical question at this point in the analysis involves determining the total extent of the disturbances to be emulated. That is, the total (additive) extent of natural disturbances and the planned emulations by means of management must not exceed the ecosystem's

inherent resilience, which is equivalent to the total potential to sustain natural disturbance (as illustrated in figure 1.1 in chapter 1). This step will justify the emulation of natural disturbance.

What are the goals of emulating natural disturbance?

The specific goals of the emulation must be explicitly formulated, and must be articulated to all stakeholders. There may be many economic and social expectations for adopting this approach that go beyond the ecological reasons addressed in step 1. Even these ecological goals must be translated into explicit and tangible objectives. This second step is an essential prerequisite to managing expectations for the management outcomes. This step requires an iterative process in which goals are set, examined, revised, and clarified so as to minimize any ambiguity and conflicts in the objectives.

Is the knowledge of the major disturbance regimes adequate?

The characteristics of the major disturbance regimes must be reasonably well understood, and their potential interactions must be clear. In particular, knowledge of the temporal, spatial, and stochastic aspects of disturbance must be readily available, so that forest practitioners can compose specific emulation practices. Using this knowledge, practitioners select the disturbance regime or regimes they should emulate. However, less-than-perfect knowledge of disturbances should not be an impediment to emulating natural disturbance if the assumptions are made explicit and the unknowns and uncertainties are noted, with the goal of progressively improving this knowledge.

How are the emulation dimensions selected?

It is important to predefine the dimensions of the planned emulation of natural disturbance. These dimensions are many, and include the spatial and temporal scales, types of disturbance, emulation criteria, degree of emulation, and templates to be used in the emulation. Although it may be impossible to exhaustively address all the dimensions of emulation, a deliberate attempt must be made to clearly describe the intended application, using as many dimensions as possible. More importantly, the scale of forest management must be congruent with the dimensions of the disturbance being emulated in terms of the planning horizon and the spatial and temporal resolution. This crucial step will decide the

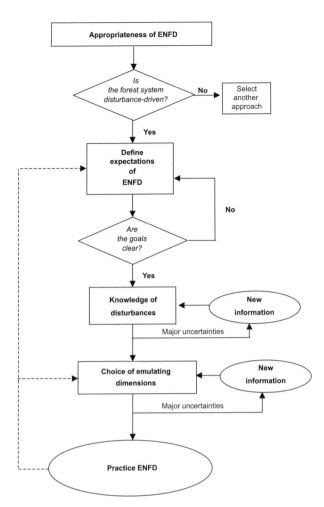

FIGURE 22.2. A schematic guideline for forest practitioners intending to emulate natural disturbance. ENFD, emulating natural forest disturbance.

success of the application; it involves describing the application, managing expectations, setting milestones for monitoring the effectiveness of the emulation practices, and progressively adapting the application. A deliberate choice must also be made on the use of an emulation template or a benchmark from among the five broad categories of template described earlier.

Directed Research Is Necessary on Many Aspects of Emulating Natural Disturbance

The most studied facet of natural disturbance emulation appears to be the individual disturbance regimes themselves. Notwithstanding the need to further the understanding of disturbance regimes and their interactions, many other aspects of this approach need additional attention from researchers. For example, we pose the following questions as the basis for potential research hypotheses.

Is the emulation of natural disturbance a sound ecological goal and an appropriate null model?

Much of the validity of emulating natural disturbance as a sensible ecological approach rests on two interrelated hypotheses: that the restoration of ecosystem resilience and the adaptation of species assemblies would occur under natural disturbance regimes. Although these premises are logical, testing those hypotheses under a variety of disturbance regimes is likely to add credibility to this approach as a null ecological model for forest management.

Is the emulation of natural disturbance feasible?

Practices designed to emulate natural disturbance must be compared with many other management alternatives in terms of their feasibility and success in meeting management goals. Even if emulating natural disturbance is the most sensible ecological approach, its popularity among practitioners is likely to rest on demonstrations of its feasibility in terms of cost-benefit analyses and social acceptability, over both the short term and the long term.

How can the emulation of
natural disturbance be implemented?

Given the diverse array of potential emulation dimensions, scrutinizing these dimensions is essential to provide specific guidance to practitioners. In particular, the appropriateness of various disturbance emulation templates, the adequacy of our emulation criteria, and the degree of emulation must all be considered by researchers. Furthermore, the effectiveness of emulation practices must be monitored and compared with that of other management alternatives to progressively assess and adapt these practices.

Conclusions

The emulation of natural disturbance is an elegant and intuitively powerful null model to guide the development of management practices and policies for forest ecosystems in which natural disturbances are the primary causal factor behind the development and maintenance of the ecosystem. The approach has garnered widespread interest among policymakers, practitioners, forest stakeholders, and researchers. Given the uncertainties surrounding long-term forest sustainability, the emulation of natural disturbance offers a conservative premise for forest management at broad scales and is fundamentally simple in its logic. However, its practice can become quite complicated when associated goals, assumptions, management approaches, and the myriad of expectations are included.

Not surprisingly, emulating natural disturbance in forest management has become the center of an active public debate in North America, focusing on both its motives and on the modus operandi. Immediate tasks awaiting researchers and practitioners are to add rigor to the terminology, goals, approaches, and methods of emulating natural disturbance so as to minimize ambiguity and manage expectations. Over the long term, the ultimate success of the approach will be determined by the answers to the questions of whether it is ecologically superior to other forest management paradigms, economically feasible for forestry practitioners, and socially acceptable.

REFERENCES

Abrams, M. D. 1984. Uneven-aged jack pine in Michigan. Journal of Forestry 82:306–309.

Abrams, M. D. 2001. Eastern white pine versatility in the presettlement forest. Bioscience 51:967–979.

Abrams, M. D., and D. I. Dickmann. 1984. Apparent heat stimulation of buried seeds of *Geranium bicknellii* on jack pine sites in northern lower Michigan. The Michigan Botanist 23:81–88.

Abrams, M. D., and C. M. Ruffner. 1995. Physiographic analysis of witness-tree distribution (1765–1798) and present forest cover through north central Pennsylvania. Canadian Journal of Forest Research 25:659–668.

Acevedo, M. F., D. L. Urban, and M. Ablan. 1995. Transition and gap models of forest dynamics. Ecological Applications 5:1040–1055.

Agee, J. K., editor. 1991. Fire history of Douglas-fir forests in the Pacific Northwest. U.S. Department of Agriculture Forest Service, Portland, Oregon, United States. General Technical Report PNW-GTR-285.

Agee, J. K. 1993. Fire ecology of Pacific Northwest forests. Island Press, Washington, D.C., United States.

Agee, J. K. 1995. Fire regimes and approaches for determining fire history. Pages 12–13 *in* C. C. Hardy and S. F. Arno, editors. The use of fire in forest restoration. General session at the annual meeting of the Society of Ecosystem Restoration "Taking a broader view." 14–16 September, 1995. University of Washington, Seattle. U.S. Department of Agriculture Forest Service, Missoula, Montana, United States. General Technical Report INT-GTR-341.

Agee, J. K. 1998. The landscape ecology of western fire regimes. Northwest Science 72:24–34.

Agee, J. K., and R. Flewelling. 1983. A fire cycle model based on climate for the Olympic Mountains, Washington. Pages 32–57 *in* Seventh conference on fire and forest meteorology. American Meteorological Society, Boston, Massachusetts, United States.

Agee, J. K., and F. Krusemark. 2001. Forest fire regime of the Bull Run watershed, Oregon. Northwest Science 75:292–306.

Agee, J. K., and L. Smith. 1984. Subalpine tree establishment after fire in the Olympic Mountains, Washington. Ecology 65:810–819.

Agee, J. K., M. Finney, and R. de Gouvenain. 1990. Forest fire history of Desolation Peak, Washington. Canadian Journal of Forest Research 20:350–356.

Agriculture and Agri-Food Canada. 1996. Soil landscapes of Canada. Version 2.2. National Soil Database, Ottawa, Ontario, Canada.

Ahl, V., and T. F. H. Allen. 1996. Hierarchy theory: A vision, vocabulary, and epistemology. Columbia University Press, New York, New York, United States.

Alatalo, R. V. 1981. Problems in the measurement of evenness in ecology. Oikos 37:199–204.

Alban, D. H., D. A. Perala, M. F. Jurgensen, M. E. Ostry, and J. R. Probst. 1991. Aspen ecosystem properties in the upper Great Lakes. U.S. Department of Agriculture Forest Service, St. Paul, Minnesota, United States. Research Paper NC-300.

Albert, D. A. 1995. Regional landscape ecosystems of Michigan, Minnesota, and Wisconsin: A working map and classification. U.S. Department of Agriculture Forest Service, St. Paul, Minnesota, United States. General Technical Report NC-178.

Alberta Forest Service. 1984. Alberta phase 3 forest inventory: Single tree volume tables, Appendix 1. Alberta Energy and Natural Resources, Edmonton, Alberta, Canada.

Allen, C. D., M. Savage, D. A. Falk, K. F. Suckling, T. W. Swetnam, T. Schulke, P. B. Stacey, P. Morgan, M. Hoffman, and J. T. Klingel. 2002. Ecological restoration of southwestern ponderosa pine ecosystems: A broad perspective. Ecological Applications 12:1418–1433.

Allen, R. B., P. J. Bellingham, and S. K. Wiser. 1999. Immediate damage by an earthquake to a temperate montane forest. Ecology 80:708–714.

Allen, T. F. H., and T. W. Hoekstra. 1992. Toward a unified ecology. Columbia University Press, New York, New York, United States.

Allen, T. F. H., and T. B. Starr. 1982. Hierarchy: Perspectives for ecological complexity. University of Chicago Press, Chicago, Illinois, United States.

Alley, W. 1984. The Palmer Drought Severity Index: Limitations and assumptions. Journal of Climate and Applied Meteorology 23:1100–1109.

Alvarado E., D. V. Sandberg, and S. G. Pickford. 1998. Modeling large forest fires as extreme events. Northwest Science 72:66–75.

Andersen, O., T. R. Crow, S. M. Lietz, and F. Stearns. 1996. Transformation of a landscape in the upper mid-west, USA: The history of the lower St. Croix river valley, 1830 to present. Landscape and Urban Planning 35:247–267.

Anderson, J. E., and W. H. Romme. 1991. Initial floristics in lodgepole pine (*Pinus contorta*) forests following the 1988 Yellowstone fires. International Journal of Wildland Fire 1:119–124.

Anderson, L., C. E. Carlson, and R. H. Wakimoto. 1987. Forest fire frequency and western spruce budworm outbreaks in western Montana. Forest Ecology and Management 22:251–260.

Anderson, R. L. 1973. A summary of white pine blister rust research in the Lake States. U.S. Department of Agriculture Forest Service, St. Paul, Minnesota, United States. General Technical Report NC-6.

Andison, D. W. 1999. Temporal patterns of age-class distribution on foothills landscapes in Alberta. Ecography 21:543–550.

Andison, D. W., and P. L. Marshall. 1999. Simulating the impact of landscape-level guidelines: A case study. The Forestry Chronicle 75:655–665.

Andren, H. 1994. Effects of habitat fragmentation on birds and mammals in landscapes with different proportions of suitable habitat: A review. Oikos 71:355–366.

Angelstäm, P. 1998. Maintaining and restoring biodiversity in European boreal forests by developing natural disturbance regimes. Journal of Vegetation Science 9:593–602.

Antonovski, M. Y., M. T. Ter-Mikaelian, and V. V. Furyaev. 1992. A spatial model of long-term forest fire dynamics and its applications to forests in western Siberia. Pages 373–403 in H. H. Shugart, R. Leemans, and G. B. Bonan, editors. A systems analysis of the global boreal forest. Cambridge University Press, Cambridge, United Kingdom.

Antos, J. A., and R. Parish. 2002. Structure and dynamics of a nearly steady-state subalpine forest in south-central British Columbia, Canada. Oecologia 130:126–135.

Arbogast, C. Jr. 1957. Marking guides for northern hardwoods under the selection system. U.S. Department of Agriculture Forest Service, St. Paul, Minnesota, United States. Station Paper 56.

Arlidge, C. 1995. Wood supply and woodland caribou on the Nakina/Geraldton forests and the Lac Seul forest. Ontario Ministry of Natural Resources, Thunder Bay, Ontario, Canada. NWST Technical Report TR-97.

Armstrong, G. W. 1999. A stochastic characterisation of the natural disturbance regime of the boreal mixedwood forest with implications for sustainable forest management. Canadian Journal of Forest Research 29:424–433.

Arno, S. F. 2002. Flames in our forest: Disaster or renewal? Island Press, Washington, D.C., United States.

Arno, S. F., and G. E. Gruell. 1986. Douglas-fir encroachment into mountain grasslands in southwestern Montana. Journal of Range Management 39:272–276.

Arno, S. F., D. G. Simmerman, and R. E. Keane. 1985. Forest succession on four habitat types in western Montana. U.S. Department of Agriculture Forest Service, Ogden, Utah, United States. General Technical Report INT-GTR-177.

Askins, R. A. 2000. Restoring North America's birds: Lessons from landscape ecology. Yale University Press, New Haven, Connecticut, United States.

Assmann, E. 1970. The principles of forest yield study. (Translated by S. H. Gardiner.) Pergamon Press, New York, New York, United States.

Attiwill, P. M. 1994. The disturbance of forest ecosystems: The ecological basis for conservative management. Forest Ecology and Management 63:247–300.

Auclair, A., H. C. Martin, and S. L. Walker. 1990. A case study of forest decline in western Canada and the adjacent United States. Water Air and Soil Pollution 53:13–31.

Babbitt, B. 1995. Return fire to its place in the West. Fire Management Notes 55:6–8.

Bachelet, D., J. M. Lenihan, C. Daly, and R. P. Neilson. 2000. Interactions between fire, grazing and climate change at Wind Cave National Park, SD. Ecological Modelling 134:229–244.

Bailey, J. K., and T. G. Whitman. 2002. Interactions among fire, aspen, and elk affect insect diversity: Reversal of a community response. Ecology 83:1701–1712.

Bailey, R. G. 1995. Description of the ecoregions of the United States, second edition. U.S. Department of Agriculture Forest Service, Washington, D.C., United States. Miscellaneous Publication 1391.

Baker, J. A. 2000. Landscape ecology and adaptive management. Pages 310–322 in A. H. Perera, D. L. Euler, and I. D. Thompson, editors. Ecology of a managed terrestrial landscape: Patterns and processes in forest landscapes of Ontario. University of British Columbia Press, Vancouver, British Columbia, Canada.

Baker, J. A., T. Clark, and I. D. Thompson. 1996. Boreal mixedwoods as wildlife habitat: Observations, questions, and concerns. Pages 41–52 in C. R. Smith and G. W. Crook, editors. Advancing boreal mixedwood management in Ontario. Canadian Forest Service and Ontario Ministry Natural Resources, Sault Ste. Marie, Ontario, Canada.

Baker, P. J., J. S. Wilson, and R. I. Gara. 1999. Silviculture around the world: Past, present and future trends. Proceedings of the long-term ecological monitoring workshop. U.S. National Parks Service, Washington, D.C., United States.

Baker, W. L. 1989. Effect of scale and spatial heterogeneity on fire-interval distributions. Canadian Journal of Forest Research 19:700–706.

Baker, W. L. 1992. Effects of settlement and fire suppression on landscape structure. Ecology 73:1879–1887.

Baker, W. L. 1993. Spatially heterogeneous multi-scale response of landscapes to fire suppression. Oikos 66:66–71.

Baker, W. L. 1994. Restoration of landscape structure altered by fire suppression. Conservation Biology 8:763–769.

Baker, W. L. 1995. Longterm response of disturbance landscape to human intervention and global change. Landscape Ecology 10:143–159.

Baker, W. L., and D. Ehle. 2001. Uncertainty in surface-fire history: The case of ponderosa pine forests in the western United States. Canadian Journal of Forest Research 31:1205–1226.

Baker, W. L., S. L. Egbert, and G. F. Frazier. 1991. A spatial model for studying the effects of climatic change on the structure of landscapes subject to large disturbances. Ecological Modelling 56:109–125.

Balling, R. C., G. A. Meyer, and S. G. Wells. 1992. Climate change in Yellowstone National Park: Is the drought-related risk of wildfires increasing? Climate Change 22:35–45.

Banner, A., J. Pojart, and G. E. Rouse. 1983. Postglacial paleoecology and successional relationships of a bog woodland near Prince Rupert, British Columbia. Canadian Journal of Forest Research 13:938–947.

Barrett, L. R. 1998. Origin and history of stump prairies in northern Michigan: Forest composition and logging practices. The Great Lakes Geographer 5:105–123.

Barrett, S. W. 1988. Fire regime classification for coniferous forests of the northwestern United States. U.S. Department of Agriculture Forest Service, Missoula, Montana, United States. Final Report RJVA-82123.

Barrett, S. W. 1994. Fire regimes on andesitic mountain terrain in northeastern Yellowstone National Park, Wyoming. International Journal of Wildland Fire 4: 65–76.

Barrett, S. W., and S. F. Arno. 1991. Classifying fire regimes and defining their topographic controls in the Selway-Bitterroot Wilderness. Pages 299–307 in P. L. Andrews and D. F. Potts, editors. Proceedings of the 11th conference on fire and forest meteorology. Society of American Foresters, Missoula, Montana, United States.

Barrett, S. W., S. F. Arno, and C. H. Key. 1991. Fire regimes of western larch–lodgepole pine forests in Glacier National Park, Montana. Canadian Journal of Forest Research 21:1711–1720.

Baskent, E. Z., and G. A. Jordan. 1995. Characterizing spatial structure of forest landscapes. Canadian Journal of Forest Research 25:1830–1849.

Baskerville, G. L. 1975. Spruce budworm: Super silviculturist. The Forestry Chronicle 51:138–140.

Baskerville, G. L. 1976. Report of the task-force for evaluation of budworm control alternatives. New Brunswick Cabinet Committee on Economic Development, Fredericton, New Brunswick, Canada.

Baskerville, G. L., and S. Kleinschmidt. 1981. A dynamic model of growth in defoliated fir stands. Canadian Journal of Forest Research 11:206–214.

Baskerville, G. L., and D. A. MacLean. 1979. Budworm-caused mortality and 20-year recovery in immature balsam fir stands. Canadian Forest Service, Fredericton, New Brunswick, Canada. Information Report M-X-102.

Batek, M. J., A. J. Rebertus, W. A. Schroeder, T. L. Haithcoat, E. Compas, and R. P. Guyette. 1999. Reconstruction of early nineteenth-century vegetation and fire regimes in the Missouri Ozarks. Journal of Biogeography 26:397–412.

Baxter, J. 1976. The fire came by: The riddle of the great Siberian explosion. Doubleday, Garden City, New Jersey, United States.

Bayley, S. E., D. W. Schindler, K. G. Beaty, B. R. Parker, and M. P. Stainton. 1992. Effects of multiple fires on nutrient yields from streams draining boreal forest and fen watersheds—nitrogen and phosphorus. Canadian Journal of Fisheries and Aquatic Sciences 49:584–596.

Beier, P. 1996. Metapopulation models, tenacious tracking and cougar conservation. Pages 293–323 in D. R. McCullough, editor. Metapopulations and wildlife conservation. Island Press, Washington, D.C., United States.

Beissinger, S. R., and M. I. Westphal. 1998. On the use of demographic models of population viability in endangered species management. Journal of Wildlife Management 62:821–841.

Bellchamber, S. B., H. S. He, S. J. Shifley, F. R. Thompson, and B. Palik. 2002. Using GIS functionalities to generate riparian buffer zones for the interpretation of best management practices. Missouri Natural Resources Conference, 30 January–1 February, 2002. Lake of Ozarks, Missouri, United States.

Benda, L., D. J. Miller, T. Dunne, G. H. Reeves, and J. K. Agee. 1998. Dynamic landscape systems. Pages 261–288 in R. J. Naiman and R. E. Bilby, editors. River ecology and management. Springer-Verlag, New York, New York, United States.

Bengston, J., S. G. Nilsson, A. Franc, and P. Menozzi. 2000. Biodiversity, disturbances, ecosystem function and management of European forests. Forest Ecology and Management 132:39–50.

Benzie, J. W. 1977. Manager's handbook for red pine in the North Central States. U.S. Department of Agriculture Forest Service, St. Paul, Minnesota, United States. General Technical Report NC-33.

Bergeron, J.-F., P. Grondin, and J. Blouin. 1999. Rapport de classification écologique: Pessière à mousse de l'ouest (révisé). Ministère des Ressources naturelles du Québec, Programme de connaissance des écosystèmes forestiers du Québec méridional, St.-Foy, Quebec, Canada.

Bergeron, Y. 1991. The influence of island and mainland lakeshore landscapes on boreal forest fire regimes. Ecology 72:1980–1992.

Bergeron, Y. 2000. Species and stand dynamics in the mixedwoods of Quebec's southern boreal forest. Ecology 81:1500–1516.

Bergeron, Y., and P. Dansereau. 1993. Predicting the composition of Canadian southern boreal forest in different fire cycles. Journal of Vegetation Science 3:827–832.

Bergeron, Y., and M. D. Flannigan. 1995. Predicting the effects of climate change on fire frequency in the southeastern Canadian boreal forest. Water Air and Soil Pollution 82:437–444.

Bergeron, Y., and B. Harvey. 1997. Basing silviculture on natural ecosystem dynamics: An approach applied to the southern boreal mixedwood forest of Quebec. Forest Ecology and Management 92:235–242.

Bergeron, Y., and A. Leduc. 1998. Relationships between change in fire frequency and mortality due to spruce budworm outbreak in the southeastern Canadian boreal forest. Journal of Vegetation Science 9:492–500.

Bergeron, Y., H. Morin, A. Leduc, and C. Joyal. 1995. Balsam fir mortality following the last spruce budworm outbreak in northwestern Quebec. Canadian Journal of Forest Research 25:1375–1384.

Bergeron, Y., P. J. H. Richard, C. Carcaillet, S. Gauthier, M. Flannigan, and Y. T. Prairie. 1998. Variability in fire frequency and forest composition in Canada's southeastern boreal forest: A challenge for sustainable forest management. Conservation Ecology 2:6. [Online, URL: <http://www.consecol.org/vol2/iss2/art6>.]

Bergeron, Y., B. Harvey, A. Leduc, and S. Gauthier. 1999. Forest management guidelines based on natural disturbance dynamics: Stand and forest-level considerations. The Forestry Chronicle 75:49–54.

Bergeron, Y., S. Gauthier, V. Kafka, P. Lefort, and D. Lesieur. 2001. Natural fire frequency for the eastern Canadian boreal forest: Consequences for sustainable forestry. Canadian Journal of Forest Research 31:384–391.

Bergeron, Y., A. Leduc, B. D. Harvey, and S. Gauthier. 2002. Natural fire regime: A guide for sustainable management of the Canadian boreal forest. Silva Fennica 36:81–95.

Bergmann, F., H.-R. Gregorius, and J. B. Larsen. 1990. Levels of genetic variation in European silver fir (*Abies alba*)—are they related to species decline? Genetica 82:1–10.

Berryman, A. A. 1988. Dynamics of forest insect populations: Patterns, causes, implications. Plenum Press, New York, New York, United States.

Bessie, W. C., and E. A. Johnson. 1995. The relative importance of fuels and weather on fire behavior in subalpine forests. Ecology 76:747–762.

Bickerstaff, A., W. L. Wallace, and F. Evert. 1981. Growth of forests in Canada. Part 2. A quantitative description of the land base and mean annual increment. Canadian Forest Service, Petawawa, Ontario, Canada. Information Report PI-X-1.

Bierregaard, R. O., T. E. Lovejoy, V. Kapos, A. A. dos Santos, and R. W. Hustings. 1992. The biological dynamics of tropical rainforest fragments. Bioscience 42:859–866.

Binkley, C. S. 1997. Preserving nature through intensive plantation forestry: The case for forestland allocation with illustrations from British Columbia. The Forestry Chronicle 73:533–559.

Binkley, C. S., M. Percy, W. A. Thompson, and I. B. Vertinsky. 1994. A general equilibrium analysis of the economic impact of a reduction in harvest levels in British Columbia. The Forestry Chronicle 70:449–454.

Black, B. A., and M. D. Abrams. 2001. Analysis of temporal variation and species–site relationships of witness tree data in southeastern Pennsylvania. Canadian Journal of Forest Research 31:419–429.

Blais, J. R. 1983. Trends in the frequency, extent, and severity of spruce budworm outbreaks in eastern Canada. Canadian Journal of Forest Research 13:539–547.

Blockstein, D. A. 2002. The passenger pigeon. Birds of North America, Philadelphia, Pennsylvania, United States.

Bloomfield, P. 1976. Fourier analysis of time series: An introduction. John Wiley and Sons, New York, New York, United States.

Blum, B. M., and D. A. MacLean. 1984. Silviculture, forest management, and the spruce budworm. Pages 83–102 *in* D. M. Schmitt, D. G. Grimble, and J. L. Searcy, technical coordinators. Managing the spruce budworm in eastern North America. U.S. Department of Agriculture Forest Service, Washington, D.C., United States. Agricultural Handbook 620.

Boecklen, W. J. 1986. Effects of habitat heterogeneity on the species area relationships of forest birds. Journal of Biogeography 13:59–68.

Bonar, R. L. 2000. Availability of pileated woodpecker cavities and use by other species. Journal of Wildlife Management 64:52–59.

Bonar, R. L. 2001. Sustainable forest management practices on the Weldwood Forest Management Area at Hinton, Alberta. The Forestry Chronicle 77:69–73.

Boose, E. R., K. E. Chamberlin, and D. R. Foster. 2001. Landscape and regional impacts of hurricanes in New England. Ecological Monographs 71:27–48.

Bork, J. L. 1985. Fire history in three vegetation types on the eastern side of the Oregon Cascades. Ph.D. dissertation. Oregon State University, Corvallis, Oregon, United States.

Bormann, B. T., H. Spaltenstein, M. H. McClellan, F. C. Ugolini, and S. M. Nay Jr. 1995. Rapid soil development after windthrow disturbance in pristine forests. Journal of Ecology 83:747–757.

Botkin, D. B. 1995. Our natural history. The lessons of Lewis and Clark. G. P. Putnam and Sons, New York, New York, United States.

Botkin, D. B., J. F. Janak, and J. R. Wallin. 1972. Some ecological consequences of a computer model of forest growth. Journal of Ecology 60:849–872.

Boucher, D., L. de Grandpré, and S. Gauthier. 2003. Développement d'un outil de classification de la structure des peuplements et comparaison de deux territoires de la pessière à mousses du Québec. The Forestry Chronicle 79(2):318–328.

Boudreault, C., S. Gauthier, and Y. Bergeron. 2000. Epiphytic lichens and bryophytes on *Populus tremuloides* along a chronosequence in the southwestern boreal forest of Québec, Canada. The Bryologist 103:725–738.

Boudreault, C., Y. Bergeron, S. Gauthier, and P. Drapeau. 2002. Bryophyte and lichen communities in mature to old-growth stands in eastern boreal forests of Canada. Canadian Journal of Forest Research 32:1080–1093.

Boutin, S., and D. Hebert. 2002. Landscape ecology and forest management: Developing an effective partnership. Ecological Applications 12:390–397.

Bowes, M. D., and J. V. Krutilla. 1989. Multiple-use management: The economics of public forest lands. Resources for the Future and Johns Hopkins University Press, Baltimore, Maryland, United States.

Boyce, M. S., and A. Haney, editors. 1997. Ecosystem management. Yale University Press, New Haven, Connecticut, United States.

Boychuk, D., and A. H. Perera. 1997. Modeling temporal variability of boreal landscape age-classes under different fire disturbance regimes and spatial scales. Canadian Journal of Forest Research 27:1083–1094.

Boychuk, D., A. H. Perera, M. T. Ter-Mikaelian, D. L. Martell, and C. Li. 1997. Modelling the effect of spatial scale and correlated fire disturbances on forest age distribution. Ecological Modelling 95:145–164.

Bradbury, R. H., D. G. Green, and E. N. Snoad. 2000. Are ecosystems complex systems? Pages 339–366 *in* R. J. Bossomaier and D. G. Green, editors. Complex systems. Cambridge University Press, Cambridge, United Kingdom.

Bradley, A. F., W. C. Fischer, and N. V. Noste. 1992a. Fire ecology of forest habitat types of eastern Idaho and western Wyoming. U.S. Department of Agriculture Forest Service, Missoula, Montana, United States. General Technical Report INT-GTR-290.

Bradley, A. F., N. V. Noste, and W. C. Fischer. 1992b. Fire ecology of forests and woodlands in Utah. U.S. Department of Agriculture Forest Service, Missoula, Montana, United States. General Technical Report INT-GTR-287.

British Columbia Ministry of Forests. 1995a. Biodiversity guidebook. British Columbia Ministry of Forests, Victoria, British Columbia, Canada.

British Columbia Ministry of Forests. 1995b. Forest practices code of British Columbia. British Columbia Ministry of Forests, Victoria, British Columbia, Canada.

Brook, B. W., D. W. Tonkyn, J. J. O'Grady, and R. Frankham. 2002. Contribution of inbreeding to extinction risk in threatened species. Conservation Ecology 6:16. [Online, URL: <http://www.consecol.org/vol6/iss1/art16>.]

Brooks, D. J., and G. E. Grant. 1992. New approaches to forest management. Journal of Forestry 90(1):25–28.

Brown, J. K. 1973. Fire cycles and community dynamics of lodgepole pine forests. Pages 23–55 in Baumgartner, D. B, editor. Management of lodgepole pine ecosystems. Volume I. Washington State University Press, Pullman, Washington, United States.

Brown, J. K. 1991. Should management ignitions be used in Yellowstone National Park? Pages 137–148 in R. B. Keiter and M. S. Boyce, editors. The Greater Yellowstone Ecosystem: Redefining America's wilderness heritage. Yale University Press, New Haven, Connecticut, United States.

Brown, J. K. 1995. Fire regimes and their relevance to ecosystem management. Pages 171–178 in Proceedings of the Society of American Foresters 1994 annual meeting. Society of American Foresters, Bethesda, Maryland, United States.

Brown, J. K., R. D. Oberheu, and C. M. Johnson. 1982. Handbook for inventorying surface fuel and biomass in the interior forest. U.S. Department of Agriculture Forest Service, Ogden, Utah, United States. Technical Report INT-129.

Brown, J. K., S. F. Arno, S. W. Barrett, and J. P. Menakis. 1994. Comparing the prescribed natural fire program with presettlement fires in the Selway–Bitterroot Wilderness. International Journal of Wildland Fire 4: 157–168.

Brown, P. M., and C. H. Sieg. 1999. Historical variability in fire at the ponderosa pine–Northern Great Plains prairie ecotone, Southeastern Black Hills, South Dakota. Ecoscience 6:539–547.

Brown, P. M., M. R. Kaufmann, and W. D. Shepperd. 1999. Long-term, landscape patterns of past fire events in a montane ponderosa pine forest of central Colorado. Landscape Ecology 14:513–532.

Brown, P., R. E. Spalding, D. O. ReVelle, E. Tagliaferri, and S. P. Worden. 2002. The flux of small near-Earth objects colliding with the Earth. Nature 420:294–296.

Bruhn, J. N., J. J. Wetteroff, J. D. Mihail, J. M. Kabrick, and J. B. Pickens. 2000. Distribution of *Armillaria* species in upland Ozark Mountain forests with respect to site, overstory species composition and oak decline. Forest Pathology 30:43–60.

Brunner, A., and J. P. Kimmins. 2003. Nitrogen fixation in coarse woody debris of *Thuja plicata* and *Tsuga heterophylla* forests on northern Vancouver Island. Canadian Journal of Forest Research 33(9):1670–1682.

Buchert, G. P., O. P. Rajora, J. V. Hood, and B. P. Dancik. 1997. Effects of harvesting on genetic diversity in old-growth eastern white pine in Ontario, Canada. Conservation Biology 11:747–758.

Buckman, R. E. 1962a. Two prescribed summer fires reduce abundance and vigor of hazel brush regrowth. U.S. Department of Agriculture Forest Service, St. Paul, Minnesota, United States. Technical Note 620.

Buckman, R. E. 1962b. Growth and yield of red pine in Minnesota. U.S. Department of Agriculture Forest Service, Washington, D.C., United States. Technical Bulletin 1272.

Buckman, R. E. 1964. Effects of prescribed burning on hazel in Minnesota. Ecology 45:626–629.

Buckman, R. E., and L. H. Blankenship. 1965. Abundance and vigor of aspen root suckering. Journal of Forestry 63:23–25.

Bunnell, F. L. 1995. Forest-dwelling vertebrate faunas and natural fire regimes in British Columbia: Patterns and implications for conservation. Conservation Biology 9:636–644.

Bunnell, F. L., and D. J. Huggard. 1999. Biodiversity across spatial and temporal scales: Problems and opportunities. Forest Ecology and Management 115:113–126.

Bunnell, F. L., L. L. Kremsater, and E. Wind. 1999. Managing to sustain vertebrate richness in forests of the Pacific Northwest: Relationships within stands. Environmental Reviews 7:97–146.

Buongiorno, J., and J. K. Gilless. 1987. Forest management and economics: A primer in quantitative methods. Macmillan, New York, New York, United States.

Burgar, R. J. 1963. Possibilities for control of the spruce budworm. Log Book (March/April):19–21.

Burns, R. M., and B. H. Honkala, technical coordinators. 1990. Silvics of North America. Volume 2: Hardwoods. U.S. Department of Agriculture Forest Service, Washington, D.C., United States. Agricultural Handbook 654.

Burton, P. J., D. D. Kneeshaw, and K. D. Coates. 1999. Managing forest harvesting to maintain old growth in boreal and sub-boreal forests. The Forestry Chronicle 75:623–631.

Busing, R. T., and E. F. Pauley. 1994. Mortality trends in a southern Appalachian red spruce population. Forest Ecology and Management 64:41–45.

Byram, G. M. 1959. Combustion of forest fuels. Pages 61–89 in K. P. Davis, editor. Forest fire control and use. McGraw-Hill, New York, New York, United States.

Cadenasso, M. L., and S. T. A. Pickett. 2001. Effect of edge structure on the flux of species into forest interiors. Conservation Biology 15:91–97.

Callicott, J. B. 1990. Whither conservation ethics? Conservation Biology 4:15–20.

Callicott, J. B., L. B. Crowder, and K. Mumford. 1999. Current normative concepts in conservation. Conservation Biology 13:22–35.

Camp, A., C. Oliver, P. Hessburg, and R. Everett. 1997. Predicting late-successional fire refugia pre-dating European settlement in the Wenatchee Mountains. Forest Ecology and Management 95:63–77.

Campbell, F. T., and S. E. Schlarbaum. 2002. Fading forests II—trading away North America's natural heritage. Healing Stones Foundation, American Lands

Alliance, and University of Tennessee, Knoxville, Tennessee, United States.

Canada Department of Crown Lands. 1862. Remarks on Upper Canada surveys, and extracts from the surveyor's reports, containing a description of the soil and timber of the townships in the Huron and Ottawa territory, and on the north shores of Lake Huron. Appendix 26. Report of the Commissioner of Crown Lands for 1861. Hunter, Rose, and Lemieux, Quebec, Quebec, Canada.

Canada Department of Crown Lands. 1867. Remarks on Upper Canada surveys, and extracts from the surveyor's reports, containing a description of the soil and timber of the townships in the Ottawa River and Georgian Bay Section and between the Spanish River, on the north shore of Lake Huron, and Goulay's Bay, on Lake Superior. Hunter and Rose, Ottawa, Ontario, Canada.

Canadian Council of Forest Ministers. 1997. Compendium of Canadian forestry statistics 1996. Canadian Council of Forest Ministers, Ottawa, Ontario, Canada.

Canadian Council of Forest Ministers. 1998. National forest strategy 1998–2003: Sustainable forests—a Canadian commitment. Canadian Council of Forest Ministers, Ottawa, Ontario, Canada.

Canadian Council of Forest Ministers. 2002. National forestry database program. Canadian Forest Service, Ottawa, Ontario, Canada. [Online, URL: <http://nfdp.ccfm.org>.]

Canadian Forest Service. 1998. State of Canada's forests, the peoples' forests, 1997–1998. Canadian Forest Service, Ottawa, Ontario, Canada.

Canadian Forest Service. 2000. The state of Canada's forests, 1999–2000. Canadian Forest Service, Ottawa, Ontario, Canada.

Canadian Parks and Wilderness Society–Wildlands League. 2002. Making forestry better, protecting the forest's critical characteristics. Wildlands League, Toronto, Ontario, Canada. Fact Sheet 4. [Online, URL: <www.wildlandsleague.org/forestry.html>.]

Candau, J.-N., R. A. Fleming, and A. Hopkin. 1998. Spatiotemporal patterns of large-scale defoliation caused by the spruce budworm in Ontario since 1941. Canadian Journal of Forest Research 28:1733–1741.

Candau, J.-N., R. A. Fleming, A. Hopkin, and A. H. Perera. 2000. Characterizing spruce budworm-caused disturbance regimes in Ontario. Pages 132–139 in S. G. Conard, editor. Proceedings of the International Boreal Forest Research Association (IBFRA) symposium on disturbance in boreal forest ecosystems: Human impacts and natural processes. U.S. Department of Agriculture Forest Service, St. Paul, Minnesota, United States. General Technical Report NC-209.

Canham, C. D., and O. L. Loucks. 1984. Catastrophic windthrow in the presettlement forests of Wisconsin. Ecology 65:803–809.

Cappuccino, N., D. Lavertu, Y. Bergeron, and J. Regnière. 1998. Spruce budworm impact, abundance and parasitism rate in a patch landscape. Oecologia 114:236–242.

Cardinale, B. J., K. Nelson, and M. A. Palmer. 2000. Linking species diversity to the functioning of ecosystems: On the importance of environmental context. Oikos 91:175–183.

Carleton, T. J. 2000. Vegetation responses to the managed forest landscape of central and northern Ontario. Pages 179–197 in A. H. Perera, D. L. Euler, and I. D. Thompson, editors. Ecology of a managed terrestrial landscape: Patterns and processes of forest landscapes in Ontario. University of British Columbia Press, Vancouver, British Columbia, Canada.

Carleton, T. J., and A. M. Gordon. 1992. Understanding old-growth red and white pine dominated forests in Ontario. Ontario Ministry of Natural Resources, Toronto, Ontario, Canada.

Carleton, T. J., and P. MacLennan. 1994. Woody vegetation responses to fire versus clear-cutting logging: A comparative survey in the central Canadian boreal forest. Ecoscience 1:141–152.

Carleton, T. J., and P. F. Maycock. 1978. Dynamics of the boreal forest south of James Bay. Canadian Journal of Botany 56:1157–1173.

Carter, N. E., and D. R. Lavigne. 1994. Protection spraying against spruce budworm in New Brunswick 1993. New Brunswick Department of Natural Resources and Energy, Fredericton, New Brunswick, Canada.

Casagrandi, R., and S. Rinaldi. 1999. A minimal model for forest fire regimes. The American Naturalist 153:527–539.

Caza, C. L. 1991. The ecology of planted Engelmann spruce (Picea engelmannii Parry) seedlings on subalpine forest cutovers. Ph.D. thesis. University of British Columbia, Vancouver, British Columbia, Canada.

Caza, C. L. 1993. Woody debris in the forests of British Columbia: A review of the literature and current research. British Columbia Ministry of Forests, Victoria, British Columbia. Land Management Report 78.

Centre for Topographic Information. 2000. Canadian digital elevation data: Standards and specifications. Natural Resources Canada, Sherbrooke, Quebec, Canada.

Chang, C. R. 1996. Ecosystem responses to fire and variations in fire regimes. Pages 1071–1099 in Sierra Nevada Ecosystem Project final report to Congress: Status of the Sierra Nevada. Volume II, Assessment and scientific basis for management options. Centers for Water and Wildland Resources, University of California, Davis, California, United States. Report 37.

Chapin, F. S., O. E. Sala, J. C. Burke, J. P. Grime, D. U. Hooper, W. K. Lauenroth, A. Lombard, H. A. Mooney, A. R. Mosier, S. Naeem, S. W. Pacala, J. Roy, W. L. Steffen, and D. Tilman. 1998. Ecosystem consequences of changing biodiversity. Bioscience 48:45–52.

Chapin, T. G., D. J. Harrison, and D. D. Katnik. 1998. Influence of landscape pattern on habitat use by American marten in an industrial forest. Conservation Biology 12:1327–1337.

Chen, J., J. F. Franklin, and T. A. Spies. 1992. Vegetation responses to edge environments in old-growth Douglas-fir forests. Ecological Applications 2:387–396.

Chew, J. D. 1997. Simulating vegetation patterns and processes at landscape scales. Pages 287–290 in 11th annual symposium on Geographic Information Systems—Integrating spatial information technologies for tomorrow. 17–20 February 1997. GIS World, Vancouver, British Columbia, Canada.

Christensen, N. L., J. K. Agee, P. F. Brussard, J. Hughes, D. H. Knight, G. W. Minshall, J. M. Peek, S. J. Pyne, F. J. Swanson, J. W. Thomas, S. Wells, S. E. Williams, and H. A. Wright. 1989. Interpreting the Yellowstone fires of 1988. BioScience 39:678–685.

Christensen, N. L., A. M. Bartuska, J. H. Brown, S. Carpenter, C. D'Antonio, R. Francis, J. F. Franklin, J. A. MacMahon, R. F. Noss, D. J. Parsons, C. H. Peterson, M. G. Turner, and R. G. Woodmansee. 1996. The report of the Ecological Society of America committee on the scientific basis for ecosystem management. Ecological Applications 6:665–691.

Cimino, G., and G. Toscano. 1998. Dissolution of trace metals from lava ash: Influence on the composition of rainwater in the Mount Etna volcanic area. Environmental Pollution 99(3):389–393.

Cissel, J. H., F. J. Swanson, W. A. McKee, and A. L. Burditt. 1994. Using the past to plan the future in the Pacific Northwest. Journal of Forestry 92:30–31.

Cissel, J. H., F. J. Swanson, and P. J. Weisberg. 1999. Landscape management using historical fire regimes: Blue River, Oregon. Ecological Applications 9:1217–1231.

Clark, D. F. 1994. Post-fire succession in the sub-boreal spruce forests of the Nechako plateau, central British Columbia. M.Sc. thesis. University of Victoria, Victoria, British Columbia, Canada.

Clark, J. S. 1988a. Stratigraphic charcoal analysis on petrographic thin sections: Application to fire history in northwestern Minnesota. Quaternary Research 30: 81–91.

Clark, J. S. 1988b. Effect of climate change on fire regimes in northwestern Minnesota. Nature 334:233–235.

Clarke, J., and G. F. Finnegan. 1984. Colonial survey records and the vegetation of Essex County, Ontario. Journal of Historical Geography 10:119–138.

Clements, F. E. 1916. Plant succession: An analysis of the development of vegetation. Carnegie Institution of Washington, Washington, D.C., United States. Publication 242.

Clowater, R., T. Erdle, J. Loo, D. MacLean, P. Neily, and V. Zelazny 1999. Categorization of natural disturbances in the Maritimes. Forest Stewardship Council, Maritimes Community Types Committee, Fredericton, New Brunswick, Canada. Unpublished report.

Coates, K. D. 2002. Tree recruitment in gaps of various size, clearcuts and undisturbed mixed forest of interior British Columbia, Canada. Forest Ecology and Management 155:387–398.

Cogbill, C. V. 1985. Dynamics of the boreal forests of the Laurentian Highlands, Canada. Canadian Journal of Forest Research 15:252–261.

Cohen, J. D. 2000. Preventing disaster: Home ignitability in the wildland–urban interface. Journal of Forestry 98:15–21.

Cohen, J. D., and J. E. Deeming. 1985. The National Fire Danger Rating System: Basic equations. U.S. Department of Agriculture Forest Service, Berkeley, California, United States. General Technical Report PSW-GTR-82.

Cohen, W. B., T. A. Spies, R. J. Alig, D. R. Oetter, T. K. Maiersperger, and M. Fiorella. 2002. Characterizing 23 years (1972–95) of stand replacement disturbance in western Oregon forests with Landsat imagery. Ecosystems 5:122–137.

Colbert, J. J., N. L. Crookston, W. P. Kemp, and N. Srivastava. 1981. Description of the combined Prognosis/Western Spruce Budworm model. U.S. Department of Agriculture Forest Service, CANUSA West, Portland, Oregon, United States.

Cole, K. L., F. Stearns, G. Guntenspeg, M. B. Davis, and K. Walker. 1998. Historical landcover changes in the Great Lakes Region. Pages 43–50 in T. D. Sisk, editor. Perspectives on the land-use history of North America: A context for understanding our changing environment. U.S. Geological Survey, Biological Resources Division, Reston, Virginia, United States. Biological Science Report USGS/BRD/BSR-1998-0003.

Congalton, R. G., and K. Green. 1999. Assessing the accuracy of remotely sensed data: Principles and practices. Lewis Publishers, CRC Press, Seattle, Washington, United States.

Connor, E. F., A. C. Courtney, and J. M. Yoder. 2000. Individuals–area relationships: The relationship between animal population density and area. Ecology 81:734–748.

Cook, E. R., and L. A. Kairiukstis. 1990. Methods of dendrochronology. Kluwer Academic, Dordrecht, the Netherlands.

Cook, E. R., D. M. Meko, D. W. Stahle, and M. K. Cleveland. 1999. Drought reconstructions for the continental United States. Journal of Climate 4:1145–1162.

Cornell, H. V., and J. H. Lawton. 1992. Species interactions, local and regional processes, and limits to richness of ecological communities: A theoretical perspective. Journal of Animal Ecology 61:1–12.

Costanza, R., and S. E. Jorgensen, editors. 2002. Understanding and solving environmental problems in the 21st century: Toward a new, integrated "hard problem science." Elsevier Science, Boston, Massachusetts, United States.

Costanza, R., B. G. Norton, and B. D. Haskell. 1992. Ecosystem health. New goals for environmental management. Island Press, Washington, D.C., United States.

Covington, W. W., P. Z. Fulé, M. M. Moore, S. C. Hart, T. E. Kolb, J. N. Mast, S. S. Sackett, and M. R. Wagner. 1997. Restoring ecosystem health in ponderosa pine forests of the southwest. Journal of Forestry 95: 23–29.

Craighead, J. J., and J. A. Mitchell. 1982. Grizzly bear. Pages 515–556 in J. A. Chapman and G. A. Feldhamer, editors. Wild mammals of North America: Biology, management and economics. The Johns Hopkins University Press, Baltimore, Maryland, United States.

Crane, M. F., and W. C. Fischer. 1986. Fire ecology of the forest habitat types of central Idaho. U.S. Department of Agriculture Forest Service, Missoula, Montana, United States. General Technical Report INT-218.

Crocker, R. L., and J. Major. 1955. Soil development in relation to vegetation and surface age of Glacier Bay, Alaska. Journal of Ecology 43:427–448.

Crooks, K. R., and M. E. Soulé. 1999. Mesopredator release and avifaunal extinctions in a fragmented system. Nature 400:563–566.

Crookston, N. L. 1991. Foliage dynamics and tree damage components of the western spruce budworm modelling system. U.S. Department of Agriculture Forest Service, Ogden, Utah, United States. General Technical Report INT-282.

Crow, T. R., D. S. Buckley, E. A. Nauertz, and J. C. Zasada. 2002. Effects of management on the composition and structure of northern hardwood forests in upper Michigan. Forest Science 48:129–145.

Crutzen, P. J., and J. G. Goldammer. 1993. Fire in the environment: The ecological, atmospheric, and climatic importance of vegetation fires. John Wiley and Sons, West Sussex, United Kingdom.

Culver, D. C. 1986. Cave faunas. Pages 427–443 *in* M. E. Soulé, editor. Conservation biology: The science of scarcity and diversity. Sinauer Associates, Sunderland, Massachusetts, United States.

Cumming, H. G. 1998. Status of woodland caribou in Ontario. Rangifer 19:99–104.

Cumming, H. G., and D. B. Beange. 1993. Survival of woodland caribou in commercial forests of northern Ontario. The Forestry Chronicle 69:579–588.

Cumming, S. G. 1997. Landscape dynamics of the boreal mixedwood forest. Ph.D. dissertation. University of British Columbia, Vancouver, British Columbia, Canada.

Cumming, S. G. 2001. A parametric model of the fire-size distribution. Canadian Journal of Forest Research 31:1297–1303.

Cumming, S. G., F. K. A. Schmiegelow, and P. J. Burton. 2000. Gap dynamics in boreal aspen stands: Is the forest older than we think? Ecological Applications 10:744–759.

Cundiff, B. 1989. New ideas about old growth. Seasons (Winter):31–35.

Cunningham, A. A., and D. L. Martell. 1973. A stochastic model for the occurrence of man-caused forest fires. Canadian Journal of Forest Research 3:282–287.

Currie, D. J. 1991. Energy and large-scale patterns of animal and plant species richness. American Naturalist 137:27–49.

Curtis, R. O. 1997. The role of extended rotations. Pages 165–170 *in* K. A. Kohm and J. F. Franklin, editors. Creating a forestry for the 21st century. Island Press, Washington, D.C., United States.

Cwynar, L. C. 1977. The recent fire history of Barron Township, Algonquin Park. Canadian Journal of Botany 55:1524–1538.

Cwynar, L. C. 1987. Fire and the forest history of the North Cascade Range. Ecology 68:791–802.

Dahir, S. E., and C. G. Lorimer. 1996. Variation in canopy gap formation among developmental stages of northern hardwood stands. Canadian Journal of Forest Research 26:1875–1892.

Dahlberg, A. 2002. Effects of fire on ectomycorrhizal fungi in Fennoscandian forests. Silva Fennica 36:69–80.

Dahlberg, A., J. Schimmel, A. F. S. Taylor, and H. Johannesson. 2001. Post-fire legacy of ectomycorrhizal fungal communities in the Swedish boreal forest in relation to fire severity and logging intensity. Biological Conservation 100:151–161.

Dahlgren, R. A., and C. T. Driscoll. 1994. The effects of whole-tree clear-cutting on soil processes at the Hubbard Brook Experimental Forest, New Hampshire, USA. Plant and Soil 158:239–262.

Dale, V. H., A. E. Lugo, J. McMahon, and S. T. A Pickett. 1998. Ecosystem management in the context of large infrequent disturbances. Ecosystems 1:546–557.

Dale, V. H., L. A. Joyce, S. McNulty, R. P. Neilson, M. P. Ayres, M. D. Flannigan, P. J. Hanson, L. C. Irland, A. E. Lugo, C. J. Peterson, D. Simberloff, F. J. Swanson, B. J. Stocks, and B. M. Wotton. 2001. Climate change and forest disturbances. BioScience 51:723–734.

Daly, C., R. P. Neilson, and D. L. Phillips. 1994. A statistical-topographical model for estimating climatological precipitation over mountainous terrain. Journal of Applied Meteorology 33:140–158.

Darby, W. R., H. R. Timmerman, J. B. Snider, K. F. Abraham, R. A. Stefanski, and C. A. Johnson. 1989. Woodland caribou in Ontario: Background to a policy. Ontario Ministry of Natural Resources, Toronto, Ontario, Canada.

Daubenmire, R. 1978. Plant geography: With special reference to North America. Academic Press, New York, New York, United States.

Davis, K. M. 1980. Fire history of a western larch/Douglas-fir forest type in northwestern Montana. Pages 69–74 *in* M. A. Stokes and J. H. Dietrich, editors. Proceedings of the fire history workshop. U.S. Department of Agriculture Forest Service, Fort Collins, Colorado, United States. General Technical Report GTR-RM-81.

de Grandpré, L., J. Morissette, and S. Gauthier. 2000. Long-term post-fire changes in the northeastern boreal forest of Quebec. Journal of Vegetation Science 11:791–800.

DeBano, L. F., D. G. Neary, and P. F. Folliott. 1998. Fire's effect on ecosystems. John Wiley and Sons, New York, New York, United States.

Deeming, J. E., R. E. Burgan, and J. D. Cohen. 1977. The National Fire Danger Rating System—1988. U.S. Department of Agriculture Forest Service, Ogden, Utah, United States. General Technical Report INT-GTR-19.

Delcourt, H. R., and P. A. Delcourt. 1996. Presettlement landscape heterogeneity: Evaluating grain of resolution using General Land Office Survey data. Landscape Ecology 11:363–381.

Delcourt, H. R., P. A. Delcourt, and T. Webb. 1983. Dynamic plant ecology: The spectrum of vegetational change in space and time. Quaternary Science Review 1:153–175.

D'Elia, J. 1998. Using geographic information systems to examine fire ignition patterns and fire danger in the arid and semi-arid western United States. University of Idaho, Moscow, Idaho, United States.

Delmelle, P., J. Stix, P. J. Baxter, J. Garcia-Alvarez, and J. Barquero. 2002. Atmospheric dispersion, environmental effects and potential health hazard associated with the low-altitude gas plume of Masaya volcano, Nicaragua. Bulletin of Vulcanology 64:423–434.

DeLong, S. C., and D. Tanner. 1996. Managing the pattern of forest harvest: Lessons from wildfire. Biodiversity and Conservation 5:1191–1205.

Dennis, F. C. undated. Fire-resistant landscaping. [Online, URL: <http://www.ext.colostate.edu/PUBS/NATRES/06303.html>.]

Densmore, F. 1974. *How Indians use wild plants for food, medicine and crafts.* Dover Publications, New York, New York, United States.

Despain, D. G. 1990. Yellowstone vegetation: Consequences of environment and history. Roberts Rinehart Publishing Company, Boulder, Colorado, United States.

Despain, D. G., and W. H. Romme. 1991. Ecology and management of high-intensity fires in Yellowstone National Park. Pages 43–58 *in* Proceedings of the 17th Tall Timbers fire ecology conference. Tall Timbers Research Station, Tallahassee, Florida, United States.

Dey, D. C., P. S. Johnson, and H. E. Garrett. 1996. Modeling the regeneration of oak stands in the Missouri Ozark Highlands. Canadian Journal of Forest Research 26:573–583.

Diaz, S., and M. Cabido. 2001. Vive la différence: Plant functional diversity matters to ecosystem processes. Trends in Ecology and Evolution 16:646–655.

Dickmann, D. I. 1993. Management of red pine for multiple benefits using prescribed fire. Northern Journal of Applied Forestry 10:53–72.

Doak, D. F., D. Bigger, E. K. Harding, M. A. Marvier, R. E. O'Malley, and D. Thomson. 1998. The statistical inevitability of stability–diversity relationships in community ecology. American Naturalist 151:264–276.

Donnelly, R. E., and J. B. Harrington. 1978. Forest fire history maps of Ontario. Canadian Forest Service, Ottawa, Ontario, Canada. Miscellaneous Report FF-Y-6.

Dorney, C. H., and J. R. Dorney. 1989. An unusual oak savanna in northeastern Wisconsin: The effect of Indian-caused fire. American Midland Naturalist 122:103–113.

Drapeau, P., A. Leduc, J. F. Giroux, J.-P. L. Savard, Y. Bergeron, and W. L. Vickery. 2000. Landscape-scale disturbances and changes in bird communities of boreal mixed-wood forests. Ecological Monographs 70:423–444.

Drapeau, P., A. Leduc, J.-P. Savard, and Y. Bergeron. 2001. Les oiseaux forestiers, des indicateurs des changements des mosaïques forestières boréales. Le Naturaliste Canadien 125:41–46.

Drapeau, P., A. Leduc, Y. Bergeron, S. Gauthier, and J.-P. Savard. 2003. Les communautés d'oiseaux des vieilles forêts de la pessière à mousses de la ceinture d'argile: Problèmes et solutions face à l'aménagement forestier. The Forestry Chronicle 79(3):531–540.

Driver, H. E. 1970. The Indians of North America, second revised edition. University of Chicago Press, Chicago, Illinois, United States.

Drobyshev, I. V. 1999. Regeneration of Norway spruce in canopy gaps in *Sphagnum–Myrtillus* old-growth forests. Forest Ecology and Management 115:71–83.

Dudley, N., and S. Stolton. 1999. Threats to forest protected areas. IUCN and World Bank–World Wildlife Foundation Alliance for Forest Conservation and Sustainable Use. IUCN (The World Conservation Union), Gland, Switzerland. Research Report.

Dunster, J., and K. Dunster. 1996. Dictionary of natural resource management. University of British Columbia Press, Vancouver, British Columbia, Canada.

Dunwiddie, P. W. 1986. A 6000-year record of forest history on Mount Rainier, Washington. Ecology 67:58–68.

Dunwiddie, P. W. 1987. Macrofossil and pollen representation of coniferous trees in modern sediments from Washington. Ecology 68:1–11.

Dutilleul, P., and P. Legendre. 1993. Spatial heterogeneity against heteroscedasticity: An ecological paradigm versus a statistical concept. Oikos 66:152–174.

Eberhart, K. E., and P. M. Woodard. 1987. Distribution of residual vegetation associated with large fires in Alberta. Canadian Journal of Forest Research 17:1207–1212.

Edenius, L., and J. Elmberg. 1996. Landscape level effects of modern forestry on bird communities in north Swedish boreal forests. Landscape Ecology 11:325–338.

Edwards, R. L., J. K. Agee, and R. I. Gara. 2000. Forest health and protection. McGraw-Hill, New York, New York, United States.

Egan, D., and E. A. Howell. 2001. The historical ecology handbook: A restorationist's guide to reference ecosystems. Island Press, Washington, D.C., United States.

Egler, F. E. 1954. Vegetation science concepts. I. Initial floristic composition—a factor in old field vegetation development. Vegetatio 4:412–418.

Eidt, D. C., and C. A. A. Weaver. 1986. Frequency of forest respraying and use of B.t. in New Brunswick, 1975–1986. Canadian Forestry Service, Fredericton, New Brunswick, Canada. Information Report M-X-158.

Eisenhart, K. S., and T. T. Veblen. 2000. Dendroecological detection of spruce bark beetle outbreak in northwestern Colorado. Canadian Journal of Forest Research 30:1788–1798.

Elbers, J. S. 1991. Changing wilderness values, 1930–1990: An annotated bibliography. Greenwood Press, New York, New York, United States.

Elkie, P. C., and R. S. Rempel. 2000. Detecting scales of pattern in boreal forest landscapes. Forest Ecology and Management 147:253–261.

Emlen, J. M. 1972. Ecology: An evolutionary approach. Addison-Wesley, Reading, Massachusetts, United States.

Environmental Systems Research Institute (ESRI). 1994. ArcInfo software. Environmental Systems Research Institute, Redlands, California, United States.

Environmental Systems Research Institute (ESRI). 1996. Arcview GIS 3.2. Environmental Systems Research Institute, San Diego, California, United States.

Environmental Systems Research Institute (ESRI). 1998. ArcInfo for Windows. Version 7.2.1. Environmental Systems Research Institute, Redlands, California, United States.

Erdle, T. A., and D. A. MacLean. 1999. Stand growth model calibration for use in forest pest impact assessment. The Forestry Chronicle 75:141–152.

Erdmann, G. G. 1986. Developing quality in second-growth stands. Pages 206–222 in G. D. Mroz and D. D. Reed, editors. Proceedings: The northern hardwood resource: Management potential. Michigan Technological University, Houghton, Michigan, United States.

Erdmann, G. G. 1987. Methods of commercial thinning in even-aged northern hardwoods. Pages 196–208 in R. D. Nyland, editor. Managing northern hardwoods: Proceedings of a silvicultural symposium. State University of New York, Syracuse, New York, United States. Faculty of Forestry Miscellaneous Publication 13 (ESF 87-002).

Erichsen-Brown, C. 1979. Medicinal and other uses of North American plants. Dover Publications, New York, New York, United States.

Eriksson, O. 1992. Evolution of seed dispersal and recruitment in clonal plants. Oikos 63:439–448.

Everett, R. L., R. Schellhaas, D. Keenum, D. Spurbeck, and P. Ohlson. 2000. Fire history in the ponderosa pine/Douglas-fir forests on the east slope of the Washington Cascades. Forest Ecology and Management 129:207–225.

Eyre, F. H., editor. 1980. Forest cover types of the United States and Canada. Society of American Foresters, Washington, D.C., United States.

Eyre, F. H., and P. Zehngraff. 1948. Red pine management in Minnesota. U.S. Department of Agriculture Forest Service, St. Paul, Minnesota, United States. Circular 778.

Fagre, D. B., D. L. Peterson, and A. E. Hessl. 2003. Taking the pulse of mountains: Ecosystem responses to climatic variability. Climatic Change 59(1–2):263–282.

Fall, J. G. 1998. Reconstructing the historical frequency of fire: A modeling approach to developing and testing methods. Simon Fraser University, Burnaby, British Columbia, Canada.

Farintosh, D. 1999. Forest management plan for Temagami Forest Management Unit 1999–2019. Ontario Ministry of Natural Resources, North Bay, Ontario, Canada.

Farnsworth, A., and P. Summerfelt. 2002. Flagstaff interface treatment prescription: Results in the wildland–urban interface. Fire Management Today 62:13–18.

Faustmann, M. 1849. On the determination of the value which forest land and immature stands possess for forestry. In M. Gane, editor. 1968. Martin Faustmann and the evolution of discounted cash flow. Commonwealth Forestry Institute, Oxford University, Oxford, United Kingdom. Institute Paper 42.

Ferguson, S. A. 1997. A climate-change scenario for the Columbia River Basin. U.S. Department of Agriculture Forest Service, Portland, Oregon, United States. Research Paper PNW-499.

Ferry, G. W., R. G. Clark, R. E. Montgomery, R. W. Mutch, W. P. Leenhouts, and G. T. Zimmerman. 1995. Altered fire regimes within fire-adapted ecosystems. U.S. Department of the Interior—National Biological Service, Washington, D.C., United States.

Finney, M. A. 1995. The missing tail and other considerations for the use of fire history models. International Journal of Wildland Fire 5:197–202.

Finney, M. A. 1998. FARSITE: Fire area simulator—model development and evaluation. U.S. Department of Agriculture Forest Service, Fort Collins, Colorado, United States. Research Paper RMRS-RP-4.

Finney, M. A. 1999. Mechanistic modeling of landscape fire patterns. Pages 186–209 in D. J. Mladenoff and W. L. Baker, editors. Spatial modeling of forest landscape change: Approaches and applications. Cambridge University Press, Cambridge, United Kingdom.

Firewise. undated. [Online, URL: <http://www.firewise.org/pubs/wnn/>.]

Fischer, W. C., and A. F. Bradley. 1987. Fire ecology of western Montana forest habitat types. U.S. Department of Agriculture Forest Service, Missoula, Montana, United States. General Technical Report GTR-INT-223.

Fisher, J. T. 2002. Adaptive boreal forestry: Anticipation and experimentation. Pages 12–15 in S. J. Song, editor. Ecological basis for stand management: A synthesis of ecological responses to wildfire and harvesting. Alberta Research Council, Vegreville, Alberta, Canada.

Fitzsimmons, A. K. 1996. Sound policy or smoke and mirrors: Does ecosystem management make sense? Water Resources Bulletin 32:217–227.

Flannigan, M. 1993. Fire regime and the abundance of red pine. International Journal of Wildland Fire 3:241–247.

Flannigan, M. D., and C. E. Van Wagner. 1991. Climate change and wildfire in Canada. Canadian Journal of Forest Research 21:66–72.

Flannigan, M. D., Y. Bergeron, O. Engelmark, and B. M. Wotton. 1998. Future wildfire in circumboreal forests in relation to global warming. Journal of Vegetation Science 9:469–476.

Fleming, K. K., and W. M. Giuliano. 1998. Effect of border-edge cuts on birds at woodlot edges in southwestern Pennsylvania. Journal of Wildlife Management 62:1430–1437.

Fleming, K. K., and W. M. Giuliano. 2001. Reduced predation of artificial nests in border-edge cuts on woodlots. Journal of Wildlife Management 65:351–355.

Fleming, R. A., C. A. Shoemaker, and J. R. Stedinger. 1984. An assessment of the impact of large-scale spraying operations on the regional dynamics of spruce budworm (Lepidoptera: Tortricidae) populations. The Canadian Entomologist 116:633–644.

Fleming, R. A., A. A. Hopkin, and J.-N. Candau. 2000. Insect and disease disturbance regimes in Ontario's forests. Pages 141–162 in A. H. Perera, D. L. Euler, and I. D. Thompson, editors. Ecology of a managed terrestrial landscape: Patterns and processes of forest landscapes in Ontario. University of British Columbia Press, Vancouver, British Columbia, Canada.

Fleming, R. A., J.-N. Candau, and R. S. McAlpine. 2002. Landscape-scale analysis of interactions between insect defoliation and forest fire in central Canada. Climate Change 55:251–272.

Flemming, S. P., G. L. Holloway, E. J. Watts, and P. S. Lawrance. 1999. Characteristics of foraging trees selected by pileated woodpeckers in New Brunswick. Journal of Wildlife Management 63:461–469.

Flieger, B. W. 1970. Forest fire and insects: The relation of fire to insect outbreak. Proceedings of the Tall Timbers Fire Ecology Conference 10:107–114.

Forest Ecosystem Management and Assessment Team. 1993. Forest ecosystem management: An ecological, economic, and social assessment. U.S. Departments of Agriculture, Commerce, and the Interior and the Environmental Protection Agency, Washington, D.C., United States.

Forestry Canada Fire Danger Group. 1992. Development and structure of the Canadian Forest Fire Behaviour Prediction System. Forestry Canada, Ottawa, Ontario, Canada. Information Report ST-X-3.

Forman, R. T. T. 1995. Land mosaics. Cambridge University Press, Cambridge, United Kingdom.

Foster, D. H., D. H. Knight, and J. F. Franklin. 1998. Landscape patterns and legacies resulting from large, infrequent forest disturbances. Ecosystems 1:497–510.

Foster, D. R. 1985. Vegetation development following fire in Picea mariana (black spruce)–Pleurozium forests of south-eastern Labrador, Canada. Journal of Ecology 73:517–534.

Foster, D. R., J. D. Aber, J. M. Melillo, R. D. Bowden, and F. Bazzaz. 1997. Forest response to disturbance and anthropogenic stress. BioScience 47:437–445.

Franklin, I. R. 1980. Evolutionary change in small populations. Pages 135–150 in M. E. Soulé and B. A. Wilcox, editors. Conservation biology: An evolutionary perspective. Sinauer Associates, Sunderland, Massachusetts, United States.

Franklin, J., A. D. Syphard, D. J. Mladenoff, H. S. He, D. K. Simons, R. P. Martin, D. Deutschman, and J. F. O'Leary. 2001. Simulating the effects of different fire regimes on plant functional groups in southern California. Ecological Modelling 142:261–283.

Franklin, J. F. 1989. Toward a new forestry. American Forests 95:37–44.

Franklin, J. F. 1993. Preserving biodiversity: Species, ecosystems or landscapes. Ecological Applications 3: 202–205.

Franklin, J. F., and C. T. Dyrness. 1988. Natural vegetation of Oregon and Washington. Oregon State University Press, Corvallis, Oregon, United States.

Franklin, J. F., K. Cromack, W. Denison, A. McKee, C. Maser, J. Sedell, F. Swanson, and G. Juday. 1981. Ecological characteristics of old-growth Douglas-fir forests. U.S. Department of Agriculture Forest Service, Portland, Oregon, United States. General Technical Report PNW-GTR-118.

Franklin, J. F., D. R. Berg, D. A. Thornburgh, and J. C. Tappeiner. 1997. Alternative silvicultural approaches to timber harvesting: Variable retention harvest systems. Pages 111–139 in K. A. Kohm and J. F. Franklin, editors. Creating a forestry for the 21st century. Island Press, Washington, D.C., United States.

Franklin, J. F., T. A. Spies, R. Van Pelt, A. B. Carey, D. A. Thornburgh, D. R. Berg, D. B. Lindenmeyer, M. E. Harmon, W. S. Keeton, D. C. Shaw, K. Bible, and J. Chen. 2002. Disturbances and structural development of natural forest ecosystems with silvicultural implications, using Douglas-fir forests as an example. Forest Ecology and Management 155:399–423.

Frelich, L. E. 2002. Forest dynamics and disturbance regimes—studies from temperate evergreen–deciduous forests. Cambridge University Press, Cambridge, United Kingdom.

Frelich, L. E., and C. G. Lorimer. 1991. Natural disturbance regimes in hemlock–hardwood forests of the Upper Great Lakes region. Ecological Monographs 61:145–164.

Frelich, L. E., and P. B. Reich. 1995. Spatial patterns and succession in a Minnesota southern-boreal forest. Ecological Monographs 65:325–346.

Friederici, P., editor. 2003. Ecological restoration of southwestern Ponderosa pine forests: A sourcebook for research and application. Island Press, Washington, D.C., United States.

Friedman, J., A. Hutchins, C. Y. Li, and D. A. Perry. 1989. Actinomycetes inducing phytotoxic or fungistatic activity in a Douglas-fir forest and in an adjacent area of repeated regeneration failure in southwestern Oregon. Biologia Plantarum 31:487–495.

Friedman, S. 1997. Forest genetics for ecosystem management. Pages 203–211 in K. A. Kohm and J. F. Franklin, editors. Creating a forestry for the 21st century. Island Press, Washington, D.C., United States.

Fritz, R., R. Suffling, and T. A. Younger. 1993. Influence of fur trade, famine and forest fires on moose and woodland caribou populations in northwestern Ontario from 1786 to 1911. Environmental Management 17: 477–489.

Frost, C. C. 1998. Presettlement fire frequency regimes of the United States: A first approximation. Tall Timbers Fire Ecology Conference 20:70–81.

Führer, E. 2000. Forest functions, ecosystem stability and management. Forest Ecology and Management 132: 29–38.

Furyaev, V. V., R. W. Wein, and D. A. MacLean. 1983. Fire influences in *Abies*-dominated forests. Pages 221–234 in R. W. Wein and D. A. MacLean, editors. The role of fire in northern circumpolar ecosystems. SCOPE 18. John Wiley and Sons, Chichester, United Kingdom.

Galatowitsch, S. M. 1990. Using the original land survey notes to reconstruct presettlement landscapes in the American west. Great Basin Naturalist 50:181–191.

Gardner, J. S. 1978. The meaning of wilderness: A problem of definition. Contact 10:7–33.

Gardner, R. H., and R. V. O'Neill. 1991. Neutral models for landscape analysis. Chapter 11 in M. G. Turner and R. H. Gardner, editors. Quantitative methods in landscape ecology. Springer-Verlag, New York, New York, United States.

Gardner, R. H., W. W. Hargrove, M. G. Turner, and W. H. Romme. 1996. Climate change, disturbances and landscape dynamics. Pages 149–172 in B. H. Walker and W. L. Steffen, editors. Global change and terrestrial ecosystems. Cambridge University Press, Cambridge, Massachusetts, United States.

Gardner, R. H., W. H. Romme, and M. G. Turner. 1999. Predicting forest fire effects at landscape scales. Pages 163–185 in D. J. Mladenoff and W. L. Baker, editors. Spatial modeling of forest landscape change: Approaches and applications. Cambridge University Press, Cambridge, United Kingdom.

Gauthier, S., A. Leduc, and Y. Bergeron. 1996. Vegetation modelling under natural fire cycles: A tool to define natural mosaic diversity for forest management. Environmental Monitoring and Assessment 39:417–434.

Gauthier, S., L. de Grandpré, and Y. Bergeron. 2000. Differences in forest composition in two ecoregions of the boreal forest of Québec. Journal of Vegetation Science 11:781–790.

Gauthier, S., P. Lefort, Y. Bergeron, and P. Drapeau. 2002. Time since fire map, age-class distribution and forest dynamics in the Lake Abitibi Model Forest. Canadian Forest Service, Ste-Foy, Quebec, Canada. Information Report LAU-X-125.

Gavin, D. G., and L. B. Brubaker 1999. A 6000-year soil pollen record of subalpine meadow vegetation in the Olympic Mountains, Washington, USA. Journal of Ecology 87:106–122.

Gedalof, Z., and D. J. Smith. 2001. Interdecadal climate variability and regime-scale shifts in Pacific North America. Geophysical Research Letters 28:1515–1518.

Gentilecore, L., and K. Donkin. 1973. Land surveys of southern Ontario. An introduction and index to the field notebooks of the Ontario Land Surveyors 1784–1859. Canadian Cartographer Volume 10, Supplement #2.

Gilbert F., A. Gonzalez, and I. Evans-Freke. 1998. Corridors maintain species richness in the fragmented landscapes of a microecosystem. Proceedings of the Royal Society of London B. 265:577–582.

Gill, A. M. 1998. A hierarchy of fire effects: impact of fire regimes on landscapes. Pages 129–144 in Third international conference on forest fire research, 14th conference on fire and forest meteorology, Luso, Portugal. Society of American Foresters, Bethesda, Maryland, United States.

Gill, A. M., and M. A. McCarthy. 1998. Intervals between prescribed fires in Australia: What intrinsic variation should apply? Biological Conservation 85:161–169.

Gillis, A. M. 1990. The new forestry: An ecosystem approach to land management. BioScience 40:558–562.

Glaubitz, J. C., J. Stark, G. F. Moran, and C. Martyas. 2000. Genetic impacts of different silvicultural practices in native eucalypt forests. Pages 183–195 in C. Martyas, editor. Forest genetics and sustainability. Kluwer Academic Publishers, Dordrecht, the Netherlands.

Gleason, H. A. 1927. Further views on the succession concept. Ecology 8:299–326.

Gleason, H. A. 1939. The individualization concept of the plant association. American Midland Naturalist 21: 92–110.

Glick, D., M. Carr, and B. Harting. 1991. An environmental profile of the Greater Yellowstone ecosystem. Greater Yellowstone Coalition, Bozeman, Montana, United States.

Gluck, M. J., and R. S. Rempel. 1996. Structural characteristics of post-wildfire and clearcut landscapes. Environmental Monitoring and Assessment 39:435–450.

Goldstein, P. Z. 1999. Functional ecosystems and biodiversity buzzwords. Conservation Biology 13:247–255.

Government of Alberta. 1997. Alberta forest conservation strategy: A new perspective on sustaining Alberta's forests. Government of Alberta, Edmonton, Alberta, Canada. [Online, URL: <http://www.borealcentre.ca/reports/afcs.html>.]

Graham, S. A. 1923. The dying balsam fir and spruce in Minnesota. University of Minnesota Agricultural Extension Division, St. Paul, Minnesota, United States. Special Bulletin 68.

Graham, S. A., and L. W. Orr. 1940. The spruce budworm in Minnesota. University of Minnesota Agricultural Experiment Station, St. Paul, Minnesota, United States. Technical Bulletin 142.

Granström, A. 2001. Fire management for biodiversity in the European boreal forest. Scandinavian Journal of Forest Research Supplement 3:62–69.

Gray, D., J. Régnière, and B. Boulet. 1998. Forecasting defoliation by spruce budworm in Quebec. Canadian Forest Service, Ste.-Foy, Quebec, Canada. Research Note Number 7.

Green, D. G. 2000. Self-organisation in complex systems. Pages 11–50 in R. J. Bossomaier and D. G. Green, editors. Complex systems. Cambridge University Press, Cambridge, United Kingdom.

Greene, D. F., and E. A. Johnson. 2000. Tree recruitment from burn edges. Canadian Journal of Forest Research 30:1264–1274.

Greene, D. F., J. C. Zasada, L. Sirois, D. Kneeshaw, H. Morin, I. Charron, and M. J. Simard. 1999. A review of the regeneration dynamics of North American boreal forest tree species. Canadian Journal of Forest Research 29:824–839.

Grieve, R., and A. Therriault. 2000. Vredefort, Sudbury, Chicxulub: Three of a kind? Annual Review of Earth and Planetary Sciences 28:305–338.

Grimm, E. C. 1984. Fire and other factors controlling the Big Woods vegetation of Minnesota in the mid-nineteenth century. Ecological Monographs 54:291–311.

Grissino-Meyer, H. D. 1995. Tree-ring reconstructions of climate and fire history at El Malpais National Monument, New Mexico. Ph.D. dissertation, University of Arizona, Tucson, Arizona, United States.

Grissino-Meyer, H. D. 1999. Modelling fire interval data from the American Southwest with the Weibull distribution. International Journal of Wildland Fire 9:37–50.

Grissino-Meyer, H. D., and T. W. Swetnam. 2000. Century-scale climatic forcing of fire regimes in the American Southwest. Holocene 10:213–220.

Grumbine, R. E. 1994. What is ecosystem management? Conservation Biology 8:27–38.

Guariguata, M. R., and M. A. Pinard. 1998. Ecological knowledge of regeneration from seed in neotropical forest areas: Implications for natural forest management. Forest Ecology and Management 112:87–99.

Gullion, G. W. 1990. Management of aspen for ruffed grouse and other wildlife—an update. Pages 133–143 in R. D. Adams, editor. Proceedings: Aspen symposium '89. U.S. Department of Agriculture Forest Service, St. Paul, Minnesota, United States. General Technical Report NC-140.

Gurd, D. B., T. D. Nudds, and D. H. Rivard. 2001. Conservation of mammals in eastern North American wildlife reserves: How small is too small? Conservation Biology 15:1355–1363.

Gustafson, E. J. 1996. Expanding the scale of forest management: Allocating timber harvests in time and space. Forest Ecology and Management 87:27–39.

Gustafson, E. J., S. R. Shifley, D. J. Mladenoff, K. K. Nimerfro, and H. S. He. 2000. Spatial simulation of forest succession and harvesting using LANDIS. Canadian Journal of Forest Research 30:32–43.

Gustafson, E. J., P. A. Zollner, H. S. He, B. R. Sturtevant, and D. J. Mladenoff. in press. Influence of forest management alternatives and land type on susceptibility to fire in northern Wisconsin, USA. Landscape Ecology.

Guyette, R. P. 1995. A tree-ring history of wildland fire in the Current River Watershed. Missouri Department of Conservation, Columbia, Missouri, United States.

Guyette, R. P., and D. C. Dey. 1997. Historic shortleaf pine (Pinus echinata) abundance and fire frequency in a mixed oak–pine forest. Pages 136–149 in B. Bookshire and S. Shifley, editors. Proceedings of the Missouri Ozark Forest Ecosystem Project symposium. U.S. Department of Agriculture Forest Service, St. Paul, Minnesota, United States. General Technical Report NC-193.

Guyette, R. P., R. M. Muzika, and D. C. Dey. 2002. Dynamics of an anthropogenic fire regime. Ecosystems 5:472–486.

Habeck, J. R. 1994. Using General Land Office records to assess forest succession in ponderosa pine–Douglas-fir forests in western Montana. Northwest Science 68: 69–78.

Haila, Y. 1994. Preserving ecological diversity in boreal forests: Ecological background, research, and management. Annales Zoologica Fennici 31:203–217.

Haila, Y., I. K. Hanski, J. Niemela, P. Punttila, S. Raivo, and H. Tukia. 1994. Forestry and the boreal fauna: Matching management with natural forest dynamics. Annales Zoologica Fennici 31:187–202.

Hann, W. J., J. L. Jones, M. G. Karl, P. F. Hessburg, R. E. Keane, D. G. Long, J. P. Menakis, C. H. McNicoll, S. G. Leonard, R. A. Gravenmeier, and B. G. Smith. 1997. An assessment of landscape dynamics of the Interior Columbia River Basin. U.S. Department of Agriculture Forest Service, Portland, Oregon, United States. General Technical Report PNW-GTR-405.

Hansen, A. J., T. A. Spies, F. J. Swanson, and J. L. Ohmann. 1991. Conserving biodiversity in managed forests. Bioscience 41:382–392.

Hansen, A. J., J. J. Rotella, M. L. Kraska, and D. Brown. 2000. Spatial patterns of primary productivity in the

Greater Yellowstone ecosystem. Landscape Ecology 15: 505–522.

Hansen, A. J., R. Rasker, B. Maxwell, J. J. Rotella, J. D. Johnson, A. W. Parmenter, U. Langner, W. B. Cohen, R. L. Lawrence, and M. P. V. Kraska. 2002. Ecological causes and consequences of demographic change in the New West. BioScience 52:151–162.

Hansen, M. H., T. Frieswyk, J. F. Glover, and J. F. Kelly. 1992. The eastwide forest inventory data base: User's manual. U.S. Department of Agriculture Forest Service, St. Paul, Minnesota, United States. General Technical Report NC-151.

Hao, Z. Q., H. S. He, L. M. Dai, D. J. Mladenoff, and G. F. Shao. 2001. Responses of individual tree species of Changbai Mountain (China) to potential global warming. Chinese Journal of Applied Ecology 12:653–658.

Hardy, C. C., and S. F. Arno, editors. 1996. The use of fire in forest restoration. U.S. Department of Agriculture Forest Service, Ogden, Utah, United States. General Technical Report INT-GTR-341.

Hardy, C. C., K. M. Schmidt, J. P. Menakis, and N. R. Sampson. 2001. Spatial data for national fire planning and fuel management. International Journal of Wildland Fire 10:353–372.

Harestad, A. S., and F. L. Bunnell. 1979. Home range and body weight—a reevaluation. Ecology 60:389–402.

Harestad, A. S., and D. G. Keisker. 1989. Nest tree use by primary cavity-nesting birds in south-central British Columbia. Canadian Journal of Zoology 67: 1067–1073.

Harestad, A. S., and G. D. Sutherland. 2001. Dispersal by mammals and forest fragmentation. Pages 93–96 in R. Field, R. J. Warren, H. Okarma, and P. R. Sievert, editors. Wildlife, land and people: Priorities for the 21st century. Proceedings of the second international wildlife management congress. The Wildlife Society, Bethesda, Maryland, United States.

Harper, J. L. 1977. Population biology of plants. Academic Press, New York, New York, United States.

Harper, K. A., and S. E. Macdonald. 2001. Structure and composition of riparian boreal forest: New methods for analyzing edge influence. Ecology 82:649–659.

Harper, K. A., Y. Bergeron, S. Gauthier, and P. Drapeau. 2002. Post-fire development of canopy structure and composition in black spruce forests of Abitibi, Québec: A landscape scale study. Silva Fennica 36: 249–263.

Harper, K., C. Boudreault, L. de Grandpré, P. Drapeau, S. Gauthier, and Y. Bergeron. 2003. Structure, composition and diversity of old-growth black spruce boreal forest of the Clay Belt region in Québec and Ontario. Environmental Reviews 11(S1):S79–S98.

Harrington, M. G. 1985. The effects of spring, summer, and fall burning on Gambel oak in a southwestern ponderosa pine stand. Forest Science 31:156–163.

Harris, L. D. 1984. The fragmented forest: Island biogeography theory and the preservation of biotic diversity. University of Chicago Press, Chicago, Illinois, United States.

Harrison, D. J., and T. G. Chapin. 1998. Extent and connectivity of habitat for wolves in eastern North America. Wildlife Society Bulletin 26:767–775.

Harrison, S., and E. Bruna. 1999. Habitat fragmentation and large-scale conservation: What do we know for sure? Ecography 22:225–232.

Hart, R. H., and W. A. Laycock. 1996. Repeat photography on range and forest lands in the western United States. Journal of Range Management 49:60–67.

Hartman, R. 1976. The harvesting decision when a standing forest has value. Economic Inquiry 14:52–58.

Harvey, B. D., A. Leduc, S. Gauthier, and Y. Bergeron. 2002. Stand–landscape integration in natural disturbance-based management of the southern boreal forest. Forest Ecology and Management 155:369–385.

Hasenstab, R. J. 1990. Agriculture, warfare, and tribalization in the Iroquois homeland of New York: A GIS analysis of late woodland settlement. Ph.D. dissertation. University of Massachusetts. UMI Dissertation Services, Ann Arbor, Michigan, United States.

He, H. S., and D. J. Mladenoff. 1999a. Spatially explicit and stochastic simulation of forest landscape fire and succession. Ecology 80:80–99.

He, H. S., and D. J. Mladenoff. 1999b. The effects of seed dispersal on the simulation of long-term forest landscape change. Ecosystems 2:308–319.

He, H. S., D. J. Mladenoff, and J. Boeder. 1999. An object-oriented forest landscape model and its representation of tree species. Ecological Modelling 119:1–19.

He, H. S., B. DeZonia, and D. J. Mladenoff. 2000. An aggregation index to quantify spatial patterns of landscapes. Landscape Ecology 15:590–602.

He, H. S., D. J. Mladenoff, and E. J. Gustafson. 2002. Study of landscape change under forest harvesting and climate warming–induced fire disturbance. Forest Ecology and Management 155:257–270.

Hearnden, K. W., S. V. Millson, and W. C. Wilson. 1992. A report on the status of forest regeneration in Ontario: Independent forest audit. Ontario Ministry of Natural Resources, Toronto, Ontario, Canada.

Hebert, D. 2002. Expectations and consequences of emulating natural forest disturbance: A forest utilization perspective. Pages 55–56 in L. J. Buse and A. H. Perera, compilers. Emulating natural forest landscape disturbances: Concepts and applications popular summaries. Ontario Ministry of Natural Resources, Sault Ste. Marie, Ontario, Canada. Forest Research Information Paper 149.

Hector, A., J. Joshi, S. P. Lawlor, E. M. Spehn, and A. Wilby. 2001. Conservation implications of the link between biodiversity and ecosystem functioning. Oecologia 129:624–628.

Hedrick, P. W. 1995. Gene flow and genetic restoration: The Florida panther as a case study. Conservation Biology 9:996–1007.

Hedrick, P. W., and P. S. Miller. 1992. Conservation genetics: Techniques and fundamentals. Ecological Applications 2:30–46.

Heidenreich, C. E. 1970. The historical geography of Huronia in the first half of the seventeenth century. Ph.D. dissertation. McMaster University, Hamilton, Ontario, Canada.

Heilman, P. E. 1966. Change in the distribution of nitrogen with forest succession on north slopes in interior Alaska. Ecology 49:825–831.

Heilman, P. E. 1968. Relationship of availability of phosphorus and cations to forest succession and bog formation in interior Alaska. Ecology 49:331–336.

Heinselman, M. L. 1973. Fire in the virgin forests of the Boundary Waters Canoe Area. Quaternary Research 3: 329–382.

Heinselman, M. L. 1981. Fire and succession in the conifer forests of North America. Pages 375-405 *in* D. C. West, H. H. Shugart, and D. B. Botkin, editors. Forest succession: Concepts and applications. Springer-Verlag, New York, New York, United States.

Heinselman, M. L. 1996. The Boundary Waters Wilderness ecosystem. University of Minnesota Press, Minneapolis, Minnesota, United States.

Hemstrom, M. A., and J. F. Franklin. 1982. Fire and other disturbances of the forests of Mt. Rainier National Park. Quaternary Research 18:32-51.

Henderson, J. A., D. H. Peter, R. D. Lesher, and D. C. Shaw. 1989. Forested plant associations of the Olympic National Forest. U.S. Department of Agriculture Forest Service, Pacific Northwest Research Station, Portland, Oregon, United States. Technical Paper 001-88 R6.

Henig-Sever, N., D. Poliakov, and M. Broza. 2001. A novel method for estimation of wildfire intensity based on ash pH and soil microarthropod community. Pedobiologia 45:98-106.

Hessburg, P. F., and J. K. Agee. in press. An environmental narrative of Inland Northwest U.S. forests, 1800-2000. Forest Ecology and Management.

Hessburg, P. F., B. G. Smith, S. D. Kreiter, C. A. Miller, R. B. Salter, C. H. McNicholl, and W. J. Hann. 1999a. Historical and current forest and range landscapes in the interior Columbia River Basin and portions of the Klamath and Great Basins. Part I: Linking vegetation patterns and landscape vulnerability to potential insect and pathogen disturbances. U.S. Department of Agriculture Forest Service, Portland, Oregon, United States. General Technical Report PNW-GTR-458.

Hessburg, P. F., B. G. Smith, and R. B. Salter. 1999b. Detecting change in forest spatial patterns. Ecological Applications 9:1232-1252.

Hessburg, P. F., R. B. Salter, M. B. Richmond, and B. G. Smith. 2000a. Ecological subregions of the Interior Columbia Basin, USA. Applied Vegetation Science 3:163-180.

Hessburg, P. F., B. G. Smith, S. D. Kreiter, C. A. Miller, C. H. McNicholl, and M. Wasienko-Holland. 2000b. Classifying plant series-level forest potential vegetation types: Methods for subbasins sampled in the midscale assessment of the interior Columbia Basin. U.S. Department of Agriculture Forest Service, Portland, Oregon, United States. Research Paper PNW-RP-524.

Hessburg, P. F., B. G. Smith, R. B. Salter, R. D. Ottmar, and E. Alvarado. 2000c. Recent changes (1930s-1990s) in spatial patterns of interior northwest forests, USA. Forest Ecology and Management 136:53-83.

Hessl, A. E., D. McKenzie, and R. Schellhaas. in press. Drought and Pacific Decadal Oscillation linked to fire occurrence in the inland Pacific Northwest. Ecological Applications.

Heyerdahl, E. K. 1997. Spatial and temporal variation in historical fire regimes of the Blue Mountains, Oregon and Washington: The influence of climate. Ph.D. dissertation. University of Washington, Seattle, Washington, United States.

Heyerdahl, E. K., D. Berry, and J. K. Agee. 1995. Fire history database of the western United States. U.S. Environmental Protection Agency, Washington, D.C., United States. Report EPA/600/R-96/081 NHEERL-COR-851.

Heyerdahl, E. K., L. B. Brubaker, and J. K. Agee. 2001. Spatial controls of historical fire regimes: A multiscale example from the interior West, USA. Ecology 82:660-678.

Heyerdahl, E. K., L. B. Brubaker, and J. K. Agee. 2002. Annual and decadal climate forcing of historical fire regimes in the interior Pacific Northwest, USA. Holocene 12:597-604.

Hill, M. O. 1973. Diversity and evenness: A unifying notation and its consequences. Ecology 54:427-431.

Hills, G. A. 1959. A ready reference to the description of the land of Ontario and its productivity. Ontario Department of Lands and Forests, Toronto, Ontario, Canada. Preliminary Research Report. (Compendium of maps, charts, tables, and brief comments.)

Hills, G. A. 1966. The ecological basis for land-use planning. Ontario Department of Lands and Forests, Maple, Ontario, Canada. Research Report 46.

Hilmo, O., and S. M. Såstad. 2001. Colonization of old-forest lichens in a young and an old boreal *Picea abies* forest: An experimental approach. Biological Conservation 102:251-259.

Hirsch, K. G. 1996. Canadian forest fire behaviour prediction (FBP) system: User's guide. Canadian Forest Service, Edmonton, Alberta, Canada. Special Report 7.

Hobbs, R. J., and L. F. Huenneke. 1992. Disturbance, diversity, and invasion: Implications for conservation. Conservation Biology 6:324-327.

Hobson, K. A., and E. Bayne. 2000. The effects of stand age on avian communities in aspen-dominated forests of central Saskatchewan, Canada. Forest Ecology and Management 136:121-134.

Hobson, K. A., and J. Schieck. 1999. Changes in bird communities in boreal mixedwood forest: Harvest and wildfire effects over 30 years. Ecological Applications 9:849-863.

Hof, J. 1993. Coactive forest management. Academic Press, San Diego, California, United States.

Hogg, E. H., J. P. Brandt, and B. Kochtubajda. 2002. Growth and dieback of aspen forests in northwestern Alberta, Canada, in relation to climate and insects. Canadian Journal of Forest Research 35:823-832.

Holling, C. S. 1973. Resilience and stability of ecological systems. Annual Review of Ecology and Systematics 4:1-23.

Holling, C. S., editor. 1978. Adaptive environmental assessment and management. International series on applied systems analysis 3. John Wiley and Sons, Toronto, Ontario, Canada.

Holling, C. S. 1981. Forest insects, forest fires, and resilience. Pages 445-463 *in* H. A. Mooney, T. M. Bonnicksen, N. L. Christensen, J. E. Lotan, and W. A. Reiners, technical coordinators. Proceedings of the conference on fire regimes and ecosystem properties. U.S. Department of Agriculture Forest Service, Washington, D.C. General Technical Report WO-26.

Holling, C. S. 1986. The resilience of terrestrial ecosystems: Local surprise and global change. Pages 292-317 *in* W. C. Clark and R. E. Munn, editors. Sustainable development of the biosphere. Cambridge University Press, Cambridge, United Kingdom.

Holling, C. S. 1992a. Cross-scale morphology, geometry, and dynamics of ecosystems. Ecological Monographs 62:447-502.

Holling, C. S. 1992b. The role of forest insects in structuring the boreal landscape. Pages 170–1991 *in* H. H. Shugart, R. Leemans, and G. B. Bonan, editors. A systems analysis of the global boreal forest. Cambridge University Press, Cambridge, United Kingdom.

Holling, C. S. 1998. Two cultures of ecology. Conservation Ecology 2:4. [Online, URL: <http://www.consecol.org/vol2/iss2/art4/>.]

Holling, C. S., and G. K. Meffe. 1996. Command and control and the pathology of natural resource management. Conservation Biology 10:328–337.

Holling, C. S., G. Peterson, P. Marples, J. Sendzimir, K. Redford, L. Gunderson, and D. Lambert. 1996. Self-organization in ecosystems: Lumpy geometries, periodicities, and morphologies. Pages 346–384 *in* B. H. Walker and W. L. Steffen, editors. Global change and terrestrial ecosystems. Cambridge University Press, Cambridge, United Kingdom.

Houston, D. B. 1973. Wildfires in northern Yellowstone National Park. Ecology 54:1111–1117.

Hubbell, S. P. 2001. The unified neutral theory of biodiversity and biogeography. Monographs in Population Biology 32. Princeton University Press, Princeton, New Jersey, United States.

Huebert, B., P. Vitousek, T. Elias, J. Heath, S. Coeppicus, S. Howell, and B. Blomquist. 1999. Volcano fixes nitrogen into plant-available forms. Biogeochemistry 47:111–118.

Huff, M. H. 1984. Postfire succession in the Olympic Mountains, Washington: Forest vegetation, fuels, and avifauna. Ph.D. dissertation. University of Washington, Seattle, Washington, United States.

Huff, M. H., R. D. Ottmar, E. Alvarado, R. E. Vihnanek, J. F. Lehmkuhl, P. F. Hessburg, and R. L. Everett. 1995. Historical and current forest landscapes of eastern Oregon and Washington. Part II: Linking vegetation characteristics to potential fire behavior and related smoke production. U.S. Department of Agriculture Forest Service, Portland, Oregon, United States. General Technical Report PNW-GTR-355.

Humphries, H. C., and P. S. Bourgeron. 2001. Methods for determining historical range of variability. Pages 273–292 *in* M. E. Jensen and P. S. Bourgeron, editors. A guide book for integrated ecological assessments. Springer-Verlag, New York, New York, United States.

Hunter, M. L. 1989. What constitutes an old-growth stand? Journal of Forestry 87:33–36.

Hunter, M. L. 1990. Wildlife, forests and forestry: Principles of managing forests for biodiversity. Prentice-Hall, Englewood Cliffs, New Jersey, United States.

Hunter, M. L. 1993. Natural fire regimes as spatial models for managing boreal forests. Biological Conservation 65:115–120.

Hunter, M. L., G. L. Jacobson Jr., and T. Webb. 1988. Paleoecology and the coarse-filter approach to maintaining biological diversity. Conservation Biology 2: 375–385.

Hunter, M. L., Jr., editor. 1999. Maintaining biodiversity in forest ecosystems. Cambridge University Press, Cambridge, United Kingdom.

Huston, M. A. 1994. Biological diversity: The coexistence of species on changing landscapes. Cambridge University Press, Cambridge, United Kingdom.

Hutchinson, M. 1988. A guide to understanding, interpreting, and using the Public Land Survey field notes in Illinois. Natural Areas Journal 8:245–255.

Hutto, R. L. 1995. Composition of bird communities following stand-replacement fires in northern Rocky Mountains (USA) conifer forests. Conservation Biology 9:1041–1058.

Hyde, W. F. 1980. Timber supply, land allocation and economic efficiency. Resources for the future. Johns Hopkins University Press, Baltimore, Maryland, United States.

Hyde, W. F. 1989. Marginal costs of managing endangered species: The case of the red-cockaded woodpecker. Journal of Agricultural Economics Research 41:12–19.

Imbeau, L., and A. Desrochers. 2002. Old-growth spruce forests (>120 years), an essential habitat for the three-toed woodpecker. Page 30 *in* Old growth forests in Canada: A science perspective. Workshop abstract booklet. 15–19 October 2001, Sault Ste. Marie, Ontario. Natural Resources Canada–Canadian Forest Service, Sault Ste. Marie, Ontario, Canada.

Imbeau, L., J.-P. L. Savard, and R. Gagnon. 1999. Comparing bird assemblages in successional black spruce stands originating from fire and logging. Canadian Journal of Zoology 77:1850–1860.

Imbeau, L., M. Monkkonen, and A. Desrochers. 2001. Long-term effects of forestry on birds of the eastern Canadian boreal forests: A comparison with Fennoscandia. Conservation Biology 15:1151–1162.

Impara, P. C. 1997. Spatial and temporal patterns of fire in the forests of the central Oregon Coast Range. Ph.D. dissertation. Oregon State University, Corvallis, Oregon, United States.

International Institute for Applied Systems Analysis. 2001. IIASA population projection results. International Institute for Applied Systems Analysis, Laxenburg, Austria. [Online, URL <http://www.iiasa.ac.at/Research/POP/index.html>.]

International Union for the Conservation of Nature and Natural Resources. 1991. A strategy for Antarctic conservation. International Union for the Conservation of Nature and Natural Resources, Gland, Switzerland.

Irwin, L. L., D. F. Rock, and G. P. Miller. 2000. Stand structures used by northern spotted owls in managed forests. Journal of Raptor Research 44:175–186.

Isebrands, J. G., E. P. McDonald, E. Kruger, G. Hendry, K. Percy, K. Pregitzer, J. Sober, and D. F. Karnosky. 2001. Growth responses of *Populus tremuloides* clones to interacting elevated carbon dioxide and tropospheric ozone. Environmental Pollution 115:359–371.

Jackson, S. M., F. Pinto, J. R. Malcolm, and E. R. Wilson. 2000. A comparison of pre-European settlement (1857) and current (1981–1995) forest composition in central Ontario. Canadian Journal of Forest Research 30:605–612.

Jakob, M. 2000. The impacts of logging on landslide activity at Clayquot Sound, British Columbia. Catena 38:279–300.

Jentsch, A., C. Beierkuhnlein, and P. S. White. 2002. Scale, the dynamic stability of forest ecosystems, and the persistence of biodiversity. Silva Fennica 36:393–400.

Johansson, P. O., and K. Lofgren. 1985. The economics of forestry and natural resources. Basil Blackwell, Oxford, United Kingdom.

Johnson, E. A. 1992. Fire and vegetation dynamics: Studies from the North American boreal forest. Cambridge University Press, Cambridge, United Kingdom.

Johnson, E. A., and S. L. Gutsell. 1994. Fire frequency models, methods, and interpretations. Advances in Ecological Research 25:239–287.

Johnson, E. A., and C. P. S. Larsen. 1991. Climatically induced change in fire frequency in the southern Canadian Rockies. Ecology 72:194–201.

Johnson, E. A., and K. Miyanishi. 1996. The need for consideration of fire behavior and effects in prescribed burning. Restoration Ecology 3:271–278.

Johnson, E. A., and C. E. Van Wagner. 1985. The theory and use of two fire history models. Canadian Journal of Forest Research 15:214–220.

Johnson, E. A., and D. R. Wowchuk. 1993. Wildfires in the southern Canadian Rocky Mountains and their relationship to mid-tropospheric anomalies. Canadian Journal of Forest Research 23:1213–1222.

Johnson, E. A., K. Miyanishi, and J. M. H. Weir. 1995. Old-growth, disturbance, and ecosystem management. Canadian Journal of Botany 73:918–926.

Johnson, E. A., K. Miyanishi, and J. M. H. Weir. 1998. Wildfires in the western Canadian boreal forest: Landscape patterns and ecosystem management. Journal of Vegetation Science 9:603–610.

Johnson, E. A., K. Miyanishi, and S. R. J. Bridge. 2001. Wildfire regime in the boreal forest and the idea of suppression and fuel build-up. Conservation Biology 15:1554–1557.

Johnson, J. E. 1995. The Lake States region. Pages 81–127 in J. W. Barrett, editor. Regional silviculture of the United States. John Wiley and Sons, New York, New York, United States.

Johnson, P. S., S. R. Shifley, and R. Rogers. 2002. The ecology and silviculture of oaks. CABI Publishing International, Oxford, United Kingdom.

Johnson, W. E., E. Eizirik, M. Roelke-Parker, and S. J. O'Brien. 2001. Applications of genetic concepts and molecular methods to carnivore conservation. Pages 335–358 in J. L. Gittleman, S. M. Funk, D. MacDonald, and R. K. Wayne, editors. Carnivore conservation. Conservation Biology 5. Cambridge University Press and the Zoological Society of London, London, United Kingdom.

Johnston, M. H., and J. A. Elliott. 1996. Impacts of logging and wildfire on an upland black spruce community in northwestern Ontario. Environmental Monitoring and Assessment 39:283–297.

Jones, T. P., and B. Lim. 2000. Extraterrestrial impacts and wildfires. Palaeogeography Palaeoclimatology Palaeoecology 164:57–66.

Jull, M. J. 1990. Long-term stand dynamics in high-elevation spruce–fir forests. M. Sc. thesis, University of British Columbia, Vancouver, British Columbia, Canada.

Kabrick, J. M., R. G. Jensen, D. R. Larsen, and S. R. Shifley. 2002. Woody vegetation following even-aged, uneven-aged, and no harvest treatments on the Missouri Forest Ecosystem Project (MOFEP). Pages 84–101 in S. R. Shifley and J. M. Kabrick, editors. Proceedings of the second Missouri Ozark Forest Ecosystem symposium: Post-treatment results of the landscape experiment. October 17–20, 2000, St. Louis, Missouri. U.S. Department of Agriculture Forest Service, St. Paul, Minnesota, United States. General Technical Report NC-227.

Kafka, V., S. Gauthier, and Y. Bergeron. 2001. Fire impacts and crowning in the boreal forest: Study of a large wildfire in western Quebec. International Journal of Wildland Fire 10:119–127.

Kasischke, E. S., D. Williams, and D. Barry. 2002. Analysis of the patterns of large fires in the boreal forest region of Alaska. International Journal of Wildland Fire 11:131–144.

Kauffman, J. B., C. Uhl, and D. L. Cummings. 1988. Fire in the Venezuelan Amazon 1: Fuel biomass and tree chemistry in the evergreen rainforest of Venezuela. Oikos 53:167–175.

Kay, J. 1985. Native Americans in the fur trade and wildlife depletion. Environmental Reviews 9:118–130.

Kay, J., and E. D. Schneider. 1994. Embracing complexity: The challenge of the ecosystem approach. Alternatives 20:32–38.

Keane, R. E. 2000. Landscape fire succession modeling: Linking ecosystem simulations for comprehensive applications. Pages 5–8 in B. C. Hawkes, and M. D. Flannigan, editors. Landscape fire modeling—challenges and opportunities. Canadian Forest Service, Edmonton, Alberta, Canada. Information Report NOR-X-371.

Keane, R. E., and M. A. Finney. 2003. The simulation design for modeling landscape fire, climate, and ecosystem dynamics. Pages 32–68 in T. T. Veblen, W. L. Baker, G. Montenegro, and T. W. Swetnam, editors. Fire and climatic change in temperate ecosystems of the western Americas. Springer-Verlag, New York, New York, USA.

Keane, R. E., and D. G. Long. 1998. A comparison of coarse-scale fire-effects simulation strategies. Northwest Science 72:76–90.

Keane, R. E., S. F. Arno, and J. K. Brown. 1990. Simulating cumulative fire effects in ponderosa pine/Douglas-fir forests. Ecology 71:189–203.

Keane, R. E., P. Morgan, and S. W. Running. 1996a. FIRE-BGC—a mechanistic ecological process model for simulating fire succession on coniferous forest landscapes of the northern Rocky Mountains. U.S. Department of Agriculture Forest Service, Missoula, Montana, United States. Research Paper INT-RP-484.

Keane, R. E., K. C. Ryan, and S. W. Running. 1996b. Simulating effects of fire on northern Rocky Mountain landscapes with the ecological process model FIRE-BCG. Tree Physiology 16:319–331.

Keane, R. E., D. G. Long, J. P. Menakis, W. J. Hann, and C. D. Bevins. 1996c. Simulating coarse-scale vegetation dynamics using the Columbia River Basin Succession Model CRBSUM. U.S. Department of Agriculture Forest Service, Fort Collins, Colorado, United States. General Technical Report INT-GTR-340.

Keane, R. E., C. C. Hardy, K. C. Ryan, and M. A. Finney. 1997. Simulating effects of fire on gaseous emissions from future landscapes of Glacier National Park, Montana, USA. World Resources Review 9:177–205.

Keane, R. E., K. C. Ryan, and M. Finney. 1998. Simulating the consequences of altered fire regimes on a complex landscape in Glacier National Park, USA. Tall Timbers Fire Ecology Conference 20:310–324.

Keane, R. E., P. Morgan, and J. D. White. 1999. Temporal patterns of ecosystem processes on simulated landscapes in Glacier National Park, Montana, USA. Landscape Ecology 14:311–329.

Keane, R. E., C. McNicoll, and M. G. Rollins. 2002a. Integrating ecosystem sampling, gradient modeling, remote sensing, and ecosystem simulation to create spatially explicit landscape inventories. U.S. Department of Agriculture Forest Service General, Fort Collins, Colorado, United States. General Technical Report RMRS-GTR-92.

Keane, R. E., R. Parsons, and P. Hessburg. 2002b. Estimating historical range and variation of landscape patch dynamics: Limitations of the simulation approach. Ecological Modelling 151:29–49.

Keane, R. E., T. Veblen, K. C. Ryan, J. Logan, C. Allen, and B. Hawkes. 2002c. The cascading effects of fire exclusion in the Rocky Mountains. Pages 133–153 in J. Baron, editor. Rocky Mountain futures: An ecological perspective. Island Press, Washington, D.C., United States.

Keenan, R., and J. P. Kimmins. 1993. Ecological effects of clearcutting. Environmental Reviews 1:121–144.

Kennedy, T. A., S. Naeem, K. M. Howe, J. N. M. Knops, D. Tilman, and P. Reich. 2002. Biodiversity as a barrier to ecological invasion. Nature 417:636–638.

Kerley, L. L., J. M. Goodrich, D. G. Miquelle, E. N. Smirnov, H. B. Quigley, and N. G. Hornocker. 2002. Effects of roads and human disturbance on Amur tigers. Conservation Biology 16:97–108.

Kertis, J. 1986. Vegetation dynamics and disturbance history of Oak Patch Natural Area Preserve, Mason County, Washington. M.Sc. thesis. University of Washington, Seattle, Washington, United States.

Kessell, S. R. 1979. Gradient modeling: Resource and fire management. Springer-Verlag, New York, New York, United States.

Kettela, E. G. 1995. Insect control in New Brunswick, 1974–1989. Pages 655–665 in J. A. Armstrong and W. G. H. Ives, editors. Forest insect pests in Canada. Canadian Forest Service, Ottawa, Ontario, Canada.

Kimmins, J. P. 1974. Sustained yield, timber mining, and the concept of ecological rotation: A British Columbian view. The Forestry Chronicle 50:27–31.

Kimmins, J. P. 1990. Modelling the sustainability of forest production and yield for a changing and uncertain future. The Forestry Chronicle 66:271–280.

Kimmins, J. P. 1993. Ecology, environmentalism and green religion. The Forestry Chronicle 69:285–289.

Kimmins, J. P. 1996. Importance of soil and role of ecosystem disturbance for sustained productivity of cool temperate and boreal forests. Soil Science Society of America Journal 60:1643–1654.

Kimmins, J. P. 1997a. Forest ecology. A foundation for sustainable management, second edition. Prentice-Hall, Upper Saddle River, New Jersey, United States.

Kimmins, J. P. 1997b. Balancing act. Environmental issues in forestry, second edition. University of British Columbia Press, Vancouver, British Columbia, Canada.

Kimmins, J. P. 1999. Biodiversity, beauty and the beast—are beautiful forests sustainable, are sustainable forests beautiful, and is "small" always ecologically desirable? The Forestry Chronicle 75:955–960.

Kimmins, J. P. 2000. Respect for nature: An essential foundation for sustainable forest management. Pages 3–24 in R. G. d'Eon, J. F. Johnson, and E. A. Ferguson, editors. Ecosystem management of forested landscapes: Directions and implementation. University of British Columbia Press, Vancouver, British Columbia, Canada.

Kimmins, J. P. 2001. Visible and non-visible indicators of forest sustainability: Beauty, beholders and belief systems. Pages 43–56 in S. R. J. Sheppard and H. W. Harshaw, editors. Forests and landscapes. Linking ecology, sustainability and aesthetics. IUFRO Research Series 6. IUFRO and CABI Publishing International, Oxford, United Kingdom.

Kimmins, J. P. 2002. Future shock in forestry. Where have we come from; where are we going: Is there a right way to manage forests? Lessons from Thoreau, Leopold, Toffler, Botkin and Nature. The Forestry Chronicle 78:263–271.

Kimmins, J. P., D. Mailly, and B. Seely. 1999a. Modelling forest ecosystem net primary production: The hybrid simulation approach used in FORECAST. Ecological Modelling 122:195–224.

Kimmins, J. P., B. Seely, D. Mailly, K. M. Tsze, K. A. Scoullar, D. W. Andison, and R. Bradley. 1999b. FORCEEing and FORECASTing the HORIZON: Hybrid simulation modelling of forest ecosystem sustainability. Pages 431–441 in A. Amaro and M. Tome, editors. Empirical and process based models for forest tree and stand growth simulation. Edicoes–Salamandra, Lisbon, Portugal.

Kirkpatrick, S., C. D. Gelatt Jr., and M. P. Vecchi. 1983. Optimization by simulated annealing. Science 220:671–680.

Kitzberger, T., and T. T. Veblen. 2002. Inter-hemispheric comparison of fire history: The Colorado Front Range, U.S.A., and the Northern Patagonian Andes, Argentina. Plant Ecology 163:187–207.

Kleindorfer, G. B., L. O'Neill, and R. Ganeshan. 1998. Validation in simulation: Various positions in the philosophy of science. Management Science 44:1087–1099.

Kletetschka, G., and S. K. Banerjee. 1995. Magnetic stratigraphy of Chinese loess as a record of natural fires. Geophysical Research Letters 22:1341–1343.

Kneeshaw, D. D., and Y. Bergeron. 1998. Canopy gap dynamics and tree replacement in the southeastern boreal forest. Ecology 79:783–794.

Kneeshaw, D. D., and P. J. Burton. 1998. A functional assessment of old-growth status: A case study in the subboreal spruce zone of British Columbia. Natural Areas Journal 18:295–310.

Kneeshaw, D. D., and S. Gauthier. 2003. Old growth in the boreal forest: A dynamic perspective at the stand and landscape level. Environmental Reviews 11(S1):S99–S114.

Knight, D. H. 1987. Parasites, lightning, and the vegetation mosaic in wilderness landscapes. Pages 59–83 in M. G. Turner, editor. Landscape heterogeneity and disturbance. Springer-Verlag, New York, New York, United States.

Knight, D. H., and L. L. Wallace. 1989. The Yellowstone fires: Issues in landscape ecology. BioScience 39:700–706.

Kohm, K. A., and J. F. Franklin, editors. 1996. Creating a forestry for the 21st century: The science of ecosystem management. Island Press, Washington, D.C., United States.

Korzukhin, M. D., M. T. Ter-Mikaelian, and R. G. Wagner. 1996. Process vs. empirical models: Which approach for ecosystem management? Canadian Journal of Forest Research 26:879–887.

Krech, S. 1999. The ecological Indian: Myth and history. W. W. Norton, New York, New York, United States.

Küchler, A. W. 1964. Potential natural vegetation of the coterminous United States. American Geographical Society, New York, New York, United States. Special Publication 36. (With separate map at 1:3,168,000.)

Küchler, A. W. 1975. Potential natural vegetation of the conterminous United States: Manual and map, second edition. American Geological Society, New York, New York, United States.

Kushla, J. D., and W. J. Ripple. 1997. The role of terrain in a fire mosaic of a temperate coniferous forest. Forest Ecology and Management 95:97–107.

Kuuluvainen, T., K. Aapala, P. Ahlroth, M. Kuusinen, T. Lindholm, T. Sallantaus, J. Siitonen, and H. Tukia. 2002. Principles of ecological restoration of boreal forested ecosystems: Finland as an example. Silva Fennica 36:409–422.

Landres, P. B., P. Morgan, and F. J. Swanson. 1999. Overview of the use of natural variability concepts in managing ecological systems. Ecological Applications 9:1179–1188.

Landsberg, J. 1990. Dieback of rural eucalypts: Response of foliar dietary quality and herbivory to defoliation. Australian Journal of Ecology 15:89–96.

Larsen, C. P. S. 1997. Spatial and temporal variations in boreal forest fire frequency in northern Alberta. Journal of Biogeography 24:663–673.

Larson, B. M., J. L. Riley, E. A. Snell, and H. G. Godschalk. 1999. Woodland heritage of southern Ontario. A study of ecological change, distribution and significance. Federation of Ontario Naturalists, Don Mills, Ontario, Canada.

Lassoie, J. P., T. M. Hinckley, and C. C. Grier. 1985. Coniferous forests of the Pacific Northwest. Pages 127–161 in B. F. Chabot and H. A. Mooney, editors. Physiological ecology of North American plant communities. Chapman and Hall, New York, New York, United States.

Laurance, W. F., S. G. Laurance, L. V. Ferreira, J. M. Rankin-de Marona, C. Gascon, and T. E. Lovejoy. 1997. Collapse of biomass in Amazonian forest fragments. Science 278:1117–1118.

Lavoie, L., and L. Sirois. 1998. Vegetation changes caused by recent fires in the northern boreal forest of eastern Canada. Journal of Vegetation Science 9:483–492.

Lawton, J. H., and V. K. Brown. 1993. Redundancy in ecosystems. Pages 255–270 in E. D. Schulze and H. A. Mooney, editors. Biodiversity and ecosystem function. Springer-Verlag, Berlin, Germany.

Leadbitter, P., D. Euler, and B. Naylor. 2002. A comparison of historical and current forest cover in selected areas of the Great Lakes–St. Lawrence forest of central Ontario. The Forestry Chronicle 78(4):522–529.

Leatherberry, E. C., and J. S. Spencer Jr. 1996. Michigan forest statistics, 1993. U.S. Department of Agriculture Forest Service, St. Paul, Minnesota, United States. Research Bulletin NC-170.

Ledig, F. T. 1992. Human impacts on genetic diversity in forest ecosystems. Oikos 63:87–108.

Lee, C. T., R. Wickneswari, M. C. Mahani, and A. H. Zakri. 2002. Effect of selective logging on the genetic diversity of Scaphium macropodum. Biological Conservation 104:107–118.

Lee, K. N. 1993. Compass and gyroscope: Integrating science and politics for the environment. Island Press, Washington, D.C., United States.

Lee, P. C., S. Crites, H. Van Nguyen, and J. B. Stelfox. 1997. Characteristics and origins of deadwood material in aspen-dominated boreal forests. Ecological Applications 7(2):691–701.

Lefort, P., S. Gauthier, and Y. Bergeron. 2003. The influence of fire weather and land use on the fire activity of the Lake Abitibi area, eastern Canada. Forest Science 49(4):509–521.

Lehmkuhl, J. F., and M. G. Raphael. 1993. Habitat patterns around northern spotted owl locations on the Olympic Peninsula, Washington. Journal of Wildlife Management 57:302–315.

Lenihan, J., C. Daly, D. Bachelet, and R. P. Neilson. 1998. Simulating broad-scale fire severity in a dynamic global vegetation model. Northwest Science 72:91–103.

Leopold, A. 1934. Some thoughts on recreational planning. Quoted on page 160 in C. Meine and R. L. Knight, editors. 1999. The essential Aldo Leopold: Quotations and commentaries. University of Wisconsin Press, Madison, Wisconsin, United States.

Leopold, A. 1949. A Sand County almanac. Oxford University Press, New York, New York, United States.

Leopold, A. 1966. A Sand County almanac, with essays on conservation from Round River. Ballantine, New York, New York, United States.

Lertzman, K., and J. Fall. 1998. From forest stands to landscapes: Spatial scales and the roles of disturbance. Pages 339–367 in D. L. Peterson and V. T. Parker, editors. Ecological scale. Columbia University Press, New York, New York, United States.

Lertzman, K., J. Fall, and B. Dorner. 1998. Three kinds of heterogeneity in fire regimes: At the crossroads of fire history and landscape ecology. Northwest Science 72:4–23.

Lertzman, K., D. Gavin, D. Hallett, L. Brubaker, D. Lepofsky, and R. Mathewes. 2002. Long-term fire regime estimated from soil charcoal in coastal temperate rainforests. Conservation Ecology 6:5. [Online, URL: <http://www.consecol.org/vol6/iss2/art5>.]

Lesieur, D., S. Gauthier, and Y. Bergeron. 2002. Fire frequency and vegetation dynamics for the south-central boreal forest of Quebec, Canada. Canadian Journal of Forest Research 32:1996–2009.

Li, C. 2000a. Reconstruction of natural fire regimes through ecological modelling. Ecological Modelling 134:129–144.

Li, C. 2000b. Modeling the influence of fire ignition source patterns on fire regimes of west-central Alberta. In Proceedings of the 4th international conference on integrating GIS and environmental modeling (GIS/EM4): Problems, prospects and research needs. [Online, URL: <http://www.Colorado.edu/research/cires/banff/upload/92>.]

Li, C. 2000c. Fire regimes and their simulation with reference to Ontario. Pages 115–140 in A. H. Perera, D. L. Euler, and I. D. Thompson, editors. Ecology of a managed terrestrial landscape: Patterns and processes of forest landscapes in Ontario. University of British Columbia Press, Vancouver, British Columbia, Canada.

Li, C. 2002. Estimation of fire frequency and fire cycle: A computational perspective. Ecological Modelling 154: 103–120.

Li, C., and H. J. Barclay. 2001. Fire disturbance patterns and forest age structure. Natural Resource Modeling 14:495–521.

Li, C., and I. G. W. Corns. 1998. An ecological knowledge-based simulation model of forest landscape dynamics. In Proceedings of the modeling of complex systems conference. [Online, URL: <http://colorado.edu/Research/cires/banff/pubapers/99/index.htm>.]

Li, C., and A. Perera. 1997. ON-FIRE: A landscape model for simulating the fire regime of northwest Ontario. Pages 369–392 in X. Chen, X. Dai, and H. Tom, editors. Ecological research and sustainable development. China Environmental Science Press, Beijing, China.

Li, C., I. G. W. Corns, and R. C. Yang. 1999. Fire frequency and size distribution under natural conditions: A new hypothesis. Landscape Ecology 14:533–542.

Li, C., M. Ter-Mikaelian, and A. Perera. 1996. Ontario Fire Regime Model: Its background, rationale, development and use. Ontario Ministry of Natural Resources, Sault Ste. Marie, Ontario, Canada. Forest Fragmentation and Biodiversity Project Report 25.

Li, C., M. Ter-Mikaelian, and A. Perera. 1997. Temporal fire disturbance patterns on a forest landscape. Ecological Modelling 99:137–150.

Li, C., M. D. Flannigan, and I. G. W. Corns. 2000. Influence of potential climate change on forest landscape dynamics of west-central Alberta. Canadian Journal of Forest Research 30:1905–1912.

Liebhold, A. M., G. Zhou, F. W. Ravlin, A. Roberts, and R. Reardon. 1998. Forecasting gypsy moth defoliation with a geographical information system. Journal of Economic Entomology 91:464–472.

Liebhold, A. M., J. Elkington, D. Williams, and R. M. Muzika. 2000. What causes outbreaks of gypsy moth in North America? Population Ecology 42:257–266.

Liedloff, A. C., M. B. Coughenour, J. A. Ludwig, and R. Dyer. 2001. Modelling the trade-off between fire and grazing in a tropical savanna landscape, Northern Australia. Environment International 27:173–180.

Lieffers, V. J., R. B. MacMillan, D. MacPherson, K. Branter, and J. D. Stewart. 1996. Semi-natural and intensive silvicultural systems for the boreal mixedwood forest. The Forestry Chronicle 72:286–292.

Lindenmayer, D., M. A. McCarthy, and A. C. Dibble. 2002. Congruence between natural and human forest disturbance: A case study from Australian montane ash forests. Forest Ecology and Management 155: 319–336.

Littell, J. S. 2002. Determinants of fire regime variability in lower elevation forests of the northern Greater Yellowstone Ecosystem. M.Sc. thesis. Montana State University, Bozeman, Montana, United States.

Liu, K.-B. 1990. Holocene paleoecology of the boreal forest and Great Lakes–St. Lawrence forest in northern Ontario. Ecological Monographs 60:179–212.

Lockwood, C., and T. Moore. 1993. Harvest scheduling with spatial constraints: A simulated annealing approach. Canadian Journal of Forest Research 23:468–478.

Long, C. J., and C. Whitlock. 2002. Fire and vegetation history from the coastal rain forest of the western Oregon Coast Range. Quaternary Research 58:215–225.

Long, C. J., C. Whitlock, P. J. Bartlein, and S. H. Millspaugh. 1998. A 9000-year fire history from the Oregon Coast Range, based on a high-resolution charcoal study. Canadian Journal of Forest Research 28: 774–787.

Long, D. G. 1998. Mapping historical fire regimes in northern Rocky Mountain landscapes. M.Sc. thesis. University of Idaho, Moscow, Idaho, United States.

Longauer, R., D. Gomoroy, P. Ladislav, D. F. Karnosky, B. Mankovska, G. Mueller-Starck, K. Percy, and R. Szaro. 2001. Selection effects of air pollution on gene pools of Norway spruce, European silver fir and European beech. Environmental Pollution 115:405–411.

Loreau, M. 2000. Biodiversity and ecosystem functioning: Recent theoretical advances. Oikos 91:3–17.

Loreau, M., and N. Mouquet. 1999. Immigration and the maintenance of local species diversity. American Naturalist 154:427–440.

Lorimer, C. G. 1977. The presettlement forest and natural disturbance cycle of northeastern Maine. Ecology 58: 139–148.

Lovejoy, J. F., R. O. Bierregaard, A. B. Rylands, J. R. Malcolm, C. E. Quintela, L. H. Harper, K. S. Brown, A. H. Powell, G. V. N. Powell, H. O. R. Schubert, and M. J. Hays. 1986. Edge and other effects of isolation on Amazon forest fragments. Pages 257–285 in M. Soulé, editor. Conservation biology: The science of scarcity and diversity. Sinauer Associates, Sunderland, Massachusetts, United States.

Loveland, T. R., J. M. Merchant, D. O. Ohlen, and J. F. Brown. 1991. Development of a land-cover characteristics database for the conterminous U.S. Photogrammetric Engineering and Remote Sensing 57:1453–1463.

Lovelock, J. E. 1987. Gaia: A new look at life on earth, second edition. Oxford University Press, Oxford, United Kingdom.

Lowman, M. D., and H. Heatwole. 1992. Spatial and temporal variability in defoliation of Australian eucalypts. Ecology 73:129–142.

Loy, W. G., S. Allan, C. P. Patton, and R. D. Plank. 1976. Atlas of Oregon. University of Oregon, Eugene, Oregon, United States.

Lundmark, J. E. 1977. The soil as part of the forest ecosystem. Pages 109–122 in The care of the soil. Properties and utilization of forest soils. [Original in Swedish]. Sveriges Skogsvardsforbunds Tidskrift Volume 75.

Lundmark, J. E. 1986. Short-term and long-term effects on site productivity by soil treatments in forestry. (Original in Swedish). Swedish University of Agricultural Sciences, Uppsala, Sweden. Skogsfacta Konferens 9:126–132.

Lundquist, J. E., and J. S. Beatty. 2002. A method for characterizing and mimicking forest canopy gaps caused by different disturbances. Forest Science 48:582–594.

Lundquist, J. E., and J. F. Negron. 2000. Endemic forest disturbances and stand structure of ponderosa pine (Pinus ponderosa) in the Upper Pine Creek Research Natural Area, South Dakota, USA. Natural Areas Journal 20:126–132.

Lutz, W., W. Sanderson, and S. Scherbov. 2001. The end of world population growth. Nature 412:543.

Lynch, D. L., W. H. Romme, and M. L. Floyd. 2000. Forest restoration in southwestern ponderosa pine. Journal of Forestry 98:17–24.

Lynch, E. A. 1998. Origin of a park–forest vegetation mosaic in the Wind River Range, Wyoming. Ecology 79:1320–1338.

Lynch, M. 1996. A quantitative genetic perspective on conservation issues. Pages 471–501 *in* J. C. Avise and J. L. Hamrick, editors. Conservation genetics: Case studies from nature. Chapman and Hall, New York, New York, United States.

Lysyk, T. J. 1990. Relationships between spruce budworm (Lepidoptera: Tortricidae) egg mass density and resultant defoliation of balsam fir and white spruce. The Canadian Entomologist 122:253–262.

MacArthur, R. H., and E. O. Wilson. 1967. The theory of island biogeography. Princeton University Press, Princeton, New Jersey, United States.

MacDonell, M. R., and A. Groot. 1996. Uneven-aged silviculture for peatland second-growth black spruce: Biological feasibility. Canadian Forest Service, Sault Ste. Marie, Ontario, Canada. NODA/NFP Technical Report TR-36.

MacLean, D. A. 1980. Vulnerability of fir–spruce stands during uncontrolled spruce budworm outbreaks: A review and discussion. The Forestry Chronicle 56:213–221.

MacLean, D. A. 1988. Effects of spruce budworm outbreaks on vegetation, structure, and succession of balsam fir forests on Cape Breton Island, Canada. Pages 253–261 *in* M. J. A. Werger, P. J. M. van der Aart, H. J. During, and J. T. A. Verhoeven, editors. Plant form and vegetation structure. SPB Academic Publishing, The Hague, The Netherlands.

MacLean, D. A. 1990. Impact of forest pests and fire on stand growth and timber yield: Implications for forest management planning. Canadian Journal of Forest Research 19:391–404.

MacLean, D. A. 1996a. The role of a stand dynamics model in the spruce budworm decision support system. Canadian Journal of Forest Research 26:1731–1741.

MacLean, D. A. 1996b. Forest management strategies to reduce spruce budworm damage in the Fundy Model Forest. The Forestry Chronicle 72:399–405.

MacLean, D. A. 1998. Landscape management for restructuring forest areas. Pages 25–45 *in* G. J. Nabuurs, T. Nuutinen, H. Bartelink, and M. Korhonen, editors. Forest scenario modelling for ecosystem management at landscape level. European Forest Institute, Joensuu, Finland. Proceedings 19.

MacLean, D. A., and W. E. MacKinnon. 1996. The accuracy of aerial sketch mapping of spruce budworm defoliation in New Brunswick. Canadian Journal of Forest Research 26:2099–2108.

MacLean, D. A., and W. E. MacKinnon. 1997. Effects of soil drainage on susceptibility and vulnerability of balsam fir to spruce budworm in New Brunswick. Canadian Journal of Forest Research 27:1859–1871.

MacLean, D. A., and H. Piene. 1995. Spatial and temporal patterns of balsam fir mortality in spaced and unspaced stands caused by spruce budworm defoliation. Canadian Journal of Forest Research 25:902–911.

MacLean, D. A., P. Etheridge, J. Pelham, and W. Emrich. 1999. Fundy Model Forest: Partners in sustainable forest management. The Forestry Chronicle 75:219–227.

MacLean, D. A., K. B. Porter, W. E. MacKinnon, and K. P. Beaton. 2000. Spruce Budworm Decision Support System: Lessons learned in development and implementation. Computers and Electronics in Agriculture 27:293–314.

MacLean, D. A., T. A. Erdle, W. E. MacKinnon, K. B. Porter, K. P. Beaton, G. Cormier, S. Morehouse, and M. Budd. 2001. The Spruce Budworm Decision Support System: Forest protection planning to sustain long-term wood supplies. Canadian Journal of Forest Research 31:1742–1757.

MacLean, D. A., K. P. Beaton, K. B. Porter, W. E. MacKinnon, and M. Budd. 2002. Potential wood supply losses to spruce budworm in New Brunswick estimated using the Spruce Budworm Decision Support System. The Forestry Chronicle 78:739–750.

MacNally, R., A. F. Bennett, G. W. Brown, L. F. Lumsden, A. Yen, S. Hinkley, P. Lillywhite, and D. Ward. 2002. How well do ecosystem-based planning units represent different components of biodiversity? Ecological Applications 12:900–912.

Madany, M. H., T. W. Swetnam, and N. E. West. 1982. Comparison of two approaches for determining fire dates from tree scars. Forest Science 28:856–861.

Maehr, D. S., E. D. Land, D. B. Shindle, O. L. Bas, and T. S. Hoctor. 2002. Florida panther dispersal and conservation. Biological Conservation 106:187–197.

Malamud, B. D., G. Morein, and D. L. Turcotte. 1998. Forest fires: An example of self-organized critical behavior. Science 281:1840–1842.

Manies, K. L., and D. J. Mladenoff. 2000. Testing methods to produce landscape-scale presettlement vegetation maps from the U.S. public land survey records. Landscape Ecology 15:741–754.

Manley, I. A., F. L. Waterhouse, and A. S. Harestad. 1999. Nesting habitat of marbled murrelets on the Sunshine Coast. British Columbia Ministry of Forests, Nanaimo, British Columbia, Canada. Forest Research Extension Note EN-002. [Online, URL: <http://www.for.gov.bc.ca/VANCOUVR/research/wildlifereports\en002.pdf>.]

Mantua, N. J., and S. R. Hare. 2002. The Pacific Decadal Oscillation. Journal of Oceanography 58:35–44.

Marshall, H. D., and K. Ritland. 2002. Genetic diversity and differentiation of Kermode bear populations. Molecular Ecology 11:686–697.

Martin, C. 1978. Keepers of the game: Indian–animal relationships and the fur trade. University of California Press, Berkeley, California, United States.

Martin, P. S. 1967. Pleistocene overkill. Natural History 76:32–38.

Masters, A. M. 1990. Changes in forest fire frequency in Kootenay National Park, Canadian Rockies. Canadian Journal of Botany 68:1763–1767.

Mattson, W. J., and N. D. Addy. 1975. Phytophagous insects as regulators of forest primary production. Science 190:515–522.

Maure, J. E., and J. P. Dawson. 1990. Interim timber management plan for the Temagami Crown Management Unit. Ontario Ministry of Natural Resources, Temagami, Ontario, Canada.

May, R. M. 1972. What is the chance a large complex system will be stable? Nature 237:413–414.

McCarthy, J. 2001. Gap dynamics of forest trees: A review with particular attention to boreal forests. Environmental Reviews 9:1–59.

McComb, W. C., M. T. McGrath, T. A. Spies, and D. Vesely. 2002. Models for mapping potential habitat at landscape scales: An example using northern spotted owls. Forest Science 48:203–216.

McComb, W. C., T. A. Spies, and W. H. Emmingham. 1993. Douglas-fir forests: Managing for timber and mature forest habitat. Journal of Forestry 91:31–42.

McCullough, D. G. 2000. A review of factors affecting the population dynamics of jack pine budworm (Choristoneura pinus pinus Freeman). Population Ecology 42:243–256.

McCullough, D. G., R. A. Werner, and D. Neuman. 1998. Fire and insect interactions in northern and boreal ecosystems of North America. Annual Review of Entomology 43:107–127.

McCune, B., and T. F. H. Allen. 1985. Will similar forests develop on similar sites? Canadian Journal of Botany 63:367–376.

McGarigal, K., and B. J. Marks. 1995. FRAGSTATS: Spatial pattern analysis program for quantifying landscape structure. U.S. Department of Agriculture Forest Service, Portland, Oregon, United States. General Technical Report PNW-GTR-351.

McGaughey, R. J. 1997. Visualizing forest stand dynamics using the stand visualization system. Pages 248–257 in Proceedings of the 1997 ACSM/ASPRS annual convention and exposition. American Society for Photogrammetry and Remote Sensing, Bethesda, Maryland, United States.

McIntire, E. 2003. Wildfire and mountain pine beetle boundary zones: Spatial pattern, boundary creation and successional consequences. Ph.D. thesis. University of British Columbia, Vancouver, British Columbia, Canada.

McKechnie, J. L., editor. 1978. Webster's new twentieth century dictionary, second edition, unabridged. Collins and World Publishing, Cleveland, Ohio, United States.

McKelvey, K. S., and K. K. Busse. 1996. Twentieth-century fire patterns on Forest Service lands. Sierra Nevada Ecosystem Project: Final Report to Congress. Volume II, Assessments and scientific basis for management options. University of California, Davis, California, United States.

McKenney, D. W., and D. B. Lindenmayer. 1994. An economic assessment of a nest-box strategy for the conservation of an endangered species. Canadian Journal of Forest Research 24:2012–2019.

McKenney, D. W., N. Beke, G. Fox, and A. Groot. 1997. Does it pay to do silviculture research on slow growing species? Forest Ecology and Management 95:141–152.

McKenney, D. W., L. Venier, I. Ball, J. McKee, H. Possingham, and B. Mackey. 2000. A "GAP" analysis of bird distributions in southern Ontario. Pages 135–143 in Proceedings of the symposium on systems analysis in forest resources. U.S. Department of Agriculture Forest Service, St. Paul, Minnesota, United States. General Technical Report NC-205.

McKenzie, D. 1998. Fire, vegetation, and scale: Toward optimal models for the Pacific Northwest. Northwest Science 72:49–65.

McKenzie, D., D. L. Peterson, and E. Alvarado. 1996a. Extrapolation problems in modelling fire effects at large spatial scales: A review. International Journal of Wildland Fire 6:165–176.

McKenzie, D., D. L. Peterson, and E. Alvarado. 1996b. Predicting the effect of fire on large-scale vegetation patterns in North America. U.S. Department of Agriculture Forest Service, Portland, Oregon, United States. Research Paper PNW-489.

McKenzie, D., D. L. Peterson, and J. K. Agee. 2000. Fire frequency in the Columbia River Basin: Building regional models from fire history data. Ecological Applications 10:1497–1516.

McKinnon, A. 1998. Biodiversity in old-growth forests. Pages 146–184 in J. Voller and S. Harrison, editors. Conservation biology principles for forested landscapes. University of British Columbia Press, Vancouver, British Columbia, Canada.

McLachlan, J. S., and L. B. Brubaker. 1995. Local and regional vegetation change on the northeastern Olympic Peninsula during the Holocene. Canadian Journal of Botany 73:1618–1627.

McNaughton, S. J. 1977. Diversity and stability of ecological communities: A comment on the role of empiricism in ecology. American Naturalist 111:515–525.

McRae, D. J., L. C. Duchesne, B. Freedman, T. J. Lynham, and S. Woodley. 2001. Comparisons between wildfire and forest harvesting and their implications in forest management. Environmental Reviews 9:223–260.

Meeker, J. E., J. E. Elias, and J. A. Heim. 1993. Plants used by the Great Lakes Ojibwa. Great Lakes Indian Fish and Wildlife Commission, Odanah, Wisconsin, United States.

Merrill, D. F., and M. E. Alexander, editors. 1987. Glossary of forest fire management terms, fourth edition. National Research Council of Canada, Ottawa, Ontario, Canada. NRCC Publication 26516.

Messier, C., M.-J. Fortin, F. Schmiegelow, F. Doyon, S. G. Cumming, J. P. Kimmins, B. Seely, C. Welham, and J. Nelson. 2003. Modelling tools to assess the sustainability of forest management scenarios. Pages 531–580 in P. J. Burton, C. Messier, D. W. Smith, and W. L. Adamowicz, editors. Towards sustainable management of the boreal forest. National Research Council Press, Ottawa, Ontario, Canada.

Methven, I. R., and U. Feunekes. 1991. Simulating process interactions on landscape attributes: Fire and spruce budworm in Pukaskwa National Park. Pages 345–352 in Proceedings of the international conference on science and management of protected areas. Elsevier, Amsterdam, the Netherlands. [Online, URL: <http://www.remsoft2.com/outgoing/freefile/sim%20proces s%2091.pdf>.]

Meyer, C. B., and S. L. Miller. 2002. Use of fragmented landscapes by marbled murrelets for nesting in southern Oregon. Conservation Biology 16:755–766.

Miles, P. D. 2001. Forest inventory mapmaker user's guide. U.S. Department of Agriculture Forest Service, St. Paul, Minnesota, United States. General Technical Report NC-221.

Miles, P. D., C. M. Chen, and E. C. Leatherberry. 1995. Minnesota forest statistics, 1990, revised. U.S. Department of Agriculture Forest Service, St. Paul, Minnesota, United States. Research Bulletin NC-158.

Miller, C. I. 1999. Genetic diversity. Pages 460–494 *in* M. L. Hunter Jr., editor. Maintaining biodiversity in forest ecosystems. Cambridge University Press, Cambridge, United Kingdom.

Miller, M. R. 1981. Ecological land classification terrestrial subsystem—a basic inventory system for planning and management on the Mark Twain National Forest. U.S. Department of Agriculture Forest Service, Mark Twain National Forest, Rolla, Missouri, United States.

Millspaugh, S. H., and C. Whitlock. 1995. A 750-year fire history based on lake sediment records in central Yellowstone National Park, USA. The Holocene 5:283–292.

Millspaugh, S. H., C. Whitlock, and P. J. Bartlein. 2000. Variations in fire frequency and climate over the past 17,000 yr in central Yellowstone National Park. Geology 28:211–214.

Milne, L. J., and M. Milne. 1960. The balance of nature. Alfred A. Knopf, New York, New York, United States.

Ministère des Ressources Naturelles du Québec. 1997. SYLVA II. Version 1.3.2. Ministère des Ressources Naturelles du Québec, Quebec, Canada.

Minnich, R. A., and Y. H. Chou. 1997. Wildland fire patch dynamics in the chaparral of southern California and northern Baja California. International Journal of Wildland Fire 7:221–248.

Miyanishi, K., and E. A. Johnson. 2001. Comment—A reexamination of the effects of fire suppression in the boreal forest. Canadian Journal of Forest Research 31:1462–1466.

Mladenoff, D. J., and W. L. Baker, editors. 1999. Advances in spatial modeling of forest landscape change: Approaches and applications. Cambridge University Press, Cambridge, United Kingdom.

Mladenoff, D. J., and H. S. He. 1999. Design and behavior of LANDIS, an object-oriented model of forest landscape disturbance and succession. Pages 125–162 *in* D. J. Mladenoff and W. L. Baker, editors. Advances in spatial modeling of forest landscape change: Approaches and applications. Cambridge University Press, Cambridge, United Kingdom.

Mladenoff, D. J., and J. Pastor. 1993. Sustainable forest ecosystems in the northern hardwood and conifer forest region: Concepts and management. Pages 145–180 *in* G. H. Aplet, N. Johnson, J. T. Olson, and V. A. Sample, editors. Defining sustainable forestry. Island Press, Washington, D.C., United States.

Mladenoff, D. J., M. A. White, J. Pastor, and T. R. Crow. 1993. Comparing spatial pattern in unaltered old-growth and disturbed forest landscapes. Ecological Applications 3:294–306.

Mladenoff, D. J., G. E. Host, J. Boeder, and T. R. Crow. 1996. LANDIS: A spatial model of forest landscape disturbance, succession, and management. Pages 175–180 *in* M. F. Goodchild, L. T. Steyaert, and B. O. Parks, editors. GIS and environmental modeling: Progress and research issues. GIS World and Oxford University Press, New York, New York, United States.

Moeur, M., and A. R. Stage. 1995. Most similar neighbor: An improved sampling inference procedure for natural resources planning. Forest Science 41:337–359.

Monserud, R. A., and N. L. Crookston. 1982. A user's guide to the combined stand prognosis and Douglas-fir tussock moth outbreak model. U.S. Department of Agriculture Forest Service, Ogden, Utah, United States. General Technical Report INT-127.

Montgomery, C. A., G. M. Brown, and D. Adams. 1994. The marginal cost of species preservation: The northern spotted owl. Journal of Environmental Economics and Management 26:111–128.

Montgomery, D. R., K. M. Schmidt, H. M. Greenberg, and W. E. Dietrich. 2000. Forest clearing and regional landsliding. Geology 28:311–314.

Mooney, H. A., and J. A. Drake, editors. 1986. Ecology of biological invasions of North America and Hawaii. Springer-Verlag, New York, New York, United States.

Morasse, J. 2000. Essais de mécanisation de coupes avec protection des sols et de la régénération de 16 cm et moins (CPRS-16) et d'assainissement. Rapport de recherche, Programme de Volet 1, Ministère des Ressources Naturelles du Québec, La Sarre, Quebec, Canada.

Morgan, P., G. H. Aplet, J. B. Haufler, H. C. Humphries, M. M. Moore, and W. D. Wilson. 1994. Historical range of variability: A useful tool for evaluating ecosystem change. Journal of Sustainable Forestry 2:87–111.

Morgan, P., S. C. Bunting, A. E. Black, T. Merrill, and S. Barrett. 1996. Past and present fire regimes in the Interior Columbia River basin. Pages 77–87 *in* K. Close and R. A. Bartlette, editors. Proceedings of the 1994 Interior West Fire Council meeting and program. Fire management under fire (adapting to change)., 1–4 November 1994. Coeur d'Alene, ID. International Association of Wildland Fire, Fairfield, Washington, United States.

Morgan, P., C. C. Hardy, T. W. Swetnam, M. G. Rollins, and D. G. Long. 2001. Mapping fire regimes across time and space: Understanding coarse and fine-scale fire patterns. International Journal of Wildland Fire 10:329–342.

Morin, H., D. Laprise, and Y. Bergeron. 1993. Chronology of spruce budworm outbreaks near Lac Duparquet, Abitibi Region, Quebec. Canadian Journal of Forest Research 23:1497–1506.

Morris, D. M., J. P. Kimmins, and D. K. Duckert. 1997. The use of soil organic matter as a criterion of the relative sustainability of forest management alternatives: A modeling approach using FORECAST. Forest Ecology and Management 94:61–78.

Morrison, P. H., and F. J. Swanson. 1990. Fire history and pattern in a Cascade Range landscape. U.S. Department of Agriculture Forest Service, Portland, Oregon, United States. General Technical Report PNW-GTR-254.

Mosseler, A., J. E. Major, D. Simpson, B. Daigle, K. Lange, K. Johnsen, Y.-S. Park, and O. P. Rajora. 2000. Indicators of population viability in red spruce: I. Reproductive traits and fecundity. Canadian Journal of Botany 78:928–940.

Mosseler, A., J. E. Major, and O. P. Rajora. 2003. Old-growth red spruce forests as reservoirs of genetic diversity and reproductive fitness. Theoretical and Applied Genetics 106:931–937.

Muir, J., and T. E. Gifford. 1996. John Muir: His life and letters and other writings. Mountaineer Books, Seattle, Washington, United States.

Murcia, C. 1995. Edge effects in fragmented forests—implications for conservation. Trends in Ecology and Evolution 10:58–62.

Mussell, A., G. Fox, and D. W. McKenney. 1996. An economic analysis of protecting woodland caribou habitat in northwestern Ontario. Canadian Forest Service, Sault Ste. Marie, Ontario, Canada. NODA/NFP File Report 49.

Mussell, R. A. 1995. Tradeoffs in protecting woodland caribou habitat in northwestern Ontario. M.Sc. dissertation. University of Guelph, Guelph, Ontario, Canada.

Mutch, R. W. 1970. Wildland fires and ecosystems—a hypothesis. Ecology 51:1046–1051.

Mutch, R. W. 1994. Fighting fire with fire: A return to ecosystem health. Journal of Forestry 92:31–33.

Muzika, R. M., and A. M. Liebhold. 1999. Effects of gypsy moth on radial growth of host and non-host tree species. Canadian Journal of Forest Research 29:1365–1373.

Muzika, R. M., and A. M. Liebhold. 2001. Effects of gypsy moth defoliation in oak–pine forests in the northeastern United States. Pages 117–123 in A. M. Liebhold, M. L. McManus, I. S. Otvos, and S. L. C. Fosbroke, editors. Proceedings of the integrated management and dynamics of forest defoliating insects symposium. U.S. Department of Agriculture Forest Service, Radnor, Pennsylvania, United States. General Technical Report NE GTR-277.

Myers, N. 1992. The primary source: Tropical forests and our future, second edition. W. W. Norton, New York, New York, United States.

Naeem, S., and S. Li. 1997. Biodiversity enhances ecosystem reliability. Nature 390:507–509.

Nakamura, F., F. J. Swanson, and S. M. Wondzell. 2000. Disturbance regimes of stream and riparian systems—a disturbance-cascade perspective. Hydrological Processes 14:2849–2860.

Nautiyal, J. C., S. Kant, and J. Williams. 1995. A mechanism for tracking the value of standing timber in an imperfect market. Canadian Journal of Forest Research 25:638–648.

Naveh, Z. 1993. Red books for threatened Mediterranean landscapes as an innovative tool for holistic landscape conservation—introduction to the Western Crete-red-book case study. Landscape and Urban Planning 24:241–247.

Needham, T., J. Kershaw, D. A. MacLean, and Q. Su. 1999. Effects of mixed stand management to reduce impacts of spruce budworm defoliation on balsam fir stand-level growth and yield. Northern Journal of Applied Forestry 16:19–24.

Neilson, R. P. 1995. A model for predicting continental-scale vegetation distribution and water balance. Ecological Applications 5:362–385.

Nelson, J. D. 2000. Reference manual for FPS-ATLAS. University of British Columbia, Vancouver, British Columbia, Canada.

Nelson, J. D. 2001. Forest and landscape models: Powerful tools or dangerous weapons? Pages 45–55 in V. LeMay and P. Marshall, editors. Proceedings of the conference on forest modelling for ecosystem management, forest certification, and sustainable management. University of British Columbia, Vancouver, British Columbia, Canada.

Nguyen, T. 2000. Développement d'une stratégie d'aménagement forestier s'inspirant de la dynamique des perturbations naturelles pour la région Nord de l'Abitibi.

Rapport de recherche effectuée dans le cadre du Volet 1 du programme de mise en valeur des ressources du milieu forestier. Année 1. [Online, URL: <http://web2.uqat.uquebec.ca/cafd/pdf/nguyen1.pdf>.]

Nguyen, T. 2001. Développement d'une stratégie d'aménagement forestier s'inspirant de la dynamique des perturbations naturelles pour la région Nord de l'Abitibi. Rapport de recherche effectuée dans le cadre du Volet 1 du programme de mise en valeur des ressources du milieu forestier. Année 2. [Online, URL: <http://web2.uqat.uquebec.ca/cafd/pdf/nguyen2.pdf>.]

Nguyen, T. 2002. Développement d'une stratégie d'aménagement forestier s'inspirant de la dynamique des perturbations naturelles pour la région Nord de l'Abitibi. Rapport de recherche effectuée dans le cadre du Volet 1 du programme de mise en valeur des ressources du milieu forestier. Année 3. [Online, URL: <http://web2.uqat.uquebec.ca/cafd/pdf/nguyen3.pdf>.]

Niemela, J. 1999. Management in relation to disturbance in the boreal forest. Forest Ecology and Management 115:127–134.

Niese, J. N., and T. F. Strong. 1992. Economic and tree diversity trade-offs in managed northern hardwoods. Canadian Journal of Forest Research 22:1807–1813.

Nigh, T. A. 1997. Landtype associations of the lower Ozark Region. Draft map and descriptions. University of Missouri, Columbia, Missouri, United States. Missouri Ecological Classification System Project.

Nigh, T. A., C. Buck, J. Grabner, J. Kabrick, and D. Meinert. 2000. An ecological classification system for the Current River Hill subsection. University of Missouri, Columbia, Missouri, United States. Working draft.

Niklasson, M., and A. Granström. 2000. Numbers and sizes of fires: Long-term spatially explicit fire history in a Swedish boreal landscape. Ecology 81:1484–1499.

Noble, I. R., and R. O. Slatyer. 1977. Post-fire succession of plants in Mediterranean ecosystems. Pages 27–63 in H. A. Mooney and C. E. Conrad, editors. Proceedings of the symposium on the environmental consequences of fire and fuel management in Mediterranean ecosystems. U.S. Department of Agriculture Forest Service, United States. General Technical Report WO-3.

Noble, I. R., and R. O. Slatyer. 1980. The use of vital attributes to predict successional changes in plant communities subject to recurrent disturbances. Vegetatio 43:5–21.

Noss, R. F. 1987. From plant communities to landscapes in conservative inventories: A look at the Nature Conservancy (USA). Biological Conservation 41:11–37.

Nyland, R. D. 2001. Silviculture: Concepts and applications, second edition. McGraw-Hill, New York, New York, United States.

O'Hara, K. L., P. A. Latham, P. F. Hessburg, and B. G. Smith. 1996. A structural classification of inland northwest forest vegetation. Western Journal of Applied Forestry 11:97–102.

Ohmann, J. L., and M. J. Gregory. 2002. Predictive mapping of forest composition and structure with direct gradient analysis and nearest-neighbor imputation in coastal Oregon, USA. Canadian Journal of Forest Research 32:725–741.

Oliver, C. D., and B. C. Larson. 1990. Forest stand dynamics. McGraw-Hill, New York, New York, United States.

Oliver, C. D., and B. C. Larson. 1996. Forest stand dynamics, updated edition. John Wiley and Sons, New York, New York, United States.

Olsen, J. 1981. Carbon balance in relation to fire regimes. Pages 327–378 *in* H. A. Mooney, T. M. Bonnicksen, N. L. Christensen, J. E. Lotan, and W. A. Reiners, technical coordinators. Proceedings of the conference on fire regimes and ecosystem properties. U.S. Department of Agriculture Forest Service, Washington D.C., United States. General Technical Report WO-26.

Olson, D. H., J. C. Hagar, A. B. Carey, J. H. Cissel, and F. J. Swanson. 2001. Wildlife of westside and high montane forests. Pages 187–212 *in* D. H. Johnson and T. A. O'Neill, editors. Wildlife–habitat relationships in Oregon and Washington. Oregon State University Press, Corvallis, Oregon, United States.

O'Neill, R. V., D. L. DeAngelis, J. B. Waide, and T. F. H. Allen. 1986. A hierarchical concept of ecosystems. Princeton University Press, Princeton, New Jersey, United States.

Ontario Environmental Assessment Board. 1994. Reasons for decision and decision. Class environmental assessment by the Ministry of Natural Resources for timber management on Crown lands in Ontario. Ontario Environmental Assessment Board, Toronto, Ontario, Canada.

Ontario Forest Policy Panel. 1992. Our future, our forests. Government of Ontario, Toronto, Ontario, Canada.

Ontario Forest Policy Panel. 1993. Diversity: Forests, people, communities. Proposed comprehensive forest policy framework for Ontario. Government of Ontario, Toronto, Ontario, Canada.

Ontario Ministry of Natural Resources. 1977. A ready reference. Ontario land inventory. (A guide to the reading and understanding of the maps produced by the Ontario land inventory.) Ontario Ministry of Natural Resources, Toronto, Ontario, Canada.

Ontario Ministry of Natural Resources. 1988. Timber management guidelines for the provision of moose habitat. Ontario Ministry of Natural Resources, Toronto, Ontario, Canada.

Ontario Ministry of Natural Resources. 1994. Management guidelines for woodland caribou habitat. Ontario Ministry of Natural Resources, Toronto, Ontario, Canada. Draft report.

Ontario Ministry of Natural Resources. 1996a. Forest management planning manual for Ontario's Crown forests. Ontario Ministry of Natural Resources, Sault Ste. Marie, Ontario, Canada.

Ontario Ministry of Natural Resources. 1996b. Forest resource inventory database manual. Ontario Ministry of Natural Resources, Toronto, Ontario, Canada.

Ontario Ministry of Natural Resources. 1997. Forest management guidelines for the emulation of fire disturbance patterns—analysis results. Ontario Ministry of Natural Resources, Thunder Bay, Ontario, Canada. Unpublished report.

Ontario Ministry of Natural Resources. 2000. Ontario Climate Model version 2.0. Ontario Ministry of Natural Resources, Sault Ste. Marie, Ontario, Canada. CD-ROM.

Ontario Ministry of Natural Resources. 2002a. Forest management guide for natural disturbance pattern emulation. Version 3.1. Ontario Ministry of Natural Resources, Toronto, Ontario, Canada.

Ontario Ministry of Natural Resources. 2002b. Forest resources of Ontario, 2001. State of the forest report, 2001. Appendix I. Government of Ontario, Toronto, Ontario, Canada.

Oregon Department of Forestry. 2001. Northwest Oregon state forests management plan. Oregon Department of Forestry, Salem, Oregon, United States.

Oregon Department of Forestry. 2002. Forest practice administrative rules. Oregon Department of Forestry, Salem, Oregon, United States.

Orlander, G., P. Gemmel, and J. Hunt. 1990. Site preparation: A Swedish overview. Canadian Forest Service, Victoria, British Columbia, Canada. FRDA Report 105.

Orton, D. 1995. The wild path forward: Left biocentrism, First Nations, park issues and forestry: A Canadian view. Wild Earth 5:45. [Online, URL <http://www.indians.org/library/wild.html#biocentrism>.]

Ostaff, D. P., and D. A. MacLean. 1995. Patterns of balsam fir foliar production and growth in relation to defoliation by spruce budworm. Canadian Journal of Forest Research 25:1128–1136.

Otten, R. H. J. M, and L. P. P. P. van Ginneken. 1989. The annealing algorithm. Kluwer Academic Publishers, Boston, Massachusetts, United States.

Ottmar, R. D., M. F. Burns, J. N. Hall, and A. D. Hanson. 1993. CONSUME users guide. U.S. Department of Agriculture Forest Service, Portland, Oregon, United States. General Technical Report PNW-GTR-304.

Ottmar, R. D., M. D. Schaaf, and E. Alvarado. 1996. Smoke considerations for using fire in maintaining healthy forest conditions. Pages 2–28 *in* C. C. Hardy and S. F. Arno, editors. The use of fire in forest restoration. U.S. Department of Agriculture Forest Service, Ogden, Utah, United States. General Technical Report INT-GTR-341.

Ottmar, R. D., E. Alvarado, P. F. Hessburg, B. G. Smith, S. G. Kreiter, C. A. Miller, and R. B. Salter. in press. Historical and current forest and range landscapes in the Interior Columbia River basin and portions of the Klamath and Great basins. Part II: Linking vegetation patterns and potential smoke production and fire behavior. U.S. Department of Agriculture Forest Service, Portland, Oregon, United States. General Technical Report PNW-GTR-XXX.

Paehlke, R., editor. 1995. Conservation and environmentalism. An encyclopedia. Garland Publishing, New York, New York, United States.

Paine, R. T., M. J. Tegner, and E. A. Johnson. 1998. Compounded perturbations yield ecological surprises. Ecosystems 1:535–545.

Palik, B. J., and R. T. Engstrom. 1999. Species composition. Pages 65–93 *in* M. L. Hunter Jr., editor. Managing forests for biodiversity. Cambridge University Press, Cambridge, United Kingdom.

Palik, B. J., and J. Zasada. 2002. An ecological context for regenerating multi-cohort, mixed-species red pine forests. Pages 54–64 *in* D. W. Gilmore and L. S. Yount, editors. Proceedings of the red pine technical conference of the Society of American Foresters Region V. University of Minnesota, St. Paul, Minnesota, United States. Staff Paper Series 157.

Palik, B. J., R. J. Mitchell, and J. K. Hiers. 2002. Modeling silviculture after natural disturbance to sustain biodiversity in the longleaf pine (*Pinus palustris*) eco-system: Balancing complexity and implementation. Forest Ecology and Management 155:347–356.

Parish, R., and J. A. Antos. 2002. Dynamics of an old-growth, fire-initiated, subalpine forest in southern interior British Columbia: Tree-ring reconstruction of 2-year cycle spruce budworm outbreaks. Canadian Journal of Forest Research 32:1947–1960.

Parish, R., T. A. Antos, and M.-J. Fortin. 1999. Stand development in an old-growth subalpine forest in southern interior British Columbia. Canadian Journal of Forest Research 29:1347–1356.

Parsons, D. J., T. W. Swetnam, and N. L. Christensen. 1999. Uses and limitations of historical variability concepts in managing ecosystems. Ecological Applications 9:1177–1178.

Partridge, L. W., N. F. Britton, and N. R. Franks. 1998. Army ant population dynamics: The effect of habitat quality and reserve size on population size and time to extinction. Proceedings of the Royal Society of London B. 265:735–741.

Pastor, J., and W. M. Post. 1985. Development of a linked forest productivity–soil process model. Oak Ridge National Laboratory, Oak Ridge, Tennessee, United States.

Payette, S., N. Bhiry, A. Delwaide, and M. Simard. 2000. Origin of the lichen woodland at its southern range limit in eastern Canada: The catastrophic impact of insect defoliators and fire on the spruce–moss forest. Canadian Journal of Forest Research 30:288–305.

Péch, G. 1993. Fire hazard in budworm-killed balsam fir stands on Cape Breton Highlands. The Forestry Chronicle 69:178–185.

Pecore, M. 1992. Menominee-sustained yield management: A successful land ethic in practice. Journal of Forestry 90:12–16.

Pennanen, J., and T. Kuuluvainen. 2002. A spatial simulation approach to natural forest landscape dynamics in boreal Fennoscandia. Forest Ecology and Management 164:157–175.

Perala, D. A. 1977. Manager's handbook for aspen in the North Central states. U.S. Department of Agriculture Forest Service, St. Paul, Minnesota, United States. General Technical Report NC-36.

Perala, D. A. 1990. Déjà vu: "Does it pay to thin aspen?" Pages 139–144 *in* V. F. Haavisto, C. R. Smith, and C. Mason, editors. Space to grow: Spacing and thinning in northern Ontario. Forestry Canada, Sault Ste. Marie, Ontario, Canada. Joint Report 15.

Perera, A. H., and D. J. B. Baldwin. 2000. Spatial patterns in the managed landscape of Ontario. Pages 74–99 *in* A. H. Perera, D. L. Euler, and I. D. Thompson, editors. Ecology of a managed terrestrial landscape: Patterns and processes of forest landscapes in Ontario. University of British Columbia Press, Vancouver, British Columbia, Canada.

Perera, A. H., D. J. B. Baldwin, F. Schnekenburger, J. E. Osborne, and R. E. Bae. 1998. Forest fires in Ontario: A spatio-temporal perspective. Ontario Ministry of Natural Resources, Sault Ste. Marie, Ontario, Canada. Forest Research Report 147.

Perera, A. H., D. J. B. Baldwin, D. Yemshanov, F. Schnekenburger, K. Weaver, and D. Boychuk. 2003. Predicting the potential for old growth forests by spatial sim-ulation of landscape aging patterns. The Forestry Chronicle 79(3):621–630.

Perera, A. H., D. Yemshanov, F. Schnekenburger, K. Weaver, D. J. B. Baldwin, and D. Boychuk. 2004. Boreal Forest Landscape Dynamics Simulator (BFOLDS): A grid-based spatially stochastic model for predicting crown fire regime and forest cover transition. Ontario Ministry of Natural Resources, Sault Ste. Marie, Ontario, Canada. Forest Research Report 152.

Perez, B., and J. M. Moreno. 1998. Methods for quantifying fire severity in shrubland-fires. Plant Ecology 139:91–101.

Perry, D. A. 1994. Forest ecosystems. The Johns Hopkins University Press, Baltimore, Maryland, United States.

Perry, D. A., and M. P. Amaranthus. 1997. Disturbance, recovery and stability. Pages 31–56 *in* K. A. Kohm and J. F. Franklin, editors. Creating a forestry for the 21st century: The science of ecosystem management. Island Press, Washington, D.C., United States.

Peterken, G. F. 1996. Natural woodland: Ecology and conservation in northern temperate regions. Cambridge University Press, Cambridge, United Kingdom.

Peterson, C. J. 2000. Catastrophic wind damage to North American forests and the potential impact of climate change. Science of the Total Environment 262:287–311.

Petraitis, P. S., R. E. Latham, and R. A. Neisenbaum. 1989. The maintenance of species diversity by disturbance. Quarterly Review of Biology 64:393–418.

Pfister, R. D., B. L. Kovalchik, S. F. Arno, and R. C. Presby. 1977. Forest habitat types of Montana. U.S. Department of Agriculture Forest Service, Missoula, Montana, United States. General Technical Report INT-34.

Phillips, J. 1935. Succession, development, the climax and the complex organism: An analysis of concepts. Journal of Ecology 23:210–246.

Pickett, S. T. A. 1989. Space for time substitution as an alternative to long-term studies. Pages 110–135 *in* G. A. Likens, editor. Long-term studies in ecology: Approaches and alternatives. Springer-Verlag, New York, New York, United States.

Pickett, S. T. A., and P. S. White, editors. 1985. The ecology of natural disturbance and patch dynamics. Academic Press, San Diego, California, United States.

Porter, K. B., D. A. MacLean, K. P. Beaton, and J. Upshall. 2001. New Brunswick permanent sample plot database (PSPDB v1.0): User's guide and analysis. Canadian Forest Service, Fredericton, New Brunswick, Canada. Information Report M-X-209.

Post, W. M., and J. Pastor. 1996. An individual-based forest ecosystem model. Climate Change 34:253–261.

Potvin, F., L. Belanger, and K. Lowell. 2000. Marten habitat selection in a clearcut landscape. Conservation Biology 14:844–857.

Power, J. M. 1991. National data on forest pest damage. Pages 119–129 *in* D. G. Brand, editor. Canada's timber resources. Canadian Forest Service, Petawawa, Ontario, Canada. Information Report PI-X-101.

Power, T. M. 1998. Lost landscapes and failed economies: The search for a value of place. Island Press, Washington, D.C., United States.

Prebble, M. L., editor. 1975. Aerial control of forest insects: A review of control projects employing chemical and biological insecticides. Canada Department of the Environment, Ottawa, Ontario, Canada.

Prescott, C. E., and G. F. Weetman, editors. 1994. Salal cedar hemlock integrated research program: A synthesis. University of British Columbia, Vancouver, British Columbia, Canada.

Preston, F. W. 1980. Noncanonical distributions of commonness and rarity. Ecology 61:88–97.

Prichard, S. J. 2003. Spatial and temporal dynamics of fire and forest succession in a mountain watershed, North Cascades National Park. Ph.D. dissertation. University of Washington, Seattle, Washington, United States.

Prigogine, I. 1980. From being to becoming. W. H. Freeman and Co., San Francisco, California, United States.

Proe, M. F., J. C. Dutch, D. G. Pyatt, and J. P. Kimmins. 1997. A strategy to develop a guide for whole-tree harvesting of Sitka spruce in Great Britain. Biomass and Bioenergy 13:289–299.

Puccia, C. J., and R. Levins. 1985. Qualitative modeling of complex systems: An introduction to loop analysis and time averaging. Harvard University Press, Cambridge, Massachusetts, United States.

Puettmann, K. J., and A. R. Ek. 1999. Status and trends of silvicultural practices in Minnesota. Northern Journal of Applied Forestry 16:203–210.

Pulliam, H. R., and B. J. Danielson. 1991. Sources, sinks and habitat selection: A landscape perspective on community dynamics. American Naturalist 137:50–66.

Pyne, S. J. 1982. Fire in America: A cultural history of wildland and rural fire. University of Washington Press, Seattle, Washington, United States.

Pyne, S. J. 1988. Fire in America. Princeton University Press, Princeton, New Jersey, United States.

Pyne, S. J., P. L. Andrews, and R. D. Laven. 1996. Introduction to wildland fire, second edition. John Wiley and Sons, New York, New York, United States.

Quigley, T. M., R. W. Haynes, and R. T. Graham. 1996. Integrated scientific assessment for ecosystem management in the Interior Columbia basin. U.S. Department of Agriculture Forest Service, Portland, Oregon, United States. General Technical Report PNW-GTR-382.

Racey, G. D., K. Abraham, W. R. Darby, H. R. Timmerman, and Q. Day. 1992. Can woodland caribou and the forest industry coexist: The Ontario scene. Rangifer 12:108–115.

Racey, G. D., A. Harris, L. Gerrish, T. Armstrong, J. McNicol, and J. A. Baker. 2002. Forest management guidelines for the conservation of woodland caribou: A landscape approach for use in northwestern Ontario. Ontario Ministry of Natural Resources, Sault Ste. Marie, Ontario, Canada.

Radeloff, V. C., D. J. Mladenoff, H. S. He, and M. S. Boyce. 1999. Forest landscape change in the northwestern Wisconsin pine barrens from pre-European settlement to the present. Canadian Journal of Forest Research 29:1649–1659.

Radeloff, V. C., D. J. Mladenoff, and M. S. Boyce. 2000. The changing relation of landscape patterns and jack pine budworm populations during an outbreak. Oikos 90:417–430.

Rains, M. T., and J. Hubbard. 2002. Protecting communities through the National Fire Plan. Fire Management Today 62:4–12.

Rajala, J. 1998. Bringing back the white pine. Jack Rajala, Rajala Company, Deer River, Minnesota, United States.

Rajora, O. P., L. DeVerno, A. Mosseler, and D. J. Innes. 1998. Genetic diversity and population structure of disjunct Newfoundland and central Ontario populations of eastern white pine. Canadian Journal of Botany 76:500–508.

Rajora, O. P., A. Mosseler, and J. E. Major. 2002. Mating system and reproductive fitness of eastern white pine (*Pinus strobus*) in large, central versus small, isolated, marginal populations. Canadian Journal of Botany 80:1173–1184.

Ramsey, G. S., and Higgins, D. G. 1991. Canadian Forest Fire Statistics: 1984–1987. Petawawa National Forestry Institute, Petawawa, Ontario. Information Report PI-X-74E.

Randall, A., and G. L. Peterson. 1984. The valuation of wildland benefits: An overview. Pages 1–52 *in* G. L. Peterson and A. Randall, editors. Valuation of wildland resource benefits. Westview Press, London, United Kingdom.

Rapport, D. J., C. Gaudet, J. R. Karr, J. S. Baron, C. Bohlen, W. Jackson, B. Jones, R. J. Naiman, B. Norton, and M. M. Pollock. 1998. Evaluating landscape health: Integrating societal goals and biophysical processes. Journal of Environmental Management 53:1–15.

Rasker, R., and A. Hansen. 2000. Natural amenities and population growth in the Greater Yellowstone region. Human Ecology Review 7:30–40.

Rasmussen, L. A., G. D. Amman, J. C. Vandygriff, R. D. Oakes, A. S. Munson, and K. E. Gibson. 1996. Bark beetle and wood borer infestation in the Greater Yellowstone area during four postfire years. U.S. Department of Agriculture Forest Service, Ogden, Utah, United States. Research Report INT-RP-487.

Rassi, P., H. Kaipiainen, I. Mannerkoski, and G. Ståhls. 1991. Report on the monitoring of threatened animals and plants in Finland. Ministry of the Environment, Helsinki, Finland. [In Finnish, with English abstract.] Committee Report 1991:30.

Reed, R. A., R. K. Peet, M. W. Palmer, and P. S. White. 1993. Scale dependence of vegetation–environment correlations—a case study of a North Carolina Piedmont woodland. Journal of Vegetation Science 4:329–340.

Reed, R. A., J. Johnson-Barnard, and W. L. Baker. 1996. Fragmentation of a forested Rocky Mountain landscape, 1950–1993. Biological Conservation 75:267–277.

Reed, R. A., M. E. Finley, W. H. Romme, and M. G. Turner. 1999. Aboveground net primary production and leaf-area index in early postfire vegetation in Yellowstone National Park. Ecosystems 2:88–94.

Reed, W. J. 1997. Estimating historical forest-fire frequencies from time-since-last-fire sample data. IMA Journal of Mathematics Applied in Medicine and Biology 14:71–83.

Reed, W. J., and K. S. McKelvey. 2002. Power-law behaviour and parametric models for the size-distribution of forest fires. Ecological Modelling 150:239–254.

Reed, W. J., C. P. S. Larsen, E. A. Johnson, and G. M. MacDonald. 1998. Estimation of temporal variation in historical fire frequency from time-since-fire map data. Forest Science 44:465–475.

Rees, D. C., and G. P. Juday. 2002. Plant species diversity on logged versus burned sites in central Alaska. Forest Ecology and Management 155:291–302.

Reeves, G. H., L. E. Benda, K. M. Burnett, P. A. Bisson, and J. R. Sedell. 1995. A disturbance-based ecosystem approach to maintaining and restoring freshwater habitats of evolutionarily significant units of anadromous salmonids in the Pacific Northwest. American Fisheries Society Symposium 17:334–349.

Reich, P. B., P. Bakken, D. Carleson, L. E. Frelich, S. K. Friedman, and D. F. Grigal. 2001. Influence of logging, fire, and forest type on biodiversity and productivity in southern boreal forests. Ecology 82:2731–2748.

Reinhardt, E. D., R. E. Keane, and J. K. Brown. 1996. First order fire effects model—FOFEM 4.0—User's Guide. U.S. Department of Agriculture, Forest Service, Ogden, Utah, United States. General Technical Report INT-GTR-344.

Reinhardt, F. 1999. Bringing the environment down to earth. Harvard Business Review 77:149–157.

Remillard, M. M., G. K. Greundling, and D. J. Booguchi. 1987. Disturbance by beaver (Castor canadenis Kuhl.) and increased landscape heterogeneity. Pages 103–122 in M. G. Turner, editor. Landscape heterogeneity and disturbance. Springer-Verlag, New York, New York, United States.

Rempel, R., P. C. Elkie, A. R. Rogers, and M. J. Gluck. 1997. Timber management and natural disturbance effects on moose habitat: Landscape evaluation. Journal of Wildlife Management 61:517–524.

Remsoft. 1996. Woodstock Forest Modelling System version 1.1 user's guide. Remsoft, Fredericton, New Brunswick, Canada.

Renkin, R. A., and D. G. Despain. 1992. Fuel moisture, forest type and lightning-caused fire in Yellowstone National Park. Canadian Journal of Forest Research 22:37–45.

Restrepo, V. R., P. M. Mace, and F. M. Serchuk. 1999. The precautionary approach: A new paradigm, or business as usual? Pages 61–70 in Our living oceans: Report on the status of U.S. living marine resources, 1999. U.S. National Marine and Fisheries Service, Seattle, Washington, United States. NOAA Technical Memo NMFS-F/SPO-41. [Online, URL: <http://spo.nwr.noaa.gov/fa1.pdf>.]

Reynolds, K. M. 1999a. EMDS users guide (version 2.0): Knowledge-based decision support for ecological assessment. U.S. Department of Agriculture, Forest Service, Portland, Oregon, United States. General Technical Report PNW-GTR-470.

Reynolds, K. M. 1999b. NetWeaver for EMDS user guide (version 1.1): A knowledge base development system. U.S. Department of Agriculture Forest Service, Portland, Oregon, United States. General Technical Report PNW-GTR-471.

Reynolds, K. M. 2001a. Using a logic framework to assess forest ecosystem sustainability. Journal of Forestry 99: 26–30.

Reynolds, K. M. 2001b. Fuzzy logic knowledge bases in integrated landscape assessment: Examples and possibilities. U.S. Department of Agriculture Forest Service, Portland, Oregon, United States. General Technical Report PNW-GTR-521.

Reynolds, K. M., and P. F. Hessburg. 2004. Decision support for integrated landscape evaluation and restoration planning. Forest Ecology and Management (invited feature).

Reynolds, K., C. Cunningham, L. Bednar, M. Saunders, M. Foster, R. Olson, D. Schmoldt, D. Latham, B. Miller, and J. Steffenson. 1996. A knowledge-based information management system for watershed analysis in the Pacific Northwest, USA. Information Applications 10:9–22.

Riebesame, W. E., H. Gosness, and D. Theobald. 1997. Atlas of the new West: Portrait of a changing region. W. W. Norton, New York, New York, United States.

Ripple, W. J., G. A. Bradshaw, and T. A. Spies. 1991. Measuring forest landscape patterns in the Cascade range of Oregon, USA. Biological Conservation 57:73–88.

Ripple, W. J., K. T. Hershey, and R. G. Anthony. 2000. Historical forest patterns of Oregon's central Coast Range. Biological Conservation 93:127–133.

Roane, M. K., G. J. Griffin, and J. R. Elkin. 1986. Chestnut blight, other Endothia diseases, and the genus Endothia. American Phytopathological Society, St. Paul, Minnesota, United States.

Roberts, D. W., and D. W. Betz. 1999. Simulating landscape vegetation dynamics of Bryce Canyon National Park with the vital attributes/fuzzy systems model VAFS.LANDSIM. Pages 99–123 in D. J. Mladenoff and W. L. Baker, editors. Spatial modeling of forest landscape change: Approaches and applications. Cambridge University Press, Cambridge, United Kingdom.

Robertson, J. M. Y., and C. P. van Shaik. 2001. Causal factors underlying the dramatic decline of Sumatran orangutan. Oryx 35:26–38.

Robitaille, A., and J.-P. Saucier. 1998. Les paysages régionaux du Québec méridional. Les Publications du Québec, Sainte-Foy, Quebec, Canada.

Rodriguez, C. E., J. J. Bommer, and R. J. Chandler. 1999. Earthquake-induced landslides: 1980–1997. Soil Dynamics and Earthquake Engineering 18:325–346.

Rogers, P. 1996. Disturbance ecology and forest management: A review of the literature. U.S. Department of Agriculture Forest Service, Ogden, Utah, United States. General Technical Report INT-GTR-336.

Roland, J. 1993. Large-scale forest fragmentation increases the duration of tent caterpillar outbreak. Oecologia 93:25–30.

Rollins, M. G., T. W. Swetnam, and P. Morgan. 2001. Evaluating a century of fire patterns in two Rocky Mountain wilderness areas using digital fire atlases. Canadian Journal of Forest Research 31:2107–2123.

Romme, W. H. 1982. Fire and landscape diversity in subalpine forests of Yellowstone National Park. Ecological Monographs 52:199–221.

Romme, W. H. 1997. Creating pseudo-rural landscapes in the mountain west. Pages 139–161 in J. Nassauer and D. Karasov, editors. Placing nature: Culture and landscape ecology. Island Press, Washington, D.C., United States.

Romme, W. H., and D. G. Despain. 1989. The Yellowstone fires. Scientific American 261:37–46.

Romme, W. H., and D. H. Knight. 1982. Landscape diversity: The concept applied to Yellowstone Park. BioScience 32:664–670.

Romme, W. H., and M. G. Turner. in press. Ten years after the Yellowstone fires: Is restoration needed? in L. L. Wallace and N. Christensen, editors. Ten years after the 1988 Yellowstone fires. Yale University Press, New Haven, Connecticut, United States.

Romme, W. H., D. H. Knight, and J. B. Yavitt. 1986. Mountain pine beetle outbreaks in the Rocky Mountains: Regulators of primary productivity? American Naturalist 127:484–494.

Romme, W. H., M. G. Turner, L. L. Wallace, and J. Walker. 1995a. Aspen, elk, and fire in northern Yellowstone National Park. Ecology 76:2097–2106.

Romme, W. H., L. Bohland, C. Persichetty, and T. Caruso. 1995b. Germination ecology of some common forest herbs in Yellowstone National Park, Wyoming, U.S.A. Arctic and Alpine Research 27:407–412.

Romme, W. H., M. G. Turner, R. H. Gardner, W. W. Hargrove, G. A. Tuskan, D. G. Despain, and R. Renkin. 1997. A rare episode of sexual reproduction in aspen (*Populus tremuloides* Michx.) following the 1988 Yellowstone fires. Natural Areas Journal 17:17–25.

Romme, W. H., L. E. Everham, M. A. Frelich, R. E. Moritz, and R. E. Sparks. 1998. Are large, infrequent disturbances qualitatively different from small, frequent disturbances? Ecosystems 1:524–534.

Romme, W. H., L. Floyd, D. Hanna, and J. S. Redders. 2000. Using natural disturbance regimes to mitigate forest fragmentation in the central Rocky Mountains. Chapter 18 *in* R. L. Knight, F. W. Smith, S. W. Buskirk, W. H. Romme, and W. L. Baker, editors. Forest fragmentation in the southern Rocky Mountains. University Press of Colorado, Niwot, Colorado, United States.

Rossi, R. E., D. J. Mulla, A. G. Journel, and E. H. Franz. 1992. Geostatistical tools for modelling and interpreting ecological spatial dependence. Ecological Monographs 62:277–314.

Rotherham, T. 1991. Timber harvest statistics: Past practice and present needs. Pages 105–110 *in* D. G. Brand, editor. Canada's timber resources. Canadian Forest Service, Petawawa, Ontario, Canada. Information Report PI-X-101.

Rothermel, R. C. 1972. A mathematical model for predicting fire spread in wildland fuels. U.S. Department of Agriculture Forest Service, Ogden, Utah, United States. Research Paper INT-RP-115.

Rothermel, R. C. 1983. How to predict the spread and intensity of forest and range fires. U.S. Department of Agriculture Forest Service, Ogden, Utah, United States. General Technical Report INT-GTR-143.

Row, C. V., H. F. Kaiser, and J. Sessions 1981. Discount rate for long-term forest service investments. Journal of Forestry 79:367–376.

Rowe, J. S. 1972. Forest regions of Canada. Canadian Forest Service, Ottawa, Ontario, Canada. Publication 1300.

Royama, T. 1984. Population dynamics of the spruce budworm *Choristoneura fumiferana*. Ecological Monographs 54:429–462.

Ruark, G. A. 1990. Evidence for the reserve shelterwood system for managing quaking aspen. Northern Journal of Applied Forestry 7:58–62.

Ruel, J. C. 2000. Factors influencing windthrow in balsam fir forests: From landscape studies to individual tree studies. Forest Ecology and Management 135:169–178.

Ruffner, C. M., and M. D. Abrams. 1998. Relating land-use history and climate to the dendroecology of a 326-year-old *Quercus prinus* talus slope forest. Canadian Journal of Forest Research 28:347–358.

Ruggerio, L. F., K. B. Aubry, A. B. Carey, and M. H. Huff. 1991. Wildlife and vegetation in unmanaged Douglas-fir forests. U.S. Department of Agriculture Forest Service, Portland, Oregon, United States. General Technical Report PNW-GTR-285.

Runkle, J. R. 1982. Patterns of disturbance in some old-growth mesic forests of eastern North America. Ecology 63:1533–1546.

Running, S. W., and S. T. Gower. 1991. FOREST-BGC. A general model of forest ecosystem processes for regional applications II. Dynamic carbon allocation and nitrogen budgets. Tree Physiology 9:147–160.

Running, S. W., and R. Hunt. 1993. Generalization of a forest ecosystem process model for other biomes, BIOME-GCG, and application for global scale models. Pages 141–158 *in* J. Ehrlinger and C. Field, editors. Scaling physiological processes: Leaf to globe. Academic Press, New York, New York, United States.

Russell, E. W. B., R. B. Davis, R. S. Anderson, T. E. Rhodes, and D. S. Anderson. 1993. Recent centuries of vegetational change in the glaciated north-eastern United States. Journal of Ecology 81:647–664.

Russell-Smith, J., P. G. Ryan, D. Klessa, G. Waight, and R. Harwood. 1998. Fire regimes, fire-sensitive vegetation and fire management of the sandstone Arnhem Plateau, monsoonal northern Australia. Journal of Applied Ecology 35:829–846.

Ryan, K. C. 2002. Dynamic interactions between forest structure and fire behavior in boreal ecosystems. Silva Fennica 36:13–39.

Ryan, K. C., and N. V. Noste. 1985. Evaluating prescribed fires. Pages 230–237 *in* J. E. Lotan, B. M. Kilgore, W. C. Fischer, and R. W. Mutch, editors. Wilderness fire symposium. U.S. Department of Agriculture Forest Service, Missoula, Montana, United States. General Technical Report INT-182.

Rykiel, E. J., Jr. 1996. Testing ecological models: The meaning of validation. Ecological Modelling 90:229–244.

Sala, O. E., F. S. Chapin III, J. J. Armesto, R. Berlow, J. Bloomfield, R. Dirzo, E. Huber-Sanwald, L. F. Huenneke, R. B. Jackson, A. Kinzig, R. Leemans, D. Lodge, H. A. Mooney, M. Oesterheld, N. L. Poff, M. T. Sykes, B. H. Walker, M. Walker, and D. H. Wall. 2000. Global biodiversity scenarios for the year 2100. Science 287: 1770–1774.

Salim, E., and O. Ullsten. 1999. Our forests, our future. Report of the World Commission on Forests and Sustainable Development. Cambridge University Press, Cambridge, United Kingdom.

Salwasser, H. 1994. Ecosystem management: Can it sustain diversity and productivity? Journal of Forestry 92: 6–10.

Samuelson, P. A. 1976. Economics of forestry in an evolving society. Economic Inquiry 14:466–492.

Sandberg, D. V., R. D. Ottmar, and G. H. Cushon. 2001. Characterizing fuels in the 21st century. International Journal of Wildland Fire 10:381–387.

Saracco, J. F., and J. A. Collazo. 1999. Predation on artificial nests along three edge types in a North Carolina bottomland hardwood forest. Wilson Bulletin 111: 541–549.

Sargent, R. G. 2000. Verification, validation, and accreditation of simulation models. Pages 50–59 *in* J. A. Joines, R. R. Barton, K. Kang, and P. A. Fishwick, editors. Proceedings of the 2000 winter simulation conference. Informs College on Simulation. [Online, URL: <http://www.informs-cs.org/wscoopapers/009.PDF>.]

Saskatchewan Environment and Resources Management. 2002. Saskatchewan forest ecosystem impacts monitoring framework. Saskatchewan Environment and Resources Management, Prince Albert, Saskatchewan, Canada.

Saucier, J.-P., J.-F. Bergeron, P. Grondin, and A. Robitaille. 1998. The land regions of southern Québec (3rd version): One element in the hierarchical land classification system developed by the Ministère des Ressources Naturelles du Québec. Ministère des Ressources Naturelles du Québec, Quebec City, Quebec, Canada. Internal report.

Schaaf, M. D. 1996. Development of the fire emission tradeoff model (FETM) and application to the Grande Ronde River basin, Oregon. CH$_2$MHill Report No. 53-82FT-03-2. (On file with: CH$_2$MHill Co., 825 NE Multinomah Building, Suite 1300, Portland, Oregon, United States, 97232.)

Schaetzl, R. J., and D. G. Brown. 1996. Forest associations and soil drainage classes in presettlement Baraga County, Michigan. The Great Lakes Geographer 3:57–74.

Schaller, G. B. 1976. The mountain gorilla: Ecology and behavior. University of Chicago Press, Chicago, Illinois, United States.

Schieck, J., M. Nietfeld, and J. B. Stelfox. 1995. Differences in bird species richness and abundance among three stages of aspen-dominated boreal forests. Canadian Journal of Zoology 73:1417–1431.

Schimmel, J., and A. Granström. 1996. Fire severity and vegetation response in the boreal Swedish forest. Ecology 77:1436–1450.

Schlapfer, F., and B. Schmid. 1999. Ecosystem effects of biodiversity: A classification of hypotheses and exploration of empirical results. Ecological Applications 9:893–912.

Schmidt, K. M., J. J. Roering, J. D. Stock, W. E. Dietrich, D. R. Montgomery, and T. Schaub. 2001. The variability of root cohesion as an influence on shallow landslide susceptibility in the Oregon Coast Range. Canadian Geotechnical Journal 38(5):995–1024.

Schmidt, K. M., J. P. Menakis, C. C. Hardy, W. J. Hann, and D. L. Bunnell. 2002. Development of coarse-scale spatial data for wildland fire management. U.S. Department of Agriculture Forest Service, Missoula, Montana, United States. General Technical Report RMRS-GTR-87.

Schmidt, T. L. 1996. Wisconsin forest statistics, 1996. U.S. Department of Agriculture Forest Service, St. Paul, Minnesota, United States. Research Bulletin NC-183.

Schmiegelow, F. K. A., and M. Monkkonen. 2002. Habitat loss and fragmentation in dynamic landscapes: Avian perspectives from the boreal forest. Ecological Applications 12:375–389.

Schmiegelow, F. K. A., C. S. Machtans, and S. J. Hannon. 1997. Are boreal birds resilient to forest fragmentation? An experimental study of short-term community responses. Ecology 78:1914–1932.

Schmoldt, D. L., and H. M. Rauscher. 1995. Building knowledge-based systems for natural resource management. Chapman and Hall, New York, New York, United States.

Schmoldt, D. L., D. L. Peterson, R. E. Keane, J. M. Lenihan, D. McKenzie, D. R. Weise, and D. V. Sandberg.

1999. Assessing the effects of fire disturbance on ecosystems: A scientific agenda for research and management. U.S. Department of Agriculture Forest Service, Portland, Oregon, United States. General Technical Report PNW-GTR-455.

Schoemaker, P. J. H. 1995. Scenario planning: A tool for strategic thinking. Sloan Management Review (Winter): 25–40.

Schorger, A. W. 1955. The passenger pigeon. University of Wisconsin Press, Madison, Wisconsin, United States.

Schowalter, T. D., W. W. Hargrove, and D. A. Crossley, Jr. 1986. Herbivory in forested ecosystems. Annual Review of Entomology 31:177–196.

Schroeder, D., and A. Perera. 2002. A comparison of large-scale spatial vegetation patterns following clearcuts and fires in Ontario's boreal forests. Forest Ecology and Management 159:217–230.

Schulte, L. A., and D. J. Mladenoff. 2001. The original U.S. public land survey records: Their use and limitations in reconstructing presettlement vegetation. Journal of Forestry 99:5–10.

Schulze, E. D., and H. A. Mooney. 1993. Biodiversity and ecosystem function. Springer-Verlag, Berlin, Germany.

Seaber, P. R., P. F. Kapinos, and G. L. Knapp. 1987. Hydrologic unit maps. United States Geological Survey, Washington, D.C., United States. Water-Supply Paper 2294.

Sedjo, R. A. 2001. The role of forest plantations in the world's future timber supply. The Forestry Chronicle 77:221–225.

Sedjo, R. A., and D. B. Botkin. 1997. Using forest plantations to spare natural forest. Environment 39:15–20, 30.

Seely, B., J. P. Kimmins, C. Welham, and K. Scoullar. 1999. Defining stand-level sustainability, exploring stand-level stewardship. Journal of Forestry 97:4–10.

Seymour, R. S., and M. L. Hunter, Jr. 1992. New forestry in eastern spruce–fir forests: Principles and applications to Maine. Maine Agricultural Experiment Station, Orono, Maine, United States. Miscellaneous Publication 716.

Seymour, R. S., and M. L. Hunter. 1999. Principles of ecological forestry. Pages 22–61 in M. L. Hunter Jr., editor. Maintaining biodiversity in forest ecosystems. Cambridge University Press, Cambridge, United Kingdom.

Seymour, R. S., A. S. White, and P. G. DeMaynadier. 2002. Natural disturbance regimes in northeastern North America—evaluating silvicultural systems using natural scales and frequencies. Forest Ecology and Management 155:357–367.

Sharov, A. A., D. Leonard, A. M. Liebhold, E. A. Roberts, and W. Dickerson. 2002. "Slow the spread": A national program to contain the gypsy moth. Journal of Forestry 100:30–36.

Shaw, C. G. III, A. R. Stage, and P. McNamee. 1991. Modeling the dynamics, behavior, and impact of Armillaria root disease. U.S. Department of Agriculture Forest Service, Washington, D.C., United States. Agricultural Handbook 691.

Sheppard, S. R. J., and H. W. Harshaw, editors. 2001. Forests and landscapes. Linking ecology, sustainability and aesthetics. IUFRO and CABI Publishing International, Oxford, United Kingdom. IUFRO Research Series 6.

Shifley, S. R., and B. L. Brookshire, editors. 2000. Missouri Ozark Forest Ecosystem Project: Site history, soils, landforms, woody and herbaceous vegetation, down wood, and inventory methods for the landscape experiment. U.S. Department of Agriculture Forest Service, St. Paul, Minnesota, United States. General Technical Report NC-GTR-208.

Shifley, S. R., and J. M. Kabrick, editors. 2002. Proceedings of the second Missouri Ozark Forest Ecosystem symposium: Post-treatment results of the landscape experiment. October 17–20, 2000, St. Louis, Missouri. U.S. Department of Agriculture Forest Service, St. Paul, Minnesota, United States. General Technical Report NC-227.

Shifley, S. R., F. R. Thompson, III, D. R. Larsen, and D. J. Mladenoff. 1997. Modeling forest landscape change in the Ozarks: Guiding principles and preliminary implementation. U.S. Department of Agriculture Forest Service, St. Paul, Minnesota, United States. General Technical Report NC-GTR-188.

Shifley, S. R., F. R. Thompson III, W. D. Dijak, and D. R. Larsen. 1999. Modeling landscape changes in the Missouri Ozarks in response to alternative management practices. Pages 267–278 in J. W. Stringer and D. L. Loftis, editors. Proceedings of the 12th central hardwood forest conference. U.S. Department of Agriculture Forest Service, Asheville, North Carolina, United States. General Technical Report SRS-GTR-24.

Shifley, S. R., F. R. Thompson III, D. R. Larsen, and W. D. Dijak. 2000. Modeling forest landscape change in the Missouri Ozarks under alternative management practices. Computers and Electronics in Agriculture 27:7–27.

Shinneman, D. J., and W. L. Baker. 1997. Nonequilibrium dynamics between catastrophic disturbances and old-growth forests in ponderosa pine landscapes of the Black Hills. Conservation Biology 11:1276–1288.

Shumway, D. L., M. D. Abrams, and C. M. Ruffner. 2001. A 400-year history of fire and oak recruitment in an old-growth oak forest in western Maryland, U.S.A. Canadian Journal of Forest Research 31:1437–1443.

Sidle, R. C. 1991. A conceptual model of changes in root cohesion in response to vegetation management. Journal of Environmental Quality 20:43–52.

Sidle, R. C. 1992. A theoretical model of the effects of timber harvesting on slope stability. Water Resources Research 28:1897–1910.

Sierra Nevada Ecosystem Project. 1996. Status of the Sierra Nevada. Final Report to Congress. Centers for Water and Wildland Resources, University of California, Davis, California, United States.

Sillett, S. C., B. McCune, J. E. Peck, T. R. Rambo, and A. Ruchty. 2000. Dispersal limitations of epiphytic lichens result in species dependent on old growth forests. Ecological Applications 10:789–799.

Silvester, R. 1986. The wilderness theme. Environment Views 9:21–24.

Simard, A. J. 1991. Fire severity, changing scales, and how things hang together. International Journal of Wildland Fire 1:23–34.

Simila, M., J. Kouki, P. Martikainen, and A. Uotila. 2002. Conservation of beetles in boreal pine forests: The effects of forest age and naturalness on species assemblages. Biological Conservation 106:19–27.

Simon, N. P. P., F. E. Schwab, and R. D. Otto. 2002. Songbird abundance in clear-cut and burned stands: A comparison of natural disturbance and forest management. Canadian Journal of Forest Research 32:1343–1350.

Simpson, E. H. 1949. Measurement of diversity. Nature 163:688.

Skinner, C. N., and C. R. Chang. 1996. Fire regimes, past and present. Sierra Nevada Ecosystem Project: Final report to Congress. Volume II. Wildland Resources Center, University of California Davis, Davis, California, United States. Report 37.

Skinner, W. R., B. J. Stocks, D. L. Martell, B. Bonsal, and A. Shabbar. 1999. The association between circulation anomalies in the mid-troposphere and area burned by wildland fire in Canada. Theoretical and Applied Climatology 63:89–105.

Skinner, W. R., M. D. Flannigan, B. J. Stocks, D. L. Martell, B. M. Wotton, J. B. Todd, J. A. Mason, K. A. Logan, and E. M. Bosch. 2002. A 500 ha synoptic wildland fire climatology for large Canadian forest fires, 1959–1996. Theoretical and Applied Climatology 71:157–169.

Skre, O., F. E. Wielgolaski, and B. Moe. 1998. Biomass and chemical composition of common forest plants in response to fire in western Norway. Journal of Vegetation Science 9:501–510.

Snow, D. R. 1994. The Iroquois. Blackwell Scientific, Cambridge, Massachusetts, United States.

Sousa, W. P. 1984. The role of disturbance in natural communities. Annual Review of Ecology and Systematics 15:353–391.

Spectranalysis. 1999. Ontario land cover database: User's manual. Revised October 1999. Spectranalysis, Oakville, Ontario, Canada. Prepared for the Natural Resources Management Branch of the Ontario Ministry of Natural Resources.

Spence, J. R., and W. J. A. Volney. 1999. EMEND—Ecosystem management emulating natural disturbance. Sustainable Forest Management Network, Edmonton, Alberta, Canada. Project Report 1999-14.

Spence, J. R., W. J. A. Volney, V. Lieffers, M. G. Weber, and T. Vinge. 1999. The Alberta EMEND project: Recipe + cooks = argument. Pages 583–590 in Sustaining the boreal forest: Science and practice. Network of Centres of Excellence, Sustainable Forest Management Network, Edmonton, Alberta, Canada.

Spies, T. A. 1997. Forest stand structure, composition, and function. Pages 11–30 in K. A. Kohm and J. F. Franklin, editors. Creating a forestry for the 21st century. Island Press, Washington, D.C., United States.

Spies, T. A. 1998. Forest structure: A key to the ecosystem. Northwest Science 72:34–39.

Spies, T. A., and J. F. Franklin. 1991. The structure of natural young, mature, and old-growth Douglas-fir forests in Oregon and Washington. Pages 91–109 in L. F. Ruggerio, K. B. Aubry, A. B. Carey, and M. H. Huff, editors. Wildlife and vegetation of unmanaged Douglas-fir forests. U.S. Department of Agriculture Forest Service, Portland, Oregon, United States. General Technical Report PNW-GTR-285.

Spies, T. A., G. H. Reeves, K. M. Burnett, W. C. McComb, K. N. Johnson, G. Grant, J. L. Ohmann, S. L. Garman, and P. Bettinger. 2002. Assessing the ecological consequences of forest policies in a multi-ownership province in Oregon. Pages 179–207 in J. Liu and W. W. Taylor, editors. Integrating landscape ecology into natural resource management. Cambridge University Press, Cambridge, United Kingdom.

Sprugel, D. G. 1991. Disturbance, equilibrium, and environmental variability: What is "natural" vegetation in a changing environment? Biological Conservation 58: 1–18.

SPSS. 1999. SPSS for Windows. Version 10.0.5. SPSS, Chicago, Illinois, United States.

Staddon, W. J., L. C. Duchesne, and J. T. Trevors. 1998. Acid phosphatase, alkaline phosphatase and arylsulfatase activities in soils from a jack pine (*Pinus banksiana* Lamb.) ecosystem after clear-cutting, prescribed burning, and scarification. Biology and Fertility of Soils 27:1–4.

Stage, A. R. 1973. Prognosis model for stand development. U.S. Department of Agriculture Forest Service, Ogden, Utah, United States. Research Paper INT-137.

Stage, A. R., N. L. Crookston, and M. R. Wiitala. 1986. Procedures for including pest management activities in forest planning using present or simplified planning models. *In* R. G. Bailey, editor. Lessons from using FORPLAN: Proceedings of a planning and implementation workshop. U.S. Department of Agriculture Forest Service, Washington, D.C., United States.

Stage, A. R., C. G. Wells, III, M. A. Marsden, J. W. Byler, D. L. Renner, B. B. Eav, P. J. McNamee, G. D. Sutherland, and T. M. Webb. 1990. User's manual for the Western Root Disease Model. U.S. Department of Agriculture Forest Service, Ogden, Utah, United States. General Technical Report INT-GTR-267.

Stahle, D. W., R. D. D'Arrigo, P. J. Krusic, M. K. Cleaveland, E. R. Cook, R. J. Allan, J. E. Cole, R. B. Dunbar, M. D. Therrell, D. A. Gay, M. D. Moore, M. A. Stokes, B. T. Burns, J. Villanueva-Diaz, and L. G. Thompson. 1998. Experimental dendroclimatic reconstruction of the Southern Oscillation. Bulletin of the American Meteorological Society 79:2137–2152.

Statistical Sciences. 1993. S-PLUS user's manual, ver. 3.2. StatSci, a division of MathSoft, Seattle, Washington, United States.

Statutes of Ontario. 1975. Environmental Assessment Act. S.O. c. 25.

Statutes of Ontario. 1990. Crown Timber Act. Revised S.O. R.S.O., c. 51.

Statutes of Ontario. 1995. Crown Forest Sustainability Act, revised. R.S.O. 1998. Chapter 25 and Ontario Regulation 167/95.

Statutes of Upper Canada. 1849. Crown Timber Act. 12 Vict., c. 30.

Stedinger, J. R. 1984. A spruce budworm–forest model and its implications for suppression programs. Forest Science 30:597–615.

Steegman, A. T. 1983. Boreal forest adaptations of northern Algonkians. Plenum Press, New York, New York, United States.

Steinman, J. R., and D. A. MacLean. 1994. Predicting effects of defoliation on spruce–fir stand development: A management-oriented growth and yield model. Forest Ecology and Management 69:283–298.

Sterner, T. E., and A. G. Davidson. 1982. Forest insect and disease conditions in Canada, 1981. Canadian Forest Service, Ottawa, Ontario, Canada.

Stine, R. A., and M. J. Baughman, editors. 1992. White pine symposium proceedings. University of Minnesota, St. Paul, Minnesota, United States. Publication NR-BU-6044-S.

Stocks, B. J. 1987. Fire potential in the spruce budworm-damaged forests of Ontario. The Forestry Chronicle 63:8–14.

Stocks, B. J. 1991. The extent and impact of forest fires in northern circumpolar countries. Pages 197–202 *in* J. S. Levine, editor. Global biomass burning: Atmospheric, climatic, and biospheric implications. MIT Press, Cambridge, Massachusetts, United States.

Stocks, B. J., M. A. Fosberg, T. J. Lynham, L. Mearns, B. M. Wotton, Q. Yang, J.-Z. Jin, K. Lawrence, G. R. Hartley, J. A. Mason, and D. W. McKenney. 1998. Climate change and forest fire potential in Russian and Canadian boreal forests. Climate Change 38:1–13.

Stone, D. M. 1997. A decision tree to evaluate silvicultural alternatives for mature aspen in the northern Lake States. Northern Journal of Applied Forestry 14:95–98.

Stone, D. M. 2001. Sustaining aspen productivity in the Lake States. Pages 47–59 *in* W. D. Shepperd, D. Binkley, D. L. Bartos, T. J. Stohlgren, and L. G. Eskew, compilers. Sustaining aspen in western landscapes: Symposium proceedings. U.S. Department of Agriculture Forest Service, Fort Collins, Colorado, United States. Proceedings RMRS-P-18.

Stone, D. M., J. D. Elioff, D. V. Potter, D. B. Peterson, and R. Wagner. 2001. Restoration of aspen-dominated ecosystems in the Lake States. Pages 137–144 *in* W. D. Shepperd, D. Binkley, D. L. Bartos, T. J. Stohlgren, and L. G. Eskew, compilers. Sustaining aspen in western landscapes: Symposium proceedings. U.S. Department of Agriculture Forest Service, Fort Collins, Colorado, United States. Proceedings RMRS-P-18.

Strauss, D., L. Bednar, and R. Mees. 1989. Do one percent of forest fires cause ninety-nine percent of the damage? Forest Science 35:319–328.

Strickland, M. A., and C. W. Douglas. 1987. Marten. Pages 531–546 *in* M. Novak, J. Baker, M. Obbard, and B. Malloch, editors. Wild furbearer management and conservation in North America. Ontario Fur Managers Federation, Sault Ste. Marie, Ontario, Canada.

Strong, T. F., G. G. Erdmann, and J. N. Niese. 1995. Forty years of alternative management in second-growth, pole-size northern hardwoods. I. Tree quality development. Canadian Journal of Forest Research 25: 1173–1179.

Stuart-Smith, K., and D. Hebert. 1995. Practicing sustainable forestry, putting sustainable forestry into practice. Canadian Pulp and Paper Association, Montreal, Quebec, Canada.

Su, Q., D. A. MacLean, and T. D. Needham. 1996. The influence of hardwood content on balsam fir defoliation by spruce budworm. Canadian Journal of Forest Research 26:1620–1628.

Suarez, A. V., K. S. Pfenning, and S. K. Robinson. 1997. Nesting success of a disturbance-dependent songbird on different kinds of edges. Conservation Biology 11: 928–935.

Suffling, R. 1983. Stability and diversity in boreal and mixed temperate forests: A demographic approach. Journal of Environmental Management 17:359–371.

Suffling, R. 1993. Induction of vertical zones in subalpine valley forests by avalanche-formed fuel breaks. Landscape Ecology 8:127–138.

Suffling, R. 1995. Can disturbance determine vegetation distribution during climate warming? A boreal test. Journal of Biogeography 22:501–508.

Suffling, R., and D. Speller. 1998. The fire roller coaster in Canadian boreal forests. Pages 19–26 in D. McIver and R. E. Meyer, editors. Proceedings of the workshop on decoding Canada's environmental past: Climate variations and biodiversity change during the last millennium. Environment Canada, Downsview, Ontario, Canada.

Suffling, R., and C. Wilson. 1994. The use of Hudson's Bay Company records in climatic and ecological research, with particular reference to the Great Lakes basin. Pages 295–319 in R. I. Macdonald, editor. Great Lakes archaeology and paleoecology: Exploring interdisciplinary initiatives for the nineties. Quaternary Sciences Institute, University of Waterloo, Waterloo, Ontario, Canada.

Suffling, R., M. Evans, and A. H. Perera. 2003. Historical old-growth in southern Ontario: Forest measured through a cultural prism. The Forestry Chronicle 79(3):485–501.

Sutherland, G. D., A. S. Harestad, K. Price, and K. P. Lertzman. 2000. Scaling of natal dispersal distances in terrestrial birds and mammals. Conservation Ecology 4:16. [Online, URL: <http://www.consecol.org/vol4/iss1/art16>.]

Suzuki, W., K. Osumi, T. Masaki, K. Takahashi, H. Daimaru, and K. Hoshizaki. 2002. Disturbance regimes and community structures of a riparian and an adjacent terrace stand in the Kanumazawa Riparian Research Forest, Northern Japan. Forest Ecology and Management 157:285–301.

Svetsov, V. V. 2002. Comment on "Extraterrestrial impacts and wildfires." Palaeogeography Palaeoclimatology Palaeoecology 185:403–405.

Swaine, J. M., and F. C. Craighead. 1924. Studies on the spruce budworm (Cacoecia fumiferana Clem.). Part I. A general account of the outbreaks, injury and associated insects. Canadian Department of Agriculture Bulletin 37:3–27.

Swanson, F. J., J. F. Franklin, and J. R. Sedell. 1997. Landscape patterns, disturbance, and management in the Pacific Northwest, USA. Pages 191–213 in I. S. Zonneveld and R. T. T. Forman, editors. Changing landscapes: An ecological perspective. Springer-Verlag, New York, New York, United States.

Swanson, F. J., J. A. Jones, D. O. Wallin, and J. H. Cissel. 1994. Natural variability—implications for ecosystem management. Pages 80–94 in M. E. Jensen and P. S. Bourgeron, editors. U.S. Department of Agriculture Forest Service, Portland, Oregon, United States. General Technical Report PNW-GTR-318.

Swetnam, T. W. 1993. Fire history and climate change in giant sequoia groves. Science 262:885–889.

Swetnam, T. W., and C. H. Baisan. 1996. Historical fire regime patterns in the southwestern United States since A.D. 1700. Pages 11–32 in C. D. Allen, editor. Fire effects in southwestern forests. Proceeding of the 2nd La Mesa fire symposium. U.S. Department of Agriculture Forest Service, Fort Collins, Colorado, United States. General Technical Report RM-GTR-286.

Swetnam, T. W., and J. L. Betancourt. 1990. Fire–southern oscillation relations in the southwestern United States. Science 249:1017–1020.

Swetnam, T. W., C. D. Allen, and J. L. Betancourt. 1999. Applied historical ecology: Using the past to manage for the future. Ecological Applications 9:1189–1206.

Tande, G. F. 1979. Fire history and vegetation pattern of coniferous forests in Jasper National Park, Alberta. Canadian Journal of Botany 57:1912–1931.

Tansley, A. G. 1935. The use and abuse of vegetational concepts and terms. Ecology 16:284–307.

Tappeiner, J. C., D. Huffman, D. Marshall, T. A. Spies, and J. D. Bailey. 1997. Density, ages, and growth rates in old-growth and young-growth forest in coastal Oregon. Canadian Journal of Forest Research 27:638–648.

Taylor, A. H., and C. B. Halpern. 1991. The structure and dynamics of Abies magnifica forests in the southern Cascade Range, USA. Journal of Vegetation Science 2:189–200.

Teensma, P. D., J. T. Rienstra, and M. A. Yeiter. 1991. Preliminary reconstruction and analysis of change in forest stand age classes of the Oregon Coast Range from 1850 to 1940. U.S. Bureau of Land Management, Portland, Oregon. Technical Note T/N OR-9.

Tellier, R., L. Duchesne, J.-C. Ruel, and R. S. McAlpine. 1995. Effets du brûlage dirigé et du scarifiage sur l'établissement des semis et sur leur interaction avec la végétation concurrente. The Forestry Chronicle 71:621–626.

Theobald, D. M. 2000. Fragmentation by inholdings and exurban development. Pages 155–174 in R. L. Knight, F. W. Smith, S. W. Buskirk, W. H. Romme, and W. L. Baker, editors. Forest fragmentation in the southern Rocky Mountains. University Press of Colorado, Niwot, Colorado, United States.

Thiel, H., and G. Schriever. 1990. Deep-sea mining, environmental impact and the DISCOL Project. Ambio 19:245–250.

Thiel, H., G. Schriever, A. Ahnert, H. Bluhm, C. Borowski, and K. Vopel. 2001. The large-scale environmental impact experiment DISCOL—reflection and foresight. Deep-Sea Research Part II—Topical Studies in Oceanography 48:3869–3882.

Thomas, J. W., technical editor. 1979. Wildlife habitats in managed forests: The Blue Mountains of Oregon and Washington. U.S. Department of Agriculture Forest Service, Washington, D.C., United States. Agricultural Handbook 553.

Thomas, J. W. 1999. Forestry at the millennium—pitfalls and opportunities. The Forestry Chronicle 75:603–606.

Thomas, J. W., E. D. Forsman, J. B. Lint, E. C. Meslow, B. R. Noon, and J. Verner. 1990. A conservation strategy for the northern spotted owl. Interagency scientific committee to address the conservation of the northern spotted owl. U.S. Government Printing Office, Washington, D.C., United States.

Thompson, I. D. 1987. The myth of integrated forest resource management. Queen's Quarterly 94:609–621.

Thompson, I. D. 1994. Marten populations in mature and logged forests in Ontario. Journal of Wildlife Management 58:272–280.

Thompson, I. D. 2000. Forest vertebrates of Ontario: Patterns of distribution. Pages 54–73 in A. H. Perera, D. L. Euler, and I. D. Thompson, editors. Ecology of a managed terrestrial landscape: Patterns and processes of forest landscapes in Ontario. University of British Columbia Press, Vancouver, British Columbia, Canada.

Thompson, I. D., and P. Angelstäm. 1999. Special species. Pages 434–459 in M. L. Hunter Jr., editor. Maintaining biodiversity in managed forests. Cambridge University Press, Cambridge, United Kingdom.

Thompson, I. D., and P. W. Colgan. 1987. Numerical responses of marten to a food shortage in northcentral Ontario. Journal of Wildlife Management 51:824–835.

Thompson, I. D., M. D. Flannigan, B. M. Wotton, and R. Suffling. 1998. The effects of climate change on landscape diversity: An example in Ontario forests. Environmental Monitoring and Assessment 49:213–233.

Thompson, I. D., H. A. Hogan, and W. A. Montevecchi. 1999. Avian communities of mature balsam fir forests in Newfoundland: Age-dependence and implications for harvesting. Condor 101:311–323.

Thomson, A. J., and R. I. Alfaro. 1990. A method to calculate yield-correction factors for the overstory component of budworm-attacked Douglas-fir. Forest Ecology and Management 31:255–267.

Thoreau, H. D. 1988. Walden. Princeton University Press, Princeton, New Jersey, United States.

Thornton, P. E., S. W. Running, and M. A. White. 1997. Generating surfaces of daily meteorological variables over large regions of complex terrain. Journal of Hydrology 190:214–251.

Thysell, D. R., and A. B. Carey. 2001. Manipulation of density of *Pseudotsuga menziesii* canopies: Preliminary effects on understory vegetation. Canadian Journal of Forest Research 31:1513–1525.

Tilman, D. 1999. The ecological consequences of changes in biodiversity: A search for general principles. Ecology 80:1455–1474.

Tilman, D., R. M. May, C. L. Lehman, and M. A. Nowak. 1994. Habitat destruction and the extinction debt. Nature 371:65–66.

Tinker, D. B., W. H. Romme, W. W. Hargrove, R. H. Gardner, and M. G. Turner. 1994. Landscape-scale heterogeneity in lodgepole pine serotiny. Canadian Journal of Forest Research 24:897–903.

Titterington, R. W., H. S. Crawford, and B. N. Burgason. 1979. Songbird responses to commercial clear-cutting in Maine spruce–fir forests. Journal of Wildlife Management 44:602–609.

Tittler, R., C. Messier, and P. J. Burton. 2001. Hierarchical forest management planning and sustainable forest management in the boreal forest. The Forestry Chronicle 77:998–1005.

Tothill, J. D. 1921. An estimate of the damage done in New Brunswick by the spruce budworm. Proceedings of the Acadian Entomological Society 7:45–49.

Trudell, A. 1996. Forest Resource Inventory manual. Ontario Ministry of Natural Resources, Peterborough, Ontario, Canada.

Tubbs, C. H. 1977. Manager's handbook for northern hardwoods in the North Central states. U.S. Department of Agriculture Forest Service, St. Paul, Minnesota, United States. General Technical Report NC-GTR-39.

Turner, M. G. 1987. Landscape heterogeneity and disturbance. Springer-Verlag, New York, New York, United States.

Turner, M. G., and V. H. Dale. 1998. Comparing large, infrequent disturbances: What have we learned? Ecosystems 1:493–496.

Turner, M. G., and W. H. Romme. 1994. Landscape dynamics in crown fire ecosystems. Landscape Ecology 9:59–77.

Turner, M. G., W. W. Hargrove, R. H. Gardner, and W. H. Romme. 1994a. Effects of fire on landscape heterogeneity in Yellowstone National Park, Wyoming. Journal of Vegetation Science 5:731–742.

Turner, M. G., Y. Wu, L. L. Wallace, W. H. Romme, and A. Brenkert. 1994b. Simulating winter interactions among ungulates, vegetation, and fire in northern Yellowstone Park. Ecological Applications 4:472–496.

Turner, M. G., V. H. Dale, and E. H. Everham, III. 1997a. Fires, hurricanes, and volcanoes: Comparing large disturbances. Bioscience 47:758–768.

Turner, M. G., W. H. Romme, R. H. Gardner, and W. W. Hargrove. 1997b. Effects of fire size and pattern on early succession in Yellowstone National Park. Ecological Monographs 67:411–433.

Turner, M. G., R. H. Gardner, and R. V. O'Neill. 2001. Landscape ecology in theory and practice. Springer-Verlag, New York, New York, United States.

Tuskan, G. A., K. E. Francis, S. L. Russ, W. H. Romme, and M. G. Turner. 1996. RAPD markers reveal diversity within and among clonal and seedling stands of aspen in Yellowstone National Park, U.S.A. Canadian Journal of Forest Research 26:2088–2098.

Tuttle, M., K. T. Law, L. Seeber, and K. Jakob. 1990. Liquefaction and ground failure induced by the Saguenay, Quebec, earthquake. Canadian Geotechnical Journal 27:580–589.

Ugolini, F. C., and D. H. Mann. 1979. Biopedological origins of peatlands in southeast Alaska. Nature 281:366–368.

Ulanova, N. G. 2000. The effects of windthrow on forests at different spatial scales: A review. Forest Ecology and Management 135:155–167.

Urban, D. L., R. V. O'Neill, and H. H. Shugart. 1987. Landscape ecology: A hierarchical perspective can help scientists understand spatial patterns. BioScience 37:119–127.

U.S. Department of Agriculture Forest Service. 1986. Interim definitions for old-growth Douglas-fir and mixed-conifer forests in the Pacific Northwest and California. Old-growth Definition Task Group. U.S. Department of Agriculture Forest Service, Portland, Oregon, United States. Research Note PNW-447.

U.S. Department of Agriculture Forest Service. 1995. National forest system land planning and resource management planning proposed rule. Federal Register 60(71) (April 13):18886–18932.

U.S. Department of Agriculture Forest Service. 2000. Boundary Waters Canoe Area Wilderness—fuel treatment. U.S. Department of Agriculture Forest Service, Milwaukee, Wisconsin, United States.

U.S. Department of Agriculture Forest Service. 2001. U.S. forest facts and historical trends. U.S. Department of Agriculture Forest Service, Washington, D.C., United States. Report FS-696.

U.S. Department of Agriculture Forest Service. 2002. Forest insect and disease conditions in the United States. U.S. Department of Agriculture Forest Service, Washington, D.C., United States. [Online, URL: <www.fs.fed.us/foresthealth/annual_i_d_conditions/index.html>.]

U.S. Department of Agriculture Forest Service and U.S. Department of the Interior. 2000. Report to the President: Managing the impact of wildfires on communities and the environment. [Online, URL: <http://www.fs.fed.us/fire/nfp/president>.] Executive summary *in* Fire Management Today 61:9–11.

Usher, M. B. 1981. Modelling ecological succession with particular reference to Markovian models. Vegetatio 40:3–14.

Usher, M. B. 1992. Statistical models of succession. Pages 215–248 *in* D. C. Glenn-Lewin, R. K. Peet, and T. T. Veblen, editors. Plant succession: Theory and prediction. Chapman and Hall, London, United Kingdom.

Vale, T. R., editor. 2002. Fire, native peoples and the natural landscape. Island Press, Washington, D.C., United States.

Van Cleve, K., L. A. Viereck, and C. T. Dyrness. 1996. State factor control of soils and forest succession along the Tanana River in interior Alaska, United States. Arctic and Alpine Research 28:388–400.

Van Delden, W. 1994. Genetic diversity and its role in the survival of species. Pages 41–56 *in* O. T. Solbrig, H. M. van Emden, and P. G. W. J. van Oordt, editors. Biodiversity and global change. CABI Publishing International and the International Union of Biological Sciences, Wallingford, United Kingdom.

Van Dyke, F. G., J. P. Dibenedetto, and S. C. Thomas. 1991. Vegetation and elk response to prescribed burning in south-central Montana. Pages 163–179 *in* R. B. Keiter and M. S. Boyce, editors. The Greater Yellowstone ecosystem: Redefining America's wilderness heritage. Yale University Press, New Haven, Connecticut, United States.

Van Kooten, G. C., C. S. Binkley, and G. Delcourt. 1995. Effect of carbon taxes and subsidies on optimal forest rotation age and supply of carbon services. American Journal of Agricultural Economics 77:365–374.

Van Wagner, C. E. 1971. Fire and red pine. Pages 211–219 *in* Proceedings of the 10th annual Tall Timbers fire ecology conference. Tall Timbers Research Station, Tallahassee, Florida, United States.

Van Wagner, C. E. 1978. Age-class distribution and the forest fire cycle. Canadian Journal of Forest Research 8:220–227.

Van Wagner, C. E. 1983. Fire behavior in northern conifer forests and shrublands. Pages 65–80 *in* R. W. Wein and D. A. MacLean, editors. The role of fire in northern circumpolar ecosystems. SCOPE 18. John Wiley and Sons, Chichester, United Kingdom.

Van Wagner, C. E. 1987. Development and structure of the Canadian Forest Fire Weather Index system. Canadian Forest Service, Ottawa, Ontario, Canada. Forest Technical Report 35.

Van Wagner, C. E. 1998. Modelling logic and the Canadian Forest Fire Behaviour Prediction System. The Forestry Chronicle 74:50–52.

Vanguard Forest Management Services. 1993. STAMAN stand growth model and calibration of STAMAN model for defoliation impacts. Pages B1–B39, C1–C18 *in* Forest protection planning to sustain long-term wood supplies. Contract report for Canadian Forest Service, Fredericton, New Brunswick, Canada.

Veblen, T. T., K. S. Hadley, M. S. Reid, and A. J. Rebertus. 1989. Blowdown and stand development in a Colorado subalpine forest. Canadian Journal of Forest Research 19:1218–1225.

Veblen, T. T., K. S. Hadley, M. S. Reid, and A. J. Rebertus. 1991. Methods of detecting past spruce beetle outbreaks in Rocky Mountain subalpine forests. Canadian Journal of Forest Research 21:242–254.

Veblen, T. T., K. S. Hadley, E. M. Nel, T. Kitzburger, M. Reid, and R. Villaba. 1994. Disturbance regimes and disturbance interactions in a Rocky Mountain subalpine forest. Journal of Ecology 82:125–135.

Veblen, T. T., T. Kitzberger, and J. Donnegan. 2000. Climatic and human influences on fire regimes in ponderosa pine forests in the Colorado Front Range. Ecological Applications 10:1178–1195.

Virkkala, R., A. Rajasarkka, R. A. Vaisansen, M. Vickholm, and E. Virolainen. 1994. Conservation value of nature reserves: Do hole-nesting birds prefer protected forests in southern Finland? Annales Zoologica Fennici 31:173–186.

Vittoz, P., G. H. Stewart, and R. P. Duncan. 2001. Earthquake impacts in old-growth *Nothofagus* forests in New Zealand. Journal of Vegetation Science 12:417–426.

Voigt, D. R, J. A. Baker, R. S. Rempel, and I. D. Thompson. 2000. Forest vertebrate responses to landscape-level changes in Ontario. Pages 198–233 *in* A. H. Perera, D. L. Euler, and I. D. Thompson, editors. Ecology of a managed terrestrial landscape: Patterns and processes of forest landscapes in Ontario. University of British Columbia Press, Vancouver, British Columbia, Canada.

Volney, W. J. A. 1996. Climate change and management of insect defoliators in boreal forest ecosystems. Pages 79–87 *in* M. J. Apps and D. T. Price, editors. Forest ecosystems, forest management and the global carbon cycle. Springer-Verlag, Berlin, Germany.

Vose, J. M., W. T. Swank, B. D. Clinton, J. D. Knoepp, and L. W. Swift. 1999. Using stand replacement fires to restore southern Appalachian pine–hardwood ecosystems: Effects on mass, carbon, and nutrient pools. Forest Ecology and Management 114:215–226.

Vuori, K. M., I. Joensuu, J. Latvala, E. Jutila, and A. Ahvonen. 1998. Forest drainage: A threat to benthic biodiversity of boreal headwater streams? Aquatic Conservation–Marine and Freshwater Ecosystems 8:745–759.

Wackernagel, M., and W. E. Rees. 1996. Our ecological footprint: Reducing human impact on Earth. New Society Publishers, Gabriola Island, British Columbia, Canada.

Wallace, L. L., M. G. Turner, W. H. Romme, R. V. O'Neill, and Y. Wu. 1995. Scale of heterogeneity of forage production and winter foraging by elk and bison. Landscape Ecology 10:75–83.

Wallin, D. O., F. J. Swanson, and B. Marks. 1994. Landscape pattern response to changes in pattern generation rules: Land-use legacies in forestry. Ecological Applications 4:569–580.

Walters, C. J. 1986. Adaptive management of renewable resources. McGraw-Hill, New York, New York, United States.

Ward, P. C., and A. G. Tithecott. 1993. The impact of fire management on the boreal landscape of Ontario. Ontario Ministry of Natural Resources, Sault Ste. Marie, Ontario, Canada. Publication 305.

Ward, P. C., A. G. Tithecott, and B. M. Wotton. 2001. Reply—a re-examination of the effects of fire suppression in the boreal forest. Canadian Journal of Forest Research 31:1467–1480.

Wardle, D. A., O. Zackrisson, and M. C. Nilsson. 1998. The charcoal effect in boreal forests: Mechanisms and ecological consequences. Oecologia 115:419–426.

Waring, R. H., and J. F. Franklin. 1979. Evergreen coniferous forests of the Pacific Northwest. Science 204:1380–1386.

Waring, R. H., and S. W. Running. 1998. Forest ecosystems: Analysis at multiple scales, second edition. Academic Press, San Diego, California, United States.

Weaver, K., and A. H. Perera. in press. Modelling land cover transitions: A solution to the problem of spatial dependence in data. Landscape Ecology.

Weber, M. G., and M. D. Flannigan. 1997. Canadian boreal forest ecosystem structure and function in a changing climate: Impact on fire regimes. Environmental Reviews 5:145–166.

Weber, M. G., and B. J. Stocks. 1998. Forest fires in the boreal forests of Canada. Pages 215–233 in J. S. Moreno, editor. Large forest fires. Backbuys, Leiden, the Netherlands.

Weir, J. M. H., E. A. Johnson, and K. Miyanishi. 2000. Fire frequency and the spatial age mosaic of the mixed-wood boreal forest in western Canada. Ecological Applications 10:1162–1177.

Weisberg, P. J., and F. J. Swanson. 2003. Regional synchroneity in fire regimes of the western cascades, USA. Forest Ecology and Management 172:17–28.

Wells, A., R. P. Duncan, and G. H. Stewart. 2002. Forest dynamics in Westland, New Zealand: The importance of large, infrequent earthquake-induced disturbance. Journal of Ecology 89:1006–1018.

Welsh, D. A., and S. C. Lougheed. 1996. Relationship of bird community structure and species distributions to two environmental gradients in the northern boreal forest. Ecography 19:194–208.

Westin, S. 1992. Wildfire in Missouri. Missouri Department of Conservation, Columbia, Missouri, United States.

Whelan, R. J. 1995. The ecology of fire. Cambridge University Press, Cambridge, United Kingdom.

Whisenant, S. G. 1990. Changing fire frequencies on Idaho's Snake River Plains: Ecological and management implications. Pages 4–10 in E. D. McArthur, editor. Proceedings of symposium on cheatgrass invasion, shrub dieoff, and other aspects of shrub biology. U.S. Department of Agriculture Forest Service, Missoula, Montana, United States. General Technical Report INT-GTR-276.

White, C. 1985. Wildland fires in Banff National Park 1880–1980. Environment Canada, Ottawa, Ontario, Canada. Occasional Paper 3.

White, J. D., S. W. Running, P. E. Thornton, R. E. Keane, K. C. Ryan, D. B. Fagre, and C. H. Key. 1998. Assessing simulated ecosystem processes for climate variability research at Glacier National Park, USA. Ecological Applications 8:805–823.

White, P. S., and S. T. Pickett. 1985. Natural disturbance and patch dynamics: An introduction. Pages 3–13 in S. T. Pickett and P. S. White, editors. The ecology of natural disturbance and patch dynamics. Academic Press, San Diego, California, United States.

White, R. 2001. New Brunswick Growth and Yield Unit database user guide. New Brunswick Department of Natural Resources and Energy, Fredericton, New Brunswick, Canada.

White Pine Strategies Working Group. 1996. Minnesota's white pine now and for the future. Report to the Minnesota Department of Natural Resources and Minnesota Forest Resources Council. White Pine Strategies Working Group, St. Paul, Minnesota, United States.

Whitlock, C., and S. H. Millspaugh. 1996. Testing the assumptions of fire-history studies: An examination of modern charcoal accumulation in Yellowstone National Park, USA. Holocene 6:7–15.

Whitmore, T. C. 1975. Tropical rainforests of the far east. Oxford University Press, Oxford, United Kingdom.

Whitney, G. G. 1987. An ecological history of the Great Lakes forest of Michigan. Journal of Ecology 75:667–684.

Whittaker, R. J., T. Partomihardjo, and S. H. Jones. 1999. Interesting times on Krakatau: Stand dynamics in the 1990s. Philosophical Transactions of the Royal Society of London Series B–Biological Sciences 354:1857–1867.

Wilcox, B. A., and D. D. Murphy. 1985. Conservation strategy: The effects of fragmentation on extinction. American Naturalist 125:879–887.

Wilkinson, C. F. 1993. Crossing the next meridian: Land, water, and the future of the West. Island Press, Washington, D.C., United States.

Wilkinson, D. M. 1997. Plant colonization: Are wind dispersed seeds really dispersed by birds at larger spatial and temporal scales? Journal of Biogeography 24:61–65.

Williams, M. 1989. Americans and their forests. Cambridge University Press, Cambridge, United Kingdom.

Wilson, S. M., and A. B. Carey. 2000. Legacy retention versus thinning: Influences on small mammals. Northwest Science 74:131–145.

Wimberly, M. C. 2002. Spatial simulation of historical landscape patterns in coastal forests of the Pacific Northwest. Canadian Journal of Forest Research 32: 1316–1328.

Wimberly, M. C., and T. A. Spies. 2002. Landscape- vs gap-level controls on the abundance of a fire-sensitive, late-successional tree species. Ecosystems 5:232–243.

Wimberly, M. C., T. A. Spies, C. J. Long, and C. Whitlock. 2000. Simulating historical variability in the amount of old forests in the Oregon Coast Range. Conservation Biology 14:167–180.

Winters, R. K. 1974. The forests and man. Vantage, New York, New York, United States.

Wisdom, M. J., R. S. Holthausen, B. C. Wales, C. D. Hargis, V. A. Saab, D. C. Lee, W. J. Hann, T. D. Rich, M. M. Rowland, W. J. Murphy, and M. R. Eames. 2000. Source habitats for terrestrial vertebrates of focus in the interior Columbia Basin: Broad-scale trends and management implications. Volume 1, Overview. U.S. Department of Agriculture Forest Service, Portland, Oregon, United States. General Technical Report PNW-GTR-485.

Wolfe, S. A., and D. S. Ponomarenko. 2001. Potential sediment transport in the prairie provinces from principal climate station data. Report submitted to Environment Canada, Ottawa, Ontario, Canada. Climate Change Action Fund (CCAF) Science Project #S00-15-09. Unpublished report.

Wolock, D. M. and G. J. McCabe. 1995. Comparison of single and multiple flow direction algorithms for computing topographic parameters in TOPMODEL. Water Resources Research 31:1315–1324.

Woodley, S., and G. Forbes, editors. 1997. Forest management guidelines to protect native biodiversity in the Fundy Model Forest. Greater Fundy Ecosystem Research Group, University of New Brunswick, Fredericton, New Brunswick, Canada.

Woodward, F. I. 1987. Climate and plant distribution. Cambridge University Press, London, United Kingdom.

Woodward, F. I., and I. F. McKee. 1991. Vegetation and climate. Environment International 17:535–546.

Wright, H. A., and A. W. Bailey. 1982. Fire ecology: United States and southern Canada. John Wiley and Sons, New York, New York, United States.

Wu, J., and O. L. Loucks. 1995. From balance of nature to hierarchical patch dynamics: A paradigm shift in ecology. Quarterly Review of Biology 70:439–466.

Wykoff, W. R., N. L. Crookston, and A. R. Stage. 1982. User's guide to the Stand Prognosis model. U.S. Department of Agriculture Forest Service, Ogden, Utah, United States. Technical Report INT-133.

Yamaguchi, D. K. 1986. The development of old growth Douglas fir forests northeast of Mt. St. Helens, Washington, following an A.D. 1480 eruption. Ph.D. dissertation. University of Washington, Seattle, Washington, United States.

Yanchuk, A. D. 2001. A quantitative framework for breeding and conservation of forest tree genetic resources in British Columbia. Canadian Journal of Forest Research 31:566–576.

Yarie, J. 1981. Forest fire cycles and life tables: A case study from interior Alaska. Canadian Journal of Forest Research 11:554–562.

Yemshanov, D. G., and A. H. Perera. 2002. A spatially explicit stochastic model to simulate forest cover transitions: General structure and properties. Ecological Modelling 150:189–209.

Zasada, J. 1990. Developing silvicultural alternatives for the boreal forest—an Alaskan perspective on regeneration of white spruce. University of Alberta, Edmonton, Alberta, Canada. Forest Industry Lecture 25.

Zehngraff, P. 1947. Possibilities of managing aspen. U.S. Department of Agriculture Forest Service, St. Paul, Minnesota, United States. Lake States Aspen Report 21.

Zhang, Q., K. S. Pregitzer, and D. D. Reed. 1999. Catastrophic disturbance in the presettlement forests of the Upper Peninsula of Michigan. Canadian Journal of Forest Research 29:106–114.

Zhong, J., and B. J. van der Kamp. 1999. Pathology of conifer seed and timing of germination in high-elevation subalpine fir and Engelmann spruce forests of the southern interior of British Columbia. Canadian Journal of Forest Research 29:187–193.

Zhu, Z. 1994. Forest density mapping in the lower 48 states: A regression procedure. U.S. Department of Agriculture Forest Service, Asheville, North Carolina, United States. Research Paper SO-280.

INDEX

The suffix *f* on a page number indicates a figure; *t* indicates a table. **Bold** numbers are color plate numbers.